MW00399606

Spectra of Graphs

This is a volume in
PURE AND APPLIED MATHEMATICS

A Series of Monographs and Textbooks

Editors: SAMUEL EILENBERG AND HYMAN BASS

A list of recent titles in this series appears at the end of this volume

Spectra of Graphs

Theory and Application

DRAGOŠ M. CVETKOVIĆ   MICHAEL DOOB   HORST SACHS

ACADEMIC PRESS   New York   San Francisco   London

A Subsidiary of Harcourt Brace Jovanovich, Publishers

ACADEMIC PRESS, INC.
111 Fifth Avenue, New York, New York 10003

Library of Congress Cataloging in Publication Data

Cvetković, Dragoš M. Doob, Michael Sachs, Horst
Spectra of Graphs

(Pure and Applied Mathematics, a series of monographs and textbooks)

ISBN 0-12-195150-2
LCCCN 79-50490

PRINTED IN THE GERMAN DEMOCRATIC REPUBLIC

To Zora, Judy, and Barbara

Preface

It has the curious feature that some of the main results, although purely combinatorial in character, seem in the present state of knowledge to be unobtainable without resorting to algebraic methods involving a consideration of eigenvalues of adjacency matrices of graphs.

CRISPIN ST. J. A. NASH-WILLIAMS, Unexplored and Semi-explored Territories in Graph Theory†

It will be apparent ... that the results achieved so far barely scratch the surface of what appears to be a rich area of investigation.

ALAN J. HOFFMAN, The Eigenvalues of the Adjacency Matrix of a Graph††

This book has been written for mathematicians working in the area of graph theory and combinatorics, for chemists who are interested in quantum chemistry, and, at least partly, for physicists and electrical engineers using graph theory in their work. The book is almost entirely self-contained; only a little familiarity with graph theory and matrix theory is assumed.

The theory of graph spectra can, in a way, be considered as an attempt to utilize linear algebra including, in particular, the well-developed theory of matrices for the purposes of graph theory and its applications. However, that does not mean that the theory of graph spectra can be reduced to the theory of matrices; on the contrary, it has its own characteristic features and specific ways of reasoning fully justifying it to be treated as a theory in its own right.

We are convinced that a book such as this should have been written. On the one hand, the standard text-books on graph theory barely mention graph spectra (N. BIGGS' excellent monograph on Algebraic Graph Theory [Big5] being an exception). On the other hand, considerable interest has been paid to graph spectra in the mathema-

---

† [Nash], p. 181; the author is speaking about "the problem of determining the smallest possible number of vertices of a regular graph of prescribed girth and valency".
†† [Hof9], p. 578.

tical as well as chemical literature. The material is spread in various journals and other publications and therefore is not well known on the whole: so it has happened that some results have been rediscovered many times (see Theorems 1.2 and 1.3).

This monograph should not be considered as a systematic treatment on graph spectra, but rather as a unifying collection of material on that subject. Nevertheless, we hope that it will be useful since it contains a lot of information, an extensive bibliography and an appendix with numerical data on graph spectra.

We are aware that some important topics related to graph spectra are only roughly outlined in the book. This lack will certainly be diminished by another book being prepared on graph spectra by J. M. Goethals, A. J. Hoffman, and J. J. Seidel.

Chapters 0, 2, 3, 7, 8, 9 have been written by D. M. Cvetković. M. Doob has written Chapter 6. Chapters 1, 4, 5 have been written by H. Sachs. The Appendix has been compiled by D. M. Cvetković and H. Sachs. In addition, all the authors have put some small inserts into all of the chapters. We have tried to improve the text of the whole book and to unify the material from different chapters. Hence all three authors are collectively responsible for the book.

We have endeavoured to find a style which is, on the one hand, concise enough to enable the extensive material to be treated in a book of limited size, and which, on the other hand, is still intuitive enough to make the book readable for the applied scientist using this material as well as for the mathematician.

We wish to thank J. J. Seidel (Eindhoven) for his valuable comments on Chapter 6 and part of Chapter 7, and I. Gutman (Kragujevac) who helped in writing Chapter 8, especially by providing references to the chemical literature. We also thank the following for their help in the preparation of this book: N. Obradović (Belgrade), W. Börner, I. Röhrs, B. Schönefeld, S. Pech, K.-H. Schwolow (Ilmenau), A. Graovac (Zagreb), J. D. Kečkić, D. Maksimović, S. K. Simić (Belgrade), A. K. Kel'mans (Moscow), L. Babai (Budapest), R. B. Mallion (Oxford) B. Preiss, E. Gushulak (Winnipeg), Th. J. M. van den Hurk, and A. M. Janson-Jansen (Eindhoven).

We are greatly indebted to the authors of several tables given in the Appendix for allowing the reproduction of their tables in this book.

We also wish to thank the University of Belgrade, Faculty of Electrical Engineering and the Mathematical Institute, Belgrade, The University of Manitoba, Winnipeg, and Technische Hochschule Ilmenau, respectively, for supporting our work on the book. D. Cvetković thanks the Technological University Eindhoven for the use of their facilities during the 1975/76 academic year. M. Doob acknowledges the support of the National Research Council of Canada whose assistence was most valuable.

Finally, we are grateful to the Publishers and their Staff, in particular Mrs. B. Mai, for their patience and permanent cooperation, and to the Printing-Office and its Staff.

Summer 1978

D. Cvetković
M. Doob
H. Sachs

Contents

0. Introduction

This introductory chapter is devoted mainly to the reader who is not familiar with graph theory to help him to enter the topic of the book. The basic definitions and facts about the spectra of graphs are given together with a description of some general graph theoretic notions and necessary facts from matrix theory. For a general introduction to graph theory the reader is referred to the books [BeCh 2], [Ber 1], [Ber 2], [Ber 3], [BoMu], [Deo], [Har 4], [Maye], [Nolt], [Sac 9], [Wi,RJ 2], [Уилс], [Xapa], a chemist may be especially interested in [Bala]†, and for a survey of matrix theory we recommend the books [Gant], [MaMi].

0.1. What the spectrum of a graph is and how it is presented in this book

By a *graph* $G = (\mathscr{X}, \mathscr{U})$ we mean a finite set $\mathscr{X}$ (whose elements are called *vertices*) together with a set $\mathscr{U}$ of two-element subsets of $\mathscr{X}$ (the elements of $\mathscr{U}$ are called *edges*). Similarly, a *digraph* (*directed graph*) $(\mathscr{X}, \mathscr{U})$ is defined to be a finite set $\mathscr{X}$ and a set $\mathscr{U}$ of ordered pairs of elements of $\mathscr{X}$ (these pairs are called *directed edges* or *arcs*). The sets of vertices and edges are sometimes denoted by $\mathscr{V}(G)$ and $\mathscr{E}(G)$, respectively.

If *multiple* undirected or directed *edges* are allowed, we shall speak of *multigraphs* or *multi-digraphs*, respectively. These two cases include the possible existence of loops (a *loop* is an edge or arc with both of its vertices identical). The terminology is that of F. Harary [Har 4] except for the fact that in this book multi-(di-)graphs are allowed to have loops. Although the term graph denotes what in many graph theoretical papers is called "a finite, undirected graph without loops or multiple edges" (or, briefly, a "*schlicht* graph"), for the sake of readability we shall sometimes (when there is no danger of confusion) use the term graph in the most general meaning, i.e., we shall mean undirected graphs, digraphs and even multigraphs and multi-digraphs.

Two vertices are called *adjacent* if they are connected by an edge (arc). The *adjacency matrix* $\boldsymbol{A}$ of a multi-(di-)graph G whose vertex set is $\{x_1, x_2, \ldots, x_n\}$ is a square matrix of order n, whose entry a_{ij} at the place (i, j) is equal to the number of edges

† Recently two books on Hückel theory have appeared (see the end of p. 359).

(arcs) starting at the vertex x_i and terminating at the vertex x_j. We shall write $\boldsymbol{A} = (a_{ij}) = (a_{ij})_1^n$; sometimes it is convenient to denote the element a_{ij} by $(\boldsymbol{A})_{ij}$.

For example, the adjacency matrix of the graph shown in Fig. 0.1 is given by

$$\boldsymbol{A} = \begin{pmatrix} 0 & 1 & 0 & 0 \\ 1 & 0 & 1 & 0 \\ 0 & 1 & 0 & 1 \\ 0 & 0 & 1 & 0 \end{pmatrix}$$

For a multigraph the entry a_{ii} is twice the number of (undirected) loops attached to the vertex x_i.†

Fig. 0.1

Sometimes it is convenient to identify a multigraph G with the multi-digraph which has the same adjacency matrix as G. A bit more generally: any undirected edge connecting distinct vertices can be considered as a pair of arcs connecting the same vertices but having opposite directions, and conversely; also, any two "directed loops" attached to the same vertex can be replaced by one undirected loop, and vice versa.

The multi-(di-)graphs $G = (\mathscr{X}, \mathscr{U})$ and $H = (\mathscr{Y}, \mathscr{V})$ are called *isomorphic* if there is a (1, 1)-mapping $y = \varphi(x)$ of $\mathscr{X}$ onto $\mathscr{Y}$ such that, for any pair of vertices $x', x'' \in \mathscr{X}$, there are in H precisely as many edges (arcs) going from $y' = \varphi(x')$ to $y'' = \varphi(x'')$ as there are in G going from x' to x''. Such an adjacency preserving (1, 1)-mapping φ is called an *isomorphism*. Obviously, G and H are isomorphic if and only if their vertices can be labelled (numbered) so that the corresponding adjacency matrices are equal. Clearly, the relation of being isomorphic is an equivalence relation partitioning the set of all multi-(di-)graphs into equivalence classes which can be considered as "abstract" multi-(di-)graphs: thus isomorphic multi-(di-)graphs represent the same abstract multi-(di-)graph. If G and H are isomorphic we write $G \cong H$ or, if G and H are regarded as abstract multi-(di-)graphs, simply $G = H$.

The determinant of a square matrix $\boldsymbol{A}$ is denoted by $|\boldsymbol{A}|$ or by det $\boldsymbol{A}$.

The characteristic polynomial $|\lambda \boldsymbol{I} - \boldsymbol{A}|$ of the adjacency matrix $\boldsymbol{A}$ of G is called the *characteristic polynomial of* G and denoted by $P_G(\lambda)$. The eigenvalues of $\boldsymbol{A}$ (i.e. the zeros of $|\lambda \boldsymbol{I} - \boldsymbol{A}|$) and the spectrum of $\boldsymbol{A}$ (which consists of the eigenvalues) are also called the *eigenvalues* and the *spectrum of* G, respectively. If $\lambda_1, \ldots, \lambda_n$ are the eigenvalues of G, the whole spectrum is denoted by $\boldsymbol{Sp}(G) = [\lambda_1, \ldots, \lambda_n]$. Clearly, isomorphic multi-(di-)graphs have the same spectrum.

The eigenvalues of $\boldsymbol{A}$ can equivalently be defined as the numbers λ satisfying $\boldsymbol{Ax} = \lambda \boldsymbol{x}$ for a non-zero vector $\boldsymbol{x}$. Each such vector $\boldsymbol{x}$ is called an *eigenvector* of the matrix $\boldsymbol{A}$ (or *of the graph* G) belonging to the eigenvalue λ.

---

† However, in some cases it is more convenient to assume that a loop contributes 1 to the trace of $\boldsymbol{A}$: We shall then say that such loops are *simply counted*. A directed loop is always simply counted.

For the graph G of Fig. 0.1 we have

$$P_G(\lambda) = \begin{vmatrix} \lambda & -1 & 0 & 0 \\ -1 & \lambda & -1 & 0 \\ 0 & -1 & \lambda & -1 \\ 0 & 0 & -1 & \lambda \end{vmatrix} = \lambda^4 - 3\lambda^2 + 1,$$

$$Sp(G) = \left[\frac{1+\sqrt{5}}{2}, \frac{-1+\sqrt{5}}{2}, \frac{1-\sqrt{5}}{2}, \frac{-1-\sqrt{5}}{2}\right].$$

The spectrum of a complete graph K_n on $n \geqq 1$ vertices (definition on p. 16) consists of the number $n - 1$ as well as $n - 1$ numbers equal to -1 (see Section 2.6).

Several other types of graph spectra will be defined in Chapter 1.

Spectra of graphs have appeared frequently in the mathematical literature since the fundamental papers of L. M. LIHTENBAUM [Лих 1] (1956) and of L. COLLATZ and U. SINOGOWITZ [CoSi 1] (1957). Even earlier, starting from the thesis of E. HÜCKEL [Hück] in 1931, theoretical chemists were interested in graph spectra, although they used different terminology. We shall now point out some reasons showing why graph spectra are important.

1. In quantum chemistry the skeletons of certain non-saturated hydrocarbons are represented by graphs. Energy levels of electrons in such a molecule are, in fact, the eigenvalues of the corresponding graph. The stability of the molecule as well as other chemically relevant facts are closely connected with the graph spectrum and the corresponding eigenvectors (see Chapter 8).

2. There are many theorems in graph theory and combinatorics in the proofs of which graph spectra appear, although the statements of the theorems do not mention the spectra explicitly. Hence, graph spectra appear here as a very important tool the use of which can be denoted by *spectral techniques* (see Chapter 7).

3. There is also a computational reason: The spectrum is a finite sequence of numerical invariants. Under the hypothesis that the information we are interested in (or an essential part of it) is contained in the spectrum, we can use the spectrum instead of the graph: a finite sequence of numbers can easily be put on a computer. This, of course, is reasonable only if we have efficient methods for wide enough classes of graphs to calculate their spectra and, conversely, to "decode" a given spectrum, i.e., to retrieve the graph properties in question from a spectrum which has been obtained as a result of algebraic computations.

It should be mentioned that chemists and physicists know the structure of the graph and they are looking for the corresponding spectrum, whereas, in general, graph theorists and combinatorialists assume that the spectrum is known and they try to say something about the graph structure.

These facts, together with a general need in graph theory to look for and to investigate graph invariants, and together with certain intriguing facts behind the whole thing, stimulate the investigation of relations between spectral and structural properties of a graph. Also, the behaviour of eigenvalues under several graph transformations is of interest, especially in some cases when the spectra of compound graphs can be expressed in terms of the spectra of simpler graphs.

The plan of the book is as follows:

In *Section* 0.2 some general graph theoretic notions are given together with some conventions used throughout this book. *Section* 0.3 contains the necessary theorems from matrix theory and describes some basic facts about graph spectra.

In *Chapters* 1, 3, 4, 5, 6 relations between spectral and structural properties of graphs are described.

Chapter 2 describes the relations between the spectrum of a graph constructed by operations on some given graphs and the spectra of these graphs themselves.

Chapters 7 *and* 8 describe the applications of the theory developed in Chapters 1 to 6. Chapter 7 is related to the applications in graph theory and combinatorics and Chapter 8 gives the applications beyond mathematics, i.e. in chemistry and physics.

Chapter 9 contains some additional material which did not fit into the classification of the other chapters. The *Appendix* contains numerical data on graph spectra and the corresponding characteristic polynomials.

The last section in each of Chapters 1—8 has the title *Miscellaneous results and problems*. At these places some additional material is reviewed, partly in form of exercises and problems. Section 9.5 gives a list of unsolved problems.

The *Bibliography* contains more than 650 references from both the mathematical and the chemical literature. Although the authors believe that all important papers from the viewpoint of this book are included, they are aware that a complete bibliography on graph spectra is almost impossible to compile because of the thousands of chemical papers where graph spectra are only mentioned in passing and the hundreds of papers on association schemes, block designs, and related combinatorial objects where eigenvalues are also involved, although not always in an important way.

0.2. Some more graph theoretic notions and conventions

We shall now give some more definitions of the graph theoretic notions frequently used throughout the book. We shall also point out some standard notations and explain some conventions used in the subsequent text.

A graph $H = (\mathcal{Y}, \mathcal{V})$ is said to be a *subgraph* of the graph $G = (\mathcal{X}, \mathcal{U})$ if $\mathcal{Y} \subset \mathcal{X}$ and $\mathcal{V} \subset \mathcal{U}$. The graph H is called a *spanning subgraph* or a *partial graph of* G if $\mathcal{Y} = \mathcal{X}$. If $\mathcal{V}$ consists of all the edges from $\mathcal{U}$ which connect the vertices from $\mathcal{Y}$, then H is called an *induced subgraph*. An induced subgraph is said to be *spanned by its vertices*, and a partial graph is sometimes said to be *spanned by its edges*.

The number of edges incident with a vertex in an undirected graph is called the *degree* or the *valency* of the vertex. Note that an undirected loop is counted twice, thus its contribution to the valency of the vertex to which it is attached is equal to 2. If all the vertices have the same valency r, the graph is called *regular of degree* r.

In digraphs we shall distinguish between the *indegree* or *rear valency* and the *outdegree* or *front valency* (of a vertex) by indicating how many arcs go into and go out from the vertex, respectively.

If there is an arc from vertex x to vertex y, we shall sometimes indicate this by writing $y \cdot x$; x and y are *neighbours* of each other, x is a *rear neighbour* of y and y is a *front neighbour* of x.

A *cycle of length n*, denoted by $\vec{C}_n$, is a digraph with the vertex set $\{x_1, \ldots, x_n\}$ having arcs (x_i, x_{i+1}), $i = 1, \ldots, n - 1$, and (x_n, x_1). A *linear directed graph* is a digraph in which each indegree and each outdegree is equal to 1, i.e., it consists of cycles.

A *spanning linear subgraph* of a multi-(di-)graph G, i.e., a linear subgraph of G which contains all vertices of G is sometimes called a *linear factor* of G. A linear factor of a multigraph consists of disjoint copies of K_2.

A regular spanning subgraph of degree s of a multigraph G is called a (*regular*) *factor of degree s* or, briefly, an *s-factor* of G.

In a multi-(di-)graph any sequence of consecutive edges (arcs) (having in mind the orientation in directed case) is called a *walk*. The *length of the walk* is the number of edges (arcs) in it. A walk can pass through the same edge (arc) more than once.

A *path* of length $n - 1$ $(n \geqq 2)$, denoted by P_n, is a graph with n vertices, say $x_1, \ldots, x_n$, and with $n - 1$ edges in which x_i and x_{i+1} are connected by an edge for $i = 1, \ldots, n - 1$.

A multi-(di-)graph is (*strongly*) *connected* if any two of its vertices are joined by a path (walk). A multigraph is *disconnected* if it is not connected, and it then consists of two or more parts called *components*, two vertices being in different components if they cannot be joined by a path. A vertex x is called a *cutpoint* and an edge u is called a *bridge* if the deletion of x or u, respectively, causes an increase of the number of components.

The length of a shortest path between two vertices is called the *distance between the vertices*. The *diameter* of a connected multigraph is the largest distance between the vertices in it.

A *circuit C_n of length n* is a regular connected graph of degree 2 on n vertices. Considered as a subgraph, C_1 is a loop, C_2 is a pair of parallel edges, C_3 is a *triangle*, C_4 is a *quadrangle*. The *girth* of a multi-(di-)graph is the length of a shortest circuit (cycle) contained in it.

A multigraph G is said to be *properly coloured* if each vertex is coloured so that adjacent vertices have different colours. G is *k-colourable* if it can be properly coloured by k colours. The *chromatic number* $\chi(G)$ is k if G is k-colourable and not $(k - 1)$-colourable. G is called *bipartite* if its chromatic number is 1 or 2. The vertex set of a bipartite multigraph G can be partitioned into two parts, say $\mathcal{X}$ and $\mathcal{Y}$, in such a way that every edge of G connects a vertex from $\mathcal{X}$ with a vertex from $\mathcal{Y}$. If $\mathcal{U}$ denotes the edge set of G, we have also the following notation: $G = (\mathcal{X}, \mathcal{Y}; \mathcal{U})$. If G is connected and has an edge, then $\mathcal{X}$ and $\mathcal{Y}$ are non-void and (up to an interchange) uniquely determined. If the vertices are labelled so that

$$\mathcal{X} = \{x_1, x_2, \ldots, x_m\}, \qquad \mathcal{Y} = \{x_{m+1}, x_{m+2}, \ldots, x_{m+n}\},$$

then the adjacency matrix of G takes the form

$$\boldsymbol{A} = \begin{pmatrix} \boldsymbol{O} & \boldsymbol{B}^{\mathsf{T}} \\ \boldsymbol{B} & \boldsymbol{O} \end{pmatrix},$$

where $\boldsymbol{B}$ is an $n \times m$ matrix and $\boldsymbol{B}^{\mathsf{T}}$ is the transpose of $\boldsymbol{B}$.

A multigraph is called *semiregular of degrees* r_1, r_2 (possibly $r_1 = r_2$) if it is bipartite, each vertex has valency r_1 or r_2, and each edge connects a vertex of valency r_1 with a vertex of valency r_2.

K_n denotes the *complete graph on n vertices* (any two distinct vertices of K_n are connected by an edge). $K_{m,n}$ is a *complete bipartite graph on* $n + m$ *vertices*; $K_{1,n}$ is called a *star*. The *complete k-partite graph on* $n_1 + n_2 + \cdots + n_k$ *vertices* is denoted by $K_{n_1, n_2, \ldots, n_k}$.

A *forest* is a graph without circuits, a *tree* is a connected forest.

The *complement* $\bar{G}$ *of a graph* G is the graph with the same vertex set as G, where any two distinct vertices are adjacent if and only if they are non-adjacent in G. Obviously, $\bar{\bar{G}} = G$. A graph without any edges is called *totally disconnected*, its complement is a complete graph.

The *subdivision graph* $S(G)$ of a graph G is obtained from G by replacing each of its edges by a path of length 2, or, equivalently, by inserting an additional vertex into each edge of G. Clearly, $S(G)$ is a bipartite graph $(\mathscr{X}, \mathscr{Y}; \mathscr{U})$ where $\mathscr{X}$ and $\mathscr{Y}$ are the sets of the original and of the additional vertices, respectively.

The *line graph* $L(G)$ of a graph G is the graph whose vertices correspond to the edges of G with two vertices being adjacent if and only if the corresponding edges in G have a vertex in common.

The *vertex-edge incidence matrix* $\boldsymbol{R}$ of a loopless multigraph $G = (\mathscr{X}, \mathscr{U})$ is defined as follows: Let

$$\mathscr{X} = \{x_1, x_2, \ldots, x_n\}, \qquad \mathscr{U} = \{u_1, u_2, \ldots, u_m\}.$$

$\boldsymbol{R} = (b_{ij})$ is an $n \times m$ matrix where $b_{ij} = 1$ if x_i is incident with (i.e., is an end vertex of) u_j, and $b_{ij} = 0$ otherwise. The *edge-vertex incidence matrix* is the transpose $\boldsymbol{R}^{\mathsf{T}}$ of $\boldsymbol{R}$.

The adjacency matrix of a multi-(di-)graph G is denoted by $\boldsymbol{A} = \boldsymbol{A}(G)$. The *valency* or *degree matrix* $\boldsymbol{D}$ of a multigraph is a diagonal matrix with the valency v_i of vertex x_i in the position (i, i).

It is not difficult to see that, for a graph G, the vertex-edge incidence matrix $\boldsymbol{R}$, the degree matrix $\boldsymbol{D}$, and the adjacency matrices of G, $L(G)$, and $S(G)$ are connected by the following formulas:

$$\boldsymbol{A}(G) = \boldsymbol{R}\boldsymbol{R}^{\mathsf{T}} - \boldsymbol{D},$$

$$\boldsymbol{A}\big(L(G)\big) = \boldsymbol{R}^{\mathsf{T}}\boldsymbol{R} - 2\boldsymbol{I},$$

$$\boldsymbol{A}\big(S(G)\big) = \begin{pmatrix} \boldsymbol{O} & \boldsymbol{R}^{\mathsf{T}} \\ \boldsymbol{R} & \boldsymbol{O} \end{pmatrix}$$

The above definitions and formulas can easily be generalized for arbitrary multigraphs.

The $(0, 1, -1)$-*incidence matrix* $\boldsymbol{V}$ of a loopless multi-digraph G with vertices $x_1, x_2, \ldots, x_n$ and arcs $u_1, u_2, \ldots, u_m$ is defined as follows: $\boldsymbol{V} = (v_{ij})$ is an $n \times m$

matrix where

$$v_{ij} = 1 \text{ if } u_j \text{ issues from } x_i,$$

$$v_{ij} = -1 \text{ if } u_j \text{ terminates in } x_i,$$

$$v_{ij} = 0 \text{ otherwise}.$$

In the majority of cases we shall use the following standard notation.

The number of vertices of a graph is denoted by n, the number of edges or arcs by m. The degree of a regular graph is denoted by r as is the index of a graph (see the next section). The symbol $\boldsymbol{I}$ means a unit matrix in general and $\boldsymbol{I}_n$ is a unit matrix of order n. The symbol $\boldsymbol{J}$ denotes a square matrix all of whose entries are equal to 1. The transpose of a matrix $\boldsymbol{X}$ is denoted by $\boldsymbol{X}^\top$, and $rk\boldsymbol{X}$ is the rank of $\boldsymbol{X}$.

The *Kronecker symbol* δ_{ij} is defined by $\delta_{ii} = 1$ and $\delta_{ij} = 0$ if $j \neq i$.

$a \mid b$ means a divides b.

In the case of undirected multigraphs the spectrum consists of real numbers. In that case, the eigenvalues $\lambda_1, \lambda_2, \ldots, \lambda_n$ are ordered so that always $\lambda_1 = r \geqq \lambda_2 \geqq \cdots \geqq \lambda_n$.

Other notations and graph theoretic concepts will be given at the place of their use.

0.3. Some theorems from matrix theory and their application to the spectrum of a graph

Some fundamental properties of spectra of graphs (or, more generally, multi-digraphs) can be established immediately by using several theorems of matrix theory. We shall formulate in this section only the most important matrix theorems. Others, which are also useful, will be given in the subsequent chapters as lemmas at the places where they are needed.

The set of eigenvectors belonging to an eigenvalue λ along with the zero vector forms the *eigenspace belonging to* λ. The *geometric multiplicity of an eigenvalue* λ is the dimension of its eigenspace. The *algebraic multiplicity of* λ is the multiplicity of λ considered as a zero of the corresponding characteristic polynomial. The geometric multiplicity is never greater than the algebraic multiplicity.

A matrix $\boldsymbol{X}$ is called *symmetric* if $\boldsymbol{X}^\top = \boldsymbol{X}$.

Theorem 0.1 (see, for example, [MaMi], p. 64): *The geometric and algebraic multiplicities of an eigenvalue of a symmetric matrix are equal.*

In the subsequent text the *multiplicity of an eigenvalue* will always mean the algebraic multiplicity.

A matrix is called *non-negative* if all its elements are non-negative numbers.

Since the adjacency matrix of a multi-(di-)graph G is non-negative, the spectrum of G has the properties of the spectrum of non-negative matrices. For non-negative matrices the following theorem holds.

Theorem 0.2 (see, for example, [Gant], vol. II, p. 66): *A non-negative matrix always has a non-negative eigenvalue r such that the moduli of all its eigenvalues do not exceed r. To this "maximal" eigenvalue there corresponds an eigenvector with non-negative coordinates.*

In the subsequent text a vector with positive (non-negative) coordinates will be called a *positive* (*non-negative*) *vector*. A matrix $\boldsymbol{A}$ is called *reducible* if there is a permutation matrix $\boldsymbol{P}$ such that the matrix $\boldsymbol{P}^{-1}\boldsymbol{AP}$ is of the form $\begin{pmatrix} \boldsymbol{X} & \boldsymbol{O} \\ \boldsymbol{Y} & \boldsymbol{Z} \end{pmatrix}$, where $\boldsymbol{X}$ and $\boldsymbol{Z}$ are square matrices. Otherwise, $\boldsymbol{A}$ is called *irreducible.*

Spectral properties of irreducible non-negative matrices are described by the following theorem of FROBENIUS.

Theorem 0.3 (see, for example, [Gant], vol. II, pp. 53–54): *An irreducible non-negative matrix $\boldsymbol{A}$ always has a positive eigenvalue r that is a simple root of the characteristic polynomial. The modulus of any other eigenvalue does not exceed r. To the "maximal" eigenvalue r there corresponds a positive eigenvector. Moreover, if $\boldsymbol{A}$ has h eigenvalues of modulus r, then these numbers are all distinct and are roots of the equation $\lambda^h - r^h = 0$. More generally: the whole spectrum $[\lambda_1 = r, \lambda_2, \ldots, \lambda_n]$ of $\boldsymbol{A}$, regarded as a system of points in the complex λ-plane, is mapped onto itself under a rotation of the plane by the angle $\frac{2\pi}{h}$. If $h > 1$, then by a permutation of rows and the same permutation of columns $\boldsymbol{A}$ can be put into the following "cyclic" form*

$$\boldsymbol{A} = \begin{bmatrix} \boldsymbol{O} & \boldsymbol{A}_{12} & \boldsymbol{O} & \cdots & \boldsymbol{O} \\ \boldsymbol{O} & \boldsymbol{O} & \boldsymbol{A}_{23} & \cdots & \boldsymbol{O} \\ \vdots & & & \ddots & \\ \boldsymbol{O} & \boldsymbol{O} & \boldsymbol{O} & \cdots & \boldsymbol{A}_{h-1,h} \\ \boldsymbol{A}_{h1} & \boldsymbol{O} & \boldsymbol{O} & \cdots & \boldsymbol{O} \end{bmatrix}, \tag{0.1}$$

where there are square blocks along the main diagonal.

If $h > 1$, the matrix $\boldsymbol{A}$ is called *imprimitive* and h is the *index of imprimitivity*. Otherwise, $\boldsymbol{A}$ is *primitive.*

According to Theorem 0.3, the spectrum of a multi-(di-)graph G lies in the circle $|\lambda| \leqq r$, where r is the greatest real eigenvalue. This eigenvalue is called the *index of G*. The algebraic multiplicity of the index can be greater than 1 and there exists a corresponding eigenvector which is non-negative.

Irreducibility of the adjacency matrix of a graph is related to the property of connectedness. *A strongly connected multi-digraph has an irreducible adjacency matrix and a multi-digraph with irreducible adjacency matrix has the property of strong connectedness* [DuMe], [Sed 1]. In undirected multigraphs the strong connectedness reduces to the property of connectedness.

According to Theorem 0.3, *the index of a strongly connected multi-digraph is a simple eigenvalue of the adjacency matrix and a positive eigenvector belongs to it.*

If the adjacency matrix is symmetric, the converse of the last statement also holds, as shown by the following theorem.

Theorem 0.4 (see, for example, [Gant], vol. II, p. 79): *If the "maximal" eigenvalue r of a non-negative matrix $\boldsymbol{A}$ is simple and if positive eigenvectors belong to r both in $\boldsymbol{A}$ and $\boldsymbol{A}^{\mathsf{T}}$, then $\boldsymbol{A}$ is irreducible.*

Theorem 0.5 (see, for example, [Gant], vol. II, p. 78): *To the "maximal" eigenvalue r of a non-negative matrix $\boldsymbol{A}$ there belongs a positive eigenvector both in $\boldsymbol{A}$ and $\boldsymbol{A}^{\mathsf{T}}$ if and only if $\boldsymbol{A}$ can be represented by a permutation of rows and by the same permutation of columns in quasi-diagonal form $\boldsymbol{A} = \operatorname{diag}(\boldsymbol{A}_1, \ldots, \boldsymbol{A}_s)$, where $\boldsymbol{A}_1, \ldots, \boldsymbol{A}_s$ are irreducible matrices each of which has r as its "maximal" eigenvalue.*

We shall now list some more theorems from the theory of matrices showing new spectral properties of graph.

Theorem 0.6 (see, for example, [Gant], vol. II, p. 69): *The "maximal" eigenvalue r' of every principal submatrix (of order less than n) of a non-negative matrix $\boldsymbol{A}$ (of order n) does not exceed the "maximal" eigenvalue r of $\boldsymbol{A}$. If $\boldsymbol{A}$ is irreducible, then $r' < r$ always holds. If $\boldsymbol{A}$ is reducible, then $r' = r$ holds for at least one principal submatrix.*

Theorem 0.7 (see, for example, [CoSi1]): *The increase of any element of a non-negative matrix $\boldsymbol{A}$ does not decrease the "maximal" eigenvalue. The "maximal" eigenvalue increases strictly if $\boldsymbol{A}$ is an irreducible matrix.*

Theorems 0.6 and 0.7 state that in a (strongly) connected multi-(di-)graph G every subgraph has the index smaller than the index of G.

Theorem 0.8 (see, for example, [MaMi], p. 64): *All the eigenvalues of a Hermitian*[†] *matrix are real numbers.*

Theorem 0.9 (see, for example, [Hof1]): *Let $\boldsymbol{A}$ be a real symmetric matrix whose greatest and smallest eigenvalues are denoted by r and q, respectively. Let $\boldsymbol{x}$ be the eigenvector belonging to r. For a principal submatrix $\boldsymbol{B}$ of $\boldsymbol{A}$, let q' be the smallest eigenvalue whose eigenvector is denoted by $\boldsymbol{y}$. Then $q' \geqq q$. If $q' = q$, vector $\boldsymbol{y}$ is orthogonal to the projection of vector $\boldsymbol{x}$ on the subspace corresponding to $\boldsymbol{B}$.*

Theorem 0.10 (see, for example, [MaMi], p. 119): *Let $\boldsymbol{A}$ be a Hermitian matrix with eigenvalues $\lambda_1, \ldots, \lambda_n$ and $\boldsymbol{B}$ be one of its principal submatrices; let $\boldsymbol{B}$ have eigenvalues $\mu_1, \ldots, \mu_m$. Then the inequalities $\lambda_{n-m+i} \leqq \mu_i \leqq \lambda_i$ $(i = 1, \ldots, m)$ hold.*

These inequalities are known as *Cauchy's inequalities* and the whole theorem is also known as *interlacing theorem.*

Theorem 0.11 (C. C. Sims, see [HeHi])[††]: *Let $\boldsymbol{A}$ be a real symmetric matrix with eigenvalues $\lambda_1, \ldots, \lambda_n$. Given a partition $\{1, \ldots, n\} = \Delta_1 \cup \Delta_2 \cup \cdots \cup \Delta_m$ with $|\Delta_i| = n_i > 0$,*

[†] The complex matrix $\boldsymbol{A} = (a_{ij})$ is called *Hermitian* if $\boldsymbol{A}^{\mathsf{T}} = \bar{\boldsymbol{A}}$, i.e. $a_{ji} = \bar{a}_{ij}$.
[††] Recently W. H. Haemers [Haem] has shown that the interlacing properties also hold for matrices $\boldsymbol{A}$ and $\boldsymbol{B}$ of this theorem.

consider the corresponding blocking $\boldsymbol{A} = (\boldsymbol{A}_{ij})$, *so that* $\boldsymbol{A}_{ij}$ *is an* $n_i \times n_j$ *block. Let* e_{ij} *be the sum of the entries in* $\boldsymbol{A}_{ij}$ *and put* $\boldsymbol{B} = (e_{ij}/n_i)$ (*i.e.*, e_{ij}/n_i *is an average row sum in* $\boldsymbol{A}_{ij}$). *Then the spectrum of* $\boldsymbol{B}$ *is contained in the segment* $[\lambda_n, \lambda_1]$.

If we assume that in each block $\boldsymbol{A}_{ij}$ from Theorem 0.11 all row sums are equal, then we can say more.

Theorem 0.12 (E. V. Haynsworth [Hayn]; M. Petersdorf, H. Sachs [PeS1]†): *Let* $\boldsymbol{A}$ *be any matrix partitioned into blocks as in Theorem* 0.11. *Let the block* $\boldsymbol{A}_{ij}$ *have constant row sums* b_{ij} *and let* $\boldsymbol{B} = (b_{ij})$. *Then the spectrum of* $\boldsymbol{B}$ *is contained in the spectrum of* $\boldsymbol{A}$ (*having in view also the multiplicities of the eigenvalues*).

The square matrices $\boldsymbol{A}$ and $\boldsymbol{B}$ are called *similar* if there is a (non-singular) square matrix $\boldsymbol{X}$ transforming $\boldsymbol{A}$ into $\boldsymbol{B}$, i.e., such that $\boldsymbol{X}^{-1}\boldsymbol{A}\boldsymbol{X} = \boldsymbol{B}$. Each symmetric matrix and each matrix which has all distinct eigenvalues is similar to a diagonal matrix. If $\boldsymbol{A}$ is the adjacency matrix of a multigraph, then $\boldsymbol{A}$ is symmetric and, consequently, similar to a diagonal matrix $\boldsymbol{D}$, namely, $\boldsymbol{D} = (\delta_{ij}\lambda_i)$.

We mention the famous *Cayley-Hamilton Theorem* which says that each square matrix $\boldsymbol{A}$ satisfies its own characteristic equation, i.e.:

$$\textit{If } f(\lambda) = |\lambda \boldsymbol{I} - \boldsymbol{A}|, \textit{ then } f(\boldsymbol{A}) = \boldsymbol{O}.$$

The *minimal polynomial* $m(\lambda)$ *of* $\boldsymbol{A}$ is the polynomial $m(\lambda) = \lambda^\mu + \cdots$ such that

(i) $m(\boldsymbol{A}) = \boldsymbol{O}$,

(ii) under condition (i), the degree μ of $m(\lambda)$ has its minimum value.

Then the following propositions hold:

(a) $m(\lambda)$ *is uniquely determined by* $\boldsymbol{A}$.

(b) *If* $F(\lambda)$ *is any polynomial with* $F(\boldsymbol{A}) = \boldsymbol{O}$, *then* $m(\lambda) \mid F(\lambda)$; *in particular,* $m(\lambda) \mid f(\lambda)$.

(c) *Let* $\{\lambda^{(1)}, \lambda^{(2)}, \ldots, \lambda^{(k)}\}$ *be the set of distinct eigenvalues of* $\boldsymbol{A}$, $\lambda^{(\varkappa)}$ *having algebraic multiplicity* $m_\varkappa$. *Then*

$$f(\lambda) = (\lambda - \lambda^{(1)})^{m_1} (\lambda - \lambda^{(2)})^{m_2} \cdots (\lambda - \lambda^{(k)})^{m_k}$$

and

$$m(\lambda) = (\lambda - \lambda^{(1)})^{q_1} (\lambda - \lambda^{(2)})^{q_2} \cdots (\lambda - \lambda^{(k)})^{q_k}$$

where the $q_\varkappa$ *satisfy*

$$0 < q_\varkappa \leqq m_\varkappa \qquad (\varkappa = 1, 2, \ldots, k).$$

(d) *If* $\boldsymbol{A}$ *is similar to a diagonal matrix, then all* $q_\varkappa$ *are equal to* 1:

$$m(\lambda) = (\lambda - \lambda^{(1)}) (\lambda - \lambda^{(2)}) \cdots (\lambda - \lambda^{(k)}).$$

(e) *Let* $\boldsymbol{A}$ *have order* n. *If* $\boldsymbol{A}$ *has all distinct eigenvalues, then*

$$m(\lambda) = f(\lambda) = (\lambda - \lambda^{(1)}) (\lambda - \lambda^{(2)}) \cdots (\lambda - \lambda^{(n)}).$$

---

† See Theorem 4.7.

A square matrix with the property that its minimal and characteristic polynomials are identical is called *non-derogatory*. Thus Proposition (e) says that *a square matrix which has all eigenvalues distinct is non-derogatory.*

We shall now describe some more basic properties of the spectrum of an undirected multigraph. The facts will be given almost without any proof for the convenience of the reader. The proofs can be found at the corresponding places in the subsequent chapters.

The adjacency matrix of an undirected multigraph G is symmetric (and, therefore, Hermitian) and the spectrum of G, containing only real numbers, according to Theorem 0.8 lies in the segment $[-r, r]$.

Let $[\lambda_1, \ldots, \lambda_n]$ be the spectrum of a multigraph. *Twice the number of loops is equal to the trace of the adjacency matrix.* Therefore, we have for multigraphs without loops $\operatorname{tr} \boldsymbol{A} = 0$, i.e., $\lambda_1 + \cdots + \lambda_n = 0$. The number of vertices is, of course, equal to n, and *for undirected graphs without loops or multiple edges the number m of edges is given by* $m = \frac{1}{2} \sum_{i=1}^{n} \lambda_i^2$ (see Section 3.2).

It is stated in [CoSi 1] that for the index r of a connected graph the inequality $2 \cos \frac{\pi}{n+1} \leqq r \leqq n - 1$ holds. The lower bound is attained by a path, and the upper bound by a complete graph. If we omit the assumption of connectedness, then for a graph without edges we have $r = 0$ and otherwise $r \geqq 1$.

For the smallest eigenvalue q of the spectrum of a graph G the inequality $-r \leqq q \leqq 0$ holds. For the graph without edges we have $q = 0$. Otherwise $q \leqq -1$. This is a consequence of Theorem 0.9, since the subgraph K_2 corresponds to a principal submatrix with least eigenvalue equal to -1. We have $q = -1$ if and only if all components of G are complete graphs (Theorem 6.4). The lower bound $q = -r$ is achieved if a component of G having the greatest index is a bipartite graph (Theorem 3.4). According to the foregoing, the following theorem describes the fundamental spectral properties of (undirected) graphs.

Theorem 0.13: *For the spectrum $[\lambda_1, \ldots, \lambda_n]$ of an (undirected) graph G the following statements hold:*

1° *The numbers $\lambda_1, \ldots, \lambda_n$ are real and $\lambda_1 + \cdots + \lambda_n = 0$.*

2° *If G contains no edges, we have $\lambda_1 = \cdots = \lambda_n = 0$.*

3° *If G contains at least one edge, we have*

$$1 \leqq r \leqq n - 1, \tag{0.2}$$

$$-r \leqq q \leqq -1. \tag{0.3}$$

In (0.2) *the upper bound is attained if and only if G is a complete graph, while the lower bound is reached if and only if the components of G consists of graphs K_2 and possibly K_1. In* (0.3) *the upper bound is reached if and only if the components of G are complete graphs, and the lower bound if and only if a component of G having the greatest index is* *a bipartite graph. If G is connected, the lower bound in* (0.2) *is replaced with* $2\cos\frac{\pi}{n+1}$. *Then equality holds if and only if G is a path.*

We shall now list some spectral properties of *regular* multigraphs. *The index is equal to the degree* [CoSi1]. It can easily be seen that this holds for disconnected multigraphs too, but then the index is not a simple eigenvalue. *The multiplicity of the index is equal to the number of components.* It can be seen immediately that *the vector having all coordinates equal to* 1 *is an eigenvector that corresponds to the index.*

The eigenvectors of the other eigenvalues are orthogonal to this vector, i.e., *the sum of their coordinates is equal to* 0.

Further spectral properties of graphs can be obtained using the fact that the coefficients of the characteristic polynomial are integers. It follows from this that the elementary symmetric functions and sums of k-th powers (k a natural number) of eigenvalues are integers, too. Since the coefficient of the highest power term of the characteristic polynomial is equal to 1, *rational eigenvalues* (if they exist) *are integers.*

1. Basic Properties of the Spectrum of a Graph

The ordinary spectrum of a (multi-di-)graph G is the spectrum of its adjacency matrix, but there are various other methods of connecting a spectrum or a characteristic polynomial with G. A general method of defining characteristic polynomials (in one or more variables) and graph spectra is outlined, the most important spectra currently used and their interrelations are discussed, and it is shown how the coefficients of the corresponding characteristic polynomials can be obtained directly from the "cyclic structure" or from the "tree structure" of G, respectively. Eventually, the generating function for the numbers of walks of length k ($k = 1, 2, \ldots$) in G is expressed in terms of the ordinary characteristic polynomial and some conclusions are drawn.

1.1. The adjacency matrix and the (ordinary) spectrum of a graph

In order to obtain an arithmetic method for describing and investigating the structural properties of a finite (directed or undirected) (multi-)graph G, it seems quite reasonable to start with the adjacency matrix $\boldsymbol{A}$ of G.

Obviously, G is uniquely determined by $\boldsymbol{A}$, but the converse statement does not, in general, hold true since the ordering (numbering) of the vertices of G is arbitrary: To each graph G there corresponds uniquely a class $\mathscr{A} = \mathscr{A}(G)$ of adjacency matrices, two adjacency matrices $\boldsymbol{A}$ and $\boldsymbol{A}^*$ belonging to the same class (i.e., determining the same graph) if and only if there is a permutation matrix $\boldsymbol{P}$ such that $\boldsymbol{A}^* = \boldsymbol{P}^{-1}\boldsymbol{A}\boldsymbol{P}$. Thus the theory of graphs G may be identified with the theory of these matrix classes $\mathscr{A}$ and their invariants. An important invariant of a class $\mathscr{A}$ is the *characteristic polynomial* $P_G(\lambda) = |\lambda \boldsymbol{I} - \boldsymbol{A}|$ with $\boldsymbol{A} \in \mathscr{A}(G)$, or, what amounts to the same thing, the *spectrum* $\mathbf{Sp}(G) = [\lambda_1, \lambda_2, \ldots, \lambda_n]$, where the λ_i's are the roots of the equation $P_G(\lambda) = 0$ (i.e., the *eigenvalues* of $\boldsymbol{A}$).[†]

The main question arising is this: how much information concerning the structure of G is contained in its spectrum, and how can this information be retrieved from the spectrum? Of course, the amount of information contained in the spectrum must

† In order to avoid confusion, this "ordinary" spectrum will later sometimes be called the *P-spectrum of G*, and it will be denoted by $\mathbf{Sp}_P(G)$.

not be overestimated, since the spectrum remains invariant not only under the group of permutations, but also under the group of all orthogonal (and even of all non-singular) transformations: Thus the spectrum reflects common properties of all those graphs the adjacency matrices of which may be transformed into one another by some non-singular matrix. Any such matrix transforming the adjacency matrix $\boldsymbol{A}$ of a graph G into the adjacency matrix $\boldsymbol{A}'$ of some graph G' not isomorphic with G is subject to stringent diophantine conditions as all entries of $\boldsymbol{A}$ and $\boldsymbol{A}'$ are required to be non-negative integers: Therefore it may be expected that the classes of *isospectral*† graphs are, in a sense, not too extensive. Isomorphic graphs are, of course, isospectral, and it has been conjectured, conversely, that any two isospectral graphs are isomorphic; however, this is not true. It is very easy indeed to find isospectral non-isomorphic digraphs, e.g., all digraphs with n vertices, containing no cycle, have the same spectrum [0, 0, ..., 0] (see 1.4, Theorem 1.2).

An essentially different situation arises if only undirected (multi-)graphs are taken into consideration, and the construction of pairs of isospectral non-isomorphic (multi-)graphs becomes more and more difficult if one passes from multigraphs to graphs and from graphs to regular graphs. Thus the spectral method may be expected to be particularly efficient, when applied to the class of regular graphs.

Nevertheless, in the theory of block designs it has been shown that, even among strongly regular graphs (which form a narrow subclass within the class of all regular graphs) with sufficiently many vertices, pairs of isospectral non-isomorphic graphs†† are in fact not uncommon; see Chapter 6.

This phenomenon may, on the one hand, be taken as an indication of the scope and the bounds of this special spectral method; on the other hand, it probably reflects a peculiarity of the theory of block designs, showing that there are indeed close relations between this theory and the spectral method.

1.2. A general method for defining different kinds of graph spectra

In this section we shall consider another very natural approach to the spectral method which, by appropriate variation, yields arbitrarily many different "spectra", i.e., systems of numerical invariants.

Let us start with the *ordinary spectrum* $\boldsymbol{Sp}_P(G)$, as an example. We consider a set of n (unspecified) variables x_k being in (1, 1)-correspondence with the set of vertices k ($k = 1, 2, \ldots, n$) of a given (multi-di-)graph $G = (\mathscr{X}, \mathscr{U})$. We try to find numerical values x_k^0 for all of the x_k, not all equal to zero and such that for each vertex i the corresponding number x_i^0 is proportional to the sum s_i^0 of all those x_k^0 corresponding to the (front) neighbours of i (i.e., such that the ratio $s_i^0 : x_i^0$ is the same for all i). In other words, the x_k^0 are to satisfy, in a non-trivial way, the system of homogeneous

---

† Graphs having the same spectrum are called *isospectral* or *cospectral*.

†† Such a Pair of Isospectral Non-isomorphic Graphs is sometimes given the acronym *PING*; more information about the construction of PINGs will be found in Chapter 6.

linear equations

$$\lambda x_i = \sum_{k \cdot i} x_k \qquad (i \in \mathscr{X}),^{\dagger} \tag{1.1}$$

the value of λ being suitably chosen; if G is a multi-(di-)graph, the multiplicity a_{ik} of the adjacency $k \cdot i$ is to be taken into account by considering x_k exactly a_{ik} times as a member of the right side sum of (1.1). Obviously, (1.1) may be given the shorter form

$$\lambda \boldsymbol{x} = \boldsymbol{A}\boldsymbol{x}, \tag{1.2}$$

$\boldsymbol{A} = (a_{ik})$ being the adjacency matrix of G and $\boldsymbol{x}$ denoting a column vector with components x_k $(k \in \mathscr{X})$. As a necessary and sufficient condition for the existence of a non-trivial solution of (1.1) or (1.2), we have

$$|\lambda \boldsymbol{I} - \boldsymbol{A}| \equiv P_G(\lambda) = 0,$$

i.e., the possible proportionality factors λ are identical with the eigenvalues of G.

This way of reasoning has the advantage of being particularly intuitive, as the components of an eigenvector may be directly interpreted as "weights" of the corresponding vertices; at a later stage we shall find that the immediate rationale of the spectrum (via equations (1.1)) by inspection of the graph itself and, particularly, simultaneous consideration of its eigenvectors, will be very useful for a series of investigations and proofs.

Certain applications necessitate the determination of the weights x_k^* of the vertices in such a way that x_i^* is proportional not to the sum (as above) but to the mean value of all those x_k^* corresponding to the (front) neighbours of i, i.e., the x_k^* are required to satisfy the system of equations

$$\lambda x_i = \frac{1}{d_i} \sum_{k \cdot i} x_k \qquad (i \in \mathscr{X}).^{\dagger\dagger} \tag{1.3}$$

(1.3) may be replaced by

$$\lambda \boldsymbol{D}\boldsymbol{x} = \boldsymbol{A}\boldsymbol{x}, \tag{1.4}$$

yielding immediately

$$|\lambda \boldsymbol{D} - \boldsymbol{A}| = 0$$

as a necessary and sufficient condition for the existence of a non-trivial solution of (1.3) or (1.4). Thus we are led to introduce as a modified characteristic polynomial

$$Q_G(\lambda) = \frac{1}{|\boldsymbol{D}|} |\lambda \boldsymbol{D} - \boldsymbol{A}| = \lambda^n + q_1 \lambda^{n-1} + \cdots + q_n \tag{1.5}$$

---

† $k \cdot i$ means that k is a (front) neighbour of i (and i is a (rear) neighbour of k).

†† d_i here denotes the (out-)degree or *(front) valency* of vertex i, i.e. the number of arcs issuing from vertex i; it is assumed here that $d_i > 0$; the diagonal matrix $\boldsymbol{D} = (\delta_{ik} d_i)$ is called the *(out-)degree* or *(front) valency matrix* of G.

with corresponding spectrum

$$\boldsymbol{Sp}_Q(G) = [\lambda_1, \lambda_2, \ldots, \lambda_n]_Q. \tag{1.6}$$

Note that

$$Q_G(\lambda) = |\lambda \boldsymbol{I} - \boldsymbol{D}^{-1}\boldsymbol{A}| = |\lambda \boldsymbol{I} - \boldsymbol{A}\boldsymbol{D}^{-1}|. \tag{1.5$'$}$$

Let $\boldsymbol{D}^{\frac{1}{2}} = \left(\delta_{ik} \sqrt{d_i}\right)$ and $\boldsymbol{A}^* = \boldsymbol{D}^{\frac{1}{2}} (\boldsymbol{D}^{-1}\boldsymbol{A}) \boldsymbol{D}^{-\frac{1}{2}}$. Then

$$\boldsymbol{A}^* = \boldsymbol{D}^{-\frac{1}{2}} \boldsymbol{A} \boldsymbol{D}^{-\frac{1}{2}} = \left(\frac{a_{ik}}{\sqrt{d_i d_k}}\right),$$

$$Q_G(\lambda) = |\lambda \boldsymbol{I} - \boldsymbol{A}^*|. \tag{1.5$''$}$$

For an undirected multigraph G, $\boldsymbol{A}^*$ is symmetric and, consequently, $\boldsymbol{Sp}_Q(G)$ is real.

In (1.5) $\boldsymbol{D}$ appears in a multiplicative manner; $\boldsymbol{D}$ may also be introduced in an additive way: starting from

$$\lambda x_i = d_i x_i + \sum_{k \cdot i} x_k = \sum_{k \cdot i} (x_i + x_k) \qquad (i \in \mathscr{X}) \tag{1.7}$$

we obtain another characteristic polynomial

$$R_G(\lambda) = |\lambda \boldsymbol{I} - \boldsymbol{D} - \boldsymbol{A}| = \lambda^n + r_1 \lambda^{n-1} + \cdots + r_n \tag{1.8}$$

with corresponding spectrum

$$\boldsymbol{Sp}_R(G) = [\lambda_1, \lambda_2, \ldots, \lambda_n]_R \tag{1.9}$$

(cf. L. M. Lihtenbaum [Лих2], E. V. Vahovskij [Вах1]).

J. J. Seidel [LiSe] defines a modified adjacency matrix $\boldsymbol{S} = (s_{ik})$ for (schlicht) graphs in the following way:

$$\left.\begin{aligned} s_{ik} &= \begin{cases} -1 & \text{if } i \text{ and } k \text{ are adjacent} \\ 1 & \text{if } i \text{ and } k \text{ are non-adjacent} \end{cases} \quad (i \neq k), \\ s_{ii} &= 0. \end{aligned}\right\} \tag{1.10}$$

Obviously,

$$\boldsymbol{S} = \boldsymbol{J} - \boldsymbol{I} - 2\boldsymbol{A}, \tag{1.11}$$

$\boldsymbol{J}$ denoting a square matrix all of whose entries are equal to 1.[†]

The system of linear equations, the characteristic polynomial, and the spectrum

---

[†] Obviously, if $\boldsymbol{S}$ is the Seidel matrix of the graph G and $\bar{\boldsymbol{S}}$ is the Seidel matrix of the graph $\bar{G}$ complementary to G, then simply $\bar{\boldsymbol{S}} = -\boldsymbol{S}$.

corresponding to $\boldsymbol{S}$ are

$$\lambda x_i = \sum_{k \in \mathscr{X}} s_{ik} x_k \qquad (i \in \mathscr{X}), \tag{1.12}$$

$$\begin{aligned} S_G(\lambda) &= |\lambda \boldsymbol{I} - \boldsymbol{S}| = |\lambda \boldsymbol{I} - \boldsymbol{J} + \boldsymbol{I} + 2\boldsymbol{A}| \\ &= \lambda^n + s_1 \lambda^{n-1} + \cdots + s_n, \end{aligned} \tag{1.13}$$

$$\boldsymbol{Sp}_S(G) = [\lambda_1, \lambda_2, \ldots, \lambda_n]_S, \tag{1.14}$$

respectively.

In this connection, two more spectra derived from the *matrix of admittance*[†], $\boldsymbol{C} = \boldsymbol{D} - \boldsymbol{A}$, should be mentioned. Some authors (W. N. ANDERSON Jr., T. D. MORLEY [AnMo]; M. FIEDLER [Fie 1]) consider the polynomial

$$C_G(\lambda) = |\lambda \boldsymbol{I} - \boldsymbol{C}| = |\lambda \boldsymbol{I} - \boldsymbol{D} + \boldsymbol{A}| = \lambda^n + c_1 \lambda^{n-1} + \cdots + c_n \tag{1.15}$$

with corresponding spectrum

$$\boldsymbol{Sp}_C(G) = [\lambda_1, \lambda_2, \ldots, \lambda_n]_C \tag{1.16}$$

(using, of course, different notation); A. K. KEL'MANS [KeJI 1] introduces a polynomial

$$B_\lambda^n(G) = \frac{1}{\lambda} |\lambda \boldsymbol{I} + \boldsymbol{C}| \tag{1.17}$$

of order $n - 1$; clearly,

$$B_\lambda^n(G) = \frac{(-1)^n}{\lambda} C_G(-\lambda),$$

so that no special symbol is required for the Kel'mans spectrum.

All the spectra considered so far — and only these[††] — are to be found in the literature; we shall return to this point in the next section.

We observe that all of the spectra dealt with up to this point may be derived from systems of linear equations the coefficients of which are connected with local structural properties of the graph in question. But the idea of obtaining systems of numerical invariants by exploiting the solvability conditions for a system of equations connected with the graph and depending on certain parameters is not at all restricted to the use of linear equations; for example, a most natural way of extending the method consists in the transition to a system of *quadratic* equations of the form

$$\lambda x_i^2 = \sum_{\substack{j \cdot i \\ k \cdot i}} x_j x_k \qquad (i \in \mathscr{X}) \tag{1.18}$$

taking the multiplicities of the adjacencies into account by summing over all pairs of different edges (arcs) which have i as a starting vertex. In terms of the adjacency

---

† The name *matrix of admittance* is taken from the theory of electrical networks: any multigraph G may be considered as corresponding to a special electrical network all branches of which have admittance (= conductivity) 1.

†† In addition, of course, mention should be made of the "distance polynomial" and corresponding spectrum; see Section 9.2.

matrix $\boldsymbol{A} = (a_{ik})$, (1.18) can be expressed in the following form:

$$\lambda x_i^2 = \sum_{1 \leq j < k \leq n} a_{ij} a_{ik} x_j x_k + \sum_{j=1}^{n} \binom{a_{ij}}{2} x_j^2 \qquad (i = 1, 2, \ldots, n). \tag{1.19}$$

(The right-hand side of (1.18) and (1.19) is nothing other than the elementary symmetric function of the second order of all x_k corresponding to the (front) neighbours of i, taking into account the multiplicities of the adjacencies.)

The set of all values of λ for which (1.18) and (1.19) have solutions consists of all zeros of the resultant $\mathbf{R}_G(\lambda)$ of the system (1.18): so the polynomial $\mathbf{R}_G(\lambda)$ and the system of the roots of the equation $\mathbf{R}_G(\lambda) = 0$ (condition of compatibility) can be considered as a characteristic polynomial "of quadratic origin" and the corresponding "quadratic" spectrum, respectively.

Instead of a system of quadratic equations a system of cubic (biquadratic, ...) equations could be taken into consideration, and if G is not regular we may connect a system of homogeneous equations depending on more than one parameter (one parameter for each degree, see next section) with the graph G thus obtaining a characteristic polynomial depending on several variables. We may even leave the field of algebra and connect with G a system of suitably chosen functional equations (boundary value problem, system of integral equations, ...),† thus obtaining also spectra with infinitely many eigenvalues: the possibilities of connecting "spectra" with graphs are many and varied.

It would be very desirable to learn something about the correlations between these different kinds of spectra and especially about the particular role which the "linear" spectra play among them: Perhaps it may be possible to specify some finite system of suitable spectra of a graph G, which, taken as a whole, completely characterize G.

Interesting as these problems are, they seem to be difficult ones,†† and, since there are at present scarcely any known results worth mentioning, we shall confine ourselves in this book to investigations concerning *linear spectra*, as described above.

1.3. Some remarks concerning current spectra

All spectra commonly used have been listed in the preceding section; it may be worth mentioning that all of them can be derived from a common source (the Seidel spectrum playing a somewhat exceptional role): Put

$$F_G(\lambda, \mu) = |\lambda \boldsymbol{I} + \mu \boldsymbol{D} - \boldsymbol{A}|. \tag{1.20}$$

---

† A first step in this direction can be found in [PeS1] (note that formulas (3) and (6) of [PeS1] are incorrect, they should be replaced by the above formula (1.19)). See also [Sac15].

†† Experimenting techniques applied to resultants of systems of non-linear algebraic equations will hopelessly fail as the orders of the resulting polynomials are in general beyond any reasonable size — even in simple cases.

Then

$$P_G(\lambda) = |\lambda \boldsymbol{I} - \boldsymbol{A}| = F_G(\lambda, 0), \tag{1.21}$$

$$Q_G(\lambda) = \frac{1}{|\boldsymbol{D}|} |\lambda \boldsymbol{D} - \boldsymbol{A}| = \frac{1}{|\boldsymbol{D}|} F_G(0, \lambda), \tag{1.22}$$

$$R_G(\lambda) = |\lambda \boldsymbol{I} - \boldsymbol{D} - \boldsymbol{A}| = F_G(\lambda, -1), \tag{1.23}$$

$$C_G(\lambda) = |\lambda \boldsymbol{I} - \boldsymbol{D} + \boldsymbol{A}| = (-1)^n F_G(-\lambda, 1). \tag{1.24}$$

As for the Seidel spectrum, we can only state

$$\begin{aligned} S_G(\lambda) = |\lambda \boldsymbol{I} - \boldsymbol{S}| &= (-1)^n \cdot 2^n F_{G^*}\left(-\frac{\lambda+1}{2}, 0\right) \\ &= (-1)^n \cdot 2^n P_{G^*}\left(-\frac{\lambda+1}{2}\right); \end{aligned} \tag{1.25}$$

here G^* stands for a "generalized graph" with weighted adjacencies having the "adjacency matrix" $\boldsymbol{A} - \frac{1}{2} \boldsymbol{J}$.

Remark. $F_G(\lambda, \mu)$ may be considered as a characteristic polynomial depending on two variables. But (1.20) is, of course, not the only possible way of introducing a characteristic polynomial depending on several variables: If, for example, G is non-regular with s different (out-)degrees $v_1, v_2, \ldots, v_s$, we make a parameter λ_σ correspond to every vertex i with (out-)degree $d_i = v_\sigma$ $(\sigma = 1, 2, \ldots, s)$. Let $\lambda^{(i)}$ denote the parameter belonging to the vertex i (i.e., $\lambda^{(i)} = \lambda_\sigma$ with σ satisfying $d_i = v_\sigma$) and put

$$\Lambda = \begin{pmatrix} \lambda^{(1)} & & 0 \\ & \lambda^{(2)} & \\ & & \ddots \\ 0 & & \lambda^{(n)} \end{pmatrix}, \tag{1.26}$$

then we may generalize $P_G(\lambda) = |\lambda \boldsymbol{I} - \boldsymbol{A}|$ to

$$P_G^*(\lambda_1, \lambda_2, \ldots, \lambda_s) = |\Lambda - \boldsymbol{A}|. \tag{1.27}$$

By the specialization

$$\lambda_\sigma = \lambda + \mu v_\sigma \qquad (\sigma = 1, 2, \ldots, s), \tag{1.28}$$

i.e.,

$$\lambda^{(i)} = \lambda + \mu d_i \qquad (i = 1, 2, \ldots, n),$$

from (1.27) the polynomial $F_G(\lambda, \mu)$ is retrieved:

$$F_G(\lambda, \mu) = P_G^*(\lambda + \mu v_1, \lambda + \mu v_2, \ldots, \lambda + \mu v_s) \tag{1.29}$$

which is also valid in the regular case.

It would certainly be an interesting though possibly difficult task to investigate the significance of these generalized characteristic polynomials, but we shall not pursue such questions in this book. (See also Section 4.5.)

We return to formulas (1.21)–(1.25) and assume G to be a (multi-)graph which is regular of a certain degree r: we shall show that in this case the four spectra $\mathbf{Sp}_P$, $\mathbf{Sp}_Q$,† $\mathbf{Sp}_R$, $\mathbf{Sp}_C$ are equivalent, i.e., contain the same amount of information about the structure of G, and that "almost the same" is also true for $\mathbf{Sp}_S$.

This is quite obvious in the first four cases: Since $\boldsymbol{D} = r\boldsymbol{I}$, we have

$$\lambda \boldsymbol{I} + \mu \boldsymbol{D} = (\lambda + r\mu)\, \boldsymbol{I}$$

and consequently

$$F_G(\lambda, \mu) = F_G(\lambda + r\mu, 0) = P_G(\lambda + r\mu). \tag{1.30}$$

So, according to (1.22)–(1.24),

$$Q_G(\lambda) = \frac{1}{r^n}\, P_G(r\lambda), \tag{1.31}$$

$$R_G(\lambda) = P_G(\lambda - r), \tag{1.32}$$

$$C_G(\lambda) = (-1)^n\, P_G(-\lambda + r), \tag{1.33}$$

and from

$$\mathbf{Sp}_P(G) = [\lambda_1, \lambda_2, \ldots, \lambda_n]^{\dagger\dagger}$$

we deduce

$$\mathbf{Sp}_Q(G) = \left[\frac{\lambda_1}{r}, \frac{\lambda_2}{r}, \ldots, \frac{\lambda_n}{r}\right], \tag{1.31'}$$

$$\mathbf{Sp}_R(G) = [\lambda_1 + r, \lambda_2 + r, \ldots, \lambda_n + r], \tag{1.32'}$$

$$\mathbf{Sp}_C(G) = [r - \lambda_n, r - \lambda_{n-1}, \ldots, r - \lambda_1]. \tag{1.33'}$$

In the case of the Seidel spectrum, by due computation making use of the eigenvectors, we obtain

$$S_G(\lambda) = (-1)^n \cdot 2^n\, \frac{\lambda + 1 + 2r - n}{\lambda + 1 + 2r}\, P_G\left(-\frac{\lambda + 1}{2}\right); \tag{1.34}$$

the eigenvalues with respect to $\boldsymbol{S}$ are $-2\lambda_{n+2-i} - 1$ ($i = 2, 3, \ldots, n$) and, in addition, $n - 2r - 1$. (See also Section 6.5, Lemma 6.6.)

If G is regular of degree r, then, as can easily be checked, $\boldsymbol{x}^0 = (1, 1, \ldots, 1)^\mathsf{T}$ is an eigenvector of its adjacency matrix $\boldsymbol{A}$ belonging to the eigenvalue r, and since all

† Here $r > 0$ is assumed.
†† Note that $\lambda_1 = r$ (see Section 0.3).

components of $\boldsymbol{x}^0$ are positive, it follows from Theorem 0.3 that r is the maximal eigenvalue contained in the P-spectrum of G.

It is worth mentioning that there is still another important class of multigraphs for which the spectra $\boldsymbol{Sp}_P(G)$ and $\boldsymbol{Sp}_Q(G)$ are equivalent, namely, the class of semiregular multigraphs of positive degrees. (Recall: a multigraph G is called *semiregular of degrees* r_1, r_2, if it is bipartite having a representation $G = (\mathcal{X}_1, \mathcal{X}_2; \mathcal{U})$ with $|\mathcal{X}_1| = n_1$, $|\mathcal{X}_2| = n_2$, $n_1 + n_2 = n$, where each vertex $x \in \mathcal{X}_1$ has valency r_1 and each vertex $x \in \mathcal{X}_2$ has valency r_2.) In this case, a straightforward calculation shows that the vector $\boldsymbol{x}^0 = \left(\sqrt{d_1}, \sqrt{d_2}, \ldots, \sqrt{d_n}\right)^{\mathsf{T}}$ (with $d_i = r_1$ or r_2) is an eigenvector of the adjacency matrix of G belonging to the eigenvalue $\sqrt{r_1 r_2}$, and since all components of $\boldsymbol{x}^0$ are positive, it follows again from Theorem 0.3 that $\sqrt{r_1 r_2}$ is the maximal eigenvalue. Recall that the maximal eigenvalue is called the *index* of G denoted by ϱ. According to (1.5)'' (Section 1.2),

$$Q_G(\lambda) = |\lambda \boldsymbol{I} - \boldsymbol{A}^*| = \left|\lambda \boldsymbol{I} - \frac{1}{\varrho} \boldsymbol{A}\right| = \frac{1}{\varrho^n} |\lambda \varrho \boldsymbol{I} - \boldsymbol{A}| = \frac{1}{\varrho^n} P_G(\varrho\lambda).$$

So we have proved

Theorem 1.1 (F. Runge [Rung]): *Let G be a multigraph either regular of positive degree r or semiregular of positive degrees r_1, r_2, and let ϱ be the index of G. Then $\varrho = r$ or $\varrho = \sqrt{r_1 r_2}$, respectively, and in either case*

$$Q_G(\lambda) = \frac{1}{\varrho^n} P_G(\varrho\lambda).$$

Note that *a connected multigraph G is regular or semiregular of positive degree(s) if and only if the line graph of G is regular.*

1.4. The coefficients of $P_G(\lambda)$

In the next three sections we shall be concerned with relating the coefficients of $P_G(\lambda)$, $C_G(\lambda)$, and $Q_G(\lambda)$, respectively, to structural properties of the graph G.

Let G be an arbitrary multi-(di-)graph and

$$P_G(\lambda) = \lambda^n + a_1\lambda^{n-1} + \cdots + a_n$$

its characteristic polynomial. It has been observed by several authors† that the values of the coefficients a_i can easily be computed if the set of all directed cycles of G (considered as a digraph) is known. The converse problem of deducing structural properties of G (for example, concerning the cycles contained in G) from the values of the a_i is much more difficult; we shall return to this problem in Section 3.1.

† See the remark on the history of the "coefficients theorem" (p. 36).

The following theorem is sometimes called the "coefficients theorem for digraphs".

Theorem 1.2 (M. MILIĆ [Mili], H. SACHS [Sac2], [Sac3], L. SPIALTER [Spia][†]): *Let*

$$P_G(\lambda) = |\lambda \boldsymbol{I} - \boldsymbol{A}| = \lambda^n + a_1\lambda^{n-1} + \cdots + a_n$$

be the characteristic polynomial of an arbitrary (directed) multigraph G. Then

$$a_i = \sum_{L \in \mathscr{L}_i} (-1)^{p(L)} \qquad (i = 1, 2, \ldots, n) \tag{1.35}$$

where $\mathscr{L}_i$ is the set of all linear directed subgraphs L of G with exactly i vertices; p(L) denotes the number of components of L (i.e., the number of cycles of which L is composed).

This statement may be given the following form:

The coefficient a_i depends only on the set of all linear directed subgraphs L of G having exactly i vertices, the contribution of L to a_i being +1 if L contains an even, and −1 if L contains an odd, number of cycles.

If G is an undirected multigraph, we may still consider G as a multi-digraph G' (see Section 0.1, p. 12); all that is necessary to observe is that to every edge of G which is not a loop there corresponds a cycle of length 2 in G', and to every circuit of G there corresponds a pair of cycles in G', oriented in opposite directions. Theorem 1.2 may now be easily reformulated for multigraphs as follows:

Theorem 1.3 (H. SACHS [Sac2], [Sac3], L. SPIALTER [Spia][†]): *Let*

$$P_G(\lambda) = |\lambda \boldsymbol{I} - \boldsymbol{A}| = \lambda^n + a_1\lambda^{n-1} + \cdots + a_n$$

be the characteristic polynomial of an arbitrary undirected multigraph G.

Call an "elementary figure"

a) *the graph K_2, or*

b) *every graph C_q ($q \geqq 1$) (loops being included with $q = 1$),*

call a "basic figure" U every graph all of whose components are elementary figures; let p(U), c(U) be the number of components and the number of circuits contained in U, respectively, and let $\mathscr{U}_i$ denote the set of all basic figures contained in G having exactly i vertices.

Then

$$a_i = \sum_{U \in \mathscr{U}_i} (-1)^{p(U)} \cdot 2^{c(U)} \qquad (i = 1, 2, \ldots, n). \tag{1.36}$$

This theorem may be given the following form:

Define the "contribution" b of an elementary figure E by

$$b(K_2) = -1, \qquad b(C_q) = (-1)^{q+1} \cdot 2$$

and of a basic figure U by

$$b(U) = \prod_{E \subset U} b(E).$$

---

† See the remark on the history of the "coefficients theorem" (p. 36).

Then

$$(-1)^i a_i = \sum_{U \in \mathcal{U}_i} b(U). \tag{1.37}$$

Proof of Theorem 1.2. Let us first consider the absolute term

$$a_n = P_G(0) = (-1)^n |\boldsymbol{A}| = (-1)^n |a_{ik}|.$$

According to the Leibniz definition of the determinant,

$$a_n = \sum_{\boldsymbol{P}} (-1)^{n+I(\boldsymbol{P})} a_{1i_1} a_{2i_2} \cdots a_{ni_n} \tag{1.38}$$

with summation taken over all permutations

$$\boldsymbol{P} = \begin{pmatrix} 1 & 2 & \cdots & n \\ i_1 & i_2 & \cdots & i_n \end{pmatrix};$$

$I(\boldsymbol{P})$ denotes, as usual, the parity of $\boldsymbol{P}$. For the sake of simplicity, let us first assume that there are no multiple arcs so that $a_{ik} = 0$ or 1 for all i, k. A term

$$S_{\boldsymbol{P}} = (-1)^{n+I(\boldsymbol{P})} a_{1i_1} a_{2i_2} \cdots a_{ni_n}$$

of the sum (1.38) is different from zero if and only if all of the arcs $(1, i_1)$, $(2, i_2), \ldots$, (n, i_n) are contained in G. $\boldsymbol{P}$ may be represented as a product

$$\boldsymbol{P} = (1 i_1 \cdots)(\cdots) \cdots (\cdots)$$

of disjoint cycles.†

Evidently, if $S_{\boldsymbol{P}} \neq 0$, then to each cycle of $\boldsymbol{P}$ there corresponds a cycle in G: thus to $\boldsymbol{P}$, there corresponds a direct sum of (non-intersecting) cycles containing all vertices of G, i.e., a linear directed subgraph $L \in \mathcal{L}_n$. Conversely: To each linear directed subgraph $L \in \mathcal{L}_n$ there corresponds a permutation $\boldsymbol{P}$ and a term $S_{\boldsymbol{P}} = \pm 1$, the sign depending only on the number $e(L)$ of even cycles (i.e., cycles of even length) among all cycles of L:

$$S_{\boldsymbol{P}} = (-1)^{n+e(L)}.$$

Obviously,

$$n + e(L) \equiv p(L) \pmod 2$$

hence

$$a_n = \sum_{\boldsymbol{P}} S_{\boldsymbol{P}} = \sum_{L \in \mathcal{L}_n} (-1)^{p(L)}. \tag{1.39}$$

Now, (1.39) remains valid even if $a_{ik} > 1$ is allowed:

Consider the set of all distinct linear directed subgraphs $L \in \mathcal{L}_n$ connecting the n vertices of G in exactly the way prescribed by the cycles of a fixed permutation $\boldsymbol{P} = (1 i_1 \cdots)(\cdots) \cdots (\cdots)$. It is clear that this set can be obtained by arbitrarily choosing for each k an arc from vertex k to vertex i_k, and doing so in every possible

† Note that $I(\boldsymbol{P}) \equiv e(\boldsymbol{P}) \pmod 2$, where $e(\boldsymbol{P})$ is the number of even cycles among all cycles of the cycle representation of $\boldsymbol{P}$ given above.

manner; and since for fixed k there are exactly a_{ki_k} possible choices, the total number of subgraphs so obtained equals $a_{1i_1} a_{2i_2} \cdots a_{ni_n}$. Thus the total contribution of all of these subgraphs to the sum $\sum_{L \in \mathcal{L}_n} (-1)^{p(L)}$ equals $(-1)^{n+I(\boldsymbol{P})} a_{1i_1} a_{2i_2} \cdots a_{ni_n}$. Summation with respect to all permutations $\boldsymbol{P}$ confirms the validity of (1.39) in the general case.

In order to complete the proof of (1.35) suppose $1 \leqq i \leqq n$ (i fixed). It is well known that $(-1)^i a_i$ equals the sum of all principal minors (subdeterminants) of order i of $\boldsymbol{A}$. Note that there is a (1, 1)-correspondence between the set of these minors and the set of all induced subgraphs of G having exactly i vertices. By applying the result obtained above to each of the $\binom{n}{i}$ minors, and summing, the validity of (1.35) is established.

Remark. If, instead of the determinant, the *permanent* of $\boldsymbol{A}$,

$$\operatorname{per} \boldsymbol{A} = \sum_{\boldsymbol{P}} a_{1i_1} a_{2i_2} \cdots a_{ni_n},$$

is considered, we obtain by means of analogous deductions the simple formulas

$$\operatorname{per} \boldsymbol{A} = \text{number of directed linear factors}^{\dagger} \text{ of } G, \tag{1.40}$$

and in the case of an undirected multigraph:

$$\operatorname{per} \boldsymbol{A} = \sum_{U \in \mathcal{U}_n} 2^{c(U)}. \tag{1.41}$$

Call *perm-polynomial* of an arbitrary square matrix $\boldsymbol{A}$ of order n the polynomial

$$\operatorname{per} (\lambda \boldsymbol{I} + \boldsymbol{A}) = \lambda^n + a_1^* \lambda^{n-1} + \cdots + a_n^*.$$

The analogues of Theorems 1.2 and 1.3 are then:

Theorem 1.2*: *Let*

$$P_G^*(\lambda) = \operatorname{per} (\lambda \boldsymbol{I} + \boldsymbol{A}) = \lambda^n + a_1^* \lambda^{n-1} + \cdots + a_n^*$$

be the perm-polynomial belonging to an arbitrary (directed) multigraph G with adjacency matrix $\boldsymbol{A}$. Then

$$a_i^* = \text{number of linear directed subgraphs of } G$$
$$\text{containing exactly } i \text{ vertices } (i = 1, 2, \ldots, n). \tag{1.35*}$$

Theorem 1.3*. *Let*

$$P_G^*(\lambda) = \operatorname{per} (\lambda \boldsymbol{I} + \boldsymbol{A}) = \lambda^n + a_1^* \lambda^{n-1} + \cdots + a_n^*$$

---

† A *directed linear factor* of a multi-(di-)graph G is a linear directed subgraph containing all vertices of G.

be the perm-polynomial belonging to an arbitrary undirected multigraph G with adjacency matrix $\boldsymbol{A}$. *Then*

$$a_i^* = \sum_{U \in \mathcal{U}_i} 2^{c(U)} \qquad (i = 1, 2, \ldots, n). \tag{1.36*}$$

Theorems 1.2 and 1.2* may be extended to digraphs with weighted adjacencies immediately:

Suppose that adjacency $k \cdot i$ has (arbitrary) weight a_{ik}†, and let $\boldsymbol{A} = (a_{ik})$ be the corresponding generalized adjacency matrix. Then Theorems 1.2 and 1.2* still hold with

$$a_i = \sum_{L \in \mathcal{L}_i} (-1)^{p(L)} \prod (L) \qquad (i = 1, 2, \ldots, n) \tag{1.35$'$}$$

and

$$a_i^* = \sum_{L \in \mathcal{L}_i} \prod (L) \qquad (i = 1, 2, \ldots, n) \tag{1.35*$'$}$$

instead of (1.35), (1.35*), respectively, $\prod (L)$ denoting the product of the weights of all arcs belonging to L.

If G is an undirected graph with weighted adjacencies and U is a basic figure contained in G, let

$$\prod (U) = \prod_{u \in E(U)} \big(w(u)\big)^{\zeta(u;U)},$$

where $E(U)$ is the set of edges of U, $w(u)$ is the weight of the edge u, and

$$\zeta(u; U) = \begin{cases} 1 \text{ if } u \text{ is contained in some circuit of } U, \\ 2 \text{ otherwise.} \end{cases}$$

Since U contains exactly $2^{c(U)}$ linear directed subgraphs L all having the same weight $\prod (L) = \prod (U)$, (1.35)$'$ takes the simple form

$$a_i = \sum_{U \in \mathcal{U}_i} (-1)^{p(U)} \, 2^{c(U)} \prod (U). \tag{1.35$''$}$$

With $i = n$, we obtain from (1.35)$'$ a simple formula for the calculation of the determinant of an arbitrary square matrix $\boldsymbol{A}$ considered as a generalized adjacency matrix of a digraph G:

$$|\boldsymbol{A}| = (-1)^n \sum_{L \in \mathcal{L}_n} (-1)^{p(L)} \prod (L) \tag{1.42}$$

(note that $\mathcal{L}_n$ is the set of all directed linear factors L of G).

If, in particular, $\boldsymbol{A}$ is the adjacency matrix of a multi-digraph or a multigraph, (1.42) reduces to

$$|\boldsymbol{A}| = (-1)^n \sum_{L \in \mathcal{L}_n} (-1)^{p(L)} \tag{1.42$'$}$$

† We may assume that for every pair i, k there is exactly one arc from i to k, and that a_{ik} is the weight of this arc (possibly equal to zero).

or

$$|A| = (-1)^n \sum_{U \in \mathcal{U}_n} (-1)^{p(U)} 2^{c(U)}, \tag{1.42''}$$

respectively.

(1.42) may be taken as an *intuitive form of the Leibniz definition of the determinant.* A theory of determinants based on this observation was outlined by D. M. CVETKOVIĆ [Cve 15].

Remark (concerning the history of the coefficients theorem). In order to show that this approach is not only of purely theoretical interest, it should be noted that there are two other fields in which determinants have been connected with graphs: electronics-cybernetics (signal flow graph theory) and chemistry (quantum chemistry, simple molecular orbital theory).

Apparently (1.42) was given for the first time by C. L. COATES [Coat] (1959) in connection with flow graph considerations†; (1.42) is therefore sometimes called Coates' formula. A simple proof is given by C. A. DESOER [Deso] (1960). F. HARARY [Har2] (1962) considers the case when A is the adjacency matrix of a digraph or of a graph. But before COATES other authors came close to formula (1.42) (see D. KÖNIG [Kön 1] (1916), [Kön2] (1936); see also T. MUIR [Mui2], footnote on p. 260 concerning Cauchy's rule for determining the sign of a summand in the expansion of a determinant).

For some small values of i, the coefficients a_i of the characteristic polynomial of an undirected graph G were already determined by C. A. COULSON [Cou2] (1949) and I. SAMUEL [Sam 1] (1949) (see also [Sam2]) in the context of molecular orbital theory, and, independently, by L. COLLATZ and U. SINOGOWITZ in their fundamental paper [CoSi 1] (1957)†† on graph spectra. COULSON [Cou2], however, does not use the concept of "basic figures" but expresses the coefficients by means of the numbers of all possible subgraphs of G with the given number of vertices. In this connection, E. HEILBRONNER's papers [Hei 1] (1953), [Hei2] (1954) should also be mentioned; he showed how, in the case of special graphs arising in the molecular orbital theory, the characteristic polynomial can easily be obtained by some intuitive "graphical" recurrence procedures.

It seems that the coefficients theorem in full generality was first published by H. SACHS [Sac3] (1964) (see also [Sac2] (1963)) and almost at the same time by L. SPIALTER [Spia] (1964) (in a terminology appropriate for chemical applications) and M. MILIĆ [Mili] (1964) (in terms of flow graph theory). Later it has been rediscovered several times: J. PONSTEIN [Pons] (1966), J. TURNER [Turn2] (1968), A. BECE [Беце] (1968), A. MOWSHOWITZ [Mow5] (1972), H. HOSOYA [Hos2] (1972), F. H. CLARKE [Clar] (1972); for trees it has also been given by L. LOVÁSZ and J. PELIKÁN [LoPe] (1973). TURNER's paper contains a somewhat more general theorem

† With regard to signal flow graph theory, see the fundamental papers of C. E. SHANNON [Shan] (1942) (which remained unnoticed for several years) and S. J. MASON [Mas1] (1953), [Mas2] (1956); for applications see C. S. LORENS [Lore] (1964). For proofs see also R. B. ASH [Ash] (1959) and A. NATHAN [Nath] (1961). A detailed treatment may be found in the book of W.-K. CHEN [Chen] (1971).

†† Note that this paper had already been prepared during World War II, see [CoSi2].

concerning the coefficients of a generalized characteristic polynomial

$$P_\chi(\lambda) = d_\chi(\boldsymbol{A} - \lambda \boldsymbol{I}),$$

d_χ being a matrix function generalizing determinant as well as permanent given by

$$d_\chi(\boldsymbol{A}) = \sum_{\boldsymbol{P}} \chi(\boldsymbol{P})\, a_{1i_1} a_{2i_2} \cdots a_{ni_n}$$

with summation over all permutations $\boldsymbol{P} = \begin{pmatrix} 1 & 2 & \cdots & n \\ i_1 & i_2 & \cdots & i_n \end{pmatrix}$; here $\chi(\boldsymbol{P})$ denotes some character defined on the symmetric group $\mathscr{S}_n$ of all permutations $\boldsymbol{P}$ considered.

Some simple consequences of Theorems 1.2 *and* 1.3

Proposition 1.1: *The number of linear subgraphs with exactly q edges contained in an undirected forest H is equal to $(-1)^q a_{2q}$. An undirected linear factor exists if and only if $a_n \neq 0$. In this case, n is even, and, as there evidently cannot be more than one linear factor,* $a_n = (-1)^{\frac{n}{2}}$.

Proposition 1.2: *The number of directed linear factors contained in a multi-digraph G is not smaller than* $|a_n|$.

The general problem

> "Let the characteristic polynomial $P_G(\lambda)$ of some multi-(di-)graph G be given, what information about the cycles (or circuits) contained in G can be retrieved from the coefficients a_i?"

will be treated in Chapter 3, Sections 3.1—3.3.

1.5. The coefficients of $C_G(\lambda)$

Next we shall express the coefficients c_i of the polynomial

$$C_G(\lambda) = |\lambda \boldsymbol{I} - \boldsymbol{C}| = \lambda^n + c_1\lambda^{n-1} + \cdots + c_n \tag{1.43}$$

in terms of the "tree structure" of G, where G is any multigraph (recall that $\boldsymbol{C} = \boldsymbol{D} - \boldsymbol{A} = (\delta_{ij}d_i - a_{ij})$ is the matrix of admittance of G; see Section 1.2).

Let $\boldsymbol{M}$ be any square matrix with rows $\boldsymbol{r}_1, \boldsymbol{r}_2, \ldots, \boldsymbol{r}_n$ and columns $\boldsymbol{c}_1, \boldsymbol{c}_2, \ldots, \boldsymbol{c}_n$, let $\mathscr{N} = \{1, 2, \ldots, n\}$ and $\mathscr{J} = \{j_1, j_2, \ldots, j_q\} \subset \mathscr{N}$; let $\boldsymbol{M}_{\mathscr{J}}$ denote the square matrix obtained from $\boldsymbol{M}$ by simultaneously cancelling rows $\boldsymbol{r}_{j_1}, \boldsymbol{r}_{j_2}, \ldots, \boldsymbol{r}_{j_q}$ and columns $\boldsymbol{c}_{j_1}, \boldsymbol{c}_{j_2}, \ldots, \boldsymbol{c}_{j_q}$. For the sake of convenience write $\boldsymbol{M}_i$ instead of $\boldsymbol{M}_{\{i\}}$, etc.; as usual, the determinant of the empty matrix (case $\mathscr{J} = \mathscr{N}$) is assumed to be 1.

If G is any multigraph with n vertices $1, 2, \ldots, n$, and if $\mathscr{J} \neq \emptyset$, let $G_{\mathscr{J}}$ denote the multigraph obtained from G by identifying (amalgamating) the vertices $j_1, j_2, \ldots, j_q$, thereby replacing the set $\{j_1, j_2, \ldots, j_q\}$ by a single new vertex i (by this process multiple edges and loops may be created); evidently, $G_1 = G_2 = \cdots = G_n = G$.

The following well-known important theorem connects the number of spanning trees of a multigraph with its matrix of admittance.

Matrix-Tree-Theorem†: *Let G be a multigraph with vertices $1, 2, \ldots, n$ and let $t(G)$ denote the number of spanning trees contained in G.*†† *Then*

$$t(G) = |C_j|, \tag{1.44}$$

where $C = D - A$ is the matrix of admittance of G and $j \in \{1, 2, \ldots, n\}$.

Corollary: *Let $\mathcal{J} \subset \mathcal{N}$, $\mathcal{J} \neq \emptyset$. Then*

$$t(G_{\mathcal{J}}) = |C_{\mathcal{J}}|. \tag{1.45}$$

Proof of the Corollary. If C' denotes the matrix of admittance of $G_{\mathcal{J}}$, then $C'_i = C_{\mathcal{J}}$, and according to the Matrix-Tree-Theorem, $t(G_{\mathcal{J}}) = |C'_i| = |C_{\mathcal{J}}|$.

In the sequel the convention $t(G_\emptyset) = 0$ will be adopted so that (1.45) holds for every $\mathcal{J} \subset \mathcal{N}$ (note that $|C_\emptyset| = |C| = 0$).

Now we are in a position to calculate the coefficients c_i of $C_G(\lambda) = |\lambda I - C|$. Since $(-1)^i c_i$ is equal to the sum of all principal minors of order i of C,

$$c_{n-k} = (-1)^{n-k} \sum_{\substack{\mathcal{J} \subset \mathcal{N} \\ |\mathcal{J}| = k}} |C_{\mathcal{J}}| \qquad (k = 0, 1, \ldots, n), \tag{1.46}$$

where, according to the corollary of the Matrix-Tree-Theorem, $|C_{\mathcal{J}}|$ equals $t(G_{\mathcal{J}})$. Thus we have proved

Theorem 1.4 (A. K. Kel'mans [Кел3]): *Let*

$$C_G(\lambda) = |\lambda I - C| = c_0\lambda^n + c_1\lambda^{n-1} + \cdots + c_n \qquad (c_0 = 1),$$

where G is an arbitrary multigraph and $C = D - A$ is its matrix of admittance. Then

$$c_i = (-1)^i \sum_{\substack{\mathcal{J} \subset \mathcal{N} \\ |\mathcal{J}| = n-i}} t(G_{\mathcal{J}}) \qquad (i = 0, 1, \ldots, n). \tag{1.47}$$

Let the forest F have k components T_i with n_i vertices $(i = 1, 2, \ldots, k)$ and put $\gamma(F) = n_1 n_2 \cdots n_k$. According to [KeCh] (see formula (2.14) on p. 203), *c_i can be given the following form:*

$$c_i = (-1)^i \sum_{F \in \mathcal{F}_{n-i}} \gamma(F) \qquad (i = 0, 1, \ldots, n-1), \qquad c_n = 0, \tag{1.47'}$$

where $\mathcal{F}_k$ is the set of all spanning forests of G with exactly k components.

---

† This theorem was proved in a paper by R. L. Brooks, C. A. B. Smith, A. H. Stone, and W. T. Tutte [BrSST] (1940), and independently by H. M. Trent [Tren] (1954), and others; an elementary proof was given by H. Hutschenreuther [Huts] (1967). Some authors hold that it is already implicitly contained in G. Kirchhoff's classic paper [Kirc] (1847). (For more details consult [Moo2] (Chapter 5).)

†† $t(G)$ is sometimes called the complexity of G. — A simple determinant formula for the complexity of a bipartite graph is due to F. Runge; see Section 1.9, no. 12.

A theorem for multi-digraphs with weighted adjacencies generalizing Theorem 1.4 was proved by M. FIEDLER and J. SEDLÁČEK [FiSe].

Remark. For $i = n - 1$, (1.47) yields $c_{n-1} = (-1)^{n-1} \sum_{j=1}^{n} t(G_j) = (-1)^{n-1} nt(G)$.

Hence

$$\text{(i)} \qquad t(G) = \frac{1}{n} (-1)^{n-1} c_{n-1}.$$

Let $\mu_1, \mu_2, \ldots, \mu_n$ (in some order) be the eigenvalues of $\boldsymbol{C}$. Since $c_n = |\boldsymbol{A} - \boldsymbol{D}| = 0$, it follows that $0 \in \boldsymbol{Sp}_C(G)$; let $\mu_n = 0$. Then $(-1)^{n-1} c_{n-1} = \prod_{i=1}^{n-1} \mu_i$, and from (i)

$$\text{(ii)} \qquad t(G) = \frac{1}{n} \prod_{i=1}^{n-1} \mu_i$$

is obtained.

If G is connected, $t(G) > 0$, i.e., $\mu_i \neq 0$ for $i = 1, 2, \ldots, n - 1$. Thus we have proved

Proposition 1.3: *Let G be a connected multigraph. Then*

$$t(G) = \frac{1}{n} \prod \mu,$$

where μ runs through all non-zero eigenvalues of $\boldsymbol{C} = \boldsymbol{D} - \boldsymbol{A}$.

In terms of the polynomial $C_G(\lambda)$ or the Kel'mans polynomial $B_\lambda^n(G)$ (see (1.17), Section 1.2), this result can also be expressed in the following form:

$$\text{(iii)} \qquad t(G) = \frac{(-1)^{n-1}}{n} C'_G(0) = \frac{1}{n} B_0^n(G).$$

If G is regular of degree r, formulas (1.33) and (1.33′) apply and we deduce from (i), (ii), and (iii) (recall that $\lambda_1 = r$)

Proposition 1.4 (H. HUTSCHENREUTHER [Huts]): *For any regular multigraph G of degree r,*

$$t(G) = \frac{1}{n} \prod_{i=2}^{n} (r - \lambda_i) = \frac{1}{n} P'_G(r),$$

where the λ_i are the ordinary eigenvalues of G.

By adding an appropriate number of (simply counted) loops, any multigraph G of maximal valency r can be made a regular multigraph G' of degree r. Since this process has no influence on the number of spanning trees, Proposition 1.4 can be applied to an arbitrary multigraph G, provided the λ_i are taken to be the eigenvalues not of G but of G'. This observation, due to D. A. WALLER ([Wal1], [Wal2], [Wal3]; see also [Mal2]), is equivalent with Proposition 1.3.

(See also Section 1.9, nos. 10, 11.)

1.6. The coefficients of $Q_G(\lambda)$

By a procedure very similar to the method used in the proof of the preceding theorem, the coefficients of $Q_G(\lambda)$ can be determined. (Recall: $Q_G(\lambda) = \frac{1}{|\boldsymbol{D}|} |\lambda \boldsymbol{D} - \boldsymbol{A}| = q_0\lambda^n + q_1\lambda^{n-1} + \cdots + q_n$ $(q_0 = 1)$; see Section 1.2.)

Let G be an arbitrary multigraph without isolated vertices. Consider $Q_G(\lambda)$ as a polynomial in $\lambda - 1$:

$$Q_G(\lambda) = |\lambda \boldsymbol{I} - \boldsymbol{D}^{-1}\boldsymbol{A}| = |(\lambda - 1)\,\boldsymbol{I} + \boldsymbol{D}^{-1}(\boldsymbol{D} - \boldsymbol{A})|$$
$$= |(\lambda - 1)\,\boldsymbol{I} + \boldsymbol{D}^{-1}\boldsymbol{C}| = \tilde{q}_0(\lambda - 1)^n + \tilde{q}_1(\lambda - 1)^{n-1} + \cdots + \tilde{q}_n,$$

where $\tilde{q}_i$ equals the sum of all principal minors of order i of $\boldsymbol{D}^{-1}\boldsymbol{C}$. Accordingly,

$$\tilde{q}_{n-k} = \sum_{\substack{\mathcal{J} \subset \mathcal{N} \\ |\mathcal{J}|=k}} |(\boldsymbol{D}^{-1}\boldsymbol{C})_{\mathcal{J}}| \qquad (k = 0, 1, \ldots, n)$$

with

$$|(\boldsymbol{D}^{-1}\boldsymbol{C})_{\mathcal{J}}| = |(\boldsymbol{D}^{-1})_{\mathcal{J}}\boldsymbol{C}_{\mathcal{J}}| = \frac{|\boldsymbol{C}_{\mathcal{J}}|}{|\boldsymbol{D}_{\mathcal{J}}|} = \frac{t(G_{\mathcal{J}})}{\prod\limits_{l \in \mathcal{N} - \mathcal{J}} d_l},$$

the last equation following from the Corollary to the Matrix-Tree-Theorem (Section 1.5) (if $\mathcal{J} = \mathcal{N}$, then $\prod\limits_{l \in \mathcal{N} - \mathcal{J}} d_l = 1$ is assumed). Thus

$$\tilde{q}_{n-k} = \sum_{\substack{\mathcal{J} \subset \mathcal{N} \\ |\mathcal{J}|=k}} \frac{t(G_{\mathcal{J}})}{\prod\limits_{l \in \mathcal{N} - \mathcal{J}} d_l} \qquad (k = 0, 1, \ldots, n),$$

and since $q_{n-k} = \sum\limits_{j=k}^{n} \binom{j}{k} (-1)^{j-k}\, \tilde{q}_{n-j}$, we obtain with $k = n - i$:

$$q_i = (-1)^{n-i} \sum_{j=n-i}^{n} \binom{j}{n-i} (-1)^j \sum_{\substack{\mathcal{J} \subset \mathcal{N} \\ |\mathcal{J}|=j}} \frac{t(G_{\mathcal{J}})}{\prod\limits_{l \in \mathcal{N} - \mathcal{J}} d_l} \qquad (i = 0, 1, \ldots, n).$$

So we have proved

Theorem 1.5 (F. Runge [Rung]): *Let*

$$Q_G(\lambda) = \frac{1}{|\boldsymbol{D}|} |\lambda \boldsymbol{D} - \boldsymbol{A}| = q_0\lambda^n + q_1\lambda^{n-1} + \cdots + q_n \qquad (q_0 = 1),$$

where G is an arbitrary multigraph without isolated vertices. Then

$$q_i = (-1)^{n-i} \sum_{j=n-i}^{n} \binom{j}{n-i} (-1)^j \sum_{\substack{\mathcal{J} \subset \mathcal{N} \\ |\mathcal{J}|=j}} \frac{t(G_{\mathcal{J}})}{\prod\limits_{l \in \mathcal{N} - \mathcal{J}} d_l} \qquad (i = 0, 1, \ldots, n), \tag{1.48}$$

where the conventions $t(G_\emptyset) = 0$ and $\prod\limits_{l \in \emptyset} d_l = 1$ are adopted.

Theorem 1.5 has also been extended to graphs and digraphs with weighted adjacencies by F. RUNGE [Rung].

Remark. In order to obtain a coefficients theorem for $Q_G(\lambda)$ based on the cyclic structure of G, recall that $Q_G(\lambda) = |\lambda \boldsymbol{I} - \boldsymbol{A}^*|$ with $\boldsymbol{A}^* = \left(\frac{a_{jk}}{\sqrt{d_j d_k}}\right)$ (see (1.5)″, Section 1.2).

Now formula (1.35)″ (Section 1.4) when applied to $\boldsymbol{A}^*$ yields immediately

$$q_i = \sum_{U \in \mathcal{U}_i} (-1)^{p(U)}\, 2^{c(U)} \prod (U)$$

with

$$\prod (U) = \prod_{(j,k) \in \mathcal{E}(U)} \left(\frac{1}{\sqrt{d_j d_k}}\right)^{\zeta((j,k);U)} = \frac{1}{\prod\limits_{h \in \mathcal{V}(U)} d_h},$$

where $\mathcal{E}(U)$, $\mathcal{V}(U)$ denote the sets of edges and of vertices of U, respectively. Thus we have proved

Theorem 1.5a: *Under the assumptions of Theorem* 1.5,

$$q_i = \sum_{U \in \mathcal{U}_i} (-1)^{p(U)} \frac{2^{c(U)}}{\prod\limits_{h \in \mathcal{V}(U)} d_h}. \tag{1.48a}$$

1.7. A formula connecting the cyclic structure and the tree structure of a regular or semiregular multigraph

There are two strong connections between structural graph theory and linear algebra: The first one consists of the fact that the most important general invariant of linear algebra, the determinant, may be given a combinatorial form (viz., the form it has in its "Leibniz definition") that has an interpretation in terms of the cyclic structure of a (di-)graph (with weighted adjacencies), and the second one is the validity of the Matrix-Tree-Theorem (see Section 1.5) which, in a very simple way, connects the tree structure of a graph with determinants formed from its matrix of admittance. Both of these connections are taken advantage of by spectral theory: the coefficients theorems for $P_G(\lambda)$ (Theorems 1.2, 1.3) are based on the first one, and for $C_G(\lambda)$ (Theorem 1.4) and $Q_G(\lambda)$ (Theorem 1.5) on the second one.

Of particular interest are those graphs G which have the property that their polynomials $P_G(\lambda)$ and $C_G(\lambda)$ or $Q_G(\lambda)$ can be transformed one into another: in this case, the coefficients can be expressed both in terms of the cyclic structure and in terms of the tree structure of G, thus linking the basic structural elements, cycle (or circuit) and tree, one to another.

According to Theorem 1.1 (Section 1.3),

$$Q_G(\lambda) = \frac{1}{\varrho^n} P_G(\varrho\lambda) \tag{1.49}$$

for any multigraph G which is regular or semiregular of positive degree(s) r or r_1, r_2, respectively, and has index (= maximal P-eigenvalue) ϱ, where $\varrho = r$ or $\varrho = \sqrt{r_1 r_2}$, respectively. From (1.49) we deduce

$$\varrho^i q_i = a_i \qquad (i = 0, 1, \ldots, n),$$

and applying Theorems 1.3 and 1.5, we obtain the following theorems.

Theorem 1.6 (F. Runge [Rung]): *Let G be a regular multigraph of positive degree r with n vertices $1, 2, \ldots, n$. Then*

$$\sum_{U \in \mathcal{U}_i} (-1)^{p(U)} 2^{c(U)} = \sum_{j=n-i}^{n} \binom{j}{n-i} (-1)^{i+j-n} r^{i+j-n} \sum_{\substack{\mathcal{J} \subset \mathcal{N} \\ |\mathcal{J}|=j}} t(G_{\mathcal{J}}) \quad (i = 0, 1, \ldots, n), \tag{1.50}$$

where for $i = 0$ the left-hand sum is taken to be 1.

Theorem 1.6a (F. Runge [Rung]): *Let $G = (\mathcal{X}, \mathcal{Y}; \mathcal{U})$ be a semiregular multigraph, where all vertices $x \in \mathcal{X} = \{1, 2, \ldots, n_1\}$ have valency $r_1 > 0$ and all vertices $y \in \mathcal{Y} = \{n_1 + 1, n_1 + 2, \ldots, n_1 + n_2 = n\}$ have valency $r_2 > 0$. Then*

for odd $i \in \mathcal{N}$,

$$\sum_{j=n-i}^{n} \binom{j}{n-i} (-1)^j \sum_{\substack{\mathcal{J} \subset \mathcal{N} \\ |\mathcal{J}|=j}} r_1^{j_1} r_2^{j_2} t(G_{\mathcal{J}}) = 0, \tag{1.51}$$

and for even $i \in \mathcal{N}$,

$$\sum_{U \in \mathcal{U}_i} (-1)^{p(U)} 2^{c(U)} = \sum_{j=n-i}^{n} \binom{j}{n-i} (-1)^{i+j-n} \sum_{\substack{\mathcal{J} \subset \mathcal{N} \\ |\mathcal{J}|=j}} r_1^{\frac{i}{2}+j_1-n_1} r_2^{\frac{i}{2}+j_2-n_2} t(G_{\mathcal{J}}), \tag{1.52}$$

where in the last sum of (1.51) and (1.52) $j_1 = |\mathcal{X} \cap \mathcal{J}|$, $j_2 = |\mathcal{Y} \cap \mathcal{J}|$ $(j_1 + j_2 = j)$.

Remark 1. For regular multigraphs of positive degree r, we may use the relation

$$C_G(\lambda) = (-1)^n P_G(-\lambda + r) \tag{1.33}$$

(Section 1.3) instead of (1.49), equate corresponding coefficients and apply Theorems 1.3 and 1.4 (instead of 1.5). The relation connecting the coefficients a_i of $P_G(\lambda)$ and c_j of $C_G(\lambda)$ is

$$a_i = (-1)^i \sum_{j=n-i}^{n} \binom{j}{n-i} r^{i+j-n} c_{n-j} \quad (i = 0, 1, \ldots, n) \tag{1.53}$$

with $a_0 = c_0 = 1$, and with (1.36) (Theorem 1.3) and (1.47) (Theorem 1.4) we arrive again at Theorem 1.6.

By inversion of (1.53) we obtain

$$c_i = (-1)^i \sum_{j=n-i}^{n} \binom{j}{n-i} r^{i+j-n} a_{n-j} \quad (i = 0, 1, \ldots, n), \tag{1.53'}$$

and (1.36) and (1.47) now yield the following system of equations equivalent with (1.50):

$$\sum_{\substack{\mathcal{J} \subset \mathcal{N} \\ |\mathcal{J}|=n-i}} t(G_{\mathcal{J}}) = \sum_{j=n-i}^{n} \binom{j}{n-i} r^{i+j-n} \sum_{U \in \mathcal{U}_{n-j}} (-1)^{p(U)} 2^{c(U)} \quad (i = 0, 1, \dots, n), \tag{1.50'}$$

where for $j = n$ the last sum is taken to be 1.

With $i = n - 1$ we obtain from (1.50′) a new formula for the number of spanning trees contained in a regular multigraph, namely

$$t(G) = \frac{1}{n} \sum_{j=1}^{n} j \cdot r^{j-1} \sum_{U \in \mathcal{U}_{n-j}} (-1)^{p(U)} 2^{c(U)} \tag{1.54}$$

(see also Proposition 1.4).

Remark 2. A general formula connecting cyclic structure and tree structure of any multigraph is, of course, contained in Theorems 1.5 and 1.5a (Section 1.6): From (1.48) and (1.48a), after multiplication by $\prod d_i$ we obtain

Theorem 1.7: *Let $G = (\mathcal{X}, \mathcal{U})$ be any multigraph without isolated vertices, where $\mathcal{X} = \mathcal{N} = \{1, 2, \dots, n\}$. Then*

$$\sum_{U \in \mathcal{U}_i} (-1)^{p(U)} 2^{c(U)} \prod_{h \in \mathcal{N} - \mathcal{V}(U)} d_h = \sum_{j=n-i}^{n} \binom{j}{n-i} (-1)^{n-i+j} \sum_{\substack{\mathcal{J} \subset \mathcal{N} \\ |\mathcal{J}|=j}} t(G_{\mathcal{J}}) \prod_{l \in \mathcal{J}} d_l$$

($i = 1, 2, \dots, n$), where $\mathcal{V}(U)$ is the set of vertices of the basic figure U and where the conventions $t(G_\emptyset) = 0$ and $\prod_{l \in \emptyset} d_l = 1$ are adopted.

By specialization Theorems 1.6 and 1.6a are obtained from Theorem 1.7, but in the general case the significance of Theorem 1.7 is constrained by the fact that the terms depending on the valencies d_h or d_l cannot be eliminated.

1.8. On the number of walks

In this section, "spectrum" always means "P-spectrum".

Let $\boldsymbol{A}$ be the adjacency matrix of a multi-digraph G with vertices $1, 2, \dots, n$. If, in addition to the spectrum of G, the eigenvectors of $\boldsymbol{A}$ are known, then, of course, more statements concerning the structure of G can be made than without this knowledge. Moreover, a multi-digraph G with a symmetric adjacency matrix — in particular, a multigraph — is completely determined by its eigenvalues and eigenvectors. For, if $\boldsymbol{v}_1, \boldsymbol{v}_2, \dots, \boldsymbol{v}_n$ is a complete system of mutually orthogonal normalized eigenvectors of $\boldsymbol{A}$ belonging to the spectrum $[\lambda_1, \lambda_2, \dots, \lambda_n]$, let $\boldsymbol{V} = (\boldsymbol{v}_1, \boldsymbol{v}_2, \dots, \boldsymbol{v}_n) = (v_{ij})$ and $\Lambda = (\delta_{ij}\lambda_i)$: then, as is well known, $\boldsymbol{V}$ is orthogonal (i.e., $\boldsymbol{V}^{-1} = \boldsymbol{V}^{\mathsf{T}}$) and

$$\boldsymbol{A} = \boldsymbol{V} \Lambda \boldsymbol{V}^{\mathsf{T}}. \tag{1.55}$$

Since G is determined by $\boldsymbol{A}$, we have proved

Theorem 1.8: *A multigraph is completely determined by its eigenvalues and corresponding eigenvectors.*

So, in principle, any multigraph problem can be treated in terms of spectra and eigenvectors. (For example, an algorithm for determining whether two graphs are isomorphic, which is based on Theorem 1.8, has been developed in [Kuhn].) From this point of view, we shall now investigate the problem of the number of walks of given length in a multi-(di-)graph G. (Recall: A *walk of length* $k \geqq 0$ is a sequence of arcs $u_1 u_2 \cdots u_k$, where the starting vertex of u_{j+1} coincides with the end vertex of u_j ($j = 1, 2, \ldots, k - 1$), repetitions and loops being allowed.) Some more problems concerning eigenvectors will be considered in Section 3.5.

The starting point of our considerations is the following well-known theorem.

Theorem 1.9: *Let $\boldsymbol{A}$ be the adjacency matrix of a multi-digraph G with vertices $1, 2, \ldots, n$, let $\boldsymbol{A}^k = (a_{ij}^{(k)})$; further, let $N_k(i, j)$ denote the number of walks of length k starting at vertex i and terminating at vertex j. Then*

$$N_k(i, j) = a_{ij}^{(k)} \qquad (k = 0, 1, 2, \ldots). \tag{1.56}$$

Note that for $k = 0$, (1.56) agrees with the convention $N_0(i, j) = \delta_{ij}$.

Now let G denote a multigraph and let $\boldsymbol{V} = (v_{ij})$ be an orthogonal matrix of eigenvectors of $\boldsymbol{A}$, as described above. Then, according to (1.55),

$$a_{ij}^{(k)} = \sum_{\nu=1}^{n} v_{i\nu} v_{j\nu} \lambda_\nu^k . \tag{1.57}$$

The number N_k of all walks of length k in G equals

$$N_k = \sum_{i,j} N_k(i, j) = \sum_{i,j} a_{ij}^{(k)} = \sum_{\nu=1}^{n} \left(\sum_{i=1}^{n} v_{i\nu} \right)^2 \lambda_\nu^k .$$

Thus we have proved

Theorem 1.10:† *The total number N_k of walks of length k in a multigraph G is given by*

$$N_k = \sum_{\nu=1}^{n} C_\nu \lambda_\nu^k \qquad (k = 0, 1, 2, \ldots), \tag{1.58}$$

where $C_\nu = \left(\sum_{i=1}^{n} v_{i\nu} \right)^2$.

In the next theorem, the generating function for the numbers N_k is expressed in terms of the characteristic polynomials of the graph G and its complement $\overline{G}$.

Theorem 1.11 (D. M. Cvetković [Cve8]): *Let G be a graph with complement $\overline{G}$, and let $H_G(t) = \sum_{k=0}^{\infty} N_k t^k$ be the generating function of the numbers N_k of walks of length k*

---

† Part of this theorem was proved in another way by D. M. Cvetković [Cve9] who also proved the theorem in the present form when preparing the manuscript of this book; the theorem was also partly used in [CvS1]. Another proof was given by F. Harary and A. J. Schwenk [HaS1].

in G $(k = 0, 1, 2, \ldots)$. *Then*

$$H_G(t) = \frac{1}{t}\left[(-1)^n \frac{P_{\bar{G}}\left(-\frac{t+1}{t}\right)}{P_G\left(\frac{1}{t}\right)} - 1\right]. \tag{1.59}$$

Proof. If $\boldsymbol{M}$ is a non-singular square matrix of order n, let $\{\boldsymbol{M}\}$ denote the matrix formed by the minors of order $n - 1$ so that $\{\boldsymbol{M}\}^{\mathsf{T}} = |\boldsymbol{M}|\, \boldsymbol{M}^{-1}$. Let sum $\boldsymbol{M}$ denote the sum of all elements of $\boldsymbol{M}$, and let $\boldsymbol{J}$ be a square matrix all entries of which are equal to 1; then, for an arbitrary number x,

$$|\boldsymbol{M} + x\boldsymbol{J}| = |\boldsymbol{M}| + x \operatorname{sum} \{\boldsymbol{M}\} \tag{1.60}$$

which can be proved by straightforward calculations. Now, according to Theorem 1.9, $N_k = \operatorname{sum} \boldsymbol{A}^k$, and since

$$\sum_{k=0}^{\infty} \boldsymbol{A}^k t^k = (\boldsymbol{I} - t\boldsymbol{A})^{-1} = |\boldsymbol{I} - t\boldsymbol{A}|^{-1} \{\boldsymbol{I} - t\boldsymbol{A}\} \qquad (|t| < (\max \lambda_i)^{-1}),$$

we obtain

$$\sum_{k=0}^{\infty} \operatorname{sum} \boldsymbol{A}^k t^k = \sum_{k=0}^{\infty} N_k t^k = |\boldsymbol{I} - t\boldsymbol{A}|^{-1} \operatorname{sum} \{\boldsymbol{I} - t\boldsymbol{A}\},$$

i.e.,

$$H_G(t) = \frac{\operatorname{sum} \{\boldsymbol{I} - t\boldsymbol{A}\}}{|\boldsymbol{I} - t\boldsymbol{A}|}. \tag{1.61}$$

With $\boldsymbol{M} = \boldsymbol{I} - t\boldsymbol{A}$, $x = t$, (1.60) yields

$$\operatorname{sum} \{\boldsymbol{I} - t\boldsymbol{A}\} = \frac{1}{t}\left(|(t + 1)\, \boldsymbol{I} + t\bar{\boldsymbol{A}}| - |\boldsymbol{I} - t\boldsymbol{A}|\right), \tag{1.62}$$

where $\bar{\boldsymbol{A}} = \boldsymbol{J} - \boldsymbol{I} - \boldsymbol{A}$ is the adjacency matrix of the complement $\bar{G}$ of G, and by inserting (1.62) into (1.61), the equation

$$H_G(t) = \frac{1}{t}\left[(-1)^n \frac{\left|-\frac{t+1}{t}\boldsymbol{I} - \bar{\boldsymbol{A}}\right|}{\left|\frac{1}{t}\boldsymbol{I} - \boldsymbol{A}\right|} - 1\right] \tag{1.63}$$

is obtained. Clearly, (1.63) implies (1.59), which proves the theorem.

Theorem 1.11 has been proved in [Cve8] by another method. P. W. KASTELEYN [Kas2] gave the expression for the generating function for numbers of walks between two prescribed vertices of a graph.

The generating function $H_G(t)$ will be used in Section 2.2. The numbers of walks for graphs of some special types will be determined in Section 7.5.

Let $\{\mu_1, \mu_2, \ldots, \mu_m\}$ be the set of distinct eigenvalues of a multigraph G. (1.58) can then be rewritten in the form

$$N_k = D_1\mu_1^k + D_2\mu_2^k + \cdots + D_m\mu_m^k \qquad (k = 0, 1, 2, \ldots), \tag{1.64}$$

where $D_1, D_2, \ldots, D_m$ are non-negative numbers uniquely determined by G; some (but not all) of them may be zero. In particular, with $k = 0$ the equation

$$C_1 + C_2 + \cdots + C_n = D_1 + D_2 + \cdots + D_m = N_0 = n \tag{1.65}$$

is obtained from (1.58) and (1.64).

D. M. Cvetković [Cve9] gave the following

Definition. The *main part of the spectrum* of a multigraph G is the set of all those eigenvalues μ_j for which in (1.64) $D_j \neq 0$ holds.

For a regular multigraph of degree r with n vertices, clearly, $N_k = nr^k$: hence, for regular multigraphs (and, in fact, only for these) the main part of the spectrum consists of the index only. In this case, $\sqrt[k]{\frac{N_k}{n}} = r$ which motivates $\sqrt[k]{\frac{N_k}{n}}$ in the general case to be considered as a certain kind of mean value of the valencies, in general depending on k. This gives rise to the following

Definition. Let G be a multigraph and $d = d(G) = \lim_{k\to\infty} \sqrt[k]{\frac{N_k}{n}} = \lim_{k\to\infty} \sqrt[k]{N_k}$ (it will be shown that the limit exists). Then d is called the *dynamic mean of the valencies* of the vertices of G.

Clearly, $N_k = O(d^k)$ $(k \to \infty)$.

Theorem 1.12 (D. M. Cvetković [Cve9]): *For a multigraph G, the dynamic mean $d(G)$ is equal to the index of G.*

Theorem 1.12, together with the existence of d, follows immediately from Theorem 1.10 and the fact that among the eigenvectors corresponding to the index of G there is a non-negative one.

An application of this theorem to chemistry is described in [CvG4].

We quote without proof

Theorem 1.13 (F. Harary, A. J. Schwenk [HaS1]): *For a multigraph G, the following statements are equivalent:*

1⁰ *$\mathcal{M}$ is the main part of the spectrum;*

2⁰ *$\mathcal{M}$ is the minimum set of eigenvalues the span of whose eigenvectors includes the vector $(1, 1, \ldots, 1)^\mathsf{T}$;*

3⁰ *$\mathcal{M}$ is the set of those eigenvalues which have an eigenvector not orthogonal to $(1, 1, \ldots, 1)^\mathsf{T}$.*

The proof can be performed by means of Theorem 1.10.

1.9. Miscellaneous results and problems

1. Let G be a multi-(di-)graph with vertex-set $\{1, 2, \ldots, n\}$ and let $N_k(i, j)$ denote the number of walks of length k in G joining i to j. If w_{ij} is the corresponding generating function (i.e., $w_{ij} = \sum_{k=0}^{\infty} N_k(i, j)\, t^k$) and $\boldsymbol{W} = (w_{ij})$, then $\boldsymbol{W} = (\boldsymbol{I} - t\boldsymbol{A})^{-1}$.

(P. W. Kasteleyn [Kas 2])

2. Let $\mathbb{R}$ be the set of the greatest eigenvalues of all graphs. Let $\tau = \frac{\sqrt{5} + 1}{2}$ (the golden mean). For $n = 1, 2, \ldots$, let β_n be the positive root of

$$P_n(x) = x^{n+1} - (1 + x + x^2 + \cdots + x^{n-1}).$$

Let $\alpha_n = \beta_n^{1/2} + \beta_n^{-1/2}$. Then $2 = \alpha_1 < \alpha_2 < \cdots$ are all limit points of $\mathbb{R}$ smaller than $\tau^{1/2} + \tau^{-1/2} = \lim_{n \to +\infty} \alpha_n$.

(A. J. Hoffman [Hof 13])

3. If a digraph G has at least one cycle then the index of G is not smaller than 1; otherwise all eigenvalues of G are equal to zero.

(J. Sedláček [Sed 1])

4. Let G be a digraph with vertices $1, \ldots, n$. For given vertices i and j $(i \neq j)$, a spanning subgraph of G in which

1° exactly one arc starts and no arc ends in i,

2° exactly one arc ends and no arc starts in j,

3° all the other vertices have in and out degrees equal to 1,

is called a *connection from i to j* and is denoted by $C(i \to j)$. For $i = j$ the vertex i is an isolated vertex of $C(i \to i)$ while all the other vertices have property 3°.

With a square matrix $\boldsymbol{A} = (a_{ij})_1^n$ we associate a weighted digraph $D_{\boldsymbol{A}}$, defined in the following way. The n vertices of $D_{\boldsymbol{A}}$ are numbered by $1, 2, \ldots, n$ and for each ordered pair of vertices i, j there exists an arc in $D_{\boldsymbol{A}}$ leading from j to i and having weight a_{ij}.

The product $W = W(L)$ of the weights of the arcs of a spanning linear subgraph L is called the weight of L. The number of cycles contained in a linear subgraph L is denoted by $c(L)$; $\mathscr{L}$ denotes the set of all spanning linear subgraphs L of $D_{\boldsymbol{A}}$.

The weight $W(C(i \to j))$ and the number of cycles $c(C(i \to j))$ of a connection $C(i \to j)$ are defined analogously.

Then the cofactor A_{ij} of the element a_{ij} is given by

$$A_{ij} = (-1)^{n-1} \sum_{C(i \to j)} (-1)^{c(C(i \to j))}\, W(C(i \to j)),$$

where the summation runs through all connections $C(i \to j)$ from i to j of the digraph $D_{\boldsymbol{A}}$.

Consider further the following system of linear algebraic equations:

$$\sum_{j=1}^{n} a_{ij} x_j = b_i \qquad (i = 1, \ldots, n).$$

With this system we associate a digraph D having vertices $0, 1, \ldots, n$ in which the vertices $1, \ldots, n$ induce the digraph $D_{\boldsymbol{A}}$, corresponding to the matrix $\boldsymbol{A} = (a_{ij})_1^n$ and in which there is an additional arc from vertex 0 to vertex i having weight $-b_i$ for every $i \in \{1, 2, \ldots, n\}$. Then

$$x_j = \frac{\sum_{C(0 \to j)} (-1)^{c(C(0 \to j))}\, W(C(0 \to j))}{\sum_{L \in \mathscr{L}} (-1)^{c(L)}\, W(L)} \qquad (j = 1, \ldots, n),$$

where in the upper sum the summation runs through all connections $C(0 \to j)$ in D.

(C. L. Coates [Coat])

5. Let G be the digraph corresponding to a square matrix $\boldsymbol{A}$ of order n (see nr. 4). Let M_{ij} be the cofactor of the i,j-element of $\lambda \boldsymbol{I} - \boldsymbol{A}$. Then

$$M_{ij} = \sum_{k=2}^{n} \lambda^{n-k} \sum_{C_k(i \to j)} (-1)^{c(C_k(i \to j))} W(C_k(i \to j)),$$

where in the second sum the summation runs through all connections $C_k(i \to j)$ from i to j which have exactly k vertices.

(J. Ponstein [Pons])

6. *Some remarks concerning the Q-spectrum of a multigraph.* If G is a multigraph without isolated vertices having components $G_1, G_2, \ldots, G_k$ then, clearly,

$$Q_G(\lambda) = \prod_{i=1}^{k} Q_{G_i}(\lambda). \tag{1.66}$$

Q_G may now be defined for multigraphs G having isolated vertices in the following way:
i) If P is the "*point graph*" having exactly one vertex and no edge, set

$$Q_P(\lambda) = \lambda - 1. \tag{1.67}$$

(ii) If G is any multigraph having components $G_1, G_2, \ldots, G_k$, set

$$Q_G(\lambda) = \prod_{i=1}^{k} Q_{G_i}(\lambda)$$

which is consistent with (1.66).

If G is a multigraph without isolated vertices, then, according to (1.5)′ and (1.5)″ (Section 1.2)

$$Q_G(\lambda) = |\lambda \boldsymbol{I} - \tilde{\boldsymbol{A}}| = |\lambda \boldsymbol{I} - \boldsymbol{A}^*|, \tag{1.68}$$

where

$$\tilde{\boldsymbol{A}} = \boldsymbol{D}^{-1}\boldsymbol{A} = \left(\frac{a_{ik}}{d_i}\right), \qquad \boldsymbol{A}^* = \boldsymbol{D}^{-\frac{1}{2}}\boldsymbol{A}\boldsymbol{D}^{-\frac{1}{2}} = \left(\frac{a_{ik}}{\sqrt{d_i d_k}}\right).$$

If G is the point graph P, set $\boldsymbol{A}^* = \tilde{\boldsymbol{A}} = (1)$, consistent with (1.67).

The matrix $\boldsymbol{A}^*$ is symmetric, $\tilde{\boldsymbol{A}}$ is stochastic, so all Q-eigenvalues of G are real, the largest one being equal to 1.

The Q-spectrum has many properties analogous or very similar to properties of the P-spectrum not to be itemized here; a few examples shall be quoted:

- *The number of components of G is equal to the multiplicity of the Q-eigenvalue 1 of G.*
- *Let G be a multigraph without isolated vertices. G is bipartite if and only if $Q_G(-\lambda) = Q_G(\lambda)$.*
- *Let G be a connected multigraph, not the point graph. G is bipartite if and only if $Q_G(-1) = 0$.*

Let G be a multigraph and let kG denote the multigraph derived from G by replacing every edge by exactly k parallel edges. Let $\boldsymbol{A}(G)$ denote the adjacency matrix of G, etc. Clearly, $\boldsymbol{A}(kG) = k\boldsymbol{A}(G)$, but $\boldsymbol{A}^*(kG) = \boldsymbol{A}^*(G)$, $\tilde{\boldsymbol{A}}(kG) = \tilde{\boldsymbol{A}}(G)$. Thus G, though uniquely determined by $\boldsymbol{A}$, is not determined by $\boldsymbol{A}^*$ or $\tilde{\boldsymbol{A}}$. Multigraphs G and kG have the same Q-spectrum. This observation is, of course, not meaningful when only graphs are considered.

7. Let G be a multigraph without isolated vertices and put $\tilde{\boldsymbol{A}}^l = (\tilde{a}_{ij}^{(l)})$ $(l = 1, 2, \ldots)$. Then $\tilde{a}_{ij}^{(l)}$ equals the probability of reaching vertex j as the last point in a random walk of length l starting at vertex i.

8. Let $G = (\mathcal{X}, \mathcal{U})$ be a graph with m edges and n non-isolated vertices having Q-spectrum $\boldsymbol{Sp}_Q(G) = [\lambda_1, \lambda_2, \ldots, \lambda_n]_Q$. Then

$$\sum_{\nu=1}^{n} \lambda_\nu^2 = 2 \sum_{(i,j)\in\mathcal{U}} \frac{1}{d_i d_j}.$$

Corollary: If G is either regular of positive degree r or semiregular of positive degrees r_1, r_2, and if ϱ is the index of G (i.e., $\varrho = r$ or $\varrho = \sqrt{r_1 r_2}$, respectively), then

$$m = \frac{1}{2} \varrho^2 \sum_{\nu=1}^{n} \lambda_\nu^2. \tag{1.69}$$

Problem. Is condition (1.69) sufficient for a graph G to be either regular or semiregular ?

(F. RUNGE [Rung])

9. The determinant of the adjacency matrix $\boldsymbol{A}$ of a multigraph G is given in terms of the tree structure of G by

$$|\boldsymbol{A}| = (-1)^n \sum_{j=1}^{n} (-1)^j \sum_{\substack{\mathcal{J} \subset \mathcal{N} \\ |\mathcal{J}|=j}} \left(\prod_{i \in \mathcal{J}} d_i\right) t(G_{\mathcal{J}}).$$

(F. RUNGE [Rung])

10. Let G be a multigraph having n vertices with positive valencies $d_1, \ldots, d_n$. Then the complexity of G is given by

$$t(G) = \frac{|\boldsymbol{D}|}{2m} Q'_G(1) = \frac{\prod d_i}{\sum d_i} \prod_{\nu=2}^{n} (1 - \lambda_\nu^*),$$

where m is the number of edges and where the λ_ν^* are the Q-eigenvalues of G.

(F. RUNGE [Rung], [RuSa]; see also [Sac12])

11. Let $G = (\mathcal{X}, \mathcal{Y}; \mathcal{U})$ be a bipartite multigraph without isolated vertices where $\mathcal{X} = \{x_1, \ldots, x_m\}$, $\mathcal{Y} = \{x_{m+1}, \ldots, x_{m+n}\}$; let $\boldsymbol{V}, \boldsymbol{W}$ be the valency matrices of the sets $\mathcal{X}$ and $\mathcal{Y}$, respectively, so that the adjacency matrix $\boldsymbol{A}$ and the valency matrix $\boldsymbol{D}$ of G are of the form

$$\boldsymbol{A} = \begin{pmatrix} \boldsymbol{O} & \boldsymbol{B} \\ \boldsymbol{B}^\top & \boldsymbol{O} \end{pmatrix}, \quad \boldsymbol{D} = \begin{pmatrix} \boldsymbol{V} & \boldsymbol{O} \\ \boldsymbol{O} & \boldsymbol{W} \end{pmatrix},$$

respectively ($\boldsymbol{B}$ is an $n \times m$ matrix). Put

$$\boldsymbol{V}^{-1}\boldsymbol{B} = \boldsymbol{M}, \quad \boldsymbol{W}^{-1}\boldsymbol{B}^\top = \tilde{\boldsymbol{M}},$$
$$\varphi_G^*(\lambda) = |\lambda \boldsymbol{I} - \boldsymbol{M}\tilde{\boldsymbol{M}}|, \quad \bar{\varphi}_G^*(\lambda) = |\lambda \boldsymbol{I} - \tilde{\boldsymbol{M}}\boldsymbol{M}|,$$

where $\boldsymbol{I}$ denotes the identity matrix of order m or n, respectively. Then, by a well-known theorem of the theory of matrices,

$$\lambda^n \varphi_G^*(\lambda) = \lambda^m \bar{\varphi}_G^*(\lambda).$$

Put

$$\varphi_G(\lambda) = \begin{cases} \varphi_G^*(\lambda) & \text{if } n \geqq m, \\ \bar{\varphi}_G^*(\lambda) & \text{if } n \leqq m. \end{cases}$$

Thus the order of the polynomial $\varphi_G(\lambda)$ is equal to $\min(m, n)$. Note that $\varphi_G(\lambda)$ is invariant under the interchange of the vertex sets $\mathcal{X}$ and $\mathcal{Y}$. The polynomial $\varphi_G(\lambda)$ is connected with $Q_G(\lambda)$ by the formula

$$\lambda^{\min(m,n)} Q_G(\lambda) = \lambda^{\max(m,n)} \varphi_G(\lambda^2)$$

so that essential information contained in $Q_G(\lambda)$ is already contained in $\varphi_G(\lambda)$. Thus, for bipartite multigraphs, it may be more convenient to use $\varphi_G(\lambda)$ (or the corresponding φ-spectrum) than $Q_G(\lambda)$ (or the Q-spectrum). For example, in terms of the φ-spectrum the complexity of G is given by

$$t(G) = \frac{|\boldsymbol{V}| \cdot |\boldsymbol{W}|}{l} \prod_{\varkappa=2}^{k} (1 - \tilde{\lambda}_\varkappa) = 2 \frac{\prod d_i}{\sum d_i} \prod_{\varkappa=2}^{k} (1 - \tilde{\lambda}_\varkappa),$$

where l is the number of edges of G, $k = \min(m, n)$, and where the $\tilde{\lambda}_\varkappa$ are the φ-eigenvalues of G. If the last formula is applied to the complete bipartite graph $K_{m,n}$, the well-known formula

$$t(K_{m,n}) = m^{n-1} \cdot n^{m-1}$$

(see [FiSe]) is immediately obtained. (See also Section 7.6, p. 219.)

(F. RUNGE [Rung], [RuSa], [Sac12])

12. Let $G = (\mathscr{X}, \mathscr{Y}; \mathscr{U})$ be a bipartite multigraph without isolated vertices. With the notation of no. 11, the complexity of G is given by

$$t(G) = |\boldsymbol{W}| \cdot |(\boldsymbol{V} - \boldsymbol{B}\boldsymbol{W}^{-1}\boldsymbol{B}^\mathsf{T})_i| = |\boldsymbol{V}| \cdot |(\boldsymbol{W} - \boldsymbol{B}^\mathsf{T}\boldsymbol{V}^{-1}\boldsymbol{B})_j|,$$

where i, j are arbitrary numbers taken from $\{1, 2, \ldots, m\}$ or $\{1, 2, \ldots, n\}$, respectively. (Recall that $\boldsymbol{M}_i$ denotes the matrix which is derived from the square matrix $\boldsymbol{M}$ by simultaneously deleting the i-th row and the i-th column.)

(F. RUNGE [Rung], [RuSa], [Sac12])

13. The considerations of no. 11 may be taken as a starting point for developing a spectral theory for hypergraphs: Any hypergraph H can be represented by its *incidence graph* (Levi graph) $G = L(H)$ which is a bipartite graph without isolated vertices; conversely, every connected bipartite graph G with more than one vertex uniquely determines a pair of connected hypergraphs H, $\overline{H}$ which are duals of each other (so that G is the incidence graph of H as well as of $\overline{H}$): Thus the φ-spectrum of a connected hypergraph H may be defined as the φ-spectrum of $L(H)$ — a definition which has the advantage of being invariant under dualization.

For some more results on various spectra connected with hypergraphs and graphs derived from them see [Rung].

14. A balanced incomplete block design (BIBD)† B can be considered as a special hypergraph H with the varieties and blocks of B being the vertices and hyperedges of H, respectively. So the complexity t of B may be defined as the number of spanning trees of the incidence graph corresponding to H. It turns out that $t = t(B)$ is completely determined by the parameters v, b, r, k, λ of B:

$$t(B) = k^{b-v+1}\lambda^{v-1}v^{v-2}.$$

(F. RUNGE [Rung]; see also [RuSa])

15. Show that the relation between the characteristic polynomial $P_G(\lambda)$ of a graph G and the characteristic polynomial $S_G(\lambda)$ of the Seidel adjacency matrix $\boldsymbol{S}$ of G can be written in the form

$$P_G(\lambda) = \frac{(-1)^n}{2^n} \frac{S_G(-2\lambda - 1)}{1 + \frac{1}{2\lambda} H_G\left(\frac{1}{\lambda}\right)},$$

where $H_G(t)$ is the generating function for the numbers of walks in G.

(D. M. CVETKOVIĆ [Cve18])

---

† For the definition of a BIBD see Section 6.2, pp. 165/166.

2. Operations on Graphs and the Resulting Spectra

In this chapter we shall describe some procedures for determining the spectra and/or characteristic polynomials of (directed or undirected) (multi-)graphs derived from some simpler graphs. In the majority of cases we have the following scheme. Let graphs $G_1, \ldots, G_n$ $(n = 1, 2, \ldots)$ be given and let their spectra be known. We define an n-ary operation on these graphs, resulting in a graph G. The theorems of this chapter describe the relations between the spectra of $G_1, \ldots, G_n$ and G. In particular, in some important cases, the spectrum of G is determined by the spectra of $G_1, \ldots, G_n$.

At the end of this chapter, in Section 2.6, we shall use the theory we have developed to derive the spectra and/or characteristic polynomials of several special classes of graphs.

2.1. The polynomial of a graph

Let $G_1 = (\mathcal{X}, \mathcal{U}_1)$ and $G_2 = (\mathcal{X}, \mathcal{U}_2)$ be graphs† with the (same) set of vertices $\mathcal{X} = \{x_1, \ldots, x_n\}$, where $\mathcal{U}_1$ and $\mathcal{U}_2$ are the sets of edges of these graphs. The *union* $G_1 \cup G_2$ of the graphs G_1 and G_2 is the graph $G = (\mathcal{X}, \mathcal{U})$, where $\mathcal{U} = \mathcal{U}_1 \cup \mathcal{U}_2$. It is understood that every edge from $\mathcal{U}_1$ is different from any edge from $\mathcal{U}_2$, even when the considered edges join the same pair of vertices. If $\boldsymbol{A}_1$, $\boldsymbol{A}_2$, and $\boldsymbol{A}$ are the adjacency matrices of graphs G_1, G_2, and $G_1 \cup G_2$, respectively, then $\boldsymbol{A} = \boldsymbol{A}_1 + \boldsymbol{A}_2$.

However, $G_1 \cup G_2$ depends not only on G_1 and G_2 but also on the numeration of the vertices of these graphs. Therefore, the spectrum of the graph $G_1 \cup G_2$ is, in general, not determined by the spectra of G_1 and G_2. Some information about the spectrum of the union of graphs is provided by the following theorem from general matrix theory.

Theorem 2.1 (the Courant-Weyl inequalities; see, for example, [Hof11]): *Let* $\lambda_1(\boldsymbol{X}), \ldots, \lambda_n(\boldsymbol{X})$ $\left(\lambda_1(\boldsymbol{X}) \geqq \lambda_2(\boldsymbol{X}) \geqq \cdots \geqq \lambda_n(\boldsymbol{X})\right)$ *be the eigenvalues of a real symmetric*

---

† The "graphs" considered in this section are, in fact, multi-(di-)graphs (loops being allowed); see the general remark in the Introduction (p. 11).

matrix $\boldsymbol{X}$. *If* $\boldsymbol{A}$ *and* $\boldsymbol{B}$ *are real symmetric matrices of order* n, *and if* $\boldsymbol{C} = \boldsymbol{A} + \boldsymbol{B}$, *then*

$$\lambda_{i+j+1}(\boldsymbol{C}) \leqq \lambda_{i+1}(\boldsymbol{A}) + \lambda_{j+1}(\boldsymbol{B}),$$

$$\lambda_{n-i-j}(\boldsymbol{C}) \geqq \lambda_{n-i}(\boldsymbol{A}) + \lambda_{n-j}(\boldsymbol{B}),$$

where $0 \leqq i, j, i + j + 1 \leqq n$. *In particular,*

$$\lambda_1(\boldsymbol{C}) \leqq \lambda_1(\boldsymbol{A}) + \lambda_1(\boldsymbol{B}).$$

The *product* $G_1 \cdot G_2$ (to be distinguished from the product $G_1 \times G_2$ defined in Section 2.5) of the above-mentioned graphs G_1 and G_2 is the graph having as many (oriented) edges leading from the vertex x_i to the vertex x_j ($x_i, x_j \in \mathscr{X}$) as there are pairs of edges (u', u'') with the following properties: $u' \in \mathscr{U}_1$ and $u'' \in \mathscr{U}_2$, u' starts from x_i, u'' terminates in x_j, while u'' starts from that vertex in which u' terminates. It can easily be seen that for the adjacency matrix $\boldsymbol{A}$ of this product $\boldsymbol{A} = \boldsymbol{A}_1\boldsymbol{A}_2$. As with the union, the product $G_1 \cdot G_2$ depends on the numeration of the vertices of the graphs G_1 and G_2.

The following theorem from general matrix theory enables, in some special cases, the spectra of the union and of the product to be determined.

Theorem 2.2 (see, for example, [MaMi], p. 25): *If* $\boldsymbol{A}$ *and* $\boldsymbol{B}$ *are n-square matrices over the field of complex numbers and both* $\boldsymbol{A}$ *and* $\boldsymbol{B}$ *commute with the commutator* $\boldsymbol{AB} - \boldsymbol{BA}$, *and if* $f(\lambda_1, \lambda_2)$ *is any polynomial in the indeterminates* λ_1, λ_2 *with complex coefficients, then there is an ordering of the characteristic roots of* $\boldsymbol{A}$ *and* $\boldsymbol{B}$, α_i, β_i ($i = 1, \ldots, n$), *such that the characteristic roots of* $f(\boldsymbol{A}, \boldsymbol{B})$ *are* $f(\alpha_i, \beta_i)$ ($i = 1, \ldots, n$).

Union and product of graphs are associative operations and it is possible to define powers with respect to these operations. The n-fold union of the graph G with itself (the numbering of vertices in each copy of G remaining unchanged) will be denoted by nG and the n-fold product by G^n. If $\boldsymbol{A}$ is the adjacency matrix of the graph G, the adjacency matrices of nG and G^n are $n\boldsymbol{A}$ and $\boldsymbol{A}^n$, respectively. The operation nG will be called the *multiplication of a graph by an integer* and the operation G^n will be referred to as the *power of a graph.*

We shall now define a more general unary operation yielding nG and G^n as special cases.

Let G be a graph with the adjacency matrix $\boldsymbol{A}$ and let $P(x)$ be a polynomial in x such that all entries of the matrix $P(\boldsymbol{A})$ are non-negative integers. Then we define the polynomial $P(G)$ of the graph G as the graph whose adjacency matrix is equal to $P(\boldsymbol{A})$. If $P(x) = a_0x^n + a_1x^{n-1} + \cdots + a_n$ ($a_i \in \{0, 1, 2, \ldots\}$, $i = 0, 1, \ldots, n$), we write

$$P(G) = a_0G^n \cup a_1G^{n-1} \cup \cdots \cup a_nG^0,$$

where G^0 is the graph whose adjacency matrix is the unit matrix of the corresponding order.

Obviously, we can formulate the following theorem.

Theorem 2.3: *If the spectrum of the graph G contains the numbers $\lambda_1, \ldots, \lambda_n$, then the spectrum of the graph $P(G)$ contains the numbers $P(\lambda_1), \ldots, P(\lambda_n)$.*

The case $P(x) = x^n$ leads to the power of a graph that is already defined. In fact, G^n is the graph with the same vertex set as G, in which there are as many edges leading from the vertex x_i to the vertex x_j as there are walks of length n between vertices x_i and x_j in G.

In the special case $P(x) = x + h$ (h a positive integer) the forming of $P(G)$ consists in adding h simply counted loops to each of the vertices of G. The eigenvalues of $P(G)$ are greater by h than the corresponding eigenvalues of G.

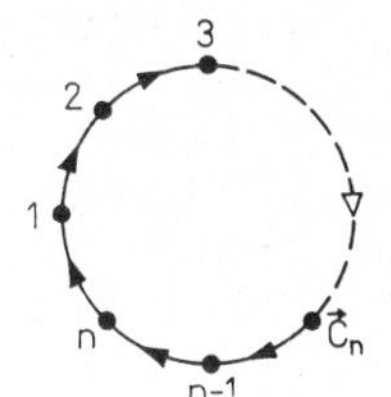

Fig. 2.1

The polynomial of a cycle $\vec{C}_n$ (Fig. 2.1) is also of interest. The adjacency matrix of this graph is the following square matrix of order n:

$$\boldsymbol{A} = \begin{pmatrix} \boldsymbol{O} & \boldsymbol{I}_{n-1} \\ \boldsymbol{I}_1 & \boldsymbol{O} \end{pmatrix}$$

It is readily seen that

$$\boldsymbol{A}^k = \begin{pmatrix} \boldsymbol{O} & \boldsymbol{I}_{n-k} \\ \boldsymbol{I}_k & \boldsymbol{O} \end{pmatrix}$$

for $k = 0, 1, \ldots, n - 1$. Therefore the adjacency matrix of $P(\vec{C}_n)$, where $P(x) = a_0 + a_1 x + \cdots + a_{n-1} x^{n-1}$, is a circulant matrix with first row $(a_0, a_1, \ldots, a_{n-1})$.

In particular, we have $C_n = \vec{C}_n \cup \vec{C}_n^{\,n-1}$, where C_n is a circuit and $\vec{C}_n$ is a cycle of n vertices.

It is well known from the theory of matrices (see, for example, [MaMi], p. 25) that the eigenvalues of $\boldsymbol{A}$ are $\lambda_j = \mathrm{e}^{\frac{2\pi j}{n} i} = \varepsilon_j$ $\left(j = 1, \ldots, n;\ i = \sqrt{-1}\right)$. (Using formula (1.35) we get simply $|\lambda \boldsymbol{I} - \boldsymbol{A}| = \lambda^n - 1$.) Therefore the eigenvalues of C_n are $\lambda_j = \varepsilon_j + \varepsilon_j^{n-1} = 2 \cos \frac{2\pi}{n} j$ $(j = 1, \ldots, n)$. The eigenvalues of the graph $P(\vec{C}_n)$ are given by $\lambda_j = P(\varepsilon_j)$, $j = 1, \ldots, n$, which is a well-known result in the theory of circulant matrices.

2.2. The spectrum of the complement, direct sum, and complete product of graphs

In this Section we shall consider three mutually related operations on graphs†. These are a unary operation: the *complementation* of a graph, and two binary operations: the *direct sum* and the *complete product* of graphs.

The *complement* $\bar{G}$ of a graph G is the graph with the same vertex set, with two (distinct) vertices, adjacent in $\bar{G}$ if and only if these vertices are non-adjacent in G.

The *direct sum* $G_1 \dot{+} G_2$ of graphs $G_1 = (\mathcal{X}_1, \mathcal{U}_1)$ and $G_2 = (\mathcal{X}_2, \mathcal{U}_2)$ $(\mathcal{X}_1 \cap \mathcal{X}_2 = \emptyset)$ is the graph $G = (\mathcal{X}, \mathcal{U})$ for which $\mathcal{X} = \mathcal{X}_1 \cup \mathcal{X}_2$ and $\mathcal{U} = \mathcal{U}_1 \cup \mathcal{U}_2$.

The *complete product* $G_1 \nabla G_2$ of graphs G_1 and G_2 is the graph obtained from $G_1 \dot{+} G_2$ by joining every vertex of G_1 with every vertex of G_2.

This sum and the product of graphs were considered in [Зык 1]. According to this paper, a graph is called *elementary*, if it is connected and cannot be represented as the complete product of two disjoint graphs. Unfortunately, according to [Зык 2], p. 508, the class of elementary graphs is very large. Nevertheless, many graphs can be represented by using the operations $\dot{+}$ and ∇, starting from a very narrow class of elementary graphs. (See Sections 2.6 and 7.6.)

The quite obvious relation

$$\overline{G_1 \nabla G_2} = \bar{G}_1 \dot{+} \bar{G}_2 \tag{2.1}$$

will be used in the subsequent discussion.

Note that both the direct sum and the ∇-product are commutative and associative operations. We shall see that the characteristic polynomial and some other matrix functions of graphs can be calculated in certain cases on the basis of these functions of elementary graphs.

We begin with the characteristic polynomials.

Consider first the direct sum of two graphs G_1 and G_2. Let $\boldsymbol{A}_i$ be the adjacency matrix of the graph G_i $(i = 1, 2)$. The adjacency matrix $\boldsymbol{A}$ of $G_1 \dot{+} G_2$ is of the form

$$\boldsymbol{A} = \begin{pmatrix} \boldsymbol{A}_1 & \boldsymbol{O} \\ \boldsymbol{O} & \boldsymbol{A}_2 \end{pmatrix},$$

i.e., $\boldsymbol{A} = \boldsymbol{A}_1 \dot{+} \boldsymbol{A}_2$, the sign $\dot{+}$ now denoting the direct sum of the matrices in question.

From the Laplacian development of the determinant, the following theorem can easily be obtained.

Theorem 2.4:

$$P_{G_1 \dot{+} G_2}(\lambda) = P_{G_1}(\lambda) \cdot P_{G_2}(\lambda). \tag{2.2}$$

Every graph is the direct sum of its components. If $G_1, \ldots, G_s$ are the components of the graph G, we have

$$P_G(\lambda) = P_{G_1}(\lambda) \cdots P_{G_s}(\lambda). \tag{2.3}$$

---

† In this section all graphs considered are supposed to be "schlicht" graphs, i.e., undirected graphs without loops or multiple edges.

This formula holds also in the case when G is an arbitrary (directed) (multi-)graph and $G_1, \ldots, G_s$ are its strong components (strongly connected components) as noticed by J. SEDLÁČEK [Sed1]. Consider the determinant $|\lambda \boldsymbol{I} - \boldsymbol{A}|$. From formula (1.42) for the development of a determinant, we see that elements of $|\lambda \boldsymbol{I} - \boldsymbol{A}|$ corresponding to edges not contained in any cycle do not affect the value of $|\lambda \boldsymbol{I} - \boldsymbol{A}|$. Accordingly, we can delete from G all the edges not forming part of a cycle without changing the value of the determinant. But upon the removal of these edges, the adjacency matrix of G (with respect to a suitable numbering of the vertices) attains the form of a direct sum of the adjacency matrices of the strong components of G so that the Laplacian development can be applied again.

This fact can be used as the basis for a graph theoretical method for the complete reduction of a matrix in order to find its eigenvalues [Har1].

We shall examine now the connection between the characteristic polynomial of a graph and that of its complement. We shall obtain this relation by considering the generating function $H_G(t)$ for the number of walks (1.59).

According to (1.58), the generating function $H_G(t)$ for the number N_k of walks of length k of an undirected graph G is

$$H_G(t) = \sum_{k=0}^{+\infty} t^k \sum_{\alpha=1}^{n} C_\alpha \lambda_\alpha^k = \sum_{\alpha=1}^{n} C_\alpha \sum_{k=0}^{+\infty} t^k \lambda_\alpha^k = \sum_{\alpha=1}^{n} \frac{C_\alpha}{1 - t\lambda_\alpha} \quad \left(|t| < \left(\max_\alpha \{\lambda_\alpha\}\right)^{-1}\right). \tag{2.4}$$

Consider the function

$$\psi(u) = (-1)^n \frac{P_{\bar{G}}(-u-1)}{P_G(u)}. \tag{2.5}$$

By means of (1.59) and (2.5) we obtain

$$\psi(u) = 1 + \frac{1}{u} H_G\left(\frac{1}{u}\right) = 1 + \sum_{\alpha=1}^{n} \frac{C_\alpha}{u - \lambda_\alpha} = \frac{H_1(u)}{H_2(u)}, \tag{2.6}$$

where $H_1(u)$, $H_2(u)$ are polynomials in u. The roots of $H_2(u)$ are all simple.

Let λ_0 be an eigenvalue of G with multiplicity p ($p \geqq 2$). Then $P_G(u) = (u - \lambda_0)^p \times Q(u)$ where $Q(\lambda_0) \neq 0$. From (2.5) and (2.6) we obtain

$$(-1)^n \frac{P_{\bar{G}}(-u-1)}{(u-\lambda_0)^p \, Q(u)} = \frac{H_1(u)}{H_2(u)}.$$

Since all roots of $H_2(u)$ are simple, the polynomial $P_{\bar{G}}(-u-1)$ must contain a factor $(u - \lambda_0)^q$, where $q \geqq p - 1$. Hence it follows that $P_{\bar{G}}(u)$ contains the factor $(u + \lambda_0 + 1)^q$. Hence, if G contains the eigenvalue λ_0 with multiplicity p ($p \geqq 2$), the complementary graph $\bar{G}$ has the eigenvalue $-\lambda_0 - 1$ with multiplicity q ($q \geqq p - 1$).

Notice that q cannot be greater than $p + 1$: suppose that $q > p + 1$, i.e., that the graph $H = \bar{G}$ contains $-\lambda_0 - 1$ as an eigenvalue with a multiplicity greater than $p + 1$. If the above statement is applied to H, we deduce that the graph $\bar{H} = G$ contains $-(-\lambda_0 - 1) - 1 = \lambda_0$ as an eigenvalue with a multiplicity greater than $p + 1 - 1 = p$, which is a contradiction. Hence, $q \leqq p + 1$.

We summarize the above facts in the following theorem.

Theorem 2.5 (D. M. CVETKOVIĆ [Cve9]): *If the spectrum of the graph G contains an eigenvalue λ_0 with multiplicity $p > 1$, then the spectrum of the complementary graph $\overline{G}$ contains an eigenvalue $-\lambda_0 - 1$ with multiplicity q, where $p - 1 \leqq q \leqq p + 1$.*

If G is a regular graph, the polynomial $P_{\overline{G}}(\lambda)$ can be expressed by means of $P_G(\lambda)$ (and vice versa). The relation is given by the following theorem.

Theorem 2.6 (H. SACHS [Sac1]): *If G is a regular graph of degree r with n vertices, then*

$$P_{\overline{G}}(\lambda) = (-1)^n \frac{\lambda - n + r + 1}{\lambda + r + 1} P_G(-\lambda - 1), \tag{2.7}$$

i.e., if the spectrum of G contains $\lambda_1 = r, \lambda_2, \ldots, \lambda_n$, then the spectrum of $\overline{G}$ contains $n - 1 - r, -\lambda_2 - 1, \ldots, -\lambda_n - 1$.

Proof. Since a walk can begin at any one of the n vertices of G, and since from any arbitrary vertex a walk may be continued in exactly r ways, for the number N_k of walks of length k in G we obtain $N_k = nr^k$. Therefore, for the generating function $H_G(t)$ we have

$$H_G(t) = \sum_{k=0}^{+\infty} N_k t^k = \sum_{k=0}^{+\infty} nr^k t^k = \frac{n}{1 - rt} \quad \left(|t| < \frac{1}{r}\right).$$

According to (1.59), we then have

$$\frac{1}{t}\left[(-1)^n \frac{P_{\overline{G}}\left(-\frac{t+1}{t}\right)}{P_G\left(\frac{1}{t}\right)} - 1\right] = \frac{n}{1 - rt}.$$

Putting $-\frac{t+1}{t} = \lambda$ in the above equality, we obtain (2.7).

This theorem was originally proved by the consideration of the eigenvectors [Sac1]. It was also proved in [Sac1] that an eigenvector $\boldsymbol{x}$ associated with the eigenvalue λ_i $(i = 2, \ldots, n)$ in the graph G is also an eigenvector belonging to the eigenvalue $-\lambda_i - 1$ in $\overline{G}$.

We proceed to the complete product of graphs. For the generating function $H_G(t)$ the relation $H_{G_1 \dotplus G_2}(t) = H_{G_1}(t) + H_{G_2}(t)$ obviously holds. By virtue of (1.59) and (2.2), this relation becomes

$$\frac{1}{t}\left[(-1)^{n_1+n_2} \frac{P_{\overline{G_1 \dotplus G_2}}\left(-\frac{1}{t} - 1\right)}{P_{G_1}\left(\frac{1}{t}\right) P_{G_2}\left(\frac{1}{t}\right)} - 1\right] = \sum_{i=1}^{2} \frac{1}{t}\left[(-1)^{n_i} \frac{P_{\overline{G}_i}\left(-\frac{1}{t} - 1\right)}{P_{G_i}\left(\frac{1}{t}\right)} - 1\right].$$

Since $\overline{G_1 \dotplus G_2} = \overline{G}_1 \nabla \overline{G}_2$, putting $-\frac{1}{t} - 1 = \lambda$ and substituting $\overline{G}_1, \overline{G}_2$ for G_1, G_2, i.e. G_1, G_2 for $\overline{G}_1, \overline{G}_2$, we arrive at the following theorem.

Theorem 2.7 (D. M. Cvetković [Cve9]): *The characteristic polynomial of the ∇-product of graphs is given by the relation*

$$P_{G_1 \nabla G_2}(\lambda) = (-1)^{n_2} P_{G_1}(\lambda) P_{\bar{G}_2}(-\lambda-1) + (-1)^{n_1} P_{G_2}(\lambda) P_{\bar{G}_1}(-\lambda-1) - (-1)^{n_1+n_2} P_{\bar{G}_1}(-\lambda-1) P_{\bar{G}_2}(-\lambda-1). \tag{2.8}$$

If G_1 and G_2 are regular graphs, Theorem 2.7 together with Theorem 2.6 gives the following result.

Theorem 2.8 (H.-J. Finck, G. Grohmann [FiGr]):† *The characteristic polynomial of the complete product of regular graphs G_1 and G_2 is given by the relation:*

$$P_{G_1 \nabla G_2}(\lambda) = \frac{P_{G_1}(\lambda) P_{G_2}(\lambda)}{(\lambda - r_1)(\lambda - r_2)} [(\lambda - r_1)(\lambda - r_2) - n_1 n_2]. \tag{2.9}$$

Let G_i ($i = 1, \ldots, k$) be a regular graph of degree r_i with n_i vertices whose characteristic polynomial $P_{G_i}(\lambda)$ is known. The graph $G_1 \nabla G_2$ is a regular graph, if and only if $r_1 + n_2 = r_2 + n_1$ (the regularity of graphs G_1, G_2 is obviously a necessary condition). If $r_1 + n_2 = r_2 + n_1$, then $G_1 \nabla G_2$ has $n^{(1)} = n_1 + n_2$ vertices and is regular of degree $r^{(1)} = r_1 + n_2 = r_2 + n_1$. Hence, the relations $n_1 - r_1 = n_2 - r_2 = n^{(1)} - r^{(1)}$ hold.

If $G_1 \nabla G_2$ is a regular graph, the polynomial

$$P_{G_1 \nabla G_2}(\lambda) = (\lambda - r^{(2)})(\lambda + n^{(1)} - r^{(1)}) \frac{P_{G_1}(\lambda) P_{G_2}(\lambda)}{(\lambda - r_1)(\lambda - r_2)} \tag{2.10}$$

determined by (2.9), can be used for the determination of $P_{(G_1 \nabla G_2) \nabla G_3}(\lambda)$. The necessary condition for $(G_1 \nabla G_2) \nabla G_3$ to be regular (of degree $r^{(2)}$ and with $n^{(2)}$ vertices) is that $n^{(1)} - r^{(1)} = n_3 - r_3 = n^{(2)} - r^{(2)}$; from (2.9) and (2.10) we have

$$P_{(G_1 \nabla G_2) \nabla G_3}(\lambda) = (\lambda - r^{(2)})(\lambda + n^{(1)} - r^{(1)}) \cdot (\lambda + n^{(2)} - r^{(2)}) \frac{P_{G_1}(\lambda) P_{G_2}(\lambda) P_{G_3}(\lambda)}{(\lambda - r_1)(\lambda - r_2)(\lambda - r_3)}.$$

Continuing this reasoning, we arrive at the following theorem.

Theorem 2.9 (H.-J. Finck, G. Grohmann [FiGr]):† *Let $G_1, \ldots, G_k$ be regular graphs; let G_i have degree r_i and n_i vertices ($i = 1, \ldots, k$), where the relations $n_1 - r_1 = n_2 - r_2 = \cdots = n_k - r_k = S$ hold. Then the graph $G = G_1 \nabla G_2 \nabla \cdots \nabla G_k$ has $n = n_1 + n_2 + \cdots + n_k$ vertices and is regular of degree $r = n - S$ so that we have*

$$P_G(\lambda) = (\lambda - r)(\lambda + n - r)^{k-1} \prod_{i=1}^{k} \frac{P_{G_i}(\lambda)}{\lambda - r_i}. \tag{2.11}$$

We shall consider now some other matrix functions which may be associated with a graph. Let $C_G(\lambda) = |\lambda \boldsymbol{I} - \boldsymbol{D} + \boldsymbol{A}|$ (see Section 1.2, (1.15) and Section 1.3, (1.24)). We shall deduce the function $C_G(\lambda)$ for the direct sum, complete product, and com-

† Theorems 2.8 aud 2.9 also follow from Theorem 2.10 (p. 59). — Note that for many applications the C-spectrum is more appropriate than the P-spectrum since it allows more general formulations valid for arbitrary multigraphs, not only for regular ones.

plement of graphs. If $G = G_1 \dot{+} G_2$, then

$$C_G(\lambda) = \begin{vmatrix} \lambda \boldsymbol{I}_1 - \boldsymbol{D}_1 + \boldsymbol{A}_1 & \boldsymbol{O} \\ \boldsymbol{O} & \lambda \boldsymbol{I}_2 - \boldsymbol{D}_2 + \boldsymbol{A}_2 \end{vmatrix},$$

where $\boldsymbol{A}_1$, $\boldsymbol{A}_2$ are the adjacency matrices and $\boldsymbol{D}_1$, $\boldsymbol{D}_2$ are the valency matrices (see footnote on p. 25) of the graphs G_1, G_2. Therefore,

$$C_{G_1 \dot{+} G_2}(\lambda) = C_{G_1}(\lambda)\, C_{G_2}(\lambda). \tag{2.12}$$

For the complementary graph $\overline{G}$ of the graph G with n vertices we have

$$C_{\overline{G}}(\lambda) = \left|\lambda \boldsymbol{I} - \big((n-1)\,\boldsymbol{I} - \boldsymbol{D}\big) + \boldsymbol{J} - \boldsymbol{I} - \boldsymbol{A}\right| = |(\lambda - n)\,\boldsymbol{I} + \boldsymbol{D} - \boldsymbol{A} + \boldsymbol{J}|. \tag{2.13}$$

If we add all remaining rows to the first row of this determinant, every entry of the first row becomes equal to λ. Taking this factor out and then subtracting the first row from all the other rows, we obtain

$$C_{\overline{G}}(\lambda) = \lambda \cdot \left|\big((\lambda - n)\,\boldsymbol{I} + \boldsymbol{D} - \boldsymbol{A}\big)^*\right|, \tag{2.14}$$

where $\big((\lambda - n)\,\boldsymbol{I} + \boldsymbol{D} - \boldsymbol{A}\big)^*$ denotes the matrix obtained from $(\lambda - n)\,\boldsymbol{I} + \boldsymbol{D} - \boldsymbol{A}$ by replacing all elements of the first row by numbers equal to 1.

Consider now the determinant

$$|(\lambda - n)\,\boldsymbol{I} + \boldsymbol{D} - \boldsymbol{A}| = (-1)^n\, C_G(n - \lambda). \tag{2.15}$$

If we add all the other rows to the first row, the entries of the first row all become equal to $\lambda - n$. Therefore,

$$|(\lambda - n)\,\boldsymbol{I} + \boldsymbol{D} - \boldsymbol{A}| = (\lambda - n)\left|\big((\lambda - n)\,\boldsymbol{I} + \boldsymbol{D} - \boldsymbol{A}\big)^*\right|. \tag{2.16}$$

According to (2.14), (2.15), and (2.16) we have

$$C_{\overline{G}}(\lambda) = (-1)^n \frac{\lambda}{\lambda - n}\, C_G(n - \lambda). \tag{2.17}$$

The following formula for the polynomial C_G of the complete product $G = G_1 \nabla G_2$ of graphs G_1 and G_2 that have n_1 and n_2 $(n_1 + n_2 = n)$ vertices, respectively, is obtained by means of (2.1), (2.12), and (2.17):

$$\begin{aligned} C_{G_1 \nabla G_2}(\lambda) &= C_{\overline{\overline{G}_1 \dot{+} \overline{G}_2}}(\lambda) = (-1)^n \frac{\lambda}{\lambda - n}\, C_{\overline{G}_1 \dot{+} \overline{G}_2}(n - \lambda) \\ &= (-1)^n \frac{\lambda}{\lambda - n}\, C_{\overline{G}_1}(n - \lambda)\, C_{\overline{G}_2}(n - \lambda) \\ &= \frac{\lambda(\lambda - n)}{(\lambda - n_1)\,(\lambda - n_2)}\, C_{G_1}(\lambda - n_2)\, C_{G_2}(\lambda - n_1). \end{aligned} \tag{2.18}$$

If we introduce the function

$$B_\lambda^n(G) = \frac{(-1)^n}{\lambda}\, C_G(-\lambda)$$

(see (1.17)) for a graph G with n vertices, formulas (2.17), (2.12), and (2.18) take the following form.

Theorem 2.10 (A. K. Kel'mans [KeЛ 1]):

$$B_\lambda^n(\overline{G}) = (-1)^{n-1} B_{-(\lambda+n)}^n(G), \tag{2.19}$$

$$B_\lambda^{n_1+n_2}(G_1 \dotplus G_2) = \lambda \cdot B_\lambda^{n_1}(G_1)\, B_\lambda^{n_2}(G_2), \tag{2.20}$$

$$B_\lambda^{n_1+n_2}(G_1 \nabla G_2) = (\lambda + n_1 + n_2)\, B_{\lambda+n_2}^{n_1}(G_1)\, B_{\lambda+n_1}^{n_2}(G_2). \tag{2.21}$$

2.3. Reduction procedures for calculating the characteristic polynomial

We shall consider now four reduction procedures which, in some cases, enable characteristic polynomials to be determined by simple numerical calculation.

1. Let x_1 be a vertex of degree 1 in the graph G and let x_2 be the vertex adjacent to x_1. Let G_1 be the induced subgraph obtained from G by deleting the vertex x_1. If x_1 and x_2 are deleted, the induced subgraph G_2 is obtained.

Theorem 2.11 (see [Hei 1], [HaKMR]):

$$P_G(\lambda) = \lambda \cdot P_{G_1}(\lambda) - P_{G_2}(\lambda). \tag{2.22}$$

The proof is obvious.

By iterating formula (2.22), the characteristic polynomial of a tree can easily be determined.

2. Let G be the graph obtained by joining the vertex x of the graph G_1 to the vertex y of the graph G_2 by an edge. Let G_1' (G_2') be the induced subgraph of G_1 (G_2) obtained by deleting the vertex x (y) from G_1 (G_2).

Theorem 2.12 (E. Heilbronner [Hei 1]):

$$P_G(\lambda) = P_{G_1}(\lambda)\, P_{G_2}(\lambda) - P_{G_1'}(\lambda)\, P_{G_2'}(\lambda). \tag{2.23}$$

The *proof* of this theorem is carried out by applying the Laplacian development to the determinant

$$P_G(\lambda) = \left| \begin{array}{c:c} \lambda I_{n_1} - A_1 & \begin{matrix} 0 & 0 & \cdots & 0 & -1 \\ & & & & 0 \\ & & 0 & & \vdots \\ & & & & 0 \end{matrix} \\ \hdashline \begin{matrix} 0 & & & & \\ \vdots & & 0 & & \\ 0 & & & & \\ -1 & 0 & \cdots & 0 & 0 \end{matrix} & \lambda I_{n_2} - A_2 \end{array} \right|$$

where A_1, A_2 are the adjacency matrices and n_1, n_2 are the numbers of vertices of the graphs G_1 and G_2, respectively.

Theorems 2.11 and 2.12 have been generalized by I. GUTMAN [Gut8].

3. Let H be the graph obtained from the graph G with vertex-set $\{x_1, x_2, \ldots, x_n\}$ in the following way:

(i) To each vertex x_i of G a set $\mathscr{V}_i$ of k new (isolated) vertices is added,

(ii) x_i is joined by an edge to each of the k vertices of $\mathscr{V}_i$ $(i = 1, 2, \ldots, n)$.

Theorem 2.13:

$$P_H(\lambda) = \lambda^{nk} P_G\left(\lambda - \frac{k}{\lambda}\right). \tag{2.24}$$

Proof. We have

$$P_H(\lambda) = \begin{vmatrix} \lambda \boldsymbol{I} - \boldsymbol{A} & -\boldsymbol{I} & -\boldsymbol{I} & \cdots & -\boldsymbol{I} \\ -\boldsymbol{I} & \lambda \boldsymbol{I} & \boldsymbol{O} & \cdots & \boldsymbol{O} \\ -\boldsymbol{I} & \boldsymbol{O} & \lambda \boldsymbol{I} & \cdots & \boldsymbol{O} \\ \vdots & \vdots & \vdots & \ddots & \vdots \\ -\boldsymbol{I} & \boldsymbol{O} & \boldsymbol{O} & \cdots & \lambda \boldsymbol{I} \end{vmatrix},$$

where there are $k + 1$ rows (and columns) and where $\boldsymbol{A}$ is the adjacency matrix of G, $\boldsymbol{I}$ is a unit matrix of order n, and $\boldsymbol{O}$ is a zero matrix of the same order. For $\lambda \neq 0$ we can multiply the rows (consisting of block-matrices) numbered $2, 3, \ldots, k + 1$ by $\frac{1}{\lambda}$ and add the resulting rows to the first row. We then have

$$P_H(\lambda) = \begin{vmatrix} \left(\lambda - \frac{k}{\lambda}\right) \boldsymbol{I} - \boldsymbol{A} & \boldsymbol{O} & \boldsymbol{O} & \cdots & \boldsymbol{O} \\ -\boldsymbol{I} & \lambda \boldsymbol{I} & \boldsymbol{O} & \cdots & \boldsymbol{O} \\ -\boldsymbol{I} & \boldsymbol{O} & \lambda \boldsymbol{I} & \cdots & \boldsymbol{O} \\ \vdots & \vdots & \vdots & \ddots & \vdots \\ -\boldsymbol{I} & \boldsymbol{O} & \boldsymbol{O} & \cdots & \lambda \boldsymbol{I} \end{vmatrix};$$

making use of the Laplacian development we obtain (2.24). For $\lambda = 0$, naturally, $P_H(0) = \lim_{\lambda \to 0} \lambda^{nk} P_G\left(\lambda - \frac{k}{\lambda}\right)$ holds, so that

$$P_H(0) = \begin{cases} (-1)^n & \text{if } k = 1, \\ 0 & \text{if } k > 1. \end{cases}$$

4. Let G be a graph with vertex-set $\{x_1, \ldots, x_n\}$. Let G_i $(i = 1, \ldots, n)$ be the subgraph induced by the vertices of $\mathscr{X} \setminus \{x_i\}$.

Theorem 2.14 (F. H. CLARKE [Clar]):

$$P'_G(\lambda) = \sum_{i=1}^{n} P_{G_i}(\lambda). \tag{2.25}$$

Proof. Row-by-row differentiation of $P_G(\lambda) = |\lambda \boldsymbol{I}_n - \boldsymbol{A}|$ yields

$$P'_G(\lambda) = \sum_{i=1}^{n} |(\lambda \boldsymbol{I}_n - \boldsymbol{A})_i| = \sum_{i=1}^{n} |\lambda \boldsymbol{I}_{n-1} - \boldsymbol{A}_i| = \sum_{i=1}^{n} P_{G_i}(\lambda).$$

Corollary. *If all subgraphs G_i $(i = 1, \ldots, n)$ are isomorphic with some graph H, then $P'_G(\lambda) = nP_H(\lambda)$.*

A formula closely related to (2.25) which connects the C-polynomial $C_G(\lambda)$ of a graph G with the C-polynomials of all those graphs that are derived from G by deleting one edge is given in [KeCh], Lemma 2.4.

The spectra of some other compositions of graphs are being considered in chemistry (see, for example, [Hei 1], [Hei 2]). See also Section 6.1.

2.4. Line graphs and total graphs

In this section we shall determine the characteristic polynomials of the line graphs of regular (and some other) graphs. Some other operations on graphs, similar to the construction of a line graph, will also be considered.

The *line graph* $L(G)$ of an undirected graph G without loops or multiple edges is the graph whose vertex-set is in one-to-one correspondence with the set of edges of the graph G, with two vertices of $L(G)$ being adjacent if and only if the corresponding edges in G have a vertex in common.

In the proofs of several theorems we shall use the following lemma from the general theory of matrices.

Lemma 2.1 (see, for example, [MaMi], p. 24): *If $\boldsymbol{A}$ is an $m \times n$ matrix, and if $P_{\boldsymbol{X}}(\lambda)$ denotes the characteristic polynomial of the square matrix $\boldsymbol{X}$, then*

$$\lambda^n P_{\boldsymbol{A}\boldsymbol{A}^\mathsf{T}}(\lambda) = \lambda^m P_{\boldsymbol{A}^\mathsf{T}\boldsymbol{A}}(\lambda). \tag{2.26}$$

Let $\boldsymbol{A}$ and $\boldsymbol{B}$ be the adjacency matrices of graphs G and $L(G)$, respectively. The $n \times m$ incidence matrix of vertices and edges of the graph G is denoted by $\boldsymbol{R}$ while $\boldsymbol{D}$ denotes the valency matrix of G. As is well known (see Section 0.2, p. 16), the following relations hold:

$$\boldsymbol{R}\boldsymbol{R}^\mathsf{T} = \boldsymbol{A} + \boldsymbol{D}, \tag{2.27}$$

$$\boldsymbol{R}^\mathsf{T}\boldsymbol{R} = \boldsymbol{B} + 2\boldsymbol{I}. \tag{2.28}$$

On the basis of (2.26) these relations yield $|\lambda\boldsymbol{I} - \boldsymbol{B} - 2\boldsymbol{I}| = \lambda^{m-n} \cdot |\lambda\boldsymbol{I} - \boldsymbol{A} - \boldsymbol{D}|$, i.e.,

$$P_{L(G)}(\lambda - 2) = \lambda^{m-n} \cdot |\lambda\boldsymbol{I} - \boldsymbol{A} - \boldsymbol{D}|. \tag{2.29}$$

For regular graphs of degree r we have $\boldsymbol{D} = r\boldsymbol{I}$ and from (2.29) we obtain

Theorem 2.15 (H. Sachs [Sac 8]): *If G is a regular graph of degree r with n vertices and $m \left(= \frac{1}{2} nr\right)$ edges, then the following relation holds*

$$P_{L(G)}(\lambda) = (\lambda + 2)^{m-n} P_G(\lambda - r + 2). \tag{2.30}$$

In [Bax 1] and [Кел 3] the analogous relation for the characteristic polynomials of the matrices $\boldsymbol{D} + \boldsymbol{A}$ and $\boldsymbol{D} - \boldsymbol{A}$ (i.e., for the polynomials $R_G(\lambda)$ and $C_G(\lambda)$; see Section 1.2) are given.

We shall show that for some other graphs (in addition to regular ones) a relation between $P_G(\lambda)$ and $P_{L(G)}(\lambda)$ can be established.

Recall that a multigraph G is called *semiregular of degrees* r_1, r_2 if it is bipartite having a representation $G = (\mathcal{X}_1, \mathcal{X}_2; \mathcal{U})$ with $|\mathcal{X}_1| = n_1$, $|\mathcal{X}_2| = n_2$, $n_1 + n_2 = n$, where each vertex $x \in \mathcal{X}_i$ has valency r_i $(i = 1, 2)$.

Theorem 2.16 (D. M. Cvetković [Cve9]): *Let G be a semiregular multigraph with $n_1 \geqq n_2$. Then the relation*

$$P_{L(G)}(\lambda) = (\lambda + 2)^\beta \sqrt{\left(-\frac{\alpha_1}{\alpha_2}\right)^{n_1 - n_2} P_G\left(\sqrt{\alpha_1\alpha_2}\right) P_G\left(-\sqrt{\alpha_1\alpha_2}\right)} \tag{2.31}$$

holds, where $\alpha_i = \lambda - r_i + 2$ $(i = 1, 2)$ and $\beta = n_1 r_1 - n_1 - n_2$.

Note that, by virtue of Theorems 1.1, 2.15, and 2.16, $Q_{L(G)}(\lambda)$, too, can be expressed in terms of $Q_G(\lambda)$ if G is regular or semiregular of positive degrees.

Proof of Theorem 2.16. Another lemma from matrix theory is needed.

Lemma 2.2 (see, for example, [Gant], vol. I, p. 45): *If $\boldsymbol{M}$ is a non-singular square matrix, we have*

$$\begin{vmatrix} \boldsymbol{M} & \boldsymbol{N} \\ \boldsymbol{P} & \boldsymbol{Q} \end{vmatrix} = |\boldsymbol{M}| \cdot |\boldsymbol{Q} - \boldsymbol{P}\boldsymbol{M}^{-1}\boldsymbol{N}|.$$

In order to prove the theorem we begin with

$$|\lambda \boldsymbol{I} - \boldsymbol{A} - \boldsymbol{D}| = \begin{vmatrix} (\lambda - r_1)\, \boldsymbol{I}_{n_1} & -\boldsymbol{K}^\mathsf{T} \\ -\boldsymbol{K} & (\lambda - r_2)\, \boldsymbol{I}_{n_2} \end{vmatrix},$$

where $\boldsymbol{K}$ is an $n_2 \times n_1$ matrix. According to Lemma 2.2, we have:

$$\begin{aligned} |\lambda \boldsymbol{I} - \boldsymbol{A} - \boldsymbol{D}| &= (\lambda - r_1)^{n_1} \cdot \left|(\lambda - r_2)\, \boldsymbol{I}_{n_2} - \boldsymbol{K} \frac{\boldsymbol{I}_{n_1}}{\lambda - r_1} \boldsymbol{K}^\mathsf{T}\right| \\ &= (\lambda - r_1)^{n_1 - n_2} \cdot |(\lambda - r_1)(\lambda - r_2)\, \boldsymbol{I}_{n_2} - \boldsymbol{K}\boldsymbol{K}^\mathsf{T}| \\ &= (\lambda - r_1)^{n_1 - n_2} P_{\boldsymbol{K}\boldsymbol{K}^\mathsf{T}}\big((\lambda - r_1)(\lambda - r_2)\big). \end{aligned} \tag{2.32}$$

The characteristic polynomial $P_{\boldsymbol{K}\boldsymbol{K}^\mathsf{T}}(\lambda)$ of the matrix $\boldsymbol{K}\boldsymbol{K}^\mathsf{T}$ can be expressed in terms of the characteristic polynomial of the adjacency matrix $\boldsymbol{A}$. We have

$$\boldsymbol{A} = \begin{pmatrix} \boldsymbol{O} & \boldsymbol{K}^\mathsf{T} \\ \boldsymbol{K} & \boldsymbol{O} \end{pmatrix}, \quad \boldsymbol{A}^2 = \begin{pmatrix} \boldsymbol{K}^\mathsf{T}\boldsymbol{K} & \boldsymbol{O} \\ \boldsymbol{O} & \boldsymbol{K}\boldsymbol{K}^\mathsf{T} \end{pmatrix}.$$

According to Lemma 2.1, $P_{\boldsymbol{K}^\mathsf{T}\boldsymbol{K}}(\lambda) = \lambda^{n_1 - n_2} P_{\boldsymbol{K}\boldsymbol{K}^\mathsf{T}}(\lambda)$. Since $P_{\boldsymbol{A}^2}(\lambda) = P_{\boldsymbol{K}^\mathsf{T}\boldsymbol{K}}(\lambda)\, P_{\boldsymbol{K}\boldsymbol{K}^\mathsf{T}}(\lambda)$ and since the eigenvalues of $\boldsymbol{A}^2$ are the squares of the eigenvalues of $\boldsymbol{A}$, i.e., $P_{\boldsymbol{A}^2}(\lambda^2) = (-1)^{n_1+n_2} P_{\boldsymbol{A}}(\lambda)\, P_{\boldsymbol{A}}(-\lambda)$,

$$P_{\boldsymbol{K}\boldsymbol{K}^\mathsf{T}}(\lambda) = \sqrt{\frac{P_{\boldsymbol{A}^2}(\lambda)}{\lambda^{n_1 - n_2}}} = \sqrt{(-1)^{n_1+n_2}\, \lambda^{n_2 - n_1} P_{\boldsymbol{A}}\left(\sqrt{\lambda}\right) P_{\boldsymbol{A}}\left(-\sqrt{\lambda}\right)}. \tag{2.33}$$

Combining expressions (2.29), (2.32), (2.33), we obtain (2.31).

This completes the proof of the theorem.

We now proceed to determine the characteristic polynomial with respect to some other unary operations on graphs.

The *subdivision graph* $S(G)$ of a graph G is the graph obtained by inserting a new vertex onto every edge of G.

Recall from the Introduction that the subdivision graph is a bipartite graph whose adjacency matrix if of the form

$$\begin{pmatrix} \boldsymbol{O} & \boldsymbol{R}^{\mathsf{T}} \\ \boldsymbol{R} & \boldsymbol{O} \end{pmatrix}.$$

By means of Lemma 2.2 and formula (2.27) we have, for a regular graph G,

$$P_{S(G)}(\lambda) = \begin{vmatrix} \lambda \boldsymbol{I}_m & -\boldsymbol{R}^{\mathsf{T}} \\ -\boldsymbol{R} & \lambda \boldsymbol{I}_n \end{vmatrix} = \lambda^m \cdot \left| \lambda \boldsymbol{I}_n - \boldsymbol{R} \frac{\boldsymbol{I}_m}{\lambda} \boldsymbol{R}^{\mathsf{T}} \right|$$
$$= \lambda^{m-n} \cdot |\lambda^2 \boldsymbol{I}_n - \boldsymbol{R}\boldsymbol{R}^{\mathsf{T}}| = \lambda^{m-n} \cdot |\lambda^2 \boldsymbol{I}_n - \boldsymbol{A} - r\boldsymbol{I}_n| = \lambda^{m-n} P_G(\lambda^2 - r).$$

We thus arrive at the following theorem.

Theorem 2.17 (D. M. Cvetković [Cve 17]): *If G is a regular graph of degree r with n vertices and m $\left(= \frac{1}{2} nr\right)$ edges, then*

$$P_{S(G)}(\lambda) = \lambda^{m-n} P_G(\lambda^2 - r). \tag{2.34}$$

Let $R(G)$ be the graph obtained from G by adding a new vertex corresponding to each edge of G and by joining each new vertex to the end points of the edge corresponding to it. The adjacency matrix of $R(G)$ is of the form

$$\begin{pmatrix} \boldsymbol{O}_m & \boldsymbol{R}^{\mathsf{T}} \\ \boldsymbol{R} & \boldsymbol{A} \end{pmatrix}.$$

Theorem 2.18 (D. M. Cvetković [Cve 17]): *If G is a regular graph of degree r with n vertices and m $\left(= \frac{1}{2} nr\right)$ edges, then*

$$P_{R(G)}(\lambda) = \lambda^{m-n} (\lambda + 1)^n \, P_G\left(\frac{\lambda^2 - r}{\lambda + 1}\right). \tag{2.35}$$

Proof.

$$P_{R(G)}(\lambda) = \begin{vmatrix} \lambda \boldsymbol{I}_m & -\boldsymbol{R}^{\mathsf{T}} \\ -\boldsymbol{R} & \lambda \boldsymbol{I}_n - \boldsymbol{A} \end{vmatrix} = \lambda^m \cdot \left| \lambda \boldsymbol{I}_n - \boldsymbol{A} - \boldsymbol{R} \frac{\boldsymbol{I}_m}{\lambda} \boldsymbol{R}^{\mathsf{T}} \right|$$
$$= \lambda^m \cdot \left| \lambda \boldsymbol{I}_n - \boldsymbol{A} - \frac{1}{\lambda} (\boldsymbol{A} + r\boldsymbol{I}_n) \right| = \lambda^{m-n} (\lambda + 1)^n \, P_G\left(\frac{\lambda^2 - r}{\lambda + 1}\right).$$

Let, further, $Q(G)$ be the graph obtained from G by inserting a new vertex into every edge of G and by joining by edges those pairs of these new vertices which lie on adjacent edges of G. The adjacency matrix of $Q(G)$ is then of the form

$$\begin{pmatrix} \boldsymbol{B} & \boldsymbol{R}^{\mathsf{T}} \\ \boldsymbol{R} & \boldsymbol{O}_n \end{pmatrix}.$$

Arguments similar to those previously used lead to the following theorem.

Theorem 2.19 (D. M. Cvetković [Cve 17]): *Let G be a graph with n vertices and m edges. Then*

$$P_{Q(G)}(\lambda) = \lambda^{n-m}(\lambda + 1)^m P_{L(G)}\left(\frac{\lambda^2 - 2}{\lambda + 1}\right). \tag{2.36}$$

Corollary: *If G is a regular graph of degree r, then according to* (2.36) *and* (2.30) *we have*

$$P_{Q(G)}(\lambda) = (\lambda + 1)^m P_G\left(\frac{\lambda^2 - (r-2)\lambda - r}{\lambda + 1}\right). \tag{2.37}$$

We proceed now to the investigation of total graphs.

The *total graph* $T(G)$ of a graph G is that graph whose set of vertices is the union of the set of vertices and of the set of edges of G, with two vertices of $T(G)$ being adjacent if and only if the corresponding elements of G are adjacent or incident.

It can easily be seen that, by a suitable numbering of the vertices, the adjacency matrix of $T(G)$ can be represented in the form

$$\begin{pmatrix} \boldsymbol{A} & \boldsymbol{R} \\ \boldsymbol{R}^{\mathrm{T}} & \boldsymbol{B} \end{pmatrix}.$$

If G is regular of degree r and has n vertices and m edges, we have

$$\begin{aligned}
P_{T(G)}(\lambda) &= \begin{vmatrix} \lambda\boldsymbol{I} + r\boldsymbol{I} - \boldsymbol{R}\boldsymbol{R}^{\mathrm{T}} & -\boldsymbol{R} \\ -\boldsymbol{R}^{\mathrm{T}} & \lambda\boldsymbol{I} + 2\boldsymbol{I} - \boldsymbol{R}^{\mathrm{T}}\boldsymbol{R} \end{vmatrix} \\
&= \begin{vmatrix} (\lambda + r)\,\boldsymbol{I} - \boldsymbol{R}\boldsymbol{R}^{\mathrm{T}} & -\boldsymbol{R} \\ -(\lambda + r + 1)\,\boldsymbol{R}^{\mathrm{T}} + \boldsymbol{R}^{\mathrm{T}}\boldsymbol{R}\boldsymbol{R}^{\mathrm{T}} & (\lambda + 2)\,\boldsymbol{I} \end{vmatrix} \\
&= \begin{vmatrix} (\lambda + r)\,\boldsymbol{I} - \boldsymbol{R}\boldsymbol{R}^{\mathrm{T}} + \dfrac{\boldsymbol{R}}{\lambda + 2}\left(-(\lambda + r + 1)\,\boldsymbol{R}^{\mathrm{T}} + \boldsymbol{R}^{\mathrm{T}}\boldsymbol{R}\boldsymbol{R}^{\mathrm{T}}\right) & \boldsymbol{O} \\ -(\lambda + r + 1)\,\boldsymbol{R}^{\mathrm{T}} + \boldsymbol{R}^{\mathrm{T}}\boldsymbol{R}\boldsymbol{R}^{\mathrm{T}} & (\lambda + 2)\,\boldsymbol{I} \end{vmatrix} \\
&= (\lambda + 2)^m \left| \lambda\boldsymbol{I} - \boldsymbol{A} + \frac{1}{\lambda + 2}(\boldsymbol{A} + r\boldsymbol{I})\left(\boldsymbol{A} - (\lambda + 1)\,\boldsymbol{I}\right) \right| \\
&= (\lambda + 2)^{m-n} \left| \boldsymbol{A}^2 - (2\lambda - r + 3)\,\boldsymbol{A} + \left(\lambda^2 - (r - 2)\,\lambda - r\right)\boldsymbol{I} \right| \\
&= (\lambda + 2)^{m-n} \prod_{i=1}^{n} \left(\lambda_i^2 - (2\lambda - r + 3)\,\lambda_i + \lambda^2 - (r - 2)\,\lambda - r\right) \\
&= (\lambda + 2)^{m-n} \prod_{i=1}^{n} \left(\lambda^2 - (2\lambda_i + r - 2)\,\lambda + \lambda_i^2 + (r - 3)\,\lambda_i - r\right),
\end{aligned}$$

λ_i $(i = 1, \ldots, n)$ being the eigenvalues of $\boldsymbol{A}$. Thus we have the following theorem.

Theorem 2.20 (D. M. Cvetković [Cve 13]): *If G is a regular graph of degree r $(r > 1)$ having n vertices and m edges, then $T(G)$ has $m - n$ eigenvalues equal to -2 and the following $2n$ eigenvalues*

$$\frac{1}{2}\left(2\lambda_i + r - 2 \pm \sqrt{4\lambda_i + r^2 + 4}\right) \quad (i = 1, \ldots, n).$$

In the subsequent discussion we shall consider connected graphs only.

Note that $-r \leqq \lambda_i \leqq r$ $(i = 1, \ldots, n)$. Consider the functions

$$f_1(x) = \frac{1}{2}\left(2x + r - 2 + \sqrt{4x + r^2 + 4}\right), \quad f_2(x) = \frac{1}{2}\left(2x + r - 2 - \sqrt{4x + r^2 + 4}\right).$$

Both are increasing on the segment $[-r, r]$ for $r \neq 2$. For $r > 2$ the first one maps this segment onto the segment $[-2, 2r]$ and the second one onto the segment $[-r, r-2]$. So the eigenvalues of $T(G)$ lie in the segment $[-r, 2r]$ (this holds also for $r = 1$). The greatest eigenvalue is, naturally, equal to $2r$. The eigenvalue $r - 2$ always appears in the spectrum. The smallest eigenvalue is equal to $-r$ if and only if G is bipartite. The multiplicity of the eigenvalue -2 in $T(G)$ is equal to $m - n + p_{-r} + p_{-1}$, where p_λ is the multiplicity of the eigenvalue λ in G, and $r > 2$.

In the case $r = 2$ the function $f_2(x)$ has a minimum for $x = -\frac{7}{4}$. Since $f_2\left(-\frac{7}{4}\right) = -\frac{9}{4}$, the smallest eigenvalue of $T(G)$ is greater than $-\frac{9}{4}$. Equality can never hold, since an eigenvalue of a graph cannot be a rational non-integral number. But, since the eigenvalues of a connected regular graph G of degree 2 with n vertices are $2 \cos \frac{2\pi}{n} i$ $(i = 1, 2, \ldots, n)$, there exist graphs G for which the smallest eigenvalue of $T(G)$ is arbitrarily close to the lower bound $-\frac{9}{4}$.

The case $r = 1$ is quite simple: G has eigenvalues $1, -1$, and $T(G)$ has eigenvalues $2, -1, -1$.

2.5. NEPS and Boolean functions

In this section we shall consider various n-ary operations defined on the set of finite, undirected graphs without loops or multiple edges; the resulting graphs shall have as vertex set the Cartesian product of the sets of vertices of those graphs on which the operation is applied. Some of the operations to be defined can be considered on other classes of graphs, too.

The known operations of this kind are the product, the sum and the p-sum of graphs ([Ber 1], p. 23 and p. 53). We now give the definitions of these operations.

Let $G_1 = (\mathscr{X}_1, \mathscr{U}_1)$ and $G_2 = (\mathscr{X}_2, \mathscr{U}_2)$ be two graphs. The *product* $G_1 \times G_2$ and the *sum* $G_1 + G_2$ of these graphs are graphs $(\mathscr{X}, \mathscr{U})$ and $(\mathscr{X}, \mathscr{V})$, respectively, where $\mathscr{X} = \mathscr{X}_1 \times \mathscr{X}_2$ and where the sets $\mathscr{U}$ and $\mathscr{V}$ are defined as follows.

Let $(x_1, x_2) \in \mathscr{X}$, $(y_1, y_2) \in \mathscr{X}$. The vertices (x_1, x_2) and (y_1, y_2) are adjacent in the product $G_1 \times G_2$ if and only if $(x_1, y_1) \in \mathscr{U}_1$ and $(x_2, y_2) \in \mathscr{U}_2$, and they are adjacent in the sum $G_1 + G_2$ if and only if either $x_1 = y_1$ and $(x_2, y_2) \in \mathscr{U}_2$ or $(x_1, y_1) \in \mathscr{U}_1$ and $x_2 = y_2$.

The *p-sum* G of graphs $G_1, \ldots, G_n$ is a graph whose set of vertices is the Cartesian product of the sets of vertices of the graphs $G_1, \ldots, G_n$. If x_i and y_i are vertices of the graphs G_i $(i = 1, \ldots, n)$, the vertices $(x_1, \ldots, x_n)$ and $(y_1, \ldots, y_n)$ of the p-sum G

are adjacent if and only if exactly p of the n pairs (x_i, y_i) $(i = 1, \ldots, n)$ are pairs of adjacent vertices in the corresponding graphs G_i, and $x_i = y_i$ for the remaining $n - p$ pairs. Thus if $p = n$, we have the product of n graphs, and if $p = 1$, we have the sum of n graphs.

The notion of the p-sum and its specializations appear in several papers under different names. So, for example, the product of two graphs is called *product* ([Ber 1], p. 23), *Cartesian product* ([Sabi]), *Kronecker's product* ([Weic]), *conjunction* ([HaWi]), *cardinal product* ([Čuli], [Mill]), *α-product* ([TeYa]), and so on. A similar statement holds for the sum of graphs.

In [БарВ] a more general operation is defined. We shall call it: *extended p-sum* or *$\mathcal{J}$-sum* of graphs $G_1, G_2, \ldots, G_n$. The set of vertices of the extended p-sum is equal to the set of vertices of the p-sum. The set of edges of the extended p-sum is the union of the sets of edges of the p-sums of graphs $G_1, G_2, \ldots, G_n$, where p takes all values from a given subset $\mathcal{J}$ of the set $\{1, \ldots, n\}$.

When $\mathcal{J} = \{1, \ldots, n\}$, we get an operation known as the *strong graph product.*

In [CvL 1] the *incomplete p-sum* and the *incomplete extended p-sum* of graphs are defined. Since the first of these operations is a special case of the second, we shall consider only the latter. The following definition of the incomplete extended p-sum, briefly: NEPS (Non-complete Extended P-Sum; originally from Serbo-Croatian: nepotpuna proširena p-suma) is given in the version of [Cve 9].

Let $\mathcal{B}$ be a set of n-tuples $(\beta_1, \ldots, \beta_n)$ of symbols 0 and 1, which does not contain the n-tuple $(0, \ldots, 0)$.

Definition 1. The *NEPS with basis $\mathcal{B}$ of the graphs $G_1, \ldots, G_n$* is the graph whose set of vertices is the Cartesian product of the sets of vertices of the graphs $G_1, \ldots, G_n$ and in which two vertices $(x_1, \ldots, x_n)$ and $(y_1, \ldots, y_n)$ are adjacent if and only if there is an n-tuple $(\beta_1, \ldots, \beta_n)$ in $\mathcal{B}$ such that $x_i = y_i$ holds exactly when $\beta_i = 0$, and x_i is adjacent to y_i in G_i exactly when $\beta_i = 1$.

If $\mathcal{B}$ consists only of n-tuples containing exactly p 1's, the operation is called *incomplete p-sum.* The *complete p-sum*, or briefly: *p-sum*, is obtained if $\mathcal{B}$ consists of all possible n-tuples with exactly p 1's. If $\mathcal{J} = \{j_1, \ldots, j_k\} \subset \{1, \ldots, n\}$, the operation represents the $\mathcal{J}$-sum under the condition that $\mathcal{B}$ consists of all n-tuples having j_i ones, $i = 1, 2, \ldots, k$.

It is of interest that in [TeYa] the special cases of NEPS' with the following bases are considered:

$$\{(1, 1)\},\ \{(0, 1), (1, 0)\},\ \{(0, 1), (1, 0), (1, 1)\},\ \{(1, 0)\},\ \{(1, 1), (1, 0)\}.$$

These operations appear in [TeYa] under the following names: *α-product, β-product, γ-product, semi β-product, semi γ-product.* The first three of these operations represent, according to our terminology, product, sum, and strong product of graphs, respectively.

We shall also note the Boolean operations on graphs. We give a definition, according to [Cve 5]. (In [HaWi] differently defined Boolean operations on graphs are also considered.)

Definition 2. Let $G_i = (\mathscr{X}_i, \mathscr{U}_i)$ $(i = 1, \ldots, n)$ be given graphs, where $\mathscr{X}_i$ and $\mathscr{U}_i$ denote the corresponding sets of vertices and of edges. If $f(p_1, \ldots, p_n)$ is an arbitrary Boolean function (f: $\{0, 1\}^n \to \{0, 1\}$), the *Boolean function* $G = f(G_1, \ldots, G_n)$ *of the graphs* $G_1, \ldots, G_n$ is the graph $G = (\mathscr{X}, \mathscr{U})$, where $\mathscr{X} = \mathscr{X}_1 \times \cdots \times \mathscr{X}_n$ and where $\mathscr{U}$ is defined in the following way. For any two vertices $(x_1, \ldots, x_n)$ and $(y_1, \ldots, y_n)$ of G the Boolean variables $p_1, \ldots, p_n$ are defined where, for every i, $p_i = 1$ if and only if x_i and y_i are adjacent in G_i. The vertices $(x_1, \ldots, x_n)$ and $(y_1, \ldots, y_n)$ are adjacent in G if and only if, for every i, $x_i \neq y_i$ and $f(p_1, \ldots, p_n) = 1$.

The set of n-tuples for which the Boolean function $f(p_1, \ldots, p_n)$ takes the value 1 will be denoted by $\mathscr{F}$. We shall also utilise the abbreviation $\beta = (\beta_1, \ldots, \beta_n)$, and sometimes β will be given an index, for example: $\beta^{(s)}$.

Definition 3. The NEPS with basis $\mathscr{B}$ of the graphs $G_1, \ldots, G_n$ *corresponds* to the Boolean function $f(G_1, \ldots, G_n)$ $\big(f(p_1, \ldots, p_n) \not\equiv 0\big)$, if $\mathscr{B} = \mathscr{F} \setminus \{(0, \ldots, 0)\}$.

The adjacency matrices of the operations on graphs considered above are expressed in terms of the adjacency matrices of the graphs on which the operations are applied by means of Kronecker's multiplication of matrices. Kronecker's multiplication of matrices will be denoted by $\otimes$. The properties of the Kronecker product of matrices used in the subsequent discussion are given below.

The *Kronecker product* $\boldsymbol{A} \otimes \boldsymbol{B}$ *of matrices* $\boldsymbol{A} = (a_{ij})_{m,n}$ and $\boldsymbol{B} = (b_{ij})_{p,q}$ is the $mp \times nq$ matrix obtained from $\boldsymbol{A}$ when every element a_{ij} is replaced by the block $a_{ij}\boldsymbol{B}$. Thus the entries of $\boldsymbol{A} \otimes \boldsymbol{B}$ consist of all the $mnpq$ possible products of an entry of $\boldsymbol{A}$ with an entry of $\boldsymbol{B}$.

The following relations are known (see, for example, [MaMi], p. 18 and 8):

$$\operatorname{tr}(\boldsymbol{A} \otimes \boldsymbol{B}) = \operatorname{tr} \boldsymbol{A} \cdot \operatorname{tr} \boldsymbol{B}, \tag{2.38}$$

$$(\boldsymbol{A} \otimes \boldsymbol{B}) \cdot (\boldsymbol{C} \otimes \boldsymbol{D}) = (\boldsymbol{AC}) \otimes (\boldsymbol{BD}). \tag{2.39}$$

(2.38) holds, if $\boldsymbol{A}$ and $\boldsymbol{B}$ are square matrices, and (2.39) holds if the products $\boldsymbol{AC}$ and $\boldsymbol{BD}$ exist.

The Kronecker products is, furthermore, an assosiative operation.

Starting from (2.39) and using induction, we obtain

$$\begin{aligned}&(\boldsymbol{A}_1 \otimes \cdots \otimes \boldsymbol{A}_n) \cdot (\boldsymbol{B}_1 \otimes \cdots \otimes \boldsymbol{B}_n) \cdots (\boldsymbol{M}_1 \otimes \cdots \otimes \boldsymbol{M}_n)\\ &= (\boldsymbol{A}_1\boldsymbol{B}_1 \cdots \boldsymbol{M}_1) \otimes \cdots \otimes (\boldsymbol{A}_n\boldsymbol{B}_n \cdots \boldsymbol{M}_n).\end{aligned} \tag{2.40}$$

It was noticed in [Čuli] that the adjacency matrix of the product $G_1 \times G_2$ of graphs G_1 and G_2 is equal to the Kronecker product $\boldsymbol{A}_1 \otimes \boldsymbol{A}_2$ of the adjacency matrices $\boldsymbol{A}_1$ and $\boldsymbol{A}_2$ of the graphs G_1 and G_2. In [Aber] the adjacency matrix of the sum $G_1 + G_2$ was determined; it is of the form $\boldsymbol{A}_1 \otimes \boldsymbol{I}_2 + \boldsymbol{I}_1 \otimes \boldsymbol{A}_2$, where $\boldsymbol{I}_1$ and $\boldsymbol{I}_2$ are unit matrices of the same orders as $\boldsymbol{A}_1$ and $\boldsymbol{A}_2$, respectively.

In [HaWi] adjacency matrices of a number of binary operations on graphs are listed.

If x and y are vertices of an arbitrary graph with adjacency matrix $\boldsymbol{A}$, we denote by $(A)_{xy}$ the element of A from the row corresponding to x and the column corresponding to y.

The adjacency matrix of the p-sum of graphs is given in [Cve2]. The following general result concerning the adjacency matrix of an NEPS is taken from [Cve9].

Theorem 2.21 (D. M. Cvetković [Cve9]): *The NEPS G with basis $\mathcal{B}$ of graphs $G_1, \ldots, G_n$ whose adjacency matrices are $A_1, \ldots, A_n$ has the following adjacency matrix:*

$$\mathbf{A} = \sum_{\beta \in \mathcal{B}} A_1^{\beta_1} \otimes \cdots \otimes A_n^{\beta_n}. \tag{2.41}$$

Proof. In each of the graphs $G_1, \ldots, G_n$ let the vertices be ordered (labelled). We shall lexicographically order the vertices of G (which represent the ordered n-tuples of vertices of graphs $G_1, \ldots, G_n$) and form the adjacency matrix $\mathbf{A}$ according to this ordering.

By virtue of the properties of the Kronecker product of matrices, the entries of $\mathbf{A}$ are

$$(\mathbf{A})_{(x_1,\ldots,x_n),(y_1,\ldots,y_n)} = \sum_{\beta \in \mathcal{B}} (A_1^{\beta_1})_{x_1 y_1} \cdots (A_n^{\beta_n})_{x_n y_n}. \tag{2.42}$$

By virtue of the lexicographic ordering,

$$(\mathbf{A})_{(x_1,\ldots,x_n),(y_1,\ldots,y_n)} = 1$$

if and only if there exists a $\beta \in \mathcal{B}$ with

$$(A_i^{\beta_i})_{x_i y_i} = 1 \qquad \text{for } i = 1, 2, \ldots, n.$$

This means precisely that x_i and y_i are adjacent in G_i if $\beta_i = 1$, and equal if $\beta_i = 0$ (i.e. $A_i^{\beta_i} = I$).

This completes the proof of the theorem.

The adjacency matrix of the p-sum is obtained from the above, if $\mathcal{B}$ contains all n-tuples with p entries equal to one. By further specialization, we get the product of graphs for $n = 2$, $p = 2$ and the sum of graphs for $n = 2$, $p = 1$.

The following theorem establishes the adjacency matrix of a Boolean function of graphs.

Theorem 2.22 (D. M. Cvetković [Cve5]): *The Boolean function $G = f(G_1, \ldots, G_n)$ of graphs $G_1, \ldots, G_n$ having adjacency matrices $A_1, \ldots, A_n$ has the adjacency matrix*

$$A = \sum_{\beta \in \mathcal{F}} A_1^{[\beta_1]} \otimes \cdots \otimes A_n^{[\beta_n]}, \tag{2.43}$$

where, for matrices A_i, the convention $A_i^{[1]} = A_i$, $A_i^{[0]} = \bar{A}_i$ is adopted, $\bar{A}_i$ being the adjacency matrix of the complement $\overline{G}_i$ of the graph G_i.

Proof. If $\mathcal{F}$ is empty, i.e. if $f(\beta_1, \ldots, \beta_n)$ is always equal to zero, the graph G has no edges, and the corresponding adjacency matrix is the zero matrix, a result consistent with formula (2.43). Consider further the case when $\mathcal{F}$ is non-empty.

As in Theorem 2.21, we use the lexicographic ordering and ask when $(A)_{(x_1,\ldots,x_n),(y_1,\ldots,y_n)}$ is equal to one. In this case the equality is attained if and only if there is some $\beta \in \mathcal{F}$

such that $(A_i^{[\beta_i]})_{x_i y_i} = 1$ for $i = 1, 2, \ldots, n$. This says precisely that x_i and y_i are adjacent if $\beta_i = 1$, and x_i and y_i are unequal and non-adjacent if $\beta_i = 0$, which in turn is the definition for adjacency in the Boolean function of the graphs.

This completes the proof of the theorem.

The following two theorems (see [Cve9], [Cve5]) describe the relation between the spectrum of the NEPS or the Boolean function and the spectra of the graphs on which the operations are performed. For the Boolean function the relation is obtained only when $G_1, \ldots, G_n$ are regular graphs.

Theorem 2.23 (D. M. Cvetković [Cve9]): *For $i = 1, 2, \ldots, n$, let G_i be a graph with n_i vertices, and let $\lambda_{i1}, \ldots, \lambda_{in_i}$ be the spectrum of G_i. Then the spectrum of the NEPS with basis $\mathscr{B}$ of the graphs $G_1, \ldots, G_n$ consists of all possible values of $\Lambda_{i_1,\ldots,i_n}$, where*

$$\Lambda_{i_1,\ldots,i_n} = \sum_{\beta \in \mathscr{B}} \lambda_{1i_1}^{\beta_1} \cdots \lambda_{ni_n}^{\beta_n} \tag{2.44}$$

$(i_k = 1, 2, \ldots, n_k;\ k = 1, 2, \ldots, n)$.

Proof.† We denote the summands of (2.41) by $\boldsymbol{B}_1, \ldots, \boldsymbol{B}_q$, where $q = |\mathscr{B}|$. The order of the summands is arbitrary. To each $\beta = (\beta_1, \ldots, \beta_n)$ from $\mathscr{B}$ there corresponds a summand $\boldsymbol{B}_s$. We define the expression R_s by $R_s = \lambda_{1i_1}^{\beta_1} \cdots \lambda_{ni_n}^{\beta_n}$.

Clearly,

$$\boldsymbol{B}_1^{s_1} \cdots \boldsymbol{B}_q^{s_q} = \boldsymbol{A}_1^{l_1} \otimes \cdots \otimes \boldsymbol{A}_n^{l_n},$$

$$R_1^{s_1} \cdots R_q^{s_q} = \lambda_{1i_1}^{l_1} \cdots \lambda_{ni_n}^{l_n},$$

where l_r is the sum of those numbers $s_1, \ldots, s_q$ which are, in the expression $\boldsymbol{B}_1^{s_1} \cdots \boldsymbol{B}_q^{s_q}$, exponents of those summands $\boldsymbol{B}_s$ which contain $\boldsymbol{A}_r$.

Hence, we have $\boldsymbol{A} = \boldsymbol{B}_1 + \cdots + \boldsymbol{B}_q$. Summands $\boldsymbol{B}_s$ commute with respect to matrix multiplication. So we have

$$(\boldsymbol{B}_1 + \cdots + \boldsymbol{B}_q)^k = \sum_{s_1,\ldots,s_q} \frac{k!}{s_1! \cdots s_q!} \boldsymbol{B}_1^{s_1} \cdots \boldsymbol{B}_q^{s_q}$$

$$= \sum_{s_1,\ldots,s_q} \frac{k!}{s_1! \cdots s_q!} \boldsymbol{A}_1^{l_1} \otimes \cdots \otimes \boldsymbol{A}_n^{l_n} \tag{2.45}$$

† This proof is referred to later. The formulation of the following short proof is due to M. Doob: Since $\boldsymbol{A}_i$, the adjacency matrix of G_i, is normal, there exist vectors $\boldsymbol{x}_{ij}$ such that $\boldsymbol{A}_i \boldsymbol{x}_{ij} = \lambda_{ij} \boldsymbol{x}_{ij}$ $(i = 1, 2, \ldots, n;\ j = 1, 2, \ldots, n_i)$. Consider the vector

$$\boldsymbol{x} = \boldsymbol{x}_{1i_1} \otimes \cdots \otimes \boldsymbol{x}_{ni_n}.$$

Applying property (2.40) to equation (2.41), we see that

$$\mathbf{A}\boldsymbol{x} = \Lambda_{i_1,\ldots,i_n} \boldsymbol{x}.$$

This yields $n_1 \cdot n_2 \cdots n_k$ eigenvalues and hence all possible values have been determined.

$(s_1 + \cdots + s_q = k)$. Further,

$$\operatorname{tr}(\boldsymbol{B}_1 + \cdots + \boldsymbol{B}_q)^k = \sum_{s_1,\ldots,s_q} \frac{k!}{s_1! \cdots s_q!} \operatorname{tr}(\boldsymbol{A}_1^{l_1} \otimes \cdots \otimes \boldsymbol{A}_n^{l_n})$$

$$= \sum_{s_1,\ldots,s_q} \frac{k!}{s_1! \cdots s_q!} \operatorname{tr} \boldsymbol{A}_1^{l_1} \cdots \operatorname{tr} \boldsymbol{A}_n^{l_n}$$

$$= \sum_{s_1,\ldots,s_q} \frac{k!}{s_1! \cdots s_q!} \sum_{i_1} \lambda_{1i_1}^{l_1} \cdots \sum_{i_n} \lambda_{ni_n}^{l_n}$$

$$= \sum_{i_1,\ldots,i_n} \sum_{s_1,\ldots,s_n} \frac{k!}{s_1! \cdots s_q!} \lambda_{1i_1}^{l_1} \cdots \lambda_{ni_n}^{l_n}$$

$$= \sum_{i_1,\ldots,i_n} \sum_{s_1,\ldots,s_q} \frac{k!}{s_1! \cdots s_q!} R_1^{s_1} \cdots R_q^{s_q}. \tag{2.46}$$

Thus,

$$\operatorname{tr}(\boldsymbol{B}_1 + \cdots + \boldsymbol{B}_q)^k = \sum_{i_1,\ldots,i_n} (R_1 + \cdots + R_q)^k \qquad (k = 1, 2, \ldots),$$

which proves the theorem.

We note some special cases of this theorem.

If λ_{1i_1} $(i_1 = 1, \ldots, m_1)$ *and* λ_{2i_2} $(i_2 = 1, \ldots, m_2)$ *are the eigenvalues of graphs* G_1 *and* G_2, *then*

1⁰ *the product* $G_1 \times G_2$ *has eigenvalues* $\lambda_{1i_1}\lambda_{2i_2}$ $(i_1 = 1, \ldots, m_1;\ i_2 = 1, \ldots, m_2)$;

2⁰ *the sum* $G_1 + G_2$ *has eigenvalues* $\lambda_{1i_1} + \lambda_{2i_2}$ $(i_1 = 1, \ldots, m_1;\ i_2 = 1, \ldots, m_2)$;

3⁰ *the strong product of* G_1 *and* G_2 *has eigenvalues* $\lambda_{1i_1}\lambda_{2i_2} + \lambda_{1i_1} + \lambda_{2i_2}$ $(i_1 = 1, \ldots, m_1;\ i_2 = 1, \ldots, m_2)$.

The result concerning the product of graphs appeared in [Mow2], but the determination of the eigenvalues of $\boldsymbol{A}_1 \otimes \boldsymbol{A}_2$ in terms of eigenvalues of $\boldsymbol{A}_1$ and $\boldsymbol{A}_2$ is well known in matrix theory (see, for example, [MaMi], p. 24). For the sum of graphs we refer to [Ruth]: in this paper, the spectrum of $\boldsymbol{A}_1 \otimes \boldsymbol{I}_2 + \boldsymbol{I}_1 \otimes \boldsymbol{A}_2$ was determined without explicitly mentioning the concept of the sum of graphs.

The result concerning the product of graphs can be specified for the bipartite product of a graph with itself ("bipartite square") $G \circ G$: Let $P_G(\lambda) = \prod_{i=1}^{n} (\lambda - \lambda_i)$. Since $G \circ G = K_2 \times G$ and since the spectrum of K_2 is $\{1, -1\}$,

$$P_{G \circ G}(\lambda) = \prod_{i=1}^{n} (\lambda - \lambda_i) \cdot \prod_{i=1}^{n} (\lambda + \lambda_i) = (-1)^n P_G(\lambda) P_G(-\lambda),$$

which appeared in [Sac7].

Note that, according to (2.44), the spectrum of the p-sum is equal to the set of all values of the elementary symmetric function of order p of variables $\lambda_{1i_1}, \ldots, \lambda_{ni_n}$.

We proceed to the determination of the spectrum of a Boolean function of graphs when each factor is regular. Let G_i be a regular graph of degree r_i with n_i vertices and with spectrum $\{\lambda_{i1} = r_i, \lambda_{i2}, \ldots, \lambda_{in_i}\}$. Then, according to Theorem 2.6, the spectrum of $\overline{G}_i$ contains $n_i - r_i - 1 = n_i - \lambda_{i1} - 1, -\lambda_{i2} - 1, \ldots, -\lambda_{in_i} - 1$; put $\bar{\lambda}_{i1} = n_i - \lambda_{i1} - 1$, $\bar{\lambda}_{ij} = -\lambda_{ij} - 1$ $(j = 2, \ldots, n_i)$. Further, the adjacency matrices of G_i and $\overline{G}_i$, $\boldsymbol{A}_i$ and $\bar{\boldsymbol{A}}_i$, have a simultaneous set of eigenvectors, i.e., there exist vectors $\boldsymbol{x}_{ij}$ such that $\boldsymbol{A}_i\boldsymbol{x}_{ij} = \lambda_{ij}\boldsymbol{x}_{ij}$ and $\bar{\boldsymbol{A}}_i\boldsymbol{x}_{ij} = \bar{\lambda}_{ij}\boldsymbol{x}_{ij}$ (see [Sac1]).

Theorem 2.24 (D. M. Cvetković [Cve5]): *For $i = 1, 2, \ldots, n$, let G_i be a regular graph of degree r_i with n_i vertices and spectrum equal to $\{\lambda_{i1} = r_i, \lambda_{i2}, \ldots, \lambda_{in_i}\}$, and let $\boldsymbol{x}_{i1}, \boldsymbol{x}_{i2}, \ldots, \boldsymbol{x}_{in_i}$ be a system of independent eigenvectors corresponding to the spectrum. Then the spectrum of the Boolean function $G = f(G_1, \ldots, G_n)$ consists exactly of all possible values of $\Lambda_{i_1,\ldots,i_n}$ where $\lambda_{ij}^{[0]} = \lambda_{ij}$, $\lambda_{ij}^{[1]} = \bar{\lambda}_{ij}$, and*

$$\Lambda_{i_1,\ldots,i_n} = \sum_{\beta\in\mathcal{F}} \lambda_{1i_1}^{[\beta_1]} \cdots \lambda_{ni_n}^{[\beta_n]}, \tag{2.47}$$

and $\boldsymbol{x}_{i_1,\ldots,i_n} := \boldsymbol{x}_{1i_1} \otimes \cdots \otimes \boldsymbol{x}_{ni_n}$ is an eigenvector of G corresponding to the eigenvalue $\Lambda_{i_1,\ldots,i_n}$.

Proof. According to the foregoing and according to the properties of Kronecker products of matrices we have

$$\begin{aligned}
\boldsymbol{A}\boldsymbol{x}_{i_1,\ldots,i_n} &= \Big(\sum_{\beta\in\mathcal{F}} \boldsymbol{A}_1^{[\beta_1]} \otimes \cdots \otimes \boldsymbol{A}_n^{[\beta_n]}\Big)(\boldsymbol{x}_{1i_1} \otimes \cdots \otimes \boldsymbol{x}_{ni_n}) \\
&= \sum_{\beta\in\mathcal{F}} (\boldsymbol{A}_1^{[\beta_1]} \otimes \cdots \otimes \boldsymbol{A}_n^{[\beta_n]})(\boldsymbol{x}_{1i_1} \otimes \cdots \otimes \boldsymbol{x}_{ni_n}) \\
&= \sum_{\beta\in\mathcal{F}} (\boldsymbol{A}_1^{[\beta_1]}\boldsymbol{x}_{1i_1}) \otimes \cdots \otimes (\boldsymbol{A}_n^{[\beta_n]}\boldsymbol{x}_{ni_n}) \\
&= \sum_{\beta\in\mathcal{F}} \big(\lambda_{1i_1}^{[\beta_1]}\boldsymbol{x}_{1i_1}\big) \otimes \cdots \otimes \big(\lambda_{ni_n}^{[\beta_n]}\boldsymbol{x}_{ni_n}\big) \\
&= \Big(\sum_{\beta\in\mathcal{F}} \lambda_{1i_1}^{[\beta_1]} \cdots \lambda_{ni_n}^{[\beta_n]}\Big)(\boldsymbol{x}_{1i_1} \otimes \cdots \otimes \boldsymbol{x}_{ni_n}) \\
&= \Lambda_{i_1,\ldots,i_n} \cdot \boldsymbol{x}_{i_1,\ldots,i_n},
\end{aligned}$$

which proves the theorem.

Several other operations on graphs related to the NEPS and the Boolean function have also been defined in the literature. Let us consider the lexicographic product of graphs.

The *lexicographic product* (or *composition*) $G_1[G_2]$ *of graphs* G_1 *and* G_2 has as its set of vertices, as in the cases considered above, the Cartesian product of the sets of vertices of graphs G_1 and G_2, but the vertices (x_1, y_1) and (x_2, y_2) of $G_1[G_2]$ are adjacent if and only if either x_1 is adjacent to x_2 in G_1 or $x_1 = x_2$ and y_1 is adjacent to y_2 in G_2. It was pointed out in [HaWi] that the adjacency matrix of $G_1[G_2]$ is of the form $(\boldsymbol{A}_1 \otimes \boldsymbol{J}_2) + (\boldsymbol{I}_1 \otimes \boldsymbol{A}_2)$, where $\boldsymbol{A}_1$ and $\boldsymbol{A}_2$ are the adjacency matrices of G_1 and G_2, $\boldsymbol{J}_2$ is a square matrix of the same order as $\boldsymbol{A}_2$ all of whose entries are equal to 1, and $\boldsymbol{I}_1$ is a unit matrix of the same order as $\boldsymbol{A}_1$.

We quote without proof the following result.

Let $\lambda_1, \ldots, \lambda_n$ ($\lambda_1 \geqq \lambda_2 \geqq \cdots \geqq \lambda_n$) *be the eigenvalues of the graph* G_1 *and let* $\mu_1, \ldots, \mu_m$ ($\mu_1 \geqq \mu_2 \geqq \cdots \geqq \mu_m$) *be the eigenvalues of the regular graph* G_2. *Then the spectrum of* $G_1[G_2]$ *consists of the numbers* $\lambda_1 m + \mu_1, \ldots, \lambda_n m + \mu_1$ *and the numbers* $\mu_2, \mu_3, \ldots, \mu_m$ *whose multiplicities are equal to* n.

Various "sums" and "products" of graphs, their spectra and their complexities are investigated also in [Far 1], [FaW 1], [FaW 2], [GoM 1], [Kers], and [Šoka]. In the last paper a generalized direct product is investigated which covers the NEPS and the Boolean function as well as the lexicographic product.

2.6. The determination of characteristic polynomials and spectra of graphs of some particular types

In this section we shall determine the characteristic polynomials and spectra of certain graphs making use of results described in this chapter. Some of the results of this section are well known in matrix theory, but we will deduce them using methods more consistent with the previously developed theory.

1⁰ For the *graph G without edges* (or loops) with n vertices $P_G(\lambda) = \lambda^n$ obviously holds, i.e., its spectrum consists of n numbers all equal to 0.

2⁰ The *complete graph* K_n with n vertices is the complement of the graph of the previous example, and by (2.7) we have $P_{K_n}(\lambda) = (\lambda - n + 1)(\lambda + 1)^{n-1}$, i.e., the spectrum of K_n consists of the number $n - 1$ and $n - 1$ numbers all equal to -1.

3⁰ Each component of a *regular graph G of degree* 1 is isomorphic with the graph K_2. Since from 2⁰ we have $P_{K_2}(\lambda) = \lambda^2 - 1$, according to (2.2) the relation $P_G(\lambda) = (\lambda^2 - 1)^k$ holds if G contains $2k$ vertices.

4⁰ By proceeding to the complementary graph of the graph from 3⁰ we obtain the characteristic polynomial of the *regular graph H of degree* $n - 2$ with $n = 2k$ vertices in the form $P_H(\lambda) = (\lambda - 2k + 2) \cdot \lambda^k \cdot (\lambda + 2)^{k-1}$. This graph is sometimes denoted by $CP(k)$ (the "*cocktail-party graph*").

5⁰ For the *complete bipartite graph* K_{n_1,n_2} the relation $K_{n_1,n_2} = G_1 \nabla G_2$ holds, where G_1 and G_2 are graphs which consist of n_1 or n_2 isolated vertices, respectively. Since $P_{G_1}(\lambda) = \lambda^{n_1}$ and $P_{G_2}(\lambda) = \lambda^{n_2}$, according to (2.9), we have $P_{K_{n_1,n_2}}(\lambda) = (\lambda^2 - n_1 n_2) \cdot \lambda^{n_1+n_2-2}$, i.e., the spectrum of the graph K_{n_1,n_2} contains the numbers $\sqrt{n_1 n_2}$, $-\sqrt{n_1 n_2}$, and $n_1 + n_2 - 2$ numbers all equal to 0. If $n_1 = n$ and $n_2 = 1$, we obtain a star with $n + 1$ vertices the characteristic polynomial of which is $P_{K_{n,1}}(\lambda) = (\lambda^2 - n)\,\lambda^{n-1}$.

6⁰ As already determined in Section 2.1, the spectrum of a *circuit* C_n contains the numbers $2 \cos \frac{2\pi}{n} i$ ($i = 1, \ldots, n$). It can easily be seen that the following relation holds:

$$P_{C_n}(\lambda) = 2\left(T_n\left(\frac{\lambda}{2}\right) - 1\right) = -2 + \sum_{k=0}^{[n/2]} (-1)^k \frac{n}{n-k} \binom{n-k}{k} \lambda^{n-2k}$$
$$= 2 \cos\left(n \arccos \frac{\lambda}{2}\right) - 2,$$

where

$$T_n(x) = \sum_{k=0}^{[n/2]} (-1)^k \frac{n}{n-k} \binom{n-k}{k} 2^{n-2k-1} x^{n-2k} = \cos(n \arccos x)$$

is the Chebyshev polynomial of the first kind.

7⁰ By applying Theorem 2.14 we can deduce from the previous result the characteristic polynomial and spectrum of the *path* P_n with n vertices. All subgraphs of the circuit C_n induced by $n-1$ vertices are isomorphic with the path P_{n-1}. Therefore, $P_{P_{n-1}}(\lambda) = \frac{1}{n} P'_{C_n}(\lambda)$. If we introduce the Chebyshev polynomials of the second kind,

$$U_n(x) = \frac{\sin[(n+1)\arccos x]}{\sqrt{1-x^2}},$$

we obtain

$$P_{P_n}(\lambda) = U_n\left(\frac{\lambda}{2}\right) = \sum_{k=0}^{[n/2]} (-1)^k \binom{n-k}{k} \lambda^{n-2k}.$$

It then easily follows that the spectrum of the path P_n consists of the numbers $2 \cos \frac{\pi}{n+1} i \ (i = 1, \ldots, n)$.

8⁰ The *complete multipartite graph* $K_{n_1,\ldots,n_k}$, where $n_1 = n_2 = \cdots = n_k = \frac{n}{k}$ (n being the number of vertices), is the complementary graph of the direct sum of k complete graphs each having $\frac{n}{k}$ vertices. Applying Theorem 2.6, we get

$$P_{K_{\frac{n}{k},\ldots,\frac{n}{k}}}(\lambda) = \lambda^{n-k}\left(\lambda + \frac{n}{k} - n\right)\left(\lambda + \frac{n}{k}\right)^{k-1}.$$

If the numbers $n_1, \ldots, n_k$ are not all equal, the characteristic polynomial of $K_{n_1,\ldots,n_k}$ can be determined by means of the generating function for the numbers of walks. Since $\overline{K}_{n_1,\ldots,n_k}$ is the direct sum of complete graphs $G_1, \ldots, G_k$ with $n_1, \ldots, n_k$ vertices, respectively, we have

$$P_{\overline{K}_{n_1,\ldots,n_k}}(\lambda) = (\lambda+1)^{n-k} \prod_{j=1}^{k} (\lambda - n_j + 1). \tag{2.48}$$

Formula (1.59) can be written in the form

$$P_G(\lambda) = (-1)^n \frac{\lambda P_{\overline{G}}(-\lambda-1)}{H_G\left(\frac{1}{\lambda}\right) + \lambda}. \tag{2.49}$$

Using the generating function $H_G(t)$ determined in Example 3 of Section 7.5 (p. 211), according to (2.48) and (2.49) we have

$$P_{K_{n_1,\dots,n_k}}(\lambda) = \lambda^{n-k}\left(1 - \sum_{i=1}^{k} \frac{n_i}{\lambda + n_i}\right) \prod_{j=1}^{k} (\lambda + n_j) \tag{2.50}$$

or

$$P_{K_{n_1,\dots,n_k}}(\lambda) = \sum_{i=1}^{k} (1 - i)\, S_i \lambda^{n-i}, \tag{2.51}$$

where S_i ($i = 1, \dots, k$) is the elementary symmetric function of order i of the numbers $n_1, \dots, n_k$ and $S_0 = 1$ ([Cve9], [HaKMR]).

9^0 Interesting graphs can be obtained if we consider the *sum of two paths or of a path and a circuit or of two circuits.*

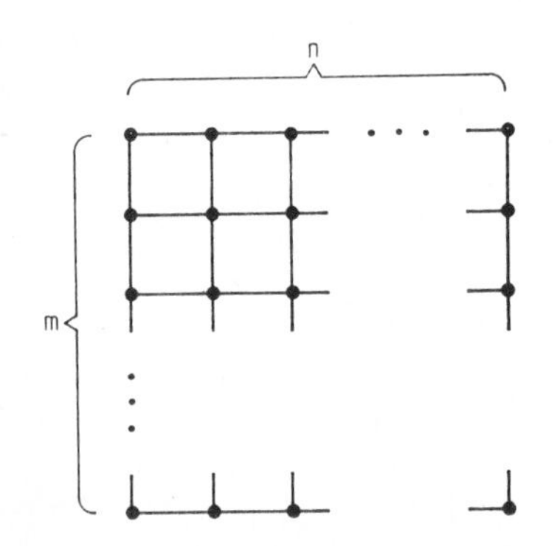

Fig. 2.2

The *sum of two paths* having m and n vertices, respectively, is the graph of an $m \times n$ square lattice, represented in Fig. 2.2. According to Theorem 2.23, the spectrum of this graph consists of all numbers of the form

$$2 \cos \frac{\pi}{m+1} i + 2 \cos \frac{\pi}{n+1} j \quad (i = 1, \dots, m;\ j = 1, \dots, n).$$

The *sum of the circuit C_m and of the path P_n* gives the graph of an analogous lattice on a cylinder which can be obtained from the graph of Fig. 2.2 (with $m + 1$ instead of m) by identifying the vertices of the first row with the corresponding vertices of the last row. The spectrum of this graph comprises the numbers

$$2 \cos \frac{2\pi}{m} i + 2 \cos \frac{\pi}{n+1} j \quad (i = 1, \dots, m;\ j = 1, \dots, n).$$

The *sum of two circuits* is the graph of a square lattice on the torus.

Consider a circular torus. A circle of the torus lying in a plane normal to the axis of the torus is called a horizontal circle. A vertical circle is obtained by cutting the torus by means of a semi-plane passing through the axis of the torus. A system of some horizontal and some vertical circles generates a square lattice on the torus. The graph G of such a square lattice has, as its vertices, intersections of horizontal and vertical circles. Any two vertices which are immediately linked by an arc of one of the circles are considered to be "adjacent".

The spectrum of such a toroidal square lattice comprises the numbers

$$2 \cos \frac{2\pi}{m} i + 2 \cos \frac{2\pi}{n} j \quad (i = 1, \ldots, m;\ j = 1, \ldots, n).$$

If we consider the strong product instead of the sum, we obtain the graphs corresponding to modified square lattices, where to every "square" belong also its "diagonals". In these cases, also, the spectra can be easily determined.

10⁰ The *graph G of the k-dimensional (finite) lattice* is the graph whose vertices are all k-tuples of numbers $1, \ldots, n$, where two k-tuples are adjacent if and only if they differ in exactly one coordinate. For $n = 2$, G reduces to the *graph of the k-dimensional unit cube*. For $k = 2$ the graph G represents the *graph* $L(K_{n,n})$. For $k = 3$ we get the *cubic lattice graph* whose spectral characterizations are considered in Section 6.4. The graph G can obviously be represented as the sum $G_1 + \cdots + G_k$ of k complete graphs ($G_j \cong K_n$, $j = 1, \ldots, k$). The sum of k graphs is, in fact, a NEPS the basis of which contains all k-tuples of numbers 0, 1 in which exactly one number 1 appears. According to Theorem 2.23, the spectrum of the sum consists of all numbers $\lambda_{1i_1} + \cdots + \lambda_{ki_k}$, where λ_{ji_j} is an eigenvalue of G_j. Since the spectrum of K_n is known, we can easily establish the spectrum of G consisting of all numbers $\lambda_j = n(k - j) - k$ $(j = 0, 1, \ldots, k)$ with multiplicities $p_j = \binom{k}{j} (n - 1)^j$.

11⁰ The *Möbius ladder* M_n is the graph with $2n$ vertices $1, 2, \ldots, 2n$ in which the following pairs of vertices are adjacent:

$$\begin{aligned} &(i, i + 1), && i = 1, 2, \ldots, 2n - 1, \\ &(1, 2n), && \\ &(i, i + n), && i = 1, 2, \ldots, n. \end{aligned}$$

It can easily be seen that $M_n = \vec{C}_{2n} \cup \vec{C}_{2n}^{\,n} \cup \vec{C}_{2n}^{\,2n-1}$, i.e., the adjacency matrix of M_n is a circulant matrix of order $2n$, whose entries in the first row are equal to 0 except for the entries in the second, $(n + 1)$-th, and $(2n)$-th columns, which are equal to 1. Therefore, the spectrum of M_n consists of the numbers

$$\lambda_j = e^{\frac{2\pi j}{2n} i} + \left(e^{\frac{2\pi j}{2n} i}\right)^n + \left(e^{\frac{2\pi j}{2n} i}\right)^{2n-1}, \quad j = 1, \ldots, 2n;$$

i.e.,

$$\lambda_j = 2 \cos \frac{\pi}{n} j + (-1)^j, \quad j = 1, \ldots, 2n.$$

12⁰ *Graphs* naturally arise *in the theory of error-correcting codes* [Cve3]. One such graph will be denoted by G_{nbl}. The vertices of the graph G_{nbl} are in one-to-one correspondence with the set of all n-tuples $(x_1, \ldots, x_n)$, $x_j = 1, \ldots, b$, $j = 1, \ldots, n$. Two vertices are adjacent if and only if the corresponding n-tuples differ in at least one and in at most $2l$ coordinates.

The graph G_{nbl} is equal to the $\mathcal{J}$-sum of n copies of the complete graph K_b, where $\mathcal{J} = \{1, \ldots, 2l\}$. As has already been proved, the eigenvalues of K_b are $\lambda_1 = b - 1$, $\lambda_2 = \lambda_3 = \cdots = \lambda_b = -1$.

Let $\sum\limits_{a_1,\ldots,a_k} \lambda_{i_{a_1}} \cdots \lambda_{i_{a_k}}$ be the symmetric function which is equal to the sum of all products of k factors of the variables $\lambda_{i_1}, \ldots, \lambda_{i_n}$. The indices $i_1, \ldots, i_n$ each take any of the values $1, \ldots, b$.

According to Theorem 2.23, the spectrum of G_{nbl} is

$$\Lambda_{i_1,\ldots,i_n} = \sum_{a_1} \lambda_{i_{a_1}} + \sum_{a_1,a_2} \lambda_{i_{a_1}}\lambda_{i_{a_2}} + \cdots + \sum_{a_1,\ldots,a_{2l}} \lambda_{i_{a_1}} \cdots \lambda_{i_{a_{2l}}}. \tag{2.52}$$

Expression (2.52) is symmetric with respect to $\lambda_{i_1}, \ldots, \lambda_{i_n}$. Therefore, the value of $\Lambda_{i_1,\ldots,i_n}$ is determined only by how many of the quantities $\lambda_{i_1}, \ldots, \lambda_{i_n}$ are equal to -1, and how many of them are equal to $b - 1$. Denote by Λ_f $(f = 0, 1, \ldots, n)$ those values $\Lambda_{i_1,\ldots,i_n}$ for which f quantities $\lambda_{i_1}, \ldots, \lambda_{i_n}$ are equal to -1 and $n - f$ quantities are equal to $b - 1$. The algebraic multiplicity of the characteristic value Λ_f is $p_f = \binom{n}{f} (b - 1)^f$.

We shall complete the calculation for $l = 1$. Then

$$\Lambda_{i_1,\ldots,i_n} = \lambda_{i_1} + \cdots + \lambda_{i_n} + \lambda_{i_1}\lambda_{i_2} + \lambda_{i_1}\lambda_{i_3} + \cdots + \lambda_{i_{n-1}}\lambda_{i_n}.$$

For Λ_f we have $\lambda_{i_1} + \cdots + \lambda_{i_n} = f \cdot (-1) + (n - f) \cdot (b - 1) = nb - bf - n$. Then among the products $\lambda_{i_{a_1}}\lambda_{i_{a_2}}$ there are $\binom{f}{2}$ products with the value $(-1) \cdot (-1) = 1$, $f(n - f)$ products with the value $(-1) \cdot (b - 1) = 1 - b$, and $\binom{n-f}{2}$ products with the value $(b - 1)^2$. Therefore,

$$\begin{aligned} \Lambda_f &= nb - bf - n + \binom{f}{2} + f(n - f)(1 - b) + \binom{n-f}{2}(b - 1)^2 \\ &= \frac{1}{2} b^2 f^2 - \frac{1}{2} bf(2nb - 2n - b + 4) + (b - 1)n + \frac{1}{2} n(n - 1)(b - 1)^2. \end{aligned} \tag{2.53}$$

Quantities Λ_f with different indices f can, in special cases, be equal. The maximal value $\Lambda_0 = (b - 1)n + \binom{n}{2}(b - 1)^2$ is always simple, and equal to the degree of the graph G_{nbl}.

In the general case the value of the symmetric function $\sum\limits_{a_1,\ldots,a_k} \lambda_{i_{a_1}} \cdots \lambda_{i_{a_k}}$ is equal to the coefficient of the term $(-1)^k x^{n-k}$ of the polynomial

$$(x + 1)^f (x + 1 - b)^{n-f} = \sum_{i=0}^{f} \binom{f}{i} x^i \sum_{j=0}^{n-f} \binom{n-f}{j} (1 - b)^{n-f-j} x^j,$$

i.e.,

$$\sum_{a_1,\ldots,a_k} \lambda_{i_{a_1}} \cdots \lambda_{i_{a_k}} = (-1)^k \sum_{i=0}^{n-k} \binom{f}{i} \binom{n-f}{n-k-i} (1 - b)^{k-f+i}.$$

The spectrum of the graph G_{nbl} is given by

$$\Lambda_f = \sum_{k=1}^{2l} (-1)^k \sum_{i=0}^{n-k} \binom{f}{i} \binom{n-f}{n-k-i} (1-b)^{k-f+i} \tag{2.54}$$

($f = 0, 1, \ldots, n$), where $p_f = \binom{n}{f} (b-1)^f$ is the multiplicity of Λ_f.

2.7. Miscellaneous results and problems

1. Show that relation (2.7) holds if and only if G is a regular graph.

(D. M. Cvetković [Cve9])

2. If the spectrum of a graph G_1 contains the eigenvalue λ with the multiplicity p ($p > 1$), then the spectrum of the complete product $G_1 \nabla G_2$ of the graph G_1 with an arbitrary graph G_2 contains λ as an eigenvalue with a multiplicity $q \geqq p - 1$.

(D. M. Cvetković [Cve9])

3. If G is a regular graph of degree r with n vertices and with the eigenvalues $\lambda_1 = r, \lambda_2, \ldots, \lambda_n$, then ([Nos1])

$$r - n \leqq \lambda_i \leqq n - 2 - r, \qquad i = 2, 3, \ldots, n. \tag{2.55}$$

Proof. According to Theorem 2.6 the eigenvalues of $\overline{G}$ are $-\lambda_i - 1$ ($i = 2, 3, \ldots, n$) and $n - 1 - r$ (the degree of $\overline{G}$). So

$$-(n - 1 - r) \leqq -\lambda_i - 1 \leqq n - 1 - r,$$

which implies (2.55).

Note that for $r > \frac{n-1}{2}$ the length of the interval $(r - n, n - 2 - r)$ is smaller than the length of the interval $(-r, r)$.

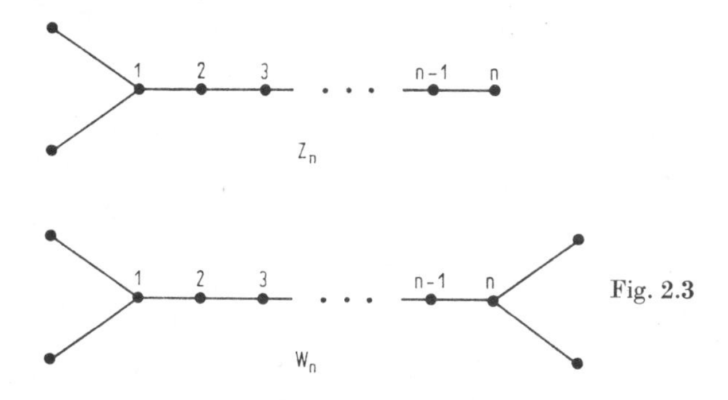

Fig. 2.3

4. The spectrum of the graph Z_n (see Fig. 2.3) having $n + 2$ vertices consists of the numbers $2 \cos \frac{2i+1}{2n+2} \pi$, $i = 0, 1, \ldots, n$, and the number 0. The spectrum of the graph W_n (see Fig. 2.3) having $n + 4$ vertices is the union (paying attention to the multiplicities of the eigenvalues) of the spectra of the circuit C_4 and the path P_n.

(See [CoSi1], [GuT7], [CvGT3])

5. Let $G_{l,k}$ be the graph obtained by inserting $k - 1$ new vertices into each edge of a star $K_{1,l}$. Let, further, H_k be the graph obtained from k copies of $K_{1,2}$ and an isolated vertex x by joining with edges the vertex x to all vertices of degree 1. Then [CoSi1]

$$P_{G_{l,k}}(\lambda) = (\Phi_{k+1} - (l-1)\,\Phi_{k-1})\Phi_k^{l-1},$$
$$P_{H_k}(\lambda) = \Phi_3^{k-1}(\lambda\Phi_3 - 2k\lambda^2),$$

where Φ_k is the characteristic polynomial of the path P_k with k vertices, given in Section 2.6.
The first result was generalized in [Mow5].

6. Let x be a vertex of a graph G. Denote by H the subgraph of G obtained by removing the vertex x from G. Let G_n be the graph obtained from G and a path P_n with n vertices by joining with an edge the vertex x to a vertex of valency 1 of P_n. Let r_n be the greatest eigenvalue of G_n. Assume further that $\lim_{n\to+\infty} r_n > 2$. Then $\lim_{n\to+\infty} r_n$ is equal to the greatest positive root of

$$\frac{1}{2}\left(\lambda + \sqrt{\lambda^2 - 4}\right) P_G(\lambda) - P_H(\lambda).$$

(A. J. Hoffman [Hof13])

7. Show that the Möbius ladder M_n, $n \equiv 1 \pmod 2$, can be represented as the NEPS with basis $\mathscr{B} = \{(0,1), (1,1)\}$ of the graphs C_n and K_2.

8. Applying Theorem 2.12 to a path P_n, show that for the Chebyshev polynomials of the second kind, $U_n(x)$, the following relations hold:

$$U_n(x) = U_k(x)\,U_{n-k}(x) - U_{k-1}(x)\,U_{n-k-1}(x), \qquad k = 1, \ldots, n-1.$$

9. Let uv be the edge whose end points are the vertices u and v. For a graph G, let $\mathscr{C}(u)$ and $\mathscr{C}(uv)$ denote the sets of all circuits Z containing u or uv, respectively. Then

$$P_G(\lambda) = \lambda P_{G-v}(\lambda) - \sum_{u' \cdot v} P_{G-v-u'}(\lambda) - 2\sum_{Z\in\mathscr{C}(v)} P_{G-\mathscr{V}(Z)}(\lambda),$$
$$P_G(\lambda) = P_{G-uv}(\lambda) - P_{G-u-v}(\lambda) - 2\sum_{Z\in\mathscr{C}(uv)} P_{G-\mathscr{V}(Z)}(\lambda),$$

where $G - \mathscr{V}(Z)$ is the graph obtained from G by removing the vertices belonging to Z (in the first sum of the first formula the summation goes over all vertices u' adjacent to v).

(A. J. Schwenk [Schw3])

10. Prove or disprove that the minimum of the least eigenvalue of $K_{n_1,\ldots,n_k}$ is attained when $n_1 = \left[\frac{n-k+1}{2}\right] + 1$, $n_2 = \left[\frac{n-k}{2}\right] + 1$, $n_3 = n_4 = \cdots = n_k = 1$ if $n \geqq 3k - 4$, and when $n_1 = n - k + 1$, $n_2 = n_3 = \cdots = n_k = 1$ if $n \leqq 3k - 4$, where $n_1 + \cdots + n_k = n$.

(M. Doob)

11. Let $\mathscr{S}$ be the set of all graphs whose index is not greater than 2. In [CvG3] the spectrum of each graph from $\mathscr{S}$ is determined. All eigenvalues are of the form $2\cos\frac{p}{q}\pi$, where p and q are integers and $q \neq 0$, and conversely, each number of this form is an eigenvalue of a graph from $\mathscr{S}$.

12. Let G be a graph with index $\varrho = \lambda_1$. Then $\varrho \leqq 2$ ($\varrho < 2$) if and only if each component of G is a subgraph (proper subgraph) of one of the graphs depicted in Fig. 2.4 which all have an index equal to 2.

(J. H. Smith [Sm,J])

13. Among all connected graphs with n vertices, the path has the smallest index which is equal to $2\cos\frac{\pi}{n+1}$ (for the last statement see 7^0 in Section 2.6).

(L. Collatz [Col1] (see also [CoSi1]); L. Lovász, J. Pelikán [LoPe])

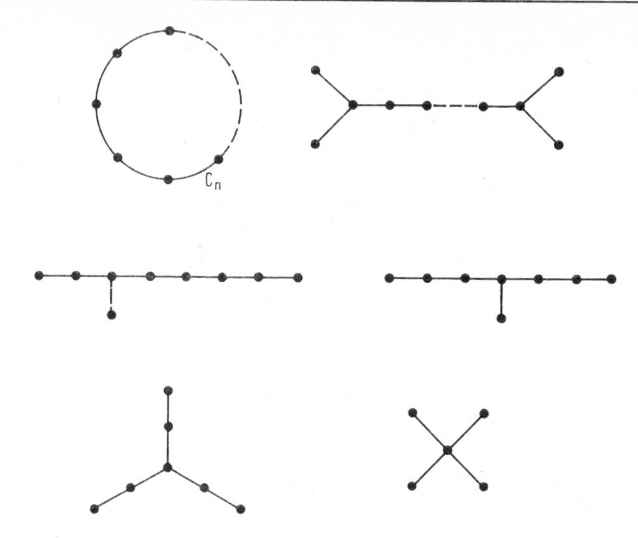

Fig. 2.4

14. An internal path of a graph G is a sequence of vertices $x_1, \ldots, x_k$ such that all x_i are distinct (except possibly $x_1 = x_k$), the vertex degrees $d(x_i)$ satisfy $d(x_1) \geqq 3$, $d(x_2) = \cdots = d(x_{k-1}) = 2$ (unless $k = 2$), $d(x_k) \geqq 3$, x_i is adjacent to x_{i+1}, $i = 1, \ldots, k-1$. Let $G_{x,y}$ be the graph obtained from G by introducing a new vertex on the edge (x, y) of G and let $r(G)$ denote the index of G. If (x, y) is an edge on an internal path of G and $G \neq W_n$ (see Fig. 2.3), then $r(G_{x,y}) < r(G)$.

(A. J. Hoffman, J. H. Smith [HoSm])

15. Let G be a connected graph and $\mathcal{T}(G)$ the set of all graphs topologically equivalent to G. Let G_T be the graph from $\mathcal{T}(G)$ with the smallest number of vertices. $r(H)$ denotes the index of a graph H. Define $r_T(G) = \inf_{H \in \mathcal{T}(G)} r(H)$, $r^T(G) = \sup_{H \in \mathcal{T}(G)} r(H)$.

1° If G_T has one vertex, $r^T(G) = 0$. If G_T has two vertices, $r^T(G) = 2$. If G_T is a triangle, $r^T(G) = 2$. If G_T is any other graph, $r^T(G) = \Theta_0 + 1/\Theta_0$, where Θ_0 is the largest root of $|\boldsymbol{D}(\Theta) - \boldsymbol{A}| = 0$, $\boldsymbol{A}$ is the adjacency matrix of G_T and $\boldsymbol{D}(\Theta)$ is a diagonal matrix, with $\Theta + 1/\Theta$ for a vertex of degree > 1, Θ for a vertex of degree 1. If there is no vertex of degree 1, then $r^T(G) = r(G_T)$.

2° If G_T has one vertex, then $r_T(G) = 0$. If G_T has two vertices, $r_T(G) = 1$. If G_T is a triangle, $r_T(G) = 2$. In all other cases

$$r_T(G) = \max_{\substack{i \\ d(i) \geqq 3}} \left(\alpha^{\frac{1}{2}} + \alpha^{-\frac{1}{2}}\right),$$

where $\alpha = \frac{1}{2}\left(d(i) - 2 + \sqrt{d^2(i) - 4f(i)}\right)$, $d(i)$ is the degree of the vertex i and $f(i)$ is the number of vertices of degree 1 adjacent to i in G_T.

(A. J. Hoffman, J. H. Smith [HoSm])

16. Let $\boldsymbol{A} = (a_{ij})_1^n$ be the adjacency matrix of a graph G and let the regular graphs $G_1, \ldots, G_n$ have $m_1, \ldots, m_n$ vertices, respectively, and degrees $r_1, \ldots, r_n$. Let H be the graph obtained from $G_1, \ldots, G_n$ by joining each vertex of G_i to each vertex of G_j whenever $a_{ij} = 1$. The spectrum of H contains all eigenvalues of all graphs $G_1, \ldots, G_n$ except for $r_1, \ldots, r_n$. The remaining n eigenvalues of H are the eigenvalues of the matrix $\boldsymbol{B} = (b_{ij})_1^n$, where $b_{ij} = a_{ij}m_j + r_j\delta_{ij}$. This was proved in [FaW1], where the described operation on graphs is called the *local join* of graphs. If all graphs $G_1, \ldots, G_n$ are isomorphic to a graph F, we get the lexicographic product $G[F]$ of graphs G and F. If $G = K_n$, we have the ∇-product of graphs $G_1, \ldots, G_n$ [Schw3]. Note that in this case the matrix $\boldsymbol{B}$ is the adjacency matrix of a front divisor of H (see Chapter 4).

3. Relations Between Spectral and Structural Properties of Graphs

In this chapter we shall describe only a part of the known relations between the spectra and the structure of (multi-)(di-)graphs. These relations represent, in fact, the main topic of this book, and they can be encountered in all other chapters.

As is well known, there are some structural properties that are not uniquely determined by the spectrum, but even in these cases we can, on the basis of the spectrum, frequently specify a range of variation of these properties. Therefore, many inequalities for various numerical characteristics (chromatic number, diameter, etc.) appear in this chapter.

In all theorems of this chapter we assume that either the spectrum or the eigenvectors of the adjacency matrix of a graph, or both, are given and that a certain class to which the graph belongs is specified. If the spectrum of the graph is given, we assume that its characteristic polynomial is also known, and conversely. The algebraic and numerical problems which appear here are assumed to be solved. Note that in some cases the class of graphs to which the graph with the given spectrum belongs can be determined by means of the spectrum.

Let, as usual, $\boldsymbol{A}$ denote the adjacency matrix, let

$$P_G(\lambda) = |\lambda \boldsymbol{I} - \boldsymbol{A}| = \lambda^n + a_1\lambda^{n-1} + \cdots + a_n \tag{3.1}$$

be the characteristic polynomial, and $[\lambda_1, \ldots, \lambda_n]$ the spectrum of the graph G.

3.1. Digraphs

First we shall assume that multiple oriented edges and loops are allowed in the digraphs to be considered. Before formulating some theorems we shall note a few simple facts.

The number of vertices of G is equal to the degree n of its characteristic polynomial, i.e. to the number of eigenvalues of G.

The number of directed loops is equal to the trace of the adjacency matrix, i.e. to the sum $\lambda_1 + \cdots + \lambda_n$, i.e. to the quantity $-a_1$.

If every vertex of G has the same number of loops, then the characteristic polynomial $P_H(\lambda)$ of the digraph H obtained from G by deleting all of its loops is completely determined by $P_G(\lambda)$: If every vertex of G has exactly h directed loops, then $h = -\frac{a_1}{n}$ and $P_H(\lambda) = P_G(\lambda + h)$.

If G is a digraph without loops, then no pair of vertices of G is joined by two edges of opposite orientation if and only if $a_2 = 0$. This fact can be easily realized by considering all principal minors of the second order of the adjacency matrix.

From Theorem 1.2 we deduce immediately: A digraph G contains no cycle if and only if all the coefficients a_i $(i = 1, \ldots, n)$ are equal to zero, i.e., if and only if the spectrum of G contains no eigenvalue different from zero (J. Sedláček [Sed1]).

According to Theorem 1.9, the number of closed walks of given length k contained in a digraph G can be determined by means of the spectrum of G; this number is equal to $\operatorname{tr} \boldsymbol{A}^k = \sum_{i=1}^{n} \lambda_i^k$.

Using the Cayley-Hamilton Theorem, we deduce from the characteristic polynomial (3.1) the following relations:

$$\boldsymbol{A}^{n+k} + a_1\boldsymbol{A}^{n+k-1} + \cdots + a_n\boldsymbol{A}^k = \boldsymbol{O} \qquad (k = 0, 1, \ldots). \tag{3.2}$$

By means of Theorem 1.9, we can obtain from (3.2) some information concerning the digraph structure.

Now we shall establish some theorems concerning the cycle structure of a digraph G without multiple edges. Some statements given in the foregoing are special cases of these theorems.

The length $g(G)$ of a shortest cycle in a digraph G (if such a cycle exists) is called the *girth* of G. If G has no cycles, then $g(G) = +\infty$. Obviously, each linear directed subgraph of G with less than $2g$ vertices, where $g = g(G)$, is necessarily a cycle. From Theorem 1.2 we deduce

$$a_i = \sum_{L \in \mathscr{L}_i} (-1)^{c(L)} = - \sum_{\vec{C}_i \subset G} 1 \qquad (i < 2g).$$

Thus $-a_i$ is the number of cycles of length i contained in G.

Theorem 3.1 (H. Sachs [Sac3]): *Let G be a digraph with the characteristic polynomial* (3.1) *and let $g(G) = g$. Let further $i \leqq \min(2g - 1, n)$. Then the number of cycles of length i contained in G is equal to $-a_i$. The girth g of G is equal to the smallest index i for which $a_i \neq 0$.*

This result can be extended so that the number of cycles of length i for some $i > 2g - 1$ can also be determined. We shall introduce a new notion: the d-girth of a digraph. For an arbitrary integer $d > 1$, the d-girth $g_d(G)$ of a digraph G is defined as the length of a shortest cycle among those cycles the lengths of which are not divisible by d. If there are no such cycles, then $g_d(G) = +\infty$.

Theorem 3.2 (H. Sachs [Sac3]): *Let G be a digraph with the characteristic polynomial* (3.1) *and let $g(G) = g$ and $g_d(G) = g_d$. Let further $i \leqq \min(g + g_d - 1, n)$, $i \not\equiv 0 \pmod{d}$. Then the number of cycles of length i contained in G is equal to $-a_i$. The d-girth g_d of G is equal to the smallest index not divisible by d for which $a_i \neq 0$.*

Remark. If g is not divisible by d, then, trivially, $g_d = g$ and Theorem 3.2 states less than Theorem 3.1. But in the opposite case, when d is a factor of g, we certainly have $g_d > g$. If, further, $g_d > g + 1$, Theorem 3.2 yields new information that is not obtainable from Theorem 3.1.

Example. Let $g = 9$, $g_9 = 15$, $g_3 = 20$, Theorem 3.1 yields the numbers of cycles of length c for $c \leqq 17$. In addition, with $d = 9$ Theorem 3.2 provides these numbers for $c = 19, 20, 21, 22, 23$, and with $d = 3$ also for $c = 25, 26, 28$.

With $d = 2$ we have the following corollary.

Corollary: *The length g_2 of the shortest odd cycle in G is equal to the index of the first non-vanishing coefficient among $a_1, a_3, a_5, \ldots$; the number of shortest odd cycles is equal to $-a_{g_2}$.*

Proof of Theorem 3.2. Let $i \leqq \min(g + g_d - 1, n)$, $i \not\equiv 0 \pmod{d}$. Then each linear directed subgraph in G with i vertices is necessarily a cycle. As in the above argument,

$$a_i = \sum_{L \in \mathscr{L}_i} (-1)^{c(L)} = -\sum_{\vec{C}_i \subset G} 1$$

which completes the proof.

From the corollary of Theorem 3.2 we can easily deduce the following theorem.

Theorem 3.3 (H. Sachs [Sac3]): *A digraph G has no odd cycles if and only if its characteristic polynomial has the following form:*

$$P_G(\lambda) = \lambda^n + a_2\lambda^{n-2} + a_4\lambda^{n-4} + \cdots = \lambda^p \cdot Q(\lambda^2),$$

where Q is a polynomial and $p = 0$ for n even, and $p = 1$ otherwise.

The following theorem can also easily be proved.

Theorem 3.4: *A strongly connected digraph G with greatest eigenvalue r has no odd cycles if and only if $-r$ is also an eigenvalue of G.*

Proof. If G has no odd cycles then, by Theorem 3.3, $-r$ is also an eigenvalue of G.

Conversely, if $-r$ belongs to the spectrum of G then the adjacency matrix of G is imprimitive. According to Theorem 0.3 (Section 0.3), the index of imprimitivity h can in that case be only an even number. By the same theorem, there exists a permutation matrix $\boldsymbol{P}$ such that $\boldsymbol{PAP}^{-1}$ has the form (0.1). Since h is even, G obviously contains no odd cycles.

This completes the proof.

A digraph G is said to be *cyclically k-partite* if its vertex set $\mathscr{X}$ can be partitioned into non-empty mutually disjoint sets $\mathscr{X}_1, \ldots, \mathscr{X}_k$ so that, if (x, y) $(x \in \mathscr{X}_i, y \in \mathscr{X}_j)$ is an arc of G, then $j - i \equiv 1 \pmod{k}$. Note that a cyclically k-partite digraph is also cyclically l-partite if k is divisible by l. The adjacency matrix of a cyclically h-partite digraph has the form (0.1). According to [DuMe] we can formulate the following theorem.

Theorem 3.5: *The characteristic polynomial of a cyclically k-partite digraph G has the form*

$$P_G(\lambda) = \lambda^p \cdot Q(\lambda^k), \tag{3.3}$$

where Q is a monic, $Q(0) \neq 0$, and p is a non-negative integer.

If G is a strongly connected digraph and if its characteristic polynomial is of the form (3.3), *then G is cyclically k-partite.*

The following theorem is taken directly from the theory of matrices (see, for example, [Gant], vol. II, p. 63).

Theorem 3.6: *Let $d_1^-, \ldots, d_n^-$ and $d_1^+, \ldots, d_n^+$ be the indegrees and outdegrees, respectively, of the vertices of a digraph G. Then, for the index r of G, the following inequalities hold:*

$$\min_i d_i^- \leqq r \leqq \max_i d_i^-, \tag{3.4}$$

$$\min_i d_i^+ \leqq r \leqq \max_i d_i^+. \tag{3.5}$$

If G is strongly connected, then equality on the left-hand side or on the right-hand side of (3.4) (*or of* (3.5)) *holds if and only if all the quantities $d_1^-, \ldots, d_n^-$ (or $d_1^+, \ldots, d_n^+$) are equal.*

Theorem 3.7 (A. J. Hoffman, M. H. McAndrew [HoMc]): *For a digraph G with the adjacency matrix **A**:*

1⁰ *There exists a polynomial $P(x)$ such that*

$$\boldsymbol{J} = P(\boldsymbol{A}), \tag{3.6}$$

if and only if G is strongly connected and regular.

2⁰ *The unique polynomial $P(x)$ of least degree such that* (3.6) *is satisfied is $nS(x)/S(d)$ where $(x - d)\, S(x)$ is the minimal polynomial of **A** and d is the degree of G.*

3⁰ *If $P(x)$ is the polynomial of least degree such that* (3.6) *is satisfied, then the degree of G is the greatest real root of $P(x) = n$.*

Proof. Assume that (3.6) holds. Let i, j be distinct vertices of G. By (3.6), there is some integer k such that $\boldsymbol{A}^k$ has a positive entry in position (i, j), i.e., there is some walk of length k from i to j. So G is strongly connected. Further, from (3.6) follows that $\boldsymbol{J}$ commutes with $\boldsymbol{A}$. Let e_i, d_j be the outdegree and the indegree of vertex i and vertex j, respectively. Now the (i, j) entry of $\boldsymbol{AJ}$ is e_i, and the (i, j) entry of $\boldsymbol{JA}$ is d_j. Thus $e_i = d_j$ for all i and j, so G is regular, i.e., all row and column sums of $\boldsymbol{A}$ are equal ($\boldsymbol{A}$ being not necessarily symmetric).

To prove the converse assume G to be strongly connected and regular. Due to the regularity, $\boldsymbol{u} = (1, 1, \ldots, 1)^\mathsf{T}$ is an eigenvector of both $\boldsymbol{A}$ and $\boldsymbol{A}^\mathsf{T}$, corresponding to the eigenvalue d. Hence, if d has multiplicity greater than 1, it must have at least one more eigenvector associated with it. But because of the strong connectedness, $\boldsymbol{u}$ is the only eigenvector corresponding to d. It follows that, if $R(x)$ is the minimal polynomial of $\boldsymbol{A}$, and if $S(x) = R(x)/(x - d)$, then $S(d) \neq 0$. We then have

$$O = R(\boldsymbol{A}) = (\boldsymbol{A} - d\boldsymbol{I})\, S(\boldsymbol{A}). \tag{3.7}$$

Let $\boldsymbol{o}$ be the zero-vector. Since $R(\boldsymbol{A})\, \boldsymbol{v} = \boldsymbol{o}$ for all vectors $\boldsymbol{v}$, it follows from (3.7) that

$$(\boldsymbol{A} - d\boldsymbol{I})\, S(\boldsymbol{A})\, \boldsymbol{v} = \boldsymbol{o},$$

so $S(\boldsymbol{A})\, \boldsymbol{v} = \alpha \boldsymbol{u}$ for some α.

Let $\langle \boldsymbol{u}, \boldsymbol{v} \rangle$ be the scalar product of vectors $\boldsymbol{u}$ and $\boldsymbol{v}$. If we take $\langle \boldsymbol{v}, \boldsymbol{u} \rangle = 0$, then $\langle \boldsymbol{A}^k \boldsymbol{v}, \boldsymbol{u} \rangle = \langle \boldsymbol{v}, (\boldsymbol{A}^{\mathsf{T}})^k \boldsymbol{u} \rangle = d^k \langle \boldsymbol{v}, \boldsymbol{u} \rangle = 0$ for every k and so $\langle S(\boldsymbol{A})\, \boldsymbol{v}, \boldsymbol{u} \rangle = 0$. Therefore, $0 = \langle S(\boldsymbol{A})\, \boldsymbol{v}, \boldsymbol{u} \rangle = \langle \alpha \boldsymbol{u}, \boldsymbol{u} \rangle = n\alpha$, i.e., $\alpha = 0$.

Thus $S(\boldsymbol{A})\, \boldsymbol{v} = \boldsymbol{o}$ for all $\boldsymbol{v}$ such that $\langle \boldsymbol{v}, \boldsymbol{u} \rangle = 0$; further, $S(\boldsymbol{A})\, \boldsymbol{u} = S(d)\, \boldsymbol{u}$. Hence $nS(\boldsymbol{A})/S(d) = \boldsymbol{J}$, i.e., (3.6) is satisfied with

$$P(x) = \frac{n}{S(d)}\, S(x). \tag{3.8}$$

This completes the proof of 1^0; part 2^0 follows since the polynomial (3.8) has smaller degree than the minimal polynomial of $\boldsymbol{A}$. To prove 3^0 we note that $\boldsymbol{A}$ is non-negative and has row and column sums all equal to d. Thus, the eigenvalues of $\boldsymbol{A}$ are all of absolute value not greater than d. The roots of $P(x)$ are eigenvalues of $\boldsymbol{A}$ and hence, for real $x > d$, $|P(x)|$ is an increasing function in x. From (3.8), $P(d) = n$ and so, since $P(x)$ is a real polynomial, $P(x) > n$ for $x > d$.

This completes the proof of Theorem 3.7. We call (3.8) the *polynomial belonging to G* and also say that *G belongs to the polynomial.*

Note that some non-regular graphs can have a polynomial with similar properties [Bri1].

3.2. Graphs

If a multi-digraph H has a symmetric adjacency matrix $\boldsymbol{A}$ with even entries on the diagonal, then the matrix $\boldsymbol{A}$ can be understood as the adjacency matrix of an (undirected) (multi-)graph G. In such a way we can apply the result from Section 3.1 to graphs. But now, due to the symmetry of the adjacency matrix, we have some further results.

The eigenvalues of a graph are real numbers, and we can order them so that the sequence $\lambda_1, \ldots, \lambda_n$ is non-increasing. This convention will always be adopted.

In the sequel we shall consider only undirected graphs without multiple edges or loops.

The following theorem can be proved using arguments directly from matrix theory.

Theorem 3.8 (L. Collatz, U. Sinogowitz [CoSi1]): *Let $\bar{d}$ be the mean value of the valencies and r the greatest eigenvalue of a graph G. Then*

$$\bar{d} \leqq r, \tag{3.9}$$

where equality holds if and only if G is regular.

Proof. As is well known, since the adjacency matrix $A = (a_{ij})_1^n$ of G is Hermitian, the problem of finding the maximal value of Rayleigh's quotient

$$R = \frac{\sum_{i=1}^{n} \sum_{j=1}^{n} a_{ij} x_i x_j}{\sum_{i=1}^{n} x_i^2} \tag{3.10}$$

(the x_i being arbitrary real numbers not all equal to zero) has the solution $R = r$. The maximum is attained if and only if the x_i ($i = 1, \ldots, n$) are the components of an eigenvector of A belonging to r.

If we put $x_i = 1$ ($i = 1, \ldots, n$) in (3.10), we have

$$R = \bar{d} = \frac{1}{n} \sum_{i=1}^{n} d_i,$$

where $d_i = \sum_{j=1}^{n} a_{ij}$ is the valency of vertex i. So, $\bar{d}$ is a particular value of Rayleigh's quotient establishing (3.9).

For regular graphs equality holds in (3.9), since in that case the greatest eigenvalue of G is equal to the degree of G. Let, conversely, equality hold in (3.9). Then the values $x_i = 1$ ($i = 1, \ldots, n$) constitute an eigenvector for A belonging to r, and $\sum_{j=1}^{n} a_{ij} x_j = r x_i$ ($i = 1, \ldots, n$) implies $d_i = \sum_{j=1}^{n} a_{ij} = r$ ($i = 1, \ldots, n$). Thus, G is regular.

This completes the proof of the theorem.

Applying Theorem 3.6 to graphs and using Theorem 3.8, we get

$$d_{\min} \leqq \bar{d} \leqq r \leqq d_{\max},$$

where $d_{\min}$ and $d_{\max}$ are the minimal and maximal values, respectively, of the valencies in G.

We continue with some more propositions relating the coefficients a_i of $P_G(\lambda)$ to some structural properties of G.

Due to the absence of loops, we always have $a_1 = 0$.

The number of closed walks of length 2 is obviously equal to twice the number m of edges, therefore $m = \frac{1}{2} \sum_{i=1}^{n} \lambda_i^2$. In a similar way the formula $t = \frac{1}{6} \sum_{i=1}^{n} \lambda_i^3$ for the number t of triangles can be obtained. Now, Theorem 1.3 gives $m = -a_2$ and $t = -\frac{1}{2} a_3$. According to the same theorem, the coefficient a_4 is equal to the number of pairs of non-adjacent edges minus twice the number of circuits C_4 of length 4 contained in G.

In a similar way the coefficient a_5 is equal to twice the number of figures consisting of a triangle and an edge (triangle and edge being disjoint) minus twice the number of circuits C_5 of length 5. These facts were noted in [CoSi1].

An interesting conclusion can be drawn from formula (1.36) for the coefficients of the characteristic polynomial [Sac 3]. For $i = n$ the 1-factors of G represent one type of basic figures. The contribution of a 1-factor to a_n is either 1 or -1, while other basic figures contribute an even number to a_n. Therefore, the number of 1-factors of G is congruent to a_n modulo 2. If a_n is odd, then there exists at least one 1-factor.

If G is a forest then, obviously, the number of 1-factors is equal to $|a_n|$ with $a_n = (-1)^{\frac{n}{2}}$ if there is a 1-factor, and $a_n = 0$ otherwise.†

For the proof of the following theorem we need a simple lemma which we state without proof. Both, Lemma 3.1 and Theorem 3.9, will be used in Section 7.7.

Lemma 3.1: *Let $\alpha_1, \ldots, \alpha_k$ be real numbers and let r, s (r even, $r < s$) be non-negative integers. Then for $a > 0$ the following implication holds:*

$$\alpha_1^r + \cdots + \alpha_k^r \leqq a^r \Rightarrow |\alpha_1^s + \cdots + \alpha_k^s| \leqq a^s.$$

Equality on the right-hand side of the implication holds if and only if the absolute value of exactly one of the quantities $\alpha_1, \ldots, \alpha_k$ is equal to a, the other quantities being all equal to zero. Strict inequality on the left-hand side implies strict inequality on the right-hand side of the implication.

Theorem 3.9 (E. Nosal [Nos 1]): *Let $[\lambda_1, \ldots, \lambda_n]$ be the spectrum of a graph G. Then the inequality*

$$\lambda_1^2 > \lambda_2^2 + \lambda_3^2 + \cdots + \lambda_n^2 \tag{3.11}$$

implies that G contains at least one triangle.

Proof. According to Lemma 3.1, (3.11) implies

$$\left| \sum_{i=2}^{n} \lambda_i^3 \right| < \lambda_1^3$$

and we obtain for the number t of triangles

$$t = \frac{1}{6} \lambda_1^3 + \frac{1}{6} \sum_{i=2}^{n} \lambda_i^3 \geq \frac{1}{6} \lambda_1^3 - \frac{1}{6} \left| \sum_{i=2}^{n} \lambda_i^3 \right| > 0.$$

This completes the proof.

Since $\sum_{i=1}^{n} \lambda_i^2 = 2m$, where m is the number of edges of G, we get the following corollary.

Corollary: *If $\lambda_1 > \sqrt{m}$, then G contains at least one triangle.*

The corollary of Theorem 3.2 can be reformulated for (undirected) graphs in the following way. Let us consider, together with a graph G, the digraph H which has the same adjacency matrix as G. To each shortest odd circuit of G there correspond

† From (1.35) follows: The number of directed 1-factors (= linear directed subgraphs with n vertices) of any digraph G is not smaller than $|a_n|$.

exactly two shortest odd cycles (with opposite orientations) of H and therefore the number of shortest odd circuits in G is half the number of the shortest odd cycles in H. Thus, we have the following theorem.

Theorem 3.10 (H. Sachs [Sac3]): *Let G be a graph*† *with the characteristic polynomial* (3.1). *Then the length f of a shortest odd circuit in G is equal to the index of the first non-vanishing coefficient among $a_1, a_3, a_5, \ldots$ The number of shortest odd circuits is equal to* $-\frac{1}{2} a_f$.

An immediate consequence of this theorem is the following:

Theorem 3.11: *A graph*† *containing at least one edge is bipartite if and only if its spectrum, considered as a set of points on the real axis, is symmetric with respect to the zero point.*

Theorem 3.11 is one of the best-known theorems making evident a close connection between the structure and spectra of graphs. It seems that the necessity part of this theorem was first recorded in chemical literature by C. A. Coulson, G. S. Rushbrooke [CoRu] (chemists usually call it the "pairing theorem").

The entire theorem was proved by H. Sachs [Sac7] in the form of Theorem 3.3. It is of interest that this theorem has been rediscovered several times. Various versions of the theorem can be found in [CoSi1], [Hof3], [Cve1], [CoLo], [Rou1], [Mari], [Sac3].

The characterization of connected bipartite graphs by Theorem 3.4 is also possible.

We shall now consider the problem of determining the girth of a graph. As in digraphs, the *girth of a graph* G is the length of the shortest circuits of G.

If we try to formulate a theorem similar to Theorem 3.1 for graphs, we encounter the following difficulties: Together with the graph G, consider the digraph H which has the same adjacency matrix as G. If G contains at least one edge, then $g(H) = 2$, while $g(G)$ can at the same time be arbitrarily large. Thus the girths of G and H are not related.

But it is easy to see the following. For $i < g(G)$ there exist basic figures only for $i = 2q$ even, and each basic figure U_{2q} consists of q non-adjacent edges, so that $p(U_{2q}) = q$ and $c(U_{2q}) = 0$. Therefore,

$$a_i = \begin{cases} 0 & \text{for odd } i \\ (-1)^q b_q & \text{for } i = 2q \end{cases} \quad \big(i < g(G)\big),$$

where b_q is the number of basic figures consisting of exactly q non-adjacent edges.

For $i = g(G)$ basic figures can be of the described type (consisting of non-adjacent edges; only for even i), or they can be circuits of length $g(G)$. In the second case the contribution of each such basic figure to a_i is -2. If

$$\tilde{a}_i = \begin{cases} a_i & \text{for odd } i, \\ a_i - (-1)^q b_q & \text{for } i = 2q, \end{cases}$$

† Theorems 3.10, 3.11 hold for multigraphs, too

then $\tilde{a}_i = 0$ for $i < g(G)$ and $-\tilde{a}_{g(G)}$ is equal to twice the number of circuits of length $g(G)$.

So we have the following

Theorem 3.12 (H. SACHS [Sac3]): *Let G be a (multi-)graph with the characteristic polynomial* (3.1) *and let b_q be the number of basic figures consisting of exactly q non-adjacent edges. Let further*

$$\tilde{a}_i = \begin{cases} a_i & \text{for odd } i, \\ a_i - (-1)^q b_q & \text{for } i = 2q. \end{cases}$$

Then $g(G)$ is equal to the index of the first non-vanishing number among $\tilde{a}_1, \tilde{a}_2, \ldots,$ and the number of circuits of length $g(G)$ is equal to $-\frac{1}{2}\tilde{a}_{g(G)}$.

(For regular graphs see also Theorems 3.26, 3.27.)

Since the adjacency matrix $\boldsymbol{A}$ of a graph is symmetric, we can determine its minimal polynomial on the basis of the spectrum. As is well known, if $\{\lambda^{(1)}, \ldots, \lambda^{(m)}\}$ is the set of all distinct eigenvalues of $\boldsymbol{A}$, the corresponding minimal polynomial $\varphi(\lambda)$ is given by

$$\varphi(\lambda) = (\lambda - \lambda^{(1)}) \cdots (\lambda - \lambda^{(m)}).$$

Let $\varphi(\lambda) = \lambda^m + b_1\lambda^{m-1} + \cdots + b_m$. Then the following relations hold:

$$\boldsymbol{A}^{m+k} + b_1\boldsymbol{A}^{m+k-1} + \cdots + b_m\boldsymbol{A}^k = \boldsymbol{O} \qquad (k = 0, 1, \ldots). \tag{3.12}$$

Using these relations we can prove the following theorem ([Nos1], [Cve9]; see also, for example, [MaMi], p. 123).

Theorem 3.13: *If a connected graph G has exactly m distinct eigenvalues, then its diameter D satisfies the inequality $D \leqq m - 1$.*

Proof. Assume the theorem to be false. Then for some connected graph G we have $D = s \geqq m$. By the definition of the diameter, for some i and j the elements $a_{ij}^{(k)}$ from the i-th row and from the j-th column of the matrices $\boldsymbol{A}^k$ $(k = 1, 2, \ldots)$ are equal to zero for $k < s$, whereas $a_{ij}^{(s)} \neq 0$.

In (3.12), put $k = s - m$. Making use of the relation so obtained, from $a_{ij}^{(k)} = 0$ $(k = 1, \ldots, s - 1)$ we deduce $a_{ij}^{(s)} = 0$, which is a contradiction.

This completes the proof of the theorem.

The interior stability number $\alpha(G)$ of the graph G is defined as the maximum number of vertices which can be chosen in G so that no pair of them is joined by an edge of G.

Theorem 3.14 (D. M. CVETKOVIĆ [Cve9], [Cve12]): *The interior stability number $\alpha(G)$ of the graph G satisfies the inequality*

$$\alpha(G) \leqq p_0 + \min(p_-, p_+), \tag{3.13}$$

where p_-, p_0, p_+ denote the number of eigenvalues of G smaller than, equal to, or greater than zero, respectively.

There are graphs for which equality holds in (3.13).

Proof. Let $s = p_0 + \min(p_-, p_+)$. Suppose that there is a graph G for which $\alpha(G) > s$ holds. Then there is an induced subgraph of G with $\alpha(G)$ vertices containing no edges. Thus, a principal submatrix, of the order $\alpha(G)$, of the adjacency matrix of G is equal to the zero-matrix. Since all eigenvalues of a zero-matrix are equal to zero, Theorem 0.10 gives for the eigenvalues $\lambda_1, \ldots, \lambda_n$ of G the inequalities

$$\lambda_i \geqq 0, \quad \lambda_{n-\alpha(G)+i} \leqq 0 \qquad \big(i = 1, \ldots, \alpha(G)\big).$$

However, this contradicts the assumption $\alpha(G) > s$. Thus, (3.13) holds.

Equality holds in (3.13), for example, for complete graphs.

This completes the proof of the Theorem 3.14.

A special case of this result is noted in [Bax1]. This paper deals with an adjacency matrix of a somewhat different structure.

Theorem 3.15 (See [Cve9], [Cve12], [AmHa], [Hof16]): *Let p_{-1}^-, p_{-1}, and p_{-1}^+ denote the number of eigenvalues of the graph G which are smaller than, equal to, or greater than -1. Let λ^* represent the smallest eigenvalue greater than -1. Let further $p = p_{-1}^- + p_{-1} + 1$ and $s = \min(p, p_{-1}^+ + p_{-1}, r + 1)$, where r is the index (= maximum eigenvalue) of G, and let*

$$\alpha = \begin{cases} 0 & \text{if} \quad \lambda^* \leqq p - 1, \\ 1 & \text{if} \quad \lambda^* > p - 1. \end{cases}$$

If $K(G)$ denotes the maximum number of vertices in a complete subgraph of G, then

$$K(G) \leqq \begin{cases} s & \text{if} \quad s < p, \\ s - \alpha & \text{if} \quad s = p. \end{cases} \tag{3.14}$$

There are graphs for which equality holds in (3.14).

Proof. If G contains a complete subgraph with k vertices, then Theorem 0.10, in a way similar to the proof of Theorem 3.14, yields the following inequalities:

$$\lambda_{n-k+1} \leqq k - 1 \leqq \lambda_1 = r,$$

$$\lambda_{n-k+i} \leqq -1 \leqq \lambda_i \qquad (i = 2, \ldots, k).$$

The greatest value of k satisfying these inequalities is given by the expression on the right-hand side of inequality (3.14). Equality holds in (3.14), for example, for complete multipartite graphs. This completes the proof of Theorem 3.15.

In the paper [Лих1] the author deals, among other things, with the connection between the spectrum of a graph and the maximum number of vertices in a complete subgraph. The proofs of the results announced in [Лих1] have, as far as we know, not yet been published.

Now we shall discuss the relations between the spectrum of a graph and its chromatic number. It is surprising that on the basis of the spectrum, some information about the chromatic number (a quantity which in general cannot easily be determined) can be obtained. For some special classes of graphs the chromatic number can even be calculated exactly from the spectrum (for example, for bichromatic graphs: see Theorem 3.11; for regular graphs of degree $n - 3$, where n is the number of vertices: see Section 3.6). Nevertheless, in the majority of cases, we have some inequalities for the chromatic number. In general, these inequalities are not too sharp, but for each inequality there are graphs for which this inequality yields a good (lower or upper) estimate of the chromatic number. Therefore, all known estimates should be applied to the given graph, and then the best one should be chosen.

In general, however, the chromatic number is not determined by the spectrum. Moreover, A. J. HOFFMAN has proved that there is, in a certain sense, an essential irrelevance between the spectrum and the chromatic number of a graph (see Section 6.1).

We shall now present some theorems concerning the topic under consideration. We begin with a theorem due to H. S. WILF.

Theorem 3.16 (H. S. WILF [Wil2]): *Let $\chi(G)$ be the chromatic number and r the index (= maximum eigenvalue) of a connected graph G. Then*

$$\chi(G) \leqq r + 1. \tag{3.15}$$

Equality holds if and only if G is a complete graph or a circuit of odd length.

Proof. Let $d_{\min}(H)$ and $d_{\max}(H)$ denote the smallest and the greatest vertex degree in a graph H and let $\lambda_1(H)$ be the index of H. Since $\chi(G)$ is the chromatic number of G, there exists an induced subgraph H of G with $d_{\min}(H) \geqq \chi(G) - 1$. By Theorems 0.6 and 3.8 we obtain

$$\lambda_1(G) \geqq \lambda_1(H) \geqq d_{\min}(H) \geqq \chi(G) - 1, \tag{3.16}$$

and therefrom (3.15). Let equality now hold in (3.15) and, consequently, also in (3.16). Then $\lambda_1(G) = \lambda_1(H)$ implies $G = H$, since G is connected. Further $\lambda_1(G) = d_{\min}(G)$, which implies, according to Theorem 3.8, that G is regular. Thus $\chi(G) = 1 + r = 1 + d_{\max}(G)$. The well-known Brooks Theorem (see, e.g., [Sac9]) now implies that G is a complete graph or a circuit of odd length. This completes the proof.

Before we quote a generalization of this theorem we shall give some definitions.

A graph G is *k-degenerate*, for some non-negative integer k, if $d_{\min}(H) \leqq k$ for each induced subgraph H of G. The *point partition number* $\varrho_k(G)$ of the graph G is the smallest number of sets into which the vertex set of G can be partitioned so that each set induces a k-degenerate subgraph of G. Since 0-degenerate graphs are exactly those which are totally disconnected, we see that $\varrho_0(G)$ is the chromatic number of G. $\varrho_1(G)$ is called the *point arboricity* of G, since 1-degenerate graphs are forests.

It can be proved (see [LiWh]) that every graph G contains an induced subgraph H with $d_{\min}(H) \geqq (k+1)\left(\varrho_k(G) - 1\right)$. On the basis of this fact the following theorem [Lick] can be proved in a manner similar to the last one (for the special case $k = 1$, see also [Mitc]).

Theorem 3.17 (D. R. LICK [Lick]): *For any graph G with index r and any non-negative integer k,*

$$\varrho_k(G) \leqq 1 + \left[\frac{r}{k+1}\right].$$

For the proof of the following theorem we quote a lemma without proof; both the lemma and theorem have been proved in [Hof 16].

Lemma 3.2: *Let A be a real symmetric matrix of order n, and let $\mathcal{S}_1 \cup \cdots \cup \mathcal{S}_t$ ($t \geqq 2$) be a partition of $\{1, \ldots, n\}$ into non-empty subsets. $\boldsymbol{A}_{kk}$ denotes the submatrix of $\boldsymbol{A}$ with row and column indices from $\mathcal{S}_k$. If $0 \leqq i_k \leqq |\mathcal{S}_k|$, $k = 1, \ldots, t$, then*

$$\lambda_{i_1+i_2+\cdots+i_t+1}(\boldsymbol{A}) + \sum_{i=1}^{t-1} \lambda_{n-i+1}(\boldsymbol{A}) \leqq \sum_{k=1}^{t} \lambda_{i_k+1}(\boldsymbol{A}_{kk}), \tag{3.17}$$

where $\lambda_i(\boldsymbol{X})$, $i = 1, 2, \ldots$, are the eigenvalues of the matrix $\boldsymbol{X}$ in decreasing order.

Theorem 3.18 (A. J. HOFFMAN [Hof 16]): *If r ($r \neq 0$) and q are the greatest and the smallest eigenvalues of the graph G, then its chromatic number $\chi(G)$ satisfies the inequality*

$$\chi(G) \geqq \frac{r}{-q} + 1. \tag{3.18}$$

Proof. Let $\chi(G) = t$ and let the vertices of G be labelled by $1, \ldots, n$. Then there exists a partition $\mathcal{S}_1 \cup \cdots \cup \mathcal{S}_t$ such that each of the subgraphs of G induced by $\mathcal{S}_i$ contains no edges. With $i_k = 0$ ($k = 1, \ldots, t$), (3.17) yields for the eigenvalues $\lambda_1 = r, \lambda_2, \ldots, \lambda_n = q$ of G

$$r + \sum_{i=1}^{t-1} \lambda_{n-i+1} \leqq 0. \tag{3.19}$$

Since $\sum_{i=1}^{t-1} \lambda_{n-i+1} \geqq (t-1)\, q$, (3.18) follows from (3.19). This completes the proof of the theorem.

Note that from (3.19) there can be drawn more information about the chromatic number than from (3.18). Actually, (3.19) yields the following bound:

$$\chi(G) \geqq 1 + \min_K \left\{K \mid \lambda_1 + \sum_{i=1}^{K} \lambda_{n-i+1} \leqq 0\right\}.$$

The following theorem provides another lower bound for the chromatic number.

Theorem 3.19 (D. M. Cvetković [Cve11]): *If G is a graph with n vertices, with index r and chromatic number $\chi(G)$, then the following inequality holds*

$$\chi(G) \geqq \frac{n}{n-r}.$$

Proof. Consider the characteristic polynomial of a k-complete graph $K_{n_1,\ldots,n_k}$ which is given by (2.50) or (2.51). The polynomial (2.50) has a single positive root which is simple. Indeed, as is seen in Theorem 6.7, complete multipartite graphs are precisely those connected graphs with a single positive eigenvalue. Thus for $x > 0$, $P_{K_{n_1,\ldots,n_k}}(x) \geqq 0$ if and only if $x \geqq \lambda_1$.

Now consider the values of

$$\sum_{i=1}^{k} \frac{n_i}{\lambda + n_i}, \quad \lambda > 0, \quad \sum_{i=1}^{k} n_i = n. \tag{3.20}$$

Assume for the moment that the n_i's can assume positive real values. Then (3.20) attains its maximum when all the n_i's are equal. Indeed, if $n_i \neq n_j$, then by letting $n_i' = n_j' = \frac{1}{2}(n_i + n_j)$ and by leaving all other values unchanged, (3.20) is increased. For the particular value $\lambda = \frac{k-1}{k} n$, (3.20) is equal to 1 when the n_i's are equal. Thus when the n_i's are positive integers (2.50) is non-negative and hence

$$\lambda_1 \leqq \frac{k-1}{k} n \tag{3.21}$$

with equality only when the graph is regular.

So we have proved

Lemma 3.3: *The index r of $K_{n_1,\ldots,n_k}$ satisfies*

$$r \leqq \frac{k-1}{k} n \quad \textit{where} \quad n = \sum_{i=1}^{k} n_i.$$

(In the Appendix the spectra of some k-complete graphs are given.)

If $\chi(G) = k$, the set of vertices of G can be partitioned into k non-empty subsets so that the subgraph induced by any one of these subsets contains no edges. If the mentioned subsets contain $n_1, \ldots, n_k$ $(n_1 + \cdots + n_k = n)$ vertices, respectively, then by adding new edges to G we can obtain $K_{n_1,\ldots,n_k}$. It is known (see Theorem 0.7) that the index of a graph does not decrease when new edges are added to the graph. Therefore, the index of G is not greater than the index of $K_{n_1,\ldots,n_k}$. According to this and the foregoing, $r \leqq \frac{k-1}{k} n$, which implies $k \geqq \frac{n}{n-r}$.

This completes the proof of the theorem.

The following theorem of Hoffman and Howes establishes the existence of an upper bound of another type for the chromatic number.

Theorem 3.20 (A. J. Hoffman, L. Howes [HoHo]): *Let $m(G)$ be the number of eigenvalues of a graph G not greater than -1. Then there exists a function f such that $\chi(G) \leqq f(m(G))$.*

Proof. Let $e = e(G)$ be the largest number such that G contains a set of $2e$ vertices $1, \ldots, e, 1', \ldots, e'$ with i and i' adjacent ($i = 1, \ldots, e$), other pairs of vertices being not adjacent. Let $K(G)$ be the maximum number of vertices in a complete subgraph of G. Using Theorem 0.10 we simply obtain $K(G) \leqq 1 + m(G)$, $e(G) \leqq m(G)$ (see also Theorem 3.15). Hence, we have only to prove that $\chi(G)$ is bounded by some function of $K(G)$ and $e(G)$. This will be done by induction on $K(G)$. If $K(G) = 1$, then $\chi(G) = 1$. For a given graph G, let G_i ($G_{i'}$), $i = 1, \ldots, e$, be the subgraph of G induced by the set of vertices adjacent to i (i'). Since $K(G_i)$, $K(G_{i'}) < K(G)$ and $e(G_i)$, $e(G_{i'}) \leqq e(G)$, the induction hypothesis can be applied to G_i ($G_{i'}$). But the set of vertices of G contained in no G_i or $G_{i'}$ induces a subgraph without edges. This fact is sufficient to prove the theorem.

It was conjectured in [Hof16] that $f(m(G)) = 1 + m(G)$. But as was observed in [HoHo], this is false since $\overline{C}_7$, the complement of a circuit of length 7, provides a counter example. In [AmHa] it was mentioned that the inequality $\chi(G) \leqq p + 1$, where p is the number of non-positive eigenvalues of G, might possibly be valid. It can be shown that at least one of the inequalities $\chi(G) \leqq \text{rk}\,(G)$, $\chi(\overline{G}) \leqq \text{rk}\,(\overline{G})$ ($\text{rk}\,(G)$ being the rank of the adjacency matrix of G) holds [Nuf2]. On the basis of this and some other facts, it can be conjectured that, except for $\overline{K}_n$, $\chi(G) \leqq \text{rk}\,(G)$, where equality holds if and only if the non-isolated vertices of G form a complete multipartite graph [Nuf2].

Some other bounds for the chromatic number will be given in Sections 3.3 and 3.6.

Now we shall give some bounds for certain quantities connected with the partitioning of the edges of a graph G.

Let $\delta(G)$ be the smallest integer k such that there exists a partition

$$\mathscr{E}_1 \cup \cdots \cup \mathscr{E}_k = \mathscr{U}, \qquad \mathscr{E}_i \cap \mathscr{E}_j = \emptyset \qquad (i, j = 1, 2, \ldots, k;\ i \neq j) \tag{3.22}$$

of the set $\mathscr{U}$ of edges of G and such that the subgraph G_i of G, induced by $\mathscr{E}_i$, is a complete graph for each $i = 1, \ldots, k$. Let $\varepsilon(G)$ be the smallest integer k such that (3.22) holds and each G_i is a complete multipartite graph. Further, let $\vartheta(G)$ be the smallest integer k such that (3.22) holds and each G_i is a bicomplete graph.

Theorem 3.21 (A. J. Hoffman [Hof11]): *Let $\lambda_1, \ldots, \lambda_n$ be the eigenvalues of a graph G. Let p_+, p_-, p be the number of eigenvalues which are positive, negative, different from both -1 and 0, respectively. Then*

$$\varepsilon(G) \geqq p_+, \quad \vartheta(G) \geqq p_-, \quad \delta(G) \geqq -\lambda_n, \quad \binom{1 + \delta(G)}{2} \geqq p.$$

This theorem was proved by means of the Courant-Weyl inequalities (see Theorem 2.1). In the proof several lemmata appear. They are included in Section 3.6.

The essential irrelevance (in a sense) of the graph spectrum with respect to $\delta(G)$ and $\varepsilon(G)$ has also been shown in [Hof11]; $\vartheta(G)$, however, is closely related to the spectrum of G.

3.3. Regular graphs

In the theory of regular graphs, numerous new theorems are valid that do not hold for non-regular graphs. Naturally, all theorems of section 3.2 hold also for regular graphs.

We shall start with the question: How can it be decided by means of its spectrum whether or not a given graph is regular?

Theorem 3.22: *Let $\lambda_1 = r, \lambda_2, \ldots, \lambda_n$ be the spectrum of a graph G, r being the index of G. G is regular if and only if*

$$\frac{1}{n} \sum_{i=1}^{n} \lambda_i^2 = r. \tag{3.23}$$

If (3.23) *holds, then G is regular of degree r.*

Proof. Since the mean value $\bar{d}$ of the vertex degrees in G is given by $\bar{d} = \frac{2m}{n} = \frac{1}{n} \sum_{i=1}^{n} \lambda_i^2$ (m is the number of edges), Theorem 3.22 is a corollary of Theorem 3.8.

This theorem is implicitely contained in [CoSi1]. See also [Cve7]. It can be easily modified for the case when the existence of loops in some vertices of G is allowed. If G contains multiple edges or multiple loops, Theorem 3.33 can be applied for the establishment of regularity.

The following theorem is obvious.

Theorem 3.23: *The number of components of a regular graph G is equal to the multiplicity of its index.*

Theorems 3.22 and 3.23 will be used several times in this book. In many theorems a graph G is required to be 1^0 regular, or 2^0 regular and connected. These conditions can be replaced by the following ones: 1^0 The spectrum of G satisfies (3.23), 2^0 The spectrum of G satisfies (3.23), and r is a simple eigenvalue. Thus, in such theorems the assumptions concerning the general graph structure are only seemingly of a non-spectral nature.

The following theorem is taken from [Finc].

Theorem 3.24 (H.-J. Finck [Finc]): *Let z be the number of circuits C_2 of length 2 in a regular multi-graph G of degree r with n vertices and without loops. Then $4z = -2a_2 - nr$, where a_2 is the coefficient of λ^{n-2} in the characteristic polynomial of G.*

Proof. Let a_{ij} be the elements of the adjacency matrix. Then a_2 is given by

$$a_2 = \sum_{i<j} \begin{vmatrix} 0 & a_{ij} \\ a_{ji} & 0 \end{vmatrix} = -\frac{1}{2} \sum_{i=1}^{n} \sum_{j=1}^{n} a_{ij}^2 .$$

If z_{ij} is the number of circuits C_2 containing the vertices i and j, then $z_{ij} = \binom{a_{ij}}{2}$ and we get

$$4z = 4 \cdot \frac{1}{2} \sum_{i=1}^{n} \sum_{j=1}^{n} z_{ij} = \sum_{i=1}^{n} \sum_{j=1}^{n} (a_{ij}^2 - a_{ij}) = -2a_2 - nr .$$

Corollary: *G has no multiple edges if and only if $2a_2 = -nr$.*

Since for graphs the minimal polynomial is obtainable from the spectrum, Theorem 3.7 takes now the following form.

Theorem 3.25 (A. J. Hoffman [Hof3]): *For a graph G with adjacency matrix $\boldsymbol{A}$ there exists a polynomial $P(x)$, such that $P(\boldsymbol{A}) = \boldsymbol{J}$, if and only if G is regular and connected. In this case we have*

$$P(x) = \frac{n(x - \lambda^{(2)}) \cdots (x - \lambda^{(m)})}{(r - \lambda^{(2)}) \cdots (r - \lambda^{(m)})},$$

where n is the number of vertices, r is the index, and $\lambda^{(1)} = r, \lambda^{(2)}, \ldots, \lambda^{(m)}$ are all distinct eigenvalues of G.

This important theorem provides great possibilities for the investigation of the structure of graphs by means of spectra. It will be used many times in the sequel.

We proceed now to the investigation of the circuit structure of regular graphs. We shall apply Theorem 3.12 and a result from [Sac4]. Consider a regular graph G. According to [Sac4], in regular graphs the number b_q occurring in Theorem 3.12 can for $q < g(G)$ be expressed in terms of q, the number of vertices n, and the degree r of G. Since n and r are obtainable from the characteristic polynomial of G, the following result is immediately obtained.

Theorem 3.26 (H. Sachs [Sac3]): *The girth g and the number of circuits of length g of a regular graph G are determined by the corresponding characteristic polynomial $P_G(\lambda)$.*

Now we can go further and extend the whole of Theorem 3.12 to the case of regular graphs. We shall again use a result from [Sac4].

Consider the basic figures U_i with i ($g \leq i < 2g$) vertices contained in G. Let U_i^0 be those basic figures which contain no circuits (i.e. which contain only graphs K_2 as components) and let A_i be their number. For odd i there are obviously no basic figures U_i^0. For $i = 2q$ we have $A_i = b_q$ (numbers b_q being defined in Theorem 3.12), and the contribution of U_i^0 to the corresponding coefficient $(-1)^i a_i$ of the characteristic polynomial of G is $(-1)^q = (-1)^{\frac{i}{2}}$. Let us further consider those basic figures U_i^c which contain a circuit of length c. Clearly, $g \leq c \leq i$; U_i^c contains exactly one circuit, since, by hypothesis, the number i of vertices of U_i^c is smaller than $2g$. Now, $i - c$ vertices, not belonging to that circuit, are vertices belonging to $\frac{i-c}{2}$ graphs K_2; thus, $c \equiv i \pmod 2$ must be valid. The contribution of a basic figure U_i^c to $(-1)^i a_i$ is, according to Theorem 1.3, equal to

$$(-1)^{\frac{i-c}{2}} \cdot (-1)^{c+1} \cdot 2 = 2 \cdot (-1)^{1+\frac{i+c}{2}}.$$

The last formula holds also in the case $c = i$, when U_i^i reduces to a circuit of length i.

Now, for each c with $g \leqq c \leqq i$, $i \equiv c \pmod 2$ the number B_i^c of different basic figures U_i^c must be determined.

Let G have exactly D_c circuits of length c and let these circuits be denoted by C_c^j $(j = 1, \ldots, D_c)$. If $c = i$, we have $B_i^i = D_i$, while in the case $c < i$ we have the following situation:

Let G_c^j be the subgraph of G induced by those vertices not lying on C_c^j. Then the number of basic figures U_i^c which contain a fixed circuit C_c^j is obviously equal to the number $E_{i,c}^j$ of forests, containing exactly $\frac{i-c}{2}$ graphs K_2 as components, in G_c^j. According to [Sac4] the number $E_{i,c}^j$ depends only on i, n, r, and c, but neither on j nor on the special structure of G. Therefore we can omit the upper index j and, since the numbers n and r are directly obtainable from $P_G(\lambda)$, we can assume that the numbers $E_{i,c}$ are also given through $P_G(\lambda)$.

So we have for $c < i$

$$B_i^c = \sum_{j=1}^{D_c} E_{i,c}^j = E_{i,c} D_c .$$

If $b(U_i)$ is the contribution of the basic figure U_i to the corresponding coefficient $(-1)^i a_i$ of $P_G(\lambda)$, we obtain

$$(-1)^i a_i = \sum_{U_i^0} b(U_i^0) + \sum_{\substack{g \leqq c < i \\ c \equiv i \,(\mathrm{mod} 2)}} \sum_{U_i^c} b(U_i^c) + \sum_{U_i^i} b(U_i^i)$$

and hence for even i

$$a_i = (-1)^{\frac{i}{2}} b_{\frac{i}{2}} + \sum_{\substack{g \leqq c < i \\ c \equiv 0 \,(\mathrm{mod} 2)}} (-1)^{1+\frac{i+c}{2}} 2E_{i,c} D_c - 2D_i, \tag{3.24}$$

and for odd i

$$a_i = \sum_{\substack{g \leqq c < i \\ c \equiv 1 \,(\mathrm{mod} 2)}} (-1)^{\frac{i+c}{2}} 2E_{i,c} D_c - 2D_i . \tag{3.25}$$

These formulas hold if i is smaller than $2g$.

By a recursive procedure, equations (3.24) and (3.25) can easily be solved with respect to the desired numbers D_i. If, for example, g is even, using Theorem 3.12 we obtain in order from (3.25):

$$D_{g+1} = -\frac{1}{2}\,\tilde{a}_{g+1},$$

from (3.24):

$$D_{g+2} = -\frac{1}{2}\,(\tilde{a}_{g+2} - 2E_{g+2,g}D_g),$$

where, according to Theorem 3.26, D_g is the known number of circuits of length g, from (3.25):

$$D_{g+3} = -\frac{1}{2}\,(\tilde{a}_{g+3} - 2E_{g+3,g+1}D_{g+1}),$$

etc. The numbers $\tilde{a}_j$ defined in Theorem 3.12 are, as has already been stated, determined by r, n, and a_j, i.e. by $P_G(\lambda)$.

Thus, we have proved the following theorem.

Theorem 3.27 (H. Sachs [Sac3]): *Let G be a regular graph with girth g and with the characteristic polynomial* (3.1). *Let $h \leqq n$ be a non-negative integer not greater than $2g - 1$. Then the number of circuits of length h, which are contained in G, is determined by the largest root r and the first h coefficients $a_1, a_2, \ldots, a_h$ of the characteristic polynomial of G.*

The following theorem establishes a spectral property of self-complementary graphs. A graph G is self-complementary if it is isomorphic to its complement $\bar{G}$. Self-complementary graphs were primarily studied by G. Ringel [Ring] and H. Sachs [Sac1] (see also [Clap], [Rea1]). If G is a regular self-complementary graph, then G is connected and has $n = 4k + 1$ vertices and degree $r = 2k$ [Ring], [Sac1]. We shall assume $n > 1$, i.e. $k \geqq 1$. According to Theorem 2.6,

$$P_{\bar{G}}(\lambda) = P_G(\lambda) = -\frac{\lambda - 2k}{\lambda + 2k + 1}\,P_G(-\lambda - 1),$$

or

$$\frac{P_G(\lambda)}{\lambda - 2k} = \frac{P_G(-\lambda - 1)}{-\lambda - 1 - 2k}.$$

If λ_i $(i = 2, 3, \ldots, 4k + 1)$ are the eigenvalues of G different from the index λ_1 (i.e., $\lambda_i \neq r = 2k$), then

$$\prod_{i=2}^{4k+1} (\lambda - \lambda_i) = \prod_{i=2}^{4k+1} (-\lambda - 1 - \lambda_i) = \prod_{j=2}^{4k+1} (\lambda + 1 + \lambda_j).$$

To each eigenvalue $\lambda_i \neq 2k$ there corresponds another eigenvalue $\lambda_j = -\lambda_i - 1$, where $\lambda_j \neq \lambda_i$, since otherwise $\lambda_i = -\frac{1}{2}$ which is impossible due to the fact that λ_i

is an algebraic integer. Thus, $\lambda_{i+1} > -\frac{1}{2}$ and $\lambda_j = \lambda_{n+1-i} < -\frac{1}{2}$ for $i = 1, 2, \ldots, 2k$, and

$$\lambda_{i+1} + \lambda_{4k+2-i} = -1 \qquad (i = 1, 2, \ldots, 2k), \tag{3.26}$$

giving rise to the following theorem.

Theorem 3.28 (H. Sachs [Sac1]): *The characteristic polynomial of a regular self-complementary graph has the form*

$$P_G(\lambda) = (\lambda - 2k) \prod_{i=2}^{2k+1} (\lambda - \lambda_i)(\lambda + \lambda_i + 1) = (\lambda - 2k) \prod_{i=1}^{2k} (\lambda^2 + \lambda - \alpha_i),$$

where $\alpha_i = \lambda_{i+1}^2 + \lambda_{i+1}$.

Note that this theorem also follows from Theorem 2.10 (p. 59); see also the footnote to p. 57.

Formula (3.26) implies $\lambda_2 < 2k - 1$, since in the opposite case we should obtain the impossible relation $\lambda_{4k+1} \leqq -(2k - 1) - 1 = -r$ (note that a self-complementary graph with $n > 4$ cannot be bipartite, thus $\lambda_{4k+1} > -r$; see Theorem 3.11).

The converse of Theorem 3.28 does not hold. Namely, there are connected regular graphs with $4k + 1$ vertices which have the characteristic polynomial of Theorem 3.28 and which are not self-complementary. Such graphs will be mentioned in Chapter 6 (see examples of cospectral pairs of graphs consisting of a graph G and of its complement $\overline{G}$).

A statement similar to Theorem 3.28 can be made for non-regular self-complementary graphs. As a simple consequence of Theorem 2.5, we obtain the following statement [Cve8], [Cve9]:

Let G be a self-complementary graph. Then to each eigenvalue λ_i of G of multiplicity $p > 1$ (if there is such an eigenvalue) there corresponds another eigenvalue λ_j whose multiplicity q satisfies the inequality $p - 1 \leqq q \leqq p + 1$, where $\lambda_i + \lambda_j = -1$.

We shall now discuss some theorems which are closely connected with the concept of the ∇-product of graphs introduced in Section 2.2. A graph is called *∇-prime* if it cannot be represented as a ∇-product of two graphs.

Theorem 3.29 (H.-J. Finck, G. Grohmann [FiGr]): *Let G be a regular connected graph of degree r with n vertices. G can be represented as a ∇-product of $p + 1$ ($p \geqq 0$) ∇-prime graphs if and only if $r - n$ is a p-fold eigenvalue of G.*

Proof. G can be represented as a ∇-product of $p + 1$ ∇-prime factors if and only if $\overline{G}$ has $p + 1$ components. According to Theorem 3.23, this situation arises if and only if the graph $\overline{G}$, whose index is $\overline{r} = n - r - 1$, has the number $n - r - 1$ as a $(p + 1)$-fold eigenvalue. By virtue of Theorem 2.6, the last statement is equivalent to the statement that $r - n$ is a p-fold eigenvalue of G. This completes the proof.

This theorem enables us to calculate several lower and upper bounds for the chromatic number $\chi(G)$ of regular graphs G which are not ∇-prime.

Let us first consider lower bounds. Obviously, $\chi(G_1 \nabla G_2) = \chi(G_1) + \chi(G_2)$. Therefore,

$$\chi(G) \geqq k \tag{3.27}$$

if G can be represented as a ∇-product of k ∇-prime factors.

Let G be a connected regular graph of degree r with n vertices, $n > r + 1$ (complete graphs are thereby excluded, but this limitation is not essential). The multiplicity of an eigenvalue λ of G will be denoted by p_λ.

Let $p_{r-n} + 1 = k > 1$. According to Theorem 3.29, G can be represented as a ∇-product of k ∇-prime graphs, say, G_ν ($\nu = 1, 2, \ldots, k$). Each of the G_ν is regular and its degree r_ν and number n_ν of vertices satisfy the equation $n_\nu - r_\nu = n - r$ (see Section 2.2, p. 57).

Suppose that among the k ∇-prime factors of G there are exactly m_1 monochromatic, i.e. totally disconnected graphs. For such factors G_ν, $\chi(G_\nu) = 1$, and for the other $k - m_1$ factors G_ν, $\chi(G_\nu) \geqq 2$. So,

$$\chi(G) \geqq m_1 + 2(k - m_1) = 2k - m_1 . \tag{3.28}$$

Assume further that exactly m_2 of the factors G_ν are bichromatic, i.e. bipartite of positive degrees r_ν, and let $m = 2m_1 + m_2$. Then

$$\chi(G) \geqq m_1 + 2m_2 + 3(k - m_1 - m_2) = 3k - m . \tag{3.29}$$

So, any upper bound for m_1 or m will, by virtue of (3.28) or (3.29), automatically yield a (possibly trivial) lower bound for $\chi(G)$.

Before we can outline a method of finding upper bounds for m_1 or m, we need two more definitions.

1⁰ Let $[\lambda_1', \lambda_2', \ldots, \lambda_{n'}']$ be the spectrum of a graph G'. Then the family $[\lambda_2', \lambda_3', \ldots, \lambda_{n'}']$ is called the *reduced spectrum of* G'.

2⁰ Let $\mathcal{F}_1, \mathcal{F}_2, \ldots, \mathcal{F}_s$ be subfamilies of a finite family $\mathcal{F}_0$ and let $p_\sigma(e)$ be the multiplicity (possibly zero) with which element e is contained in $\mathcal{F}_\sigma$ ($\sigma = 0, 1, \ldots, s$). $\mathcal{F}_1, \mathcal{F}_2, \ldots, \mathcal{F}_s$ are called *independent in* $\mathcal{F}_0$ if $\sum_{\sigma=1}^{s} p_\sigma(e) \leqq p_0(e)$ for each element e of $\mathcal{F}_0$.

Now, according to Theorem 2.9, the reduced spectra of the ∇-prime factors G_ν of G constitute a family of independent subfamilies of the spectrum of G. Evaluating the conditions which the spectra of totally disconnected and regular bipartite graphs must satisfy, we can in principle easily obtain upper bounds for m_1 and m.

Suppose G^* to be a totally disconnected ∇-prime factor of G. Then $r^* = 0$ and $n^* = n - r$. The reduced spectrum of G consists of $n - r - 1 > 0$ numbers all equal to zero: so, zero is contained in the spectrum of G with a multiplicity not smaller than $m_1(n - r - 1)$. Thus $p_0 \geqq m_1(n - r - 1)$ and, consequently,

$$m_1 \leqq \left[\frac{p_0}{n - r - 1}\right]. \tag{3.30}$$

(3.30) is a fortiori true if $m_1 = 0$

Recall that $k = p_{r-n} + 1$. From (3.28) we deduce

$$\chi(G) \geqq 2p_{r-n} + 2 - \left[\frac{p_0}{n - r - 1}\right]. \tag{3.31}$$

A better estimate may be obtained by taking possible bichromatic ∇-prime factors into consideration. Denote by $\mathscr{B}$ the set of regular bipartite ∇-prime factors of G having positive degrees: then $|\mathscr{B}| = m_2$. For every $G_\nu \in \mathscr{B}$, $r_\nu \leqq \frac{1}{2} n_\nu$ (note that a regular bipartite graph of positive degree has an even number of vertices); moreover, $r_\nu \leqq \frac{1}{2} n_\nu - 1$, because a regular bipartite graph G' with $r' = \frac{1}{2} n'$ is bicomplete and therefore not ∇-prime. The above inequality, together with the relation $n_\nu - r_\nu = n - r$, implies $r_\nu \leqq n - r - 2$.

An arbitrary subfamily $\mathscr{S} = [\mu_1, \mu_2, \ldots, \mu_{n_\mathscr{S}-1}]$ of the spectrum of G can be the reduced spectrum of a regular bipartite ∇-prime graph $G_\nu \in \mathscr{B}$ of some degree $r_\nu = i > 0$ only if the following conditions be satisfied:

a) $n - r + 1 \leqq n_\mathscr{S} = n - r + i \leqq 2(n - r - 1)$;

b) $n_\mathscr{S}$ is even;

c) $-i \leqq \mu_l \leqq i \quad (l = 1, 2, \ldots, n_\mathscr{S} - 1)$;

d) the family $\mathscr{S} \cup \{i\} = [i, \mu_1, \mu_2, \ldots, \mu_{n_\mathscr{S}-1}]$ is symmetric with respect to the zero point of the real axis (μ and $-\mu$ in $\mathscr{S} \cup \{i\}$ having the same multiplicity);

e) $\sum_{l=1}^{n_\mathscr{S}-1} \mu_l^2 = i(n - r)$.

These conditions are either direct consequences of Theorems 3.11, 2.9, 3.22 or obvious.

In the cases $i = 1$ and $i = 2$ stronger conditions that $\mathscr{S}$ must satisfy can be formulated.

Case $i = 1$: In this case, G_ν has only complete graphs K_2 as components.

f') $n_\mathscr{S} = n - r + 1 \geqq 4$ (since K_2 is not ∇-prime);

g') $\mathscr{S}$ contains number 1 with multiplicity $\frac{1}{2} n_\mathscr{S} - 1 = \frac{n - r - 1}{2}$ and number -1 with multiplicity $\frac{1}{2} n_\mathscr{S} = \frac{n - r + 1}{2}$, and no other numbers.

Case $i = 2$: In this case, G_ν has only circuits of even length $\geqq 4$ as components. The characteristic polynomial is of the form

$$f_\mathscr{P}(\lambda) = \prod_{\alpha \in \mathscr{P}} \prod_{j=1}^{\alpha} \left(\lambda - 2 \cos \frac{2\pi j}{\alpha}\right), \tag{3.32}$$

where $\mathscr{P}$ may be any partition of $n - r + 2$ into even numbers $\geqq 4$, and α runs through all elements of $\mathscr{P}$ (see end of Section 2.1).

f'') $n_\mathscr{S} = n - r + 2 \geqq 6$ (since a circuit of length 4 is not ∇-prime);

g″) $\mathscr{S} \cup \{2\}$ is identical with the family of roots of the equation $f_{\mathscr{P}}(\lambda) = 0$ (see (3.32)) for some partition $\mathscr{P}$ of $n - r + 2$ into even numbers $\geqq 4$.

Note that conditions c)—e) are consequences of conditions f′), g′) or f″), g″), respectively.

Now let **S** be a family of $u_1 + u_2 \leqq k$ independent subfamilies $\mathscr{S}$ of the spectrum of G, exactly u_1 of them having zero as their only element with multiplicity $n - r - 1$, and each of the remaining u_2 families satisfying the conditions given above for some i; such a family **S** will be called *feasible*. There is, in particular, a feasible $\mathbf{S} = \mathbf{S}^*$ (with $u_1 = u_1^*$, $u_2 = u_2^*$) which is identical with the family of the reduced spectra of all monochromatic and bichromatic ∇-prime factors of G, thus $u_1^* = m_1$, $u_2^* = m_2$, $2u_1^* + u_2^* = m$. Consequently, the maximum value M of $2u_1 + u_2$, taken over all feasible **S**, is an upper bound for m, and so

$$\chi(G) \geqq 3k - M .$$

L. L. Kraus and D. M. Cvetković [KrC1] noted that all constraints to be satisfied (in particular the condition of independence) can be given the form of linear inequalities so that M may be obtained as the solution of an integer linear programming problem which we shall now formulate.

Let $\mathscr{S}_1$ be the family having zero as its only element, with multiplicity $n - r - 1$. Determine the set $\{\mathscr{S}_2, \mathscr{S}_3, \ldots, \mathscr{S}_f\}$ of all distinct (not necessarily independent) subfamilies $\mathscr{S}$ of the spectrum of G satisfying, for some i, the conditions given above.

Let the spectrum of G contain the distinct eigenvalues $\lambda^{(i)}$ $(i = 1, 2, \ldots, d)$ with multiplicities p_i $(p_1 + p_2 + \cdots + p_d = n)$. Let p_{ij} be the multiplicity in $\mathscr{S}_j$ of the eigenvalue $\lambda^{(i)}$. If $\mathscr{S}_j$ appears exactly x_j times as an element of the feasible family **S** then, with the notation used above: $x_1 = u_1$, $\sum_{j=2}^{f} x_j = u_2$, and so the following inequalities hold:

$$x_j \geqq 0 \qquad (j = 1, 2, \ldots, f), \tag{3.33}$$

$$\sum_{j=1}^{f} x_j \leqq k, \tag{3.34}$$

$$\sum_{j=1}^{f} p_{ij} x_j \leqq p_i \qquad (i = 1, 2, \ldots, d); \tag{3.35}$$

the last inequality is equivalent to the independence of the families $\mathscr{S}$ of **S** as subfamilies of the spectrum of G. Since $2u_1 + u_2 = 2x_1 + \sum_{j=2}^{f} x_j$, it is clear that the maximum value of $2x_1 + \sum_{j=2}^{f} x_j$, where the x_j are integers subject to the contraints (3.33)—(3.35), is equal to the maximum value M of $2u_1 + u_2$.

So we have proved

Theorem 3.30 (L. L. Kraus, D. M. Cvetković [KrC1]): *Under the assumptions made above, let M be the maximum value of $2x_1 + \sum_{j=2}^{f} x_j$, where the x_j $(j = 1, 2, \ldots, f)$*

are integers subject to the constraints (3.33)—(3.35). *Then*

$$\chi(G) \geqq 3k - M, \qquad \text{where } k = p_{r-n} + 1. \tag{3.36}$$

Along similar lines, some rougher (but more easily calculable) lower bounds have been obtained by H.-J. Finck [Finc] and again by L. L. Kraus and D. M. Cvetković [KrC1].

Now we proceed to the determination of an upper bound for the chromatic number of a regular graph which is not ∇-prime.

Let, as earlier, G be a connected regular graph of degree r with n vertices and let the eigenvalue $r - n$ occur in the spectrum of G with multiplicity $k - 1$ ($\geqq 0$). Then the ∇-prime factors of G are regular graphs G_ν of degree r_ν with n_ν vertices, where $r_\nu = r - n + n_\nu$ ($\nu = 1, \ldots, k$). According to the well-known theorem of Brooks (see, for example, [Sac9]),

$$\chi(G_\nu) \leqq r_\nu + 1 \qquad (\nu = 1, \ldots, k), \tag{3.37}$$

and so

$$\chi(G) \leqq \sum_{\nu=1}^{k} (r_\nu + 1) = \sum_{\nu=1}^{k} r_\nu + k = k(r + 1) - (k - 1)\, n. \tag{3.38}$$

This bound can be improved. In (3.37), equality holds only in the following four cases

(a) $r_\nu = 0$,

(b) $r_\nu = 1$,

(c) $r_\nu = 2$ and G_ν contains a component with an odd number of edges,

(d) $r_\nu \geqq 3$ and G_ν contains a complete graph with $r_\nu + 1$ vertices as a component.

If s is the number of graphs G_ν which satisfy one of these conditions, then

$$\chi(G) \leqq \sum_{\nu=1}^{k} r_\nu + s.$$

We shall now derive an upper bound for s. In order to simplify the analysis we shall assume that $n - r$ is even. In this way graphs G_ν satisfying condition (b) are excluded since such graphs, on the one hand, must have $n - r + 1$ vertices (this is an odd number) and, on the other hand, must have an even number of vertices.

The number of graphs G_ν satisfying condition (a) is not greater than $\left[\dfrac{p_0}{n - r - 1}\right]$, as mentioned earlier.

A graph G_ν satisfying (c) cannot be connected since its number of vertices $n_\nu = n - r + 2$ is even. This means that the characteristic polynomial of G_ν contains a factor $(\lambda - 2)^2$. Thus, according to Theorem 2.9, the characteristic polynomial of G contains a factor $\lambda - 2$ which stems from G_ν. So the number of such G_ν is not greater than p_2.

For each graph G_ν which satisfies (d), a factor $(\lambda + 1)^{r_\nu}$ appears in the characteristic polynomial of G. Since $r_\nu \geqq 3$, the number of such graphs G_ν is not greater than $\left[\dfrac{p_{-1}}{3}\right]$.

In summarizing we obtain

$$s \leqq \left[\frac{p_0}{n - r - 1}\right] + p_2 + \left[\frac{p_{-1}}{3}\right],$$

and, consequently,

$$\chi(G) \leqq kr - (k - 1)\, n + \left[\frac{p_0}{n - r - 1}\right] + p_2 + \left[\frac{p_{-1}}{3}\right]. \tag{3.39}$$

Having in view relations (3.38) and (3.39), we can formulate the following theorem.

Theorem 3.31 (H.-J. Finck [Finc]): *Let G be a connected regular graph of degree r with n vertices, where $n - r$ is even. Let p_λ be the multiplicity of the eigenvalue λ in the P-spectrum of G. For the chromatic number $\chi(G)$ of the graph G the following inequality holds:*

$$\chi(G) \leqq r + \min\left(p_{r-n} + 1, \left[\frac{p_0}{n - r - 1}\right] + p_2 + \left[\frac{p_{-1}}{3}\right]\right) - (n - r)\, p_{r-n}.$$

If $n - r$ is odd, a more complex analysis is necessary. It seems that a problem of integer linear programming, similar to the one treated above (see Theorem 3.30), will have to be solved.

3.4. Some remarks on strongly regular graphs

Let x and y be any two distinct vertices of a graph and let $\Delta(x, y)$ denote the number of vertices adjacent to both x and y. A regular graph G of positive degree r, not the complete graph, is called *strongly regular* if there exist non-negative integers e and f such that $\Delta(x, y) = e$ for each pair of adjacent vertices x, y and $\Delta(x, y) = f$ for each pair of (distinct) non-adjacent vertices x, y of G.

The concept of a strongly regular graph was introduced by R. C. Bose [Bos1] (1963), and at present there is already an extensive literature on this type of graph (see Sections 7.2 and 7.3).

From Theorem 1.9 (Section 1.8) we deduce immediately that a regular graph G of degree $r > 0$ — not the complete graph — is strongly regular if and only if there exist non-negative integers e and f such that the adjacency matrix $\boldsymbol{A} = (a_{ij})$ of G satisfies the following relation:

$$\boldsymbol{A}^2 = (e - f)\, \boldsymbol{A} + f\boldsymbol{J} + (r - f)\, \boldsymbol{I}. \tag{3.40}$$

Theorem 3.32 (S. S. Shrikhande, Bhagwandas [ShBh]): *A regular connected graph G of degree r is strongly regular if and only if it has exactly three distinct eigenvalues $\lambda^{(1)} = r$, $\lambda^{(2)}$, $\lambda^{(3)}$.*

If G is strongly regular, then

$$e = r + \lambda^{(2)}\lambda^{(3)} + \lambda^{(2)} + \lambda^{(3)} \quad \textit{and} \quad f = r + \lambda^{(2)}\lambda^{(3)}.$$

Proof. Let G be strongly regular. The eigenvalues of G are not all equal, for if they were, they would all be equal to zero — contradicting the hypothesis that G has an edge. Nor can the spectrum of G have exactly two distinct eigenvalues since then G would have at least one edge and, according to Theorem 3.13, its diameter would be equal to 1 — contradicting the hypothesis that G is not the complete graph. Since the relation (3.40) holds for G, the minimal polynomial of the adjacency matrix $\boldsymbol{A}$ of G has degree 3. Thus, G has exactly three distinct eigenvalues.

Let now G have exactly three distinct eigenvalues $\lambda^{(1)} = r$, $\lambda^{(2)}$, $\lambda^{(3)}$. Then, obviously, $r > 0$ and G is not the complete graph and according to Theorem 3.25, the relation

$$a\boldsymbol{A}^2 + b\boldsymbol{A} + c\boldsymbol{I} = \boldsymbol{J} \qquad (a \neq 0) \tag{3.41}$$

holds, where $\lambda^{(2)}$ and $\lambda^{(3)}$ are the roots of the equation $a\lambda^2 + b\lambda + c = 0$. A comparison of the diagonal elements of the left and right side of (3.41) yields the equation $ar + c = 1$, or $c = 1 - ar$.

If the vertices i, j are adjacent, i.e. if $a_{ij} = 1$, we deduce from (3.41) that the number of walks of length two between i and j equals $\frac{1-b}{a}$. If i, j are distinct and non-adjacent, the corresponding number of such walks is $\frac{1}{a}$. Hence, G is strongly regular. Comparing (3.41) and (3.40), we obtain $e = r + \lambda^{(2)} + \lambda^{(3)} + \lambda^{(2)}\lambda^{(3)}$ and $f = r + \lambda^{(2)}\lambda^{(3)}$.

This completes the proof.

3.5. Eigenvectors

In Chapter 1 we have seen that the eigenvectors of the adjacency matrix of a (multi-)graph G, together with the eigenvalues, provide a useful tool in the investigation of the structure of G. In this section we shall go into a bit more detail.

Sometimes valuable information about a (multi-)graph can be obtained from its eigenvectors alone. Such a result is given by

Theorem 3.33: *A multigraph G is regular if and only if its adjacency matrix has an eigenvector all of whose components are equal to* 1.

This theorem is a consequence of a well-known theorem of the theory of matrices (see, e.g., [MaMi], p. 133).

In [Ber 1], p. 131, the following result of T. H. Wei [Wei] is noted:

Let $N_k(i)$ be the number of walks of length k starting at vertex i of a connected graph G with vertices $1, 2, \ldots, n$. Let $s_k(i) = N_k(i) \cdot \left(\sum_{j=1}^{n} N_k(j)\right)^{-1}$. Then, for $k \to \infty$, the vector $\left(s_k(1), s_k(2), \ldots, s_k(n)\right)^{\mathrm{T}}$ tends towards the eigenvector of the index of G.

The question as to whether or not a given multigraph is connected can be decided by means of Theorem 0.4: combining Theorems 0.3 and 0.4, we obtain

Theorem 3.34: *A multigraph is connected if and only if its index is a simple eigenvalue with a positive eigenvector.*

Theorem 0.5 can also be translated into the language of graph theory:

Theorem 3.35: *If the index of a multigraph has multiplicity p, and if there is a positive eigenvector in the eigenspace corresponding to it, then G has exactly p components.*

Of particular interest are the eigenvectors of the line graph $L(G)$ of a connected regular multigraph G of degree r.[†] Let G have n vertices and m edges. The relation connecting the spectra of G and $L(G)$, namely

$$P_{L(G)}(\mu) = (\mu + 2)^{m-n}\, P_G(\mu - r + 2), \tag{2.30}$$

has already been given in Theorem 2.15.

Formula (2.30) establishes a (1, 1)-correspondence between the sets of eigenvalues $\lambda \neq -r$ of G and $\mu \neq -2$ of $L(G)$: If $\lambda \neq -r$ is a p-fold eigenvalue of G[††], then $\mu = \lambda + r - 2 \neq -2$ is a p-fold eigenvalue of $L(G)$, and if $\mu \neq -2$ is a p-fold eigenvalue of $L(G)$, then $\lambda = \mu - r + 2 \neq -r$ is a p-fold eigenvalue of G. Therefore, we shall call (λ, μ) a pair of corresponding eigenvalues if $\lambda \neq -r$ is an eigenvalue of G, $\mu \neq -2$ is an eigenvalue of $L(G)$, and $\lambda + r = \mu + 2$.

Denote the eigenspace belonging to the eigenvalue λ of G, or to the eigenvalue μ of $L(G)$, by $\mathbf{X}(\lambda)$ or $\mathbf{Y}(\mu)$, respectively. Then the following theorem holds.

Theorem 3.36 (H. Sachs [Sac 8]): *Let G be a connected regular multigraph of degree r with n vertices and m edges, let (λ, μ) be a pair of corresponding eigenvalues of G and $L(G)$, and let $\boldsymbol{R}$ denote the $n \times m$ vertex-edge incidence matrix of G. Then $\boldsymbol{R}^{\mathsf{T}}$ maps the eigenspace $\mathbf{X}(\lambda)$ onto the eigenspace $\mathbf{Y}(\mu)$, and $\boldsymbol{R}$ maps $\mathbf{Y}(\mu)$ onto $\mathbf{X}(\lambda)$.*

Proof. As in Section 2.4, let $\boldsymbol{A}$ and $\boldsymbol{B}$ denote the adjacency matrices of G and $L(G)$, respectively.

1. Let $\boldsymbol{x} \in \mathbf{X}(\lambda)$, $\boldsymbol{x} \neq \boldsymbol{o}$, and $\boldsymbol{y} = \boldsymbol{R}^{\mathsf{T}}\boldsymbol{x}$. Then, by virtue of (2.27) (Section 2.4),

$$\boldsymbol{R}\boldsymbol{y} = \boldsymbol{R}\boldsymbol{R}^{\mathsf{T}}\boldsymbol{x} = (\boldsymbol{A} + \boldsymbol{D})\,\boldsymbol{x} = (\boldsymbol{A} + r\boldsymbol{I})\,\boldsymbol{x} = (\lambda + r)\,\boldsymbol{x} \neq \boldsymbol{o},$$

thus $\boldsymbol{y} \neq \boldsymbol{o}$. Using (2.28) and again (2.27), we obtain

$$\begin{aligned} \boldsymbol{B}\boldsymbol{y} &= (\boldsymbol{R}^{\mathsf{T}}\boldsymbol{R} - 2\boldsymbol{I})\,\boldsymbol{y} = \boldsymbol{R}^{\mathsf{T}}\boldsymbol{R}\boldsymbol{R}^{\mathsf{T}}\boldsymbol{x} - 2\boldsymbol{y} = \boldsymbol{R}^{\mathsf{T}}(\lambda + r)\,\boldsymbol{x} - 2\boldsymbol{y} \\ &= (\lambda + r - 2)\,\boldsymbol{y} = \mu\boldsymbol{y}, \end{aligned}$$

thus $\boldsymbol{y} \in \mathbf{Y}(\mu)$, i.e.:

$$\boldsymbol{x} \in \mathbf{X}(\lambda),\ \boldsymbol{x} \neq \boldsymbol{o} \quad \text{implies} \quad \boldsymbol{R}^{\mathsf{T}}\boldsymbol{x} \in \mathbf{Y}(\mu),\ \boldsymbol{R}^{\mathsf{T}}\boldsymbol{x} \neq \boldsymbol{o}.$$

[†] The line graph $L(G)$ of a multigraph G with edges $1, 2, \ldots, m$ is a multigraph with vertices $1, 2, \ldots, m$ and adjacency matrix $\boldsymbol{B} = (b_{ik})$, where, for $i \neq k$, $b_{ik} = 0$ if the edges i, k of G have no vertex in common, $b_{ik} = 1$ if i, k are proper edges (i.e., not loops) having exactly one vertex in common, $b_{ik} = 2$ if i, k are proper edges having both their vertices in common or if one of them is a proper edge and the other one a loop having a vertex in common, $b_{ik} = 4$ if i, k are both loops attached to the same vertex, and where $b_{ii} = 0$ if i is a proper edge, $b_{ii} = 2$ if i is a loop.

[††] Recall that $\lambda = -r$ is an eigenvalue (of multiplicity 1) of G if and only if G is bipartite (see Theorem 3.4).

2. In a similar way it can be proved that

$$\boldsymbol{y} \in \mathbf{Y}(\mu),\ \boldsymbol{y} \neq \boldsymbol{o} \quad \text{entails} \quad \boldsymbol{Ry} \in \mathbf{X}(\lambda),\ \boldsymbol{Ry} \neq \boldsymbol{o}.$$

3. If $\boldsymbol{y} \in \mathbf{Y}(\mu)$, then there is a (unique) $\boldsymbol{x} \in \mathbf{X}(\lambda)$ such that $\boldsymbol{R}^\mathsf{T}\boldsymbol{x} = \boldsymbol{y}$, namely

$$\boldsymbol{x} = \frac{1}{\mu + 2} \boldsymbol{Ry}.$$

4. If $\boldsymbol{x} \in \mathbf{X}(\lambda)$, then there is a (unique) $\boldsymbol{y} \in \mathbf{Y}(\mu)$ such that $\boldsymbol{Ry} = \boldsymbol{x}$, namely

$$\boldsymbol{y} = \frac{1}{\lambda + r} \boldsymbol{R}^\mathsf{T}\boldsymbol{x}.$$

Theorem 3.36 is now proved.

Remark. The mappings $\boldsymbol{R}$ and $\boldsymbol{R}^\mathsf{T}$ considered in Theorem 3.36 can be given a more intuitive form: the matrices $\boldsymbol{R}$, $\boldsymbol{A}$, $\boldsymbol{B}$ and the eigenvectors $\boldsymbol{x}$, $\boldsymbol{y}$ all refer to a fixed labelling of the vertices and edges, respectively: for example, the component x_i of an eigenvector $\boldsymbol{x}$ of $\boldsymbol{A}$ corresponds to vertex v_i, so we may write $x(v_i)$ instead of x_i, or drop the subscript altogether and simply write $x(v)$ for the component of $\boldsymbol{x}$ that corresponds to vertex v and, similarly, $y(u)$ for the component of $\boldsymbol{y}$ that corresponds to edge u.

Now let a, b denote both vertices, or edges, or one of them a vertex and the other one an edge, and let the symbol $\sum_{a \cdot b}$ mean that, for fixed b, the summation is to be taken over the set of all a which are adjacent to b or incident with b, respectively. Then

$\boldsymbol{y} = \boldsymbol{R}^\mathsf{T}\boldsymbol{x}$ is equivalent to

$$y(u) = \sum_{v \cdot u} x(v) \text{ for each edge } u,$$

$\boldsymbol{x} = \boldsymbol{Ry}$ is equivalent to

$$x(v) = \sum_{u \cdot v} y(u) \text{ for each vertex } v.$$

Next the eigenvectors $\boldsymbol{x}$ of G and $\boldsymbol{y}$ of $L(G)$ belonging to $-r$, -2, respectively, shall be investigated.

Theorem 3.37: *Let G be any connected multigraph. Then G is bipartite if and only if the system of equations*

$$\sum_{v \cdot u} x(v) = 0 \textit{ for each edge } u \textit{ of } G, \tag{3.42}$$

equivalent to $\boldsymbol{R}^\mathsf{T}\boldsymbol{x} = \boldsymbol{o}$,

has a (unique) non-trivial solution.

If, in particular, G is regular of degree r and bipartite, then the solution of (3.42) *equals the eigenvector belonging to $\lambda_n = -r$.*

The simple proof may be left to the reader.

Theorem 3.38 (M. Doob [Doo2], [Doo8]; for regular multigraphs see H. Sachs [Sac8]): *Let G be any connected multigraph. Then $\boldsymbol{y}$ is an eigenvector of $L(G)$ belonging to the eigenvalue -2 if and only if*

$$\sum_{u \cdot v} y(u) = 0 \text{ for each vertex } v \text{ of } G, \text{ which is equivalent to } \boldsymbol{R}\boldsymbol{y} = \boldsymbol{o}.$$

The proof is contained in the proof of Theorem 6.11 (Section 6.3).

Corollary to Theorem 3.38: *If $\boldsymbol{y}$ is an eigenvector of $L(G)$ belonging to the eigenvalue -2, then $\sum_{i=1}^{m} y_i = 0$, i.e.: the eigenspace $\mathbf{Y}(-2)$ is orthogonal to the vector $(1, 1, \ldots, 1)^{\mathsf{T}}$. Hence, in line graphs the eigenvalue -2 never belongs to the main part of the spectrum.*

Note that, in a regular multigraph, each eigenvector which does not belong to the index is orthogonal to $(1, 1, \ldots, 1)^{\mathsf{T}}$.

Recall that a regular spanning submultigraph (of degree s) of a regular multigraph G is called a factor (s-factor) of G.

We shall now establish an interesting relation between the factors of G and the eigenvectors of $L(G)$.

A spanning submultigraph G' of a multigraph G can be represented by a vector $\boldsymbol{c} = (c_1, c_2, \ldots, c_m)^{\mathsf{T}}$, with $c_j = 1$ if the j-th edge of G belongs to G', and $c_j = 0$ otherwise; $\boldsymbol{c}$ is called the *characteristic vector of G'* (with respect to G).

Let G be a connected regular multigraph of degree r. According to Theorem 2.15, the greatest eigenvalue of $L(G)$ is $2r - 2$ and the smallest eigenvalue is not smaller than -2. Let $\{\boldsymbol{y}^1, \boldsymbol{y}^2, \ldots, \boldsymbol{y}^p\}$ be a maxinal set of linearly independent eigenvectors belonging to the eigenvalues of $L(G)$ which are greater than -2 and smaller than $2r - 2$. It is easy to see that $p = n - 2$ if G is bipartite and $p = n - 1$ in the opposite case; if $n > 2$, then $p > 0$. Denote the $m \times p$ matrix with columns $\boldsymbol{y}^1, \boldsymbol{y}^2, \ldots, \boldsymbol{y}^p$ by $\boldsymbol{M}$.

The next theorem provides a means for the investigation of the existence of an s-factor in G, providing $\boldsymbol{M}$ is known.

Theorem 3.39 (H. Sachs [Sac8], [Sac14]): *Let G be a connected regular multigraph of degree r with n vertices and m edges. A vector $\boldsymbol{z}$ with m components is the characteristic vector of an s-factor of G if and only if it satisfies the following conditions:*

1^0 $\frac{1}{2}$ *sn components of $\boldsymbol{z}$ are equal to 1, and the other components are equal to zero;*

2^0 $\boldsymbol{M}^{\mathsf{T}}\boldsymbol{z} = \boldsymbol{o}$.

Proof. 1. Let $\boldsymbol{z}$ be the characteristic vector of an s-factor G_s of G. Then 1^0 holds trivially. In order to deduce condition 2^0, consider the vector $\boldsymbol{y} = r\boldsymbol{z} - s\boldsymbol{y}^0$, where $\boldsymbol{y}^0$ is a vector all of whose components are equal to 1 (note that $\boldsymbol{y}^0$ is the eigenvector of the index $2r - 2$ of $L(G)$). Then $y_j = r - s$ if the j-th edge belongs to G_s, and $y_j = -s$ otherwise. Further,

$$\sum_{u \cdot v} y(u) = s(r - s) + (r - s)(-s) = 0$$

for each vertex v of G (or, briefly, $\boldsymbol{R}\boldsymbol{y} = \boldsymbol{o}$). According to Theorem 3.38 this means that $\boldsymbol{y}$ is an eigenvector belonging to the eigenvalue -2 of $L(G)$, so $\boldsymbol{y}$ is orthogonal to $\boldsymbol{y}^1, \boldsymbol{y}^2, \ldots, \boldsymbol{y}^p$. Since $\boldsymbol{y}^0$ has the same property, $\boldsymbol{z} = \frac{1}{r}(\boldsymbol{y} + s\boldsymbol{y}^0)$ is also orthogonal to each of $\boldsymbol{y}^1, \boldsymbol{y}^2, \ldots, \boldsymbol{y}^p$. Hence, $\boldsymbol{M}^\mathsf{T}\boldsymbol{z} = \boldsymbol{o}$.

2. Let now 1^0 and 2^0 hold for a vector $\boldsymbol{z}$ with m components. Let $\boldsymbol{y} = r\boldsymbol{z} - s\boldsymbol{y}^0$. Then

$$\langle \boldsymbol{y}^0, \boldsymbol{y} \rangle = r \cdot \frac{1}{2} sn - sm = r \cdot \frac{1}{2} sn - s \cdot \frac{1}{2} rn = 0$$

and for each vector $\boldsymbol{y}^i$ $(i = 1, 2, \ldots, p)$

$$\langle \boldsymbol{y}^i, \boldsymbol{y} \rangle = r\langle \boldsymbol{y}^i, \boldsymbol{z} \rangle - s\langle \boldsymbol{y}^i, \boldsymbol{y}^0 \rangle = 0.$$

Hence, $\boldsymbol{y}$ is orthogonal to $\boldsymbol{y}^0, \boldsymbol{y}^1, \ldots, \boldsymbol{y}^p$ and therefore $\boldsymbol{y}$ is an eigenvector belonging to the eigenvalue -2 of $L(G)$. According to Theorem 3.38, $\sum_{u \cdot v} y(u) = 0$ for each vertex v of G. Since the components of $\boldsymbol{y}$ are $r - s$ or $-s$, it follows from the last equation that for exactly s of the edges which are incident with v the equation $\boldsymbol{y}(u) = r - s$ holds and that $y(u) = -s$ for the remaining $r - s$ edges. Those edges u for which $y(u) = r - s$ (these are precisely the edges for which $z(u) = 1$) form an s-factor of G.

This completes the proof of the theorem.

Corollary to Theorem 3.39: *If a vector $\boldsymbol{z}$ with m components, q of which are equal to* 1 *and the other $m - q$ of which are equal to* 0, *satisfies the condition* $\boldsymbol{M}^\mathsf{T}\boldsymbol{z} = \boldsymbol{o}$, *then* $2q \equiv 0 \pmod{n}$ *and $\boldsymbol{z}$ is the characteristic vector of an s-factor with* $s = \frac{2q}{n}$.

The proof of the Corollary is left to the reader.

Let $\mathscr{V}$ be a subset of the set $\mathscr{U}$ of edges of G. Then $\mathscr{V}$ induces a regular factor if and only if

$$\sum_{u \in \mathscr{V}} y^i(u) = 0 \qquad (i = 1, 2, \ldots, p).$$

Let $\mathbf{S}^p$ be the vector space generated by $\boldsymbol{y}^1, \boldsymbol{y}^2, \ldots, \boldsymbol{y}^p$. Clearly, each vector of $\mathbf{S}^p$ is a solution of the following system of homogeneous linear equations

$$\sum_{u \in \mathscr{V}} y(u) = 0,$$

where $\mathscr{V}$ runs through all subsets of $\mathscr{U}$ which induce a regular factor. Therefore the rank of the coefficient matrix of this system is not greater than $m - p$. The rows of this matrix are just the characteristic vectors of the regular factors of G.

Hence, the number of linearly independent characteristic vectors of regular factors is not greater than $m - p$. So, if we call a set of regular factors *independent* (*dependent*) if the corresponding characteristic vectors are linearly independent (dependent), we have

Theorem 3.40 (H. Sachs [Sac8], [Sac14]): *The number of independent regular factors of a connected regular multigraph G of degree r with n vertices and m edges is not greater than*

$$m - p = \begin{cases} \frac{1}{2} rn - n + 2 & \text{if } G \text{ is bipartite}, \\ \frac{1}{2} rn - n + 1 & \text{otherwise.}^{\dagger} \end{cases}$$

Moreover, H. Sachs [Sac8] proved by direct construction that this bound is attained for each n, r with $nr \equiv 0 \pmod 2$ by some multigraph, for n even in both cases, and for n odd — naturally — only by non-bipartite multigraphs.

Several further results concerning eigenvectors of multigraphs can be found in Sections 5.1, 5.2, 6.3.

Remark (H. S.). Many of the results stated above for regular multigraphs can be generalized to arbitrary multigraphs if, instead of the ordinary spectrum (= P-spectrum), the Q-spectrum and the corresponding eigenvectors (= Q-eigenvectors) are utilized. (Recall that $\boldsymbol{x}$ is called a Q-eigenvector belonging to the Q-eigenvalue λ if $\boldsymbol{x} \neq \boldsymbol{o}$ and $\boldsymbol{x}$ and λ satisfy $\boldsymbol{Ax} = \lambda \boldsymbol{Dx}$.)

The following propositions can easily be proved.

Let G be a connected multigraph. Then

1⁰ *all Q-eigenvalues* λ_i *are real and satisfy* $-1 \leqq \lambda_i \leqq 1$;

2⁰ *the maximal Q-eigenvalue* λ_1 (= "*Q-index*") *is a simple eigenvalue equal to* 1;

3⁰ *the Q-eigenvector belonging to* λ_1 *is* $(1, 1, \ldots, 1)^{\mathsf{T}}$;

4⁰ *any two Q-eigenvectors* $\boldsymbol{x}, \boldsymbol{x}'$ *belonging to different Q-eigenvalues are orthogonal in the following sense:* $\boldsymbol{x}^{\mathsf{T}} \boldsymbol{D} \boldsymbol{x}' = 0$;

5⁰ *the following statements are equivalent:*
  (i) *G is bipartite,*
  (ii) *the Q-spectrum of G is symmetric with respect to the zero point of the real axis,*
  (iii) -1 *is a (necessarily simple) Q-eigenvalue;*

6⁰ *if G is bipartite, then the Q-eigenvector* $\boldsymbol{x}$ *belonging to* $\lambda_n = -1$ *satisfies* $x_i + x_k = 0$ *for each pair i, k of adjacent vertices.*

In the sequel, suppose that G is an arbitrary connected multigraph with $n \geqq 1$ vertices and $m \geqq 1$ edges.

† Note that $\frac{1}{2} rn - n + 1 = m - n + 1$ is the cyclomatic number of G.

Now introduce two "modified incidence matrices"

$$\boldsymbol{R}^* = \boldsymbol{D}^{-1}\boldsymbol{R}, \qquad \boldsymbol{S}^* = \frac{1}{2}\,\boldsymbol{R}^\mathsf{T}. \tag{3.43}$$

Both of them are stochastic with respect to their rows, and so are the "modified adjacency matrices"

$$\boldsymbol{A}^* = \boldsymbol{R}^*\boldsymbol{S}^*, \qquad \boldsymbol{B}^* = \boldsymbol{S}^*\boldsymbol{R}^*.^\dagger \tag{3.44}$$

Clearly,

$$\boldsymbol{A}^* = \frac{1}{2}\,\boldsymbol{D}^{-1}\boldsymbol{R}\boldsymbol{R}^\mathsf{T} = \frac{1}{2}\,\boldsymbol{D}^{-1}(\boldsymbol{A} + \boldsymbol{D}) = \frac{1}{2}\,(\boldsymbol{D}^{-1}\boldsymbol{A} + \boldsymbol{I}), \tag{3.45}$$

$$\boldsymbol{B}^* = \frac{1}{2}\,\boldsymbol{R}^\mathsf{T}\boldsymbol{D}^{-1}\boldsymbol{R} = \boldsymbol{B}^{*\mathsf{T}}. \tag{3.46}$$

Define the characteristic polynomials††

$$f_G(t) = |t\boldsymbol{I} - \boldsymbol{A}^*|, \qquad g_G(t) = |t\boldsymbol{I} - \boldsymbol{B}^*|$$

with corresponding spectra

$$[\lambda_1^*, \lambda_2^*, \ldots, \lambda_n^*]_f, \qquad [\mu_1^*, \mu_2^*, \ldots, \mu_m^*]_g.$$

By (3.44)

$$t^m f_G(t) = t^n g_G(t), \tag{3.47}$$

and because of (3.46), all eigenvalues λ_i^*, μ_j^* are real numbers.

By virtue of (3.45), $\boldsymbol{D}^{-1}\boldsymbol{A} = 2\boldsymbol{A}^* - \boldsymbol{I}$, so

$$Q_G(t) = |t\boldsymbol{I} - \boldsymbol{D}^{-1}\boldsymbol{A}| = |t\boldsymbol{I} - 2\boldsymbol{A}^* + \boldsymbol{I}| = 2^n \left|\frac{t+1}{2}\,\boldsymbol{I} - \boldsymbol{A}^*\right| = 2^n f_G\left(\frac{t+1}{2}\right).$$

Consequently, if $[\lambda_1, \lambda_2, \ldots, \lambda_n]_Q$ is the Q-spectrum of G,

$$\lambda_i^* = \frac{\lambda_i + 1}{2}, \qquad \lambda_i = 2\lambda_i^* - 1 \qquad (i = 1, 2, \ldots, n). \tag{3.48}$$

In connection with (3.47) and (3.48), Proposition 1⁰ yields

$$0 \leqq \lambda_i^* \leqq 1 \qquad (i = 1, 2, \ldots, n), \tag{3.49}$$

$$0 \leqq \mu_j^* \leqq 1 \qquad (j = 1, 2, \ldots, m),$$

† Note that the definitions of $\boldsymbol{R}^*, \boldsymbol{S}^*, \boldsymbol{A}^*, \boldsymbol{B}^*$ may be extended to hypergraphs with $\boldsymbol{R}^* = \boldsymbol{D}_v^{-1}\boldsymbol{R}$, $\boldsymbol{S}^* = \boldsymbol{D}_e^{-1}\boldsymbol{R}^\mathsf{T}$, where $\boldsymbol{D}_v$ is the "valency matrix of the set of vertices" and $\boldsymbol{D}_e$ is the "valency matrix of the set of (hyper-)edges". These definitions lay a basis for a theory of a modified "*Q-spectrum of a hypergraph*" and at the same time explain the apparent asymmetry in the definitions of $\boldsymbol{R}^*$, $\boldsymbol{S}^*$ given above (formula (3.43)). (See also Section 1.9, nos. 11, 13.)

†† When dealing with hypergraphs, F. Runge [Rung] has used polynomials similar to $f_G(t)$ and $g_G(t)$; see Section 1.9, nos. 11, 13.

and Proposition 5^0 (iii) may now be restated as follows:

G is bipartite if and only if $\lambda_n^* = 0$.

By (3.47) a simple multiplicity preserving (1, 1)-correspondence between the sets of all non-zero (i.e., positive) eigenvalues λ^* of $\boldsymbol{A}^*$ and eigenvalues μ^* of $\boldsymbol{B}^*$ is given, namely $\lambda^* = \mu^*$.

If λ is a Q-eigenvalue of G, let $\mathbf{X}(\lambda)$ be the corresponding eigenspace; then, clearly, the eigenspace $\mathbf{X}^*(\lambda^*)$ of the eigenvalue $\lambda^* = \dfrac{\lambda + 1}{2}$ of the matrix $\boldsymbol{A}^*$ is identical with $\mathbf{X}(\lambda)$:

$$\mathbf{X}^*(\lambda^*) = \mathbf{X}(\lambda).$$

Denote the eigenspace of the eigenvalue μ^* of the matrix $\boldsymbol{B}^*$ by $\mathbf{Y}^*(\mu^*)$. Then the analogue of Theorem 3.36 holds:

Theorem 3.36′: *Let* $\lambda^* = \mu^* > 0$ *be corresponding eigenvalues of* $\boldsymbol{A}^*, \boldsymbol{B}^*$, *respectively, and let* $\lambda = 2\lambda^* - 1$ *be the corresponding* Q-*eigenvalue* > -1. *Then* $\boldsymbol{S}^*$ *maps the eigenspace* $\mathbf{X}(\lambda) = \mathbf{X}^*(\lambda^*)$ *onto the eigenspace* $\mathbf{Y}^*(\mu^*)$, *and* $\boldsymbol{R}^*$ *maps* $\mathbf{Y}^*(\mu^*)$ *onto* $\mathbf{X}^*(\lambda^*) = \mathbf{X}(\lambda)$.

Theorem 3.38 has the following analogue:

Theorem 3.38′: $\boldsymbol{y}^*$ *is an eigenvector of* B^* *belonging to the eigenvalue* $\mu^* = 0$ *if and only if* $\boldsymbol{R}^*\boldsymbol{y}^* = \boldsymbol{o}$.

Now define a generalized factor of an arbitrary multigraph $G = (\mathscr{X}, \mathscr{U})$ as follows: Let the valencies $d(v)$ of the vertices $v \in \mathscr{X}$ have the greatest common divisor δ, and let $0 < \sigma \leqq \delta$. A spanning submultigraph $G' = (\mathscr{X}, \mathscr{U}')$ of G is called a *σ-factor* if for every vertex $v \in \mathscr{X}$ the valencies $d'(v)$ with respect to G' and $d(v)$ with respect to G have the same ratio $\sigma : \delta$, i.e., if

$$d'(v) = \frac{\sigma}{\delta}\, d(v) \text{ for every vertex } v \in \mathscr{X}.$$

A non-trivial σ-factor can, of course, exist only if $\delta > 1$.

Let $\{\boldsymbol{x}^1, \boldsymbol{x}^2, \ldots, \boldsymbol{x}^p\}$ be a maximal set of linearly independent Q-eigenvectors belonging to the Q-eigenvalues of G which are greater than -1 and smaller than 1. Clearly, $p = n - 2$ if G is bipartite and $p = n - 1$ otherwise. Denote the $n \times p$ matrix with columns $\boldsymbol{x}^i$ by $\boldsymbol{E}$ and put $\boldsymbol{M}^* = 2\boldsymbol{S}^*\boldsymbol{E} = \boldsymbol{R}^\mathsf{T}\boldsymbol{E}$; then, according to Theorem 3.36′, the p columns $\boldsymbol{y}^{*i}$ of $\boldsymbol{M}^*$ constitute a maximal set of linearly independent eigenvectors belonging to the positive eigenvalues μ^* of $\boldsymbol{B}^*$ which are smaller than 1. Now the following theorem which is a generalization of Theorem 3.39 can be proved in a way analogous to the proof of Theorem 3.39.

Theorem 3.39′: *Let G be a connected multigraph with* $m \geqq 1$ *edges. A vector* $\boldsymbol{z}$ *with m components is the characteristic vector of a σ-factor of G if and only if it satisfies the* *following conditions:*

1⁰ $\frac{\sigma}{\delta}$ *m components of* z *are equal to* 1, *and the other components of* z *are equal to* 0;

2⁰ $M^{*T}z = o$.

An analogue to the Corollary to Theorem 3.39 is also valid.

The relations between the (generalized) factors of a (multi-)graph and the eigenvectors of its line graph become particularly evident when extended to hypergraphs (see also footnote on p. 110).

3.6. Miscellaneous results and problems

1. Let ϱ and $\bar{\varrho}$ be the indices of the graphs G and $\bar{G}$. Let G have n vertices. By use of Theorem 3.8 and relation (7.29), the following inequalities are easily obtained:

$$n - 1 \leqq \varrho + \bar{\varrho} \leqq \sqrt{2}(n - 1).$$

The left-hand inequality is actually an equality if and only if G is regular.

(E. Nosal [Nos1]; A. T. Amin, S. L. Hakimi [AmHa])

2. Prove that $\lambda_1 \geqq \sqrt{d_{\max}}$, where $\lambda_1 = \varrho$ is the index and $d_{\max}$ is the maximal valency of a graph.

(E. Nosal [Nos1]; L. Lovász, J. Pelikán [LoPe])

3. Let G be a regular graph of degree r with n vertices. Show that for the number $t(G)$ of spanning trees of G the following formula holds:

$$t(G) = \frac{(-1)^n}{n^2} P_{\bar{G}}(-r - 1).$$

4. If G is a connected graph, neither a tree nor a circuit, then $\varrho > \tau^{\frac{1}{2}} + \tau^{-\frac{1}{2}}$, where ϱ is the index of G and $\tau = \frac{1}{2}(\sqrt{5} + 1)$.

(A. J. Hoffman [Hof13])

5. Show that the star has the largest index among all trees with n vertices.

6. Each closed walk in a graph G can be represented as a sequence of vertices through which it passes, for example, $x_1, x_2, \ldots, x_n, x_1$. The walks $x_1, x_2, \ldots, x_{k-1}, x_k, x_1$ and $x_2, x_3, \ldots, x_k, x_1, x_2$ are different because one starts from x_1 and the other from x_2 but are considered as cyclically equivalent: Two closed walks are called *cyclically equivalent* if one is obtained from the other by rotating an initial segment to the end of the walk. Let $C_k(G)$ be the number of cyclic equivalence classes of closed walks of length k in G. Then

$$C_k(G) = \frac{1}{k} \sum_{d|k} \Phi\left(\frac{k}{d}\right) \sum_{i=1}^{n} \lambda_i^d,$$

where $\Phi(k)$ denotes the Euler phi-function and $\lambda_1, \ldots, \lambda_n$ are the eigenvalues of G. A similar formula was obtained also for the number of dihedral equivalence classes of closed walks in a graph.

(F. Harary, A. J. Schwenk [HaS1])

7. Let A be the adjacency matrix of a graph G with n vertices, let $\mathscr{I} \subset \{1, 2, \ldots, n\}$ and let $A_{\mathscr{I}}$ be defined as in Section 1.5 (p. 37). Then the number of Hamiltonian circuits of G is, given by

$$\frac{1}{2n} \sum_{s=0}^{n} (-1)^s \sum_{|\mathscr{I}|=s} \operatorname{tr} A^n_{\mathscr{I}}.$$

(L. M. Lihtenbaum [Лих4], [Лих5])

8. Let p_-, p_0, p_+ denote the number of eigenvalues of a graph G, which are smaller than, equal to, or greater than zero, respectively. Then the chromatic number $\chi(G)$ of G satisfies

$$\chi(G) \geqq \frac{n}{p_0 + \min(p_+, p_-)},$$

where n is the number of vertices of G. This inequality is sharp; equality holds, for example for complete graphs.

(D. M. Cvetković [Cve 11])

9. Let G be a regular graph of degree $r = n - 3$ with n $(n \geqq 3)$ vertices. Put

$$(-1)^n (\lambda + n - 2)^{-1} (\lambda - 2) P_G(-\lambda - 1) = \lambda^n + a_1\lambda^{n-1} + \cdots + a_n$$
$$= (\lambda - 2)^{m_2} (\lambda + 2)^{m_{-2}} \varphi(\lambda),$$

where $\varphi(2) \neq 0$ and $\varphi(-2) \neq 0$. Then the chromatic number $\chi(G)$ of G is given by

$$\chi(G) = \frac{1}{2} (n + m_2 - m_{-2} + a_3).$$

(H.-J. Finck [Finc])

10. Show that the chromatic number χ of a graph G is determined by the spectrum of G if the index ϱ of G is smaller than 3. If G is connected, the same statement holds also for $\varrho = 3$.

(D. M. Cvetković [Cve 9])

11. For a given k let a, b, c, d, e denote the numbers of eigenvalues λ in the spectrum of a graph G, which, respectively, satisfy the following relations: $\lambda < -k + 1$, $\lambda = -k + 1$, $-k + 1 < \lambda < 1$, $\lambda = 1$, $\lambda > 1$. Further, let s be the smallest of the natural numbers k $(k > 1)$ for which the inequality

$$\min(b + c + k(d + e), k(a + b) + (k - 1)(c + d)) \geqq n$$

holds. Then $\chi(G) \geqq s$.

(D. M. Cvetković [Cve 12])

12. Let $\chi(\overline{G})$ be the chromatic number of the complement $\overline{G}$ of a graph G. If G is not a complete graph, then

$$\chi(\overline{G}) \geqq \frac{n + \lambda_2 - \lambda_1}{1 + \lambda_2},$$

where n is the number of vertices of G and λ_1, λ_2 are the first two greatest eigenvalues of G.

(A. J. Hoffman [Hof 16])

13. Let $\sigma(G)$ be the smallest number of subsets into which the vertex set of G can be partitioned such that the subgraph induced by any one of the subsets is either a complete graph or a totally disconnected graph. Let k be a positive integer and let G have eigenvalues $\lambda_1, \ldots, \lambda_n$. Then there exist functions $\mathscr{A}_k$ and $\mathscr{B}_k$ such that

$$\sigma(G) \leqq \mathscr{A}_k(\lambda_2 - \lambda_{n-k+1}), \qquad \sigma(G) \leqq \mathscr{B}_k(\lambda_k - \lambda_n).$$

(A. J. Hoffman, L. Howes [HoHo])

14. If $\mathscr{V}(G)$ is the vertex set of a graph G and if $\mathscr{S} \subset \mathscr{V}(G)$, then $G^{\mathscr{S}}$ denotes the subgraph of G induced by the set of vertices of G each of which is adjacent to all vertices in $\mathscr{S}$. Let

$\lambda^*(G)$ be the smallest eigenvalue of G and let $k(H)$ denote the cliquomatic number† of a graph H. Then

1° there exists a function f such that if $x \in \mathscr{V}(G)$,

$$k(G^{\{x\}}) \leqq f(\lambda^*(G));$$

2° there exists a function g such that if $x \in \mathscr{V}(G)$,

$$\left|\{i \mid i \text{ is not adjacent to } x, |\mathscr{V}(G^{\{x,i\}})| > g(\lambda^*(G))\}\right| < g(\lambda^*(G)).$$

(A. J. Hoffman [Hof15])

15. Prove that the Seidel $(-1, 1, 0)$-spectrum (see Section 1.2) of a self-complementary graph is symmetric with respect to the zero point.

16. Let $\boldsymbol{R}$ be the vertex-edge incidence matrix of a connected multigraph G having n vertices. Then

$$\operatorname{rk} \boldsymbol{R} = \begin{cases} n-1 & \text{if } G \text{ is bipartite,} \\ n & \text{otherwise.} \end{cases}$$

(H. Sachs [Sac8]; C. van Nuffelen [Nuf1])

17. Let the edges $u_1, u_2, \ldots u_{2k}$ form a closed walk (of even length) in a regular multigraph G and let $\boldsymbol{y}$ be an eigenvector of $L(G)$ not belonging to the eigenvalue -2. Then

$$\sum_{i=0}^{2k} (-1)^i y(u_i) = 0.$$

(H. Sachs [Sac8])

18. Let G be a graph with eigenvalues $\lambda_1, \ldots, \lambda_n$. Let $\varepsilon(G)$, $\vartheta(G)$, and $\delta(G)$ be the quantities defined as in Theorem 3.21. Then

$$\varepsilon(G) = 1 \quad \text{if and only if} \quad \lambda_2 \leqq 0,$$

$$\vartheta(G) = 1 \quad \text{if and only if} \quad \lambda_{n-1} \geqq 0,$$

$$\delta(G) = 1 \quad \text{if and only if} \quad \lambda_2 \leqq 0 \quad \text{and} \quad \lambda_n = -1.$$

(A. J. Hoffman [Hof11])

19. Let G be a graph and let $\varepsilon_m(G)$ be the smallest integer k such that (3.22) holds and each G_i is a complete s-partite graph with $s \leqq m$. Then

$$\varepsilon_m(G) \geqq \frac{p_-}{m-1},$$

where p_- is the number of negative eigenvalues of G.

(D. T. Malbaški, private communication)

20. A k-partition of a graph is a division of its vertices into k disjoint subsets containing $m_1, m_2, \ldots, m_k$ vertices, respectively, where $m_1 \geqq m_2 \geqq \cdots \geqq m_k$.

Let G be a graph with adjacency matrix $\boldsymbol{A}$, let $\boldsymbol{U}$ be any diagonal matrix such that the sum of all the elements of $\boldsymbol{A} + \boldsymbol{U}$ is zero, and let $\mu_1, \mu_2, \ldots, \mu_k$ ($\mu_1 \geqq \mu_2 \geqq \cdots \geqq \mu_k$) be the largest k eigenvalues of $\boldsymbol{A} + \boldsymbol{U}$. Then, if any k-partition σ of G is given, the number E_σ of edges of G whose two vertices belong to different subsets of σ satisfies

$$E_\sigma \geqq \frac{1}{2} \sum_{\varkappa=1}^{k} (-m_\varkappa \mu_\varkappa).$$

The right-hand sum is a concave function of $\boldsymbol{U}$.

(W. E. Donath, A. J. Hoffman [DoHo], cf. also [Fie3])

---

† The cliquomatic number of a graph G is the chromatic number of the complement $\overline{G}$ of G.

21. The eigenvalues and eigenvectors of $\boldsymbol{C} = \boldsymbol{D} - \boldsymbol{A}$ ($\boldsymbol{D}$ is the valency matrix of a graph) were used by K. M. Hall [Ha, K] in a problem of minimizing the total length of the edges of a graph which is to be imbedded into a plane.

22. Define a relationship $\sim$ on the vertex set of a graph G thus: $x \sim y$ if for every $z \neq x, y$ the vertex z is adjacent to both or to none of vertices x, y. Let $e(G)$ denote the number of equivalence classes so defined. $e(G)$ is not determined by the spectrum of G (see Fig. 6.1). But A. J. Hoffman [Hof17] has proved that $e(G)$ is bounded from above and below by some functions of the number of eigenvalues of G not contained in the interval $\left(\frac{1}{2}(\sqrt{5} - 1), 1\right)$.

23. Let G be a regular graph of degree r with n vertices. Let G_1 be an induced subgraph of G having n_1 vertices and average vertex degree r_1. Then

$$\frac{n_1(r - \lambda_n)}{n} + \lambda_n \leqq r_1 \leqq \frac{n_1(r - \lambda_2)}{n} + \lambda_2.$$

This inequality was derived in [BuCS] by the use of Theorem 0.11. Specifying $r_1 = 0$ and $r_1 = n_1 - 1$ we get the following inequalities for the cardinalities $\alpha(G)$ and $K(G)$ of an internal stable set and of a complete subgraph of a regular graph G, respectively:

$$\alpha(G) \leqq \frac{-n\lambda_n}{r - \lambda_n}, \qquad K(G) \leqq \frac{(\lambda_2 + 1)\, n}{n - r + \lambda_2},$$

where λ_2 and λ_n are the second largest and the least eigenvalue of G. The first bound was found by A. J. Hoffman (unpublished). Together with $\alpha(G)\,\chi(G) \geqq n$ it gives the bound from Theorem 3.18 in the case of regular graphs. Similarly, the second inequality gives the bound for the cliquomatic number (see Section 3.6, no. 12). Specifying r_1 in some other ways we can get bounds for some more characteristics of a graph. The inequality for $\alpha(G)$ in the case of strongly regular graphs was noted in [Del1]. The inequality can be extended to non-regular graphs (W. Haemers [Haem]).

24. Further bounds for $K(G)$, defined in Theorem 3.15, can be obtained by the same technique using the Seidel adjacency matrix instead of the (0, 1)-adjacency matrix. For example, J. M. Goethals and J. J. Seidel [GoS4] found the inequality $K(G) \leqq \min(1 - \varrho_2, \mu_1, \mu_2)$, where G is a graph whose Seidel adjacency matrix has only two distinct eigenvalues ϱ_1, ϱ_2 ($\varrho_1 > \varrho_2$) with the multiplicities μ_1, μ_2, respectively. Generalize this result for arbitrary graphs and, in the case of regular graphs, express it in terms of the eigenvalues of the (0, 1)-adjacency matrix.

25. If D_i ($i = 3, 4, 5$) are the numbers of circuits of length i in a regular graph of degree r, and if a_j ($j = 0, 1, \ldots, n$) are the coefficients of the corresponding characteristic polynomial, then

$$D_3 = -\frac{1}{2}\, a_3,$$

$$D_4 = \frac{1}{4}\,(a_2^2 + 2ra_2 - a_2 - 2a_4),$$

$$D_5 = \frac{1}{2}\,(a_3a_2 + 3ra_3 - 3a_3 - a_5).$$

26. If $\lambda_1, \ldots, \lambda_n$ are the eigenvalues of a regular graph G, then the number D_4 of circuits of length 4 in G is given by

$$D_4 = \frac{1}{8}\left(\sum_{i=1}^{n} \lambda_i^4 - n\lambda_1(2\lambda_1 - 1)\right).$$

4. The Divisor of a Graph

In this chapter various concepts of a divisor of a multi-(di-)graph are defined. It is shown in which way the divisor, on the one hand, is connected with structural, in particular (generalized) symmetry properties and, on the other hand, is connected with spectral properties of a graph. The divisor is utilized for factoring the characteristic polynomial; a special "geometric" factorization procedure is developed, which in many cases proves very efficient.

4.1. The divisor concept

Let us start with an intuitive definition (see [FiSa], [PeS 1], [ReSc], [Sac 7], [Sac 10], [Sac 16]).

Let G be a connected multi-digraph and $\boldsymbol{D} = (d_{ij})$ a square matrix of order m where all numbers d_{ij} are non-negative integers. We shall say that G has a D-feasible coloration if it is possible to colour each of its vertices with one of the colours $1, 2, \ldots, m$ in such a way that from each vertex of colour i and for each colour j there issue exactly d_{ij} arcs which terminate in vertices of colour j.

The matrix $\boldsymbol{D}$ determines a multi-digraph D which has $\boldsymbol{D}$ as its adjacency matrix; trivially, D has a D-feasible coloration: just take the number of a vertex as its colour. In order to be sure that in any D-feasible coloration all of the m colours are used we shall in addition — without loss of essential generality — assume that D is strongly connected.

It may be helpful to imagine that the coloured multi-digraph G is a maze in which we are lost and in which we can see only the colours of the vertex we are in and of its front neighbours: we will never find out whether we are in G or in D.

Any D-feasible coloration of G defines a mapping of (the vertex set of) G onto (the vertex set of) D which preserves front neighbourhood; especially, a vertex of G and its image in D have the same front valency.

Definition 4.1. If G has a D-feasible coloration we shall say that D is a *front divisor of* G and write $D \mid_f G$. (In fact, the term "factor" would be more appropriate but is already used for a different concept.) D is called a *rear divisor of* G ($D \mid_r G$) if $D^{\mathsf{T}} \mid_f G^{\mathsf{T}}$ (the "*transpose*" or "converse digraph" G^{T} is obtained from G by reversing the orientations of all of its arcs); we say that D is a *divisor* of G and write $D \mid G$ if there is a coloration of the vertices of G which is simultaneously D-feasible for G and D^{T}-feasible for G^{T}. D is called a *proper* (front, rear) *divisor of* G if D is a (front, rear) divisor of G and has fewer vertices than G.

Some obvious consequences:

Proposition 4.1: *$D \mid G$ implies $D \mid_f G$ and $D \mid_r G$.*

Proposition 4.2: *For undirected multigraphs the statements $D \mid_f G$, $D \mid_r G$, and $D \mid G$ are equivalent.*

Proposition 4.3: *If $D \mid G$ and if G has a symmetric adjacency matrix, then so has D.*

Proposition 4.4: *Each of the relations $\mid_f$, $\mid_r$, and $\mid$ is a semi-order relation; for example, for $\mid$ we have*
(a) $G \mid G$,
(b) *$F \mid G$ and $G \mid H$ imply $F \mid H$,*
(c) *$G \mid H$ and $H \mid G$ imply that H is isomorphic with G.*

Examples. Consider the following *graphs of regular solids*: (a) the cube graph C, (b) the octahedron graph O, (c) the icosahedron graph I, (d) the dodecahedron graph D, each drawn regularly on a sphere. A projection from the center of the sphere identifies antipodal points and thus yields images embedded in the projective plane which are, respectively, isomorphic with (a) the tetrahedron graph K_4, (b) the multigraph $K_3^{(2)}$ on three vertices any two of which are connected by exactly two edges, (c) K_6, (d) the Petersen graph P. This means:

$$K_4 \mid C, \quad K_3^{(2)} \mid O, \quad K_6 \mid I, \quad P \mid D.$$

But there are, of course, other possibilities of obtaining (front-)divisors; for example, $K_2^{(3)}$ consisting of two vertices connected by exactly three edges is a divisor of C (in fact, in the set of all connected multigraphs G, the property of having $K_2^{(3)}$ as a divisor is equivalent with G being a bipartite cubic graph). A simple example for a multi-digraph which is a front divisor of a graph is the following: G is a path of length 2, D consists of vertices x and y together with two arcs from x to y and one arc from y to x.

4.2. Divisor and cover

If D and G are undirected multigraphs and if $D \mid G$, then, in a combinatorial sense, D is the homomorphic and locally homeomorphic image of G under any mapping defined by some D-feasible coloration of G. If, in particular, D and G are schlicht graphs, this is obviously also true in the topological sense when D and G are considered as linear complexes; the statement remains true even when multiple edges and loopes are allowed but this case needs a proof making use of factorization theorems (see [Sac 16]). This fact can be given any of the following two formulations:

Theorem 4.1 (H. Sachs [Sac 16]): *Let D and G be undirected multigraphs with $D \mid G$.*

(a) *Then G, considered as a linear complex, is an unbranched cover of D.*

(b) *Suppose also that the edges of D have pairwise distinct colours $c_1, c_2, \ldots, c_k$, say. Then there is a coloration of the edges of G with colours $c_1, c_2, \ldots, c_k$ such that for $\varkappa = 1, 2, \ldots, k$ each vertex is incident with exactly as many edges of colour $c_\varkappa$ as is its image vertex in D.*

Let G be a multi-digraph and let $n(G)$ and $k(G)$ denote the numbers of its vertices and arcs, respectively. The proof of the following theorem is left to the reader.

Theorem 4.2 (H. Sachs [Sac 7]): *Let D and G be arbitrary multi-digraphs satisfying $D \mid G$. Then $n(D) \mid n(G)$, $k(D) \mid k(G)$, and each vertex of D is the image of exactly $n(G)/n(D)$ vertices of G.*

For the construction of common multiples and theorems on the least common multiple and the greatest common divisor of multigraphs see [Sac 16]. For more information about divisors and covers consult, for example, [Big 5] (Chapter 19), [Far 2], [Gard], [Wal 4], [Wal 5] (there is a list of more references in [Wal 5]).

4.3. A generalization of the divisor concept

It is sometimes convenient to have a more general divisor concept (corresponding to some "branched cover" concept); we shall formulate the definition of a generalized divisor for multigraphs only; the concepts of front and rear divisors for multi-digraphs can be generalized in an analogous way.

***Definition* 4.2.** Let D and G be connected multigraphs, D having vertices $1, 2, \ldots, m$ and adjacency matrix (d_{ij}). Consider the set of all possible colorations C of the vertices of G with colours $1, 2, \ldots, m$. If x is any vertex of G, define the μ-valency of x under coloration C, denoted by $v_{C:\mu}(x)$, as the total number of edges (loops being counted twice) which connect x with vertices of colour μ. We shall say that D is a *generalized divisor of* G and write $D \mathrel{\vdots} G$ if there is a coloration C_0 with the following property: for each colour i, for each vertex x of G having colour i, and for each pair of colours j, k $(1 \leqq i, j, k \leqq m)$

(i) $v_{C_0:j}(x) = 0 \qquad \text{if } d_{ij} = 0,$

(ii) $v_{C_0:j}(x) : v_{C_0:k}(x) = d_{ij} : d_{ik} \qquad \text{if } d_{ij}^2 + d_{ik}^2 > 0$

(i.e., for each vertex of colour i, the ratio (j-valency) : (k-valency) $(1 \leqq j, k \leqq m)$ is in G the same as in D).

4.4. Symmetry properties and divisors of graphs

A permutation $\boldsymbol{P}$ acting on the set of vertices of a multi-digraph G is called an *automorphism of* G if it is adjacency preserving. If $\boldsymbol{P}$ is represented by a permutation matrix (also denoted by $\boldsymbol{P}$), then $\boldsymbol{P}$ is an automorphism if and only if

$$\boldsymbol{P}^{-1}\boldsymbol{A}\boldsymbol{P} = \boldsymbol{A}$$

where $\boldsymbol{A}$ is the adjacency matrix of G. The automorphisms of G form a group $\Gamma = \Gamma(G)$ called the *automorphism group of* G in terms of which the symmetries (with respect to permutations of the vertices) of G are expressed.

The connection between the automorphism group and the divisors of G is given by the following theorem. (For a discussion in terms of representation theory, see Section 5.3.)

Theorem 4.3 (see [PeS1]): *Let G be a strongly connected multi-digraph which has a non-trivial automorphism group Γ with non-trivial subgroup Σ (possibly $\Sigma = \Gamma$), and let $\Delta = \{\Omega_1, \Omega_2, \ldots, \Omega_s\}$ be the system of orbits into which the vertex set of G is partitioned by Σ. Clearly, the number of arcs issuing from any vertex of Ω_i and terminating in vertices of Ω_j depends only on i and j; denote this number by d_{ij} $(i, j = 1, 2, \ldots, s)$. Then the multi-digraph D_Σ with vertices $1, 2, \ldots, s$ and adjacency matrix (d_{ij}) is a proper front divisor of G:*

$$D_\Sigma \mid_f G.$$

The proof is immediate: just colour all vertices of Ω_σ with colour σ $(\sigma = 1, 2, \ldots, s)$.

If $\boldsymbol{P}$ is any non-trivial automorphism of G, we may take for Σ the cyclic group generated by $\boldsymbol{P}$: the orbits Ω_σ are then precisely the cycles of the permutation $\boldsymbol{P}$.

Now suppose that Σ' is a subgroup of Σ: then, under the assumption of Theorem 4.3, the orbits $\Omega'_{\sigma'}$ generated by Σ' are subsets of the orbits $\Omega_1, \Omega_2, \ldots, \Omega_s$. Consider $D_{\Sigma'}$ with vertices $\sigma' = 1', 2', \ldots, s'$ and colour vertex σ' with colour σ whenever $\Omega'_{\sigma'}$ is a subset of Ω_σ. This coloration of $D_{\Sigma'}$ is D_Σ-feasible since for any fixed σ and τ $(\sigma, \tau = 1, 2, \ldots, s)$ there are in $D_{\Sigma'}$ for every vertex x of colour σ exactly as many arcs issuing from x and terminating in vertices $y', y'', \ldots$ of colour τ as there are arcs in D_Σ issuing from the image vertex of x and terminating in the common image vertex of $y', y'', \ldots$

Thus we have proved

Theorem 4.4 (M. Petersdorf, H. Sachs [PeS1]): *Let, under the assumptions of Theorem 4.3, Σ' be a subgroup of Σ. Then*

$$D_\Sigma \mid_f D_{\Sigma'} \mid_f G.$$

With Theorem 4.4 we have established a sort of Galois correspondence: *If* $\Gamma(G) = \Gamma \supset \Sigma \supset \Sigma' \supset \cdots \supset I$ (I is the identity group) *then*

$$D_\Gamma \mid_f D_\Sigma \mid_f D_{\Sigma'} \mid_f \cdots \mid_f D_I \doteq G.$$

Example: Let G be the graph of Fig. 4.1. Here

$$\Gamma(G) = \{(1), (17)(26)(35), (13)(57)(48), (15)(26)(37)(48)\}$$

with orbits

$$\Omega_1 = \{1, 3, 5, 7\}, \qquad \Omega_2 = \{2, 6\}, \qquad \Omega_3 = \{4, 8\}.$$

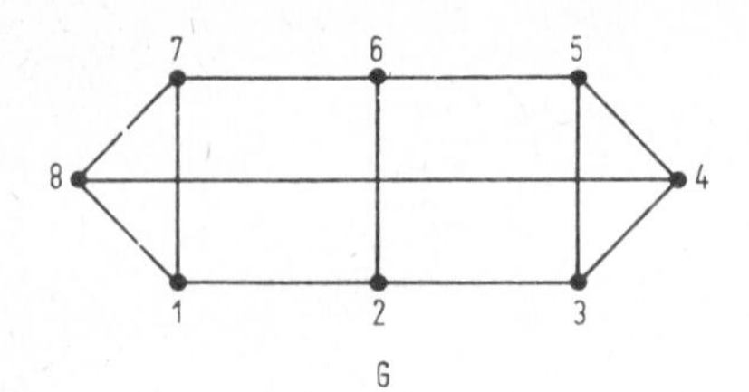

Fig. 4.1

The subgroups Σ_1, Σ_2, Σ_3 generated by the permutations $\boldsymbol{P}_1 = (17)(26)(35)$, $\boldsymbol{P}_2 = (13)(57)(48)$, $\boldsymbol{P}_3 = (15)(26)(37)(48)$, respectively, have systems of orbits

$$\Delta_1 = \big\{\{4\}, \{8\}, \{1, 7\}, \{2, 6\}, \{3, 5\}\big\},$$

$$\Delta_2 = \big\{\{2\}, \{6\}, \{1, 3\}, \{5, 7\}, \{4, 8\}\big\},$$

$$\Delta_3 = \big\{\{1, 5\}, \{2, 6\}, \{3, 7\}, \{4, 8\}\big\}$$

which give rise to front divisors D_1, D_2, D_3, respectively (see Fig. 4.2). Thus we have

$$D_\Gamma \mid_f D_i \mid_f G \qquad (i = 1, 2, 3).$$

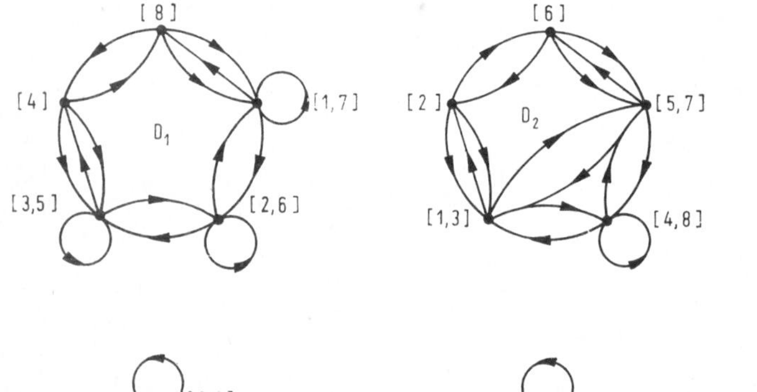

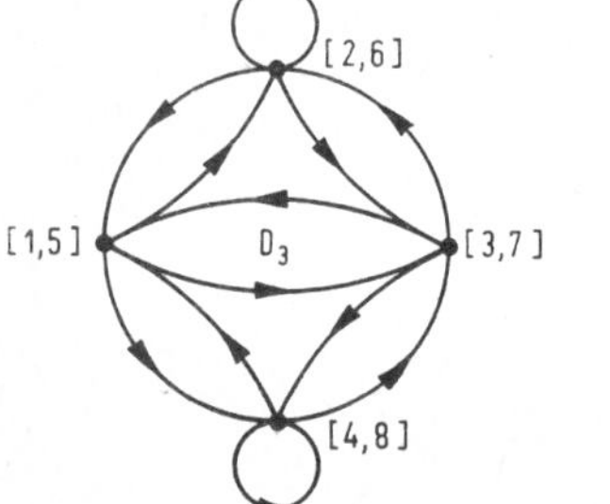

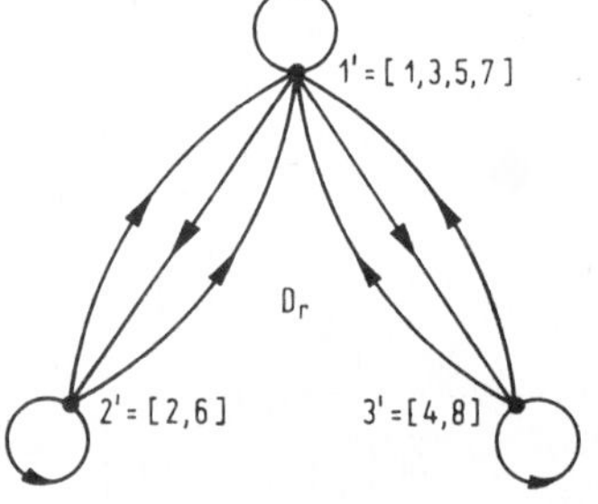

Fig. 4.2

Note that, starting from D_Γ, we find two more divisors, namely D', D'' (see Fig. 4.3) which cannot be directly obtained from $\Gamma(G)$ and its subgroups; so we have, eventually,

$$D'' \mid_f D' \mid_f D_\Gamma \mid_f D_i \mid_f G \qquad (i = 1, 2, 3).$$

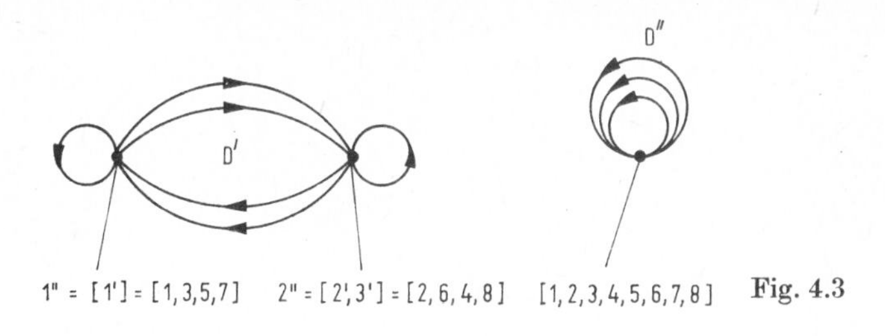

Fig. 4.3

Often there are front divisors (and even divisors) which do **not** correspond to the orbits of some subgroup of the automorphism group: thus the existence of a non-trivial (front or rear) divisor may be interpreted as a sort of (generalized) symmetry which comprises permutational symmetry but is, in general, weaker than it (see also [Schw3], pp. 159–160).

4.5. The fundamental lemma connecting divisor and spectrum

We shall first consider the case of connected multigraphs F, G. Let G have vertices $V_1, V_2, \ldots, V_n$. Suppose $F \mid G$ so that G has an F-feasible coloration. Let $\boldsymbol{x}^* = (x_1^*, x_2^*, \ldots, x_m^*)^{\mathsf{T}}$ be an eigenvector and λ^* a corresponding eigenvalue of the problem

$$\lambda x_i = \sum_{k \cdot i} x_k \qquad (i = 1, 2, \ldots, m), \tag{4.1}$$

or briefly $\lambda \boldsymbol{x} = \boldsymbol{F}\boldsymbol{x}$, defined on F, so that

$$\lambda^* x_i^* = \sum_{k \cdot i} x_k^* \qquad (i = 1, 2, \ldots, m). \tag{4.2}$$

Now "transplant" each of the values x_i^* from vertex i in F to all vertices of colour i in G: We obtain a vector $\boldsymbol{y}^* = (y_1^*, y_2^*, \ldots, y_n^*)^{\mathsf{T}}$ with $y_j^* = x_i^*$ for all j such that vertex V_j of G has colour i. By virtue of (4.2) this implies

$$\lambda^* y_p^* = \sum_{q \cdot p} y_q^* \qquad (p = 1, 2, \ldots, n)$$

where $q \cdot p$ signifies adjacency of vertices V_q, V_p in G. This means that $\boldsymbol{y}^*$ is an eigenvector of the problem

$$\lambda y_p = \sum_{q \cdot p} y_q \qquad (p = 1, 2, \ldots, n),$$

or briefly $\lambda \boldsymbol{y} = \boldsymbol{A}\boldsymbol{y}$, defined on G, with corresponding eigenvalue λ^*; thus the eigenvalue λ^* of F is also an eigenvalue of G.

If λ^* has multiplicity μ as an eigenvalue of F (i.e., of the symmetric matrix $\boldsymbol{F}$), then there is a system $\mathbb{S}$ of μ linearly independent eigenvectors $\boldsymbol{x}^*$ all belonging to λ^*; clearly, linear independence is preserved under transplantation from F to G, so we obtain from $\mathbb{S}$ a system of μ linearly independent eigenvectors $\boldsymbol{y}^*$ of G all belonging to λ^*. This implies that λ^*, as an eigenvalue of G, has multiplicity at least μ.

Thus we have proved (as a special case of Theorem 4.7)

Theorem 4.5†: *Let F and G be connected multigraphs with $F \mid G$. Then*

$$P_F(\lambda) \mid P_G(\lambda).$$

Remark. By similar arguments it can easily be proved that $F \mid G$ also implies $Q_F(\lambda) \mid Q_G(\lambda)$, $R_F(\lambda) \mid R_G(\lambda)$, and $C_F(\lambda) \mid C_G(\lambda)$. But this all follows also from the more general Theorem 4.7.

We now turn our attention to the general case.

Let us say that an eigenvalue problem $\mathbb{E}$ connected with multi-digraphs is of *local character* if it consists of a system of equations, one equation for each of the vertices, where the i-th equation is completely determined by the front neighbourhood of the i-th vertex (i.e., contains only those indeterminates x_j which belong to i or its front neighbours: $j = i$ or $j \cdot i$); so the eigenvalue problems

$$\lambda x_i = \sum_{k \cdot i} x_k \qquad (i \in \mathscr{X}), \tag{1.1}$$

$$\lambda x_i = \frac{1}{d_i} \sum_{k \cdot i} x_k \qquad (i \in \mathscr{X}), \tag{1.3}$$

$$\lambda x_i = \sum_{k \cdot i} (x_i + x_k) \qquad (i \in \mathscr{X}), \tag{1.7}$$

and

$$\lambda x_i = \sum_{k \cdot i} (x_i - x_k) \qquad (i \in \mathscr{X})$$

leading to the polynomials $P_G(\lambda)$, $Q_G(\lambda)$, $R_G(\lambda)$, and $C_G(\lambda)$, respectively, and the quadratic problem (1.18) are all of local character (and so is also the problem

$$\lambda^{(i)} x_i = \sum_{k \cdot i} x_k \qquad (i \in \mathscr{X})$$

with corresponding polynomial

$$P_G^*(\lambda_1, \lambda_2, \ldots, \lambda_s) = |\Lambda - \mathbf{A}|, \tag{1.27}$$

see Section 1.3) whereas the eigenvalue problem

$$\lambda x_i = \sum_{k \in \mathscr{X}} s_{ik} x_k \qquad (i \in \mathscr{X}) \tag{1.12}$$

which leads to the Seidel polynomial $S_G(\lambda)$, is not.

---

† [Sac7]; very probable, this simple reasoning has often been used implicitly in the theory of matrices.

In what follows, the eigenvalues and characteristic polynomials of an eigenvalue problem $\mathbb{E}$ will be called $\mathbb{E}$-eigenvalues and $\mathbb{E}$-polynomials, respectively.

For the spectra corresponding to an eigenvalue problem of local character we have the following general theorem:

Theorem 4.6 (M. Petersdorf, H. Sachs [PeS1]): *Let F and G be arbitrary multi-digraphs satisfying $F \mid_f G$; then, for any eigenvalue problem $\mathbb{E}$ of local character, each $\mathbb{E}$-eigenvalue λ^* of F is also an $\mathbb{E}$-eigenvalue of G.*

For the proof simply note that because of the local character of $\mathbb{E}$ the above transplantation argument applies (but in this case does not yield any statement as to the multiplicity of λ^* as an $\mathbb{E}$-eigenvalue of G).

In particular, Theorem 4.6 holds for the P-, Q-, R-, and C-eigenvalues. But in these cases it can be shown that the hypothesis $F \mid_f G$ in fact implies $E_F(\lambda) \mid E_G(\lambda)$ where E stands for P, Q, R, or C. We shall prove even a bit more, namely

Theorem 4.7 (Fundamental Lemma) (M. Petersdorf, H. Sachs [PeS1]†): *Let F and G be arbitrary multi-digraphs satisfying $F \mid_f G$; then*

$$P_F^*(\lambda_1, \lambda_2, \ldots, \lambda_s) \mid P_G^*(\lambda_1, \lambda_2, \ldots, \lambda_s). \tag{4.3}$$

(Recall from Section 1.3:

$$P_G^*(\lambda_1, \lambda_2, \ldots, \lambda_s) = |\Lambda - \boldsymbol{A}|.) \tag{1.27}$$

Proof. We shall follow the elementary proof given in [PeS1]. Let F have the adjacency matrix $\boldsymbol{F} = (f_{ij})$ $(i, j = 1, 2, \ldots, m)$ and assume that G has n vertices. $F \mid_f G$ implies that G has an F-feasible coloration with colours $1, 2, \ldots, m$. Denote the class of vertices of G which have colour i by $\mathscr{C}_i$ $(i = 1, 2, \ldots, m)$. From each class $\mathscr{C}_i$ select a vertex in an arbitrary way, call it V_i, and number the remaining vertices V of G from $m + 1$ through n in such a way that $i < k \leqq m$ and $V_p \in \mathscr{C}_i \setminus \{V_i\}$, $V_q \in \mathscr{C}_k \setminus \{V_k\}$ imply $p < q$. Denote the adjacency matrix of G corresponding to this labelling by $\boldsymbol{G} = (g_{pq})$ $(p, q = 1, 2, \ldots, n)$. The matrix $\Lambda = \Lambda(G)$ defined by (1.26) is of the form

$$\Lambda(G) = (\lambda^{(p)}\delta_{pq}) = \begin{pmatrix} \Lambda(F) & \boldsymbol{O} \\ \boldsymbol{O} & \Lambda' \end{pmatrix}$$

where $\lambda^{(j)} = \lambda^{(i)}$ $(i = 1, 2, \ldots, m)$ whenever vertex V_j has colour i.

Consider the determinant

$$|\boldsymbol{G} - \Lambda(G)| = (-1)^n P_G^*(\lambda_1, \lambda_2, \ldots, \lambda_s);$$

† In [PeS1] (Satz 9) only the case of the ordinary polynomial $P_G(\lambda)$ (i.e., $\lambda_1 = \lambda_2 = \ldots = \lambda_s = \lambda$) is treated. — For a general theorem on partitioned matrices see also [Gard] (Theorem 2) and [Hayn] (see Theorem 0.12, p. 20 of this book).

for $k = 1, 2, \ldots, m$, add to the k-th column all other columns belonging to vertices of colour k. Then

$$|\boldsymbol{G} - \Lambda(G)| =$$

$$\begin{vmatrix}
f_{11} - \lambda^{(1)} & f_{12} & \cdots & f_{1m} & g_{1,m+1} & \cdots & g_{1n} \\
f_{21} & f_{22} - \lambda^{(2)} & \cdots & f_{2m} & g_{2,m+1} & \cdots & g_{2n} \\
\vdots & \vdots & & \vdots & \vdots & & \vdots \\
f_{m1} & f_{m2} & \cdots & f_{mm} - \lambda^{(m)} & g_{m,m+1} & \cdots & g_{mn} \\
f_{11} - \lambda^{(1)} & f_{12} & \cdots & f_{1m} & g_{m+1,m+1} - \lambda^{(1)} & \cdots & g_{m+1,n} \\
\vdots & \vdots & & \vdots & \vdots & & \vdots \\
f_{11} - \lambda^{(1)} & f_{12} & \cdots & f_{1m} & \cdot & & \cdot \\
f_{21} & f_{22} - \lambda^{(2)} & \cdots & f_{2m} & \cdot & & \cdot \\
\vdots & \vdots & & \vdots & & & \\
f_{21} & f_{22} - \lambda^{(2)} & \cdots & f_{2m} & \cdot & & \cdot \\
\vdots & \vdots & & \vdots & \vdots & & \vdots \\
f_{m1} & f_{m2} & \cdots & f_{mm} - \lambda^{(m)} & \cdot & & \cdot \\
\vdots & \vdots & & \vdots & \vdots & & \vdots \\
f_{m1} & f_{m2} & \cdots & f_{mm} - \lambda^{(m)} & g_{n,m+1} & \cdots & g_{nn} - \lambda^{(m)}
\end{vmatrix}$$

If, for $i = 1, 2, \ldots, m$, the i-th row is subtracted from all other rows which belong to vertices of colour i, we eventually obtain

$$|\boldsymbol{G} - \Lambda(G)| = \begin{vmatrix} \boldsymbol{F} - \Lambda(F) & \boldsymbol{B} \\ \boldsymbol{O} & \boldsymbol{C} - \Lambda' \end{vmatrix} = |\boldsymbol{F} - \Lambda(F)| \cdot |\boldsymbol{C} - \Lambda'|, \tag{4.4}$$

where B and C are matrices with integer entries; so

$$P_G^*(\lambda_1, \lambda_2, \ldots, \lambda_s) = P_F^*(\lambda_1, \lambda_2, \ldots, \lambda_s) \cdot g(\lambda_1, \lambda_2, \ldots, \lambda_s)$$

with some polynomial g (with integral coefficients), which proves the theorem.

Remark. Having in view Theorem 4.7 it seems that Theorem 4.6 can be replaced by a stronger one: it is a straightforward conjecture that for every "reasonably defined" eigenvalue problem $\mathbb{E}$ of local character (with one or more variables λ_σ) connected with multi-digraphs the $\mathbb{E}$-polynomial of a front divisor F of G divides the $\mathbb{E}$-polynomial of G. But this is still unproved, even in the case of the quadratic problem (1.18).

For generalized divisors (see Section 4.3) we have

Theorem 4.8: *Let F and G be arbitrary multigraphs (or multidigraphs) satisfying $F \mid G$ (or $F \mid_f G$, respectively). Then*

$$Q_F(\lambda) \mid Q_G(\lambda).$$

The proof is quite analogous to the proof of Theorem 4.7.

4.6. The divisor — an effective tool for factoring the characteristic polynomial

Let us resume the proof of Theorem 4.7 but with all variables identified: $\lambda_1 = \lambda_2 = \cdots = \lambda_s = \lambda$. In this case (4.4) simply yields the result

$$|\lambda \boldsymbol{I} - \boldsymbol{G}| = |\lambda \boldsymbol{I} - \boldsymbol{F}| \cdot |\lambda \boldsymbol{I} - \boldsymbol{C}|$$

(where $\boldsymbol{I}$ is an identity matrix of the respective order), i.e.

$$P_G(\lambda) = P_F(\lambda) \cdot |\lambda \boldsymbol{I} - \boldsymbol{C}|. \tag{4.5}$$

The matrix $\boldsymbol{C} = (c_{ij})$ has all integral but possibly negative entries. We may interpret $\boldsymbol{C}$ as the adjacency matrix of a *generalized multi-digraph* C which can have two kinds of arcs: *positive arcs* of weight 1 and *negative arcs* of weight -1 (so that, if $c_{ij} < 0$, there are exactly $|c_{ij}|$ negative arcs from vertex i to vertex j); such generalized multi-digraphs with positive and/or negative arcs will be called *s-graphs* (*signed multi-digraphs*). With this interpretation, (4.5) becomes

$$P_G(\lambda) = P_F(\lambda) \cdot P_C(\lambda). \tag{4.6}$$

It turns out that C is in general not determined by F and G (and the adopted coloration) but depends on the set of selected vertices $V_1, V_2, \ldots, V_m$ which represent the colour classes. Clearly, for given F and G the procedure described in the proof of Theorem 4.7 yields only a finite set $\mathscr{C} = \mathscr{C}(F, G)$ of s-graphs C which, by virtue of (4.6), are all cospectral: We shall call each sample a *co-front-divisor* or, briefly, a *codivisor of F* (*with respect to G*).

At this point it is convenient to note that the whole theory developed so far can be extended to the class of s-graphs without any difficulty (the details are left to the reader). So, if we start from some s-graph G and an s-graph F which is a front divisor of G we shall obtain a class $\mathscr{C}(F, G)$ of codivisors C which again are all s-graphs. We may be lucky enough to find a new s-graph F' which is a proper front divisor of F or of one of its codivisors C and continue the factorization process; if, for example, $F' \mid_f C$, we obtain from (4.6)

$$P_G(\lambda) = P_F(\lambda) \cdot P_C(\lambda) = P_F(\lambda) \cdot P_{F'}(\lambda) \cdot P_{C'}(\lambda),$$

where C' is some codivisor of F' with respect to C; and so on.†

Of course, the bounds of the method will soon become visible: Since both polynomials $P_F(\lambda) = \lambda^m + \cdots$ and $P_C(\lambda) = \lambda^{n-m} + \cdots$ of equation (4.6) have integral rational coefficients the factorization process stops at the latest when all factors found are irreducible polynomials; this natural bound can indeed be reached, as the example of the icosahedron graph (see Fig. 4.5) shows (but it is doubtful that it is always reachable).

---

† It sometimes happens that the codivisor C is disconnected having components $C_1, C_2, \ldots, C_k$, say. In this case $P_C(\lambda) = \prod_{\varkappa=1}^{k} P_{C_\varkappa}(\lambda)$, and we may try to continue the factorization by means of a front divisor of any one of the s-graphs $C_\varkappa$.

J. Rempel and K.-H. Schwolow [ReSc] translated the procedure of finding the codivisors of a given front divisor F of G from matrix language (in which it is presented in the proof of Theorem 4.7) into label-free graph language by formulating an intuitive algorithm which starts with an F-feasible coloration of $G = (\mathscr{X}, \mathscr{U})$ and a system $\mathscr{R} = \{V_1, V_2, \ldots, V_m\}$ of representatives of the colour classes and ends with F and the corresponding codivisor C.

The steps of the algorithm are as follows (for an example compare Fig. 4.4 where negative arcs are indicated by dotted lines):

Step 1. *The hypothesis $F \mid_f G$ says that G has an F-feasible coloration; fix any such coloration with colour classes $\mathscr{C}_1, \mathscr{C}_2, \ldots, \mathscr{C}_m$.*

Step 2. *From each $\mathscr{C}_i$ select a vertex (in an arbitrary way), call it V_i and mark it with an asterisk* (in Figs. 4.4, 4.5, V_i is denoted by i*); *put* $\mathscr{R} = \{V_1, V_2, \ldots, V_m\}$, $\overline{\mathscr{C}}_i = \mathscr{C}_i \setminus \{V_i\}$ $(i = 1, 2, \ldots, m)$, $\overline{\mathscr{R}} = \overline{\mathscr{C}}_1 \cup \overline{\mathscr{C}}_2 \cup \cdots \cup \overline{\mathscr{C}}_m = \mathscr{X} \setminus \mathscr{R}$.

Step 3. *Delete all arcs which issue from $\overline{\mathscr{R}}$ and terminate in $\mathscr{R}$.*

Step 4. *Consider the set $\mathscr{S}$ of all arcs issuing from $\mathscr{R}$ and terminating in $\overline{\mathscr{R}}$. Let u be any arc of $\mathscr{S}$ going from $V_i \in \mathscr{R}$ to $V \in \overline{\mathscr{C}}_k$, say. Replace u by a set of $|\mathscr{C}_i|$ arcs consisting of*

(i) *one arc from V_i to V_k, which is positive if u is positive, and negative if u is negative;*

(ii) *one arc for each vertex of $\overline{\mathscr{C}}_i$, which goes from this vertex to V and is negative if u is positive, and positive if u is negative.*

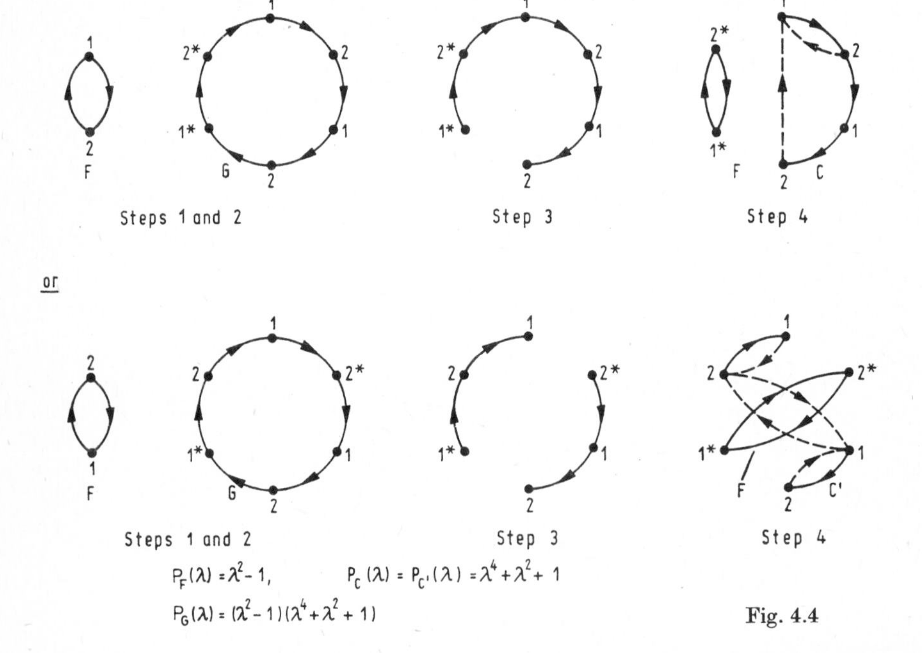

Fig. 4.4

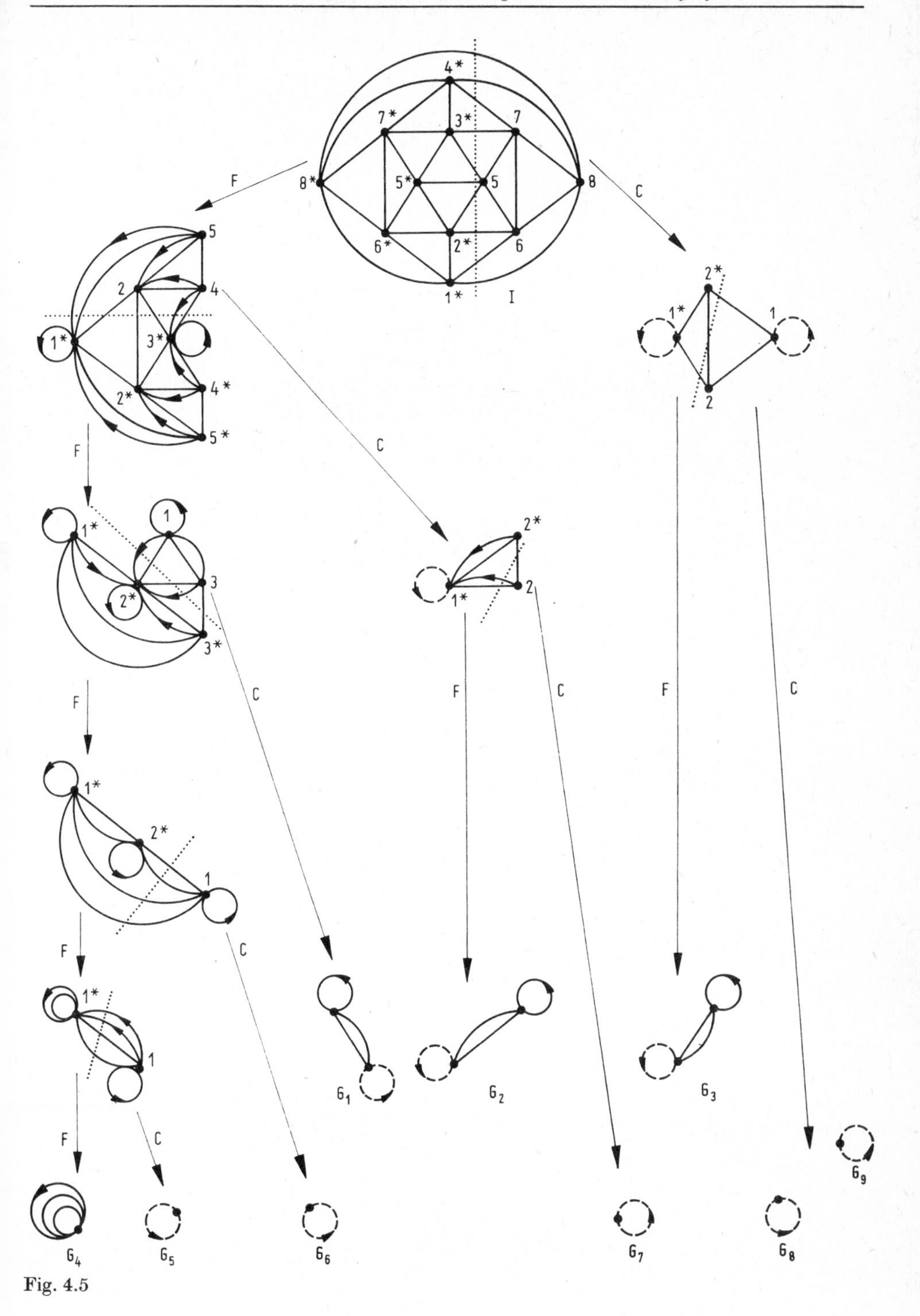

4*
7*
3*
7
8*
5*
5
8
6*
2*
6
1*
I
F
C
5
2
4
1*
3*
4*
2*
5*
2*
1*
1
2
C
F
1
1*
3
2*
3*
2*
1*
2
F
C
F
C
1*
2*
1
F
C
1*
1
F
C
G1
G2
G3
G9
G4
G5
G6
G7
G8

Fig. 4.5

Whenever there appears a pair of parallel arcs, one of them being positive and the other one negative, delete them both. Do so for all arcs of $\mathscr{S}$.

The algorithm results in a pair of disjoint s-graphs: one of them with vertex set $\mathscr{R}$ which is isomorphic to F and the other one with vertex set $\overline{\mathscr{R}}$ isomorphic to some co-divisor C of F.

In Fig. 4.5 the complete factorization of the icosahedron graph I is demonstrated: in this case the procedure results in s-graphs $G_1, G_2, \ldots, G_9$ which have irreducible characteristic polynomials, namely $\lambda^2 - 5$, $\lambda - 5$, $\lambda + 1$, respectively; thus

$$P_I(\lambda) = \prod_{i=1}^{9} P_{G_i}(\lambda) = (\lambda^2 - 5)^3 (\lambda - 5) (\lambda + 1)^5.$$

Remark 1. L. Collatz and U. Sinogowitz [CoSi1] and E. Heilbronner [Hei3] described special geometric methods for factoring the characteristic polynomial of a "symmetric" graph (i.e., a graph having an automorphism α of order 2); in [ReSc] it is shown that these methods are equivalent to factoring G by means of a front divisor F derived from α by regarding the orbits of α as colour classes (see Section 4.4, Theorem 4.3), and this also applies to a method of B. J. McClelland [McC2] which is a variant of the Heilbronner [Hei3] procedure.

Remark 2. There are transformations defined on the class of s-graphs G which leave the characteristic polynomial $P_G(\lambda)$ unchanged, for example:

(i) *transposition*$^\mathsf{T}$: G^T, the transpose of G, is the s-graph derived from G by reversing the orientation of all of its arcs (the adjacency matrix $\boldsymbol{A}$ is replaced by its transpose $\boldsymbol{A}^\mathsf{T}$);

(ii) "*switching*" S_x with respect to an arbitrary vertex x of G: G^{S_x} is the s-graph derived from G by reversing the sign of each arc, not a loop, which is incident with x (in the adjacency matrix $\boldsymbol{A}$ the row and the column belonging to x are both multiplied by -1).

It is often advantageous to perform such a transformation before factoring. Note that in case (i) this simply results in a factorization by means of a rear divisor: if, namely, $F \mid_r G$, then $F^\mathsf{T} \mid_f G^\mathsf{T}$ and

$$P_F(\lambda) = P_{F^\mathsf{T}}(\lambda) \mid P_{G^\mathsf{T}}(\lambda) = P_G(\lambda).$$

4.7. The divisor — a mediator between structure and spectrum

In Section 4.4 we have seen that the symmetries of a multi-digraph G imply the existence of corresponding front divisors, and this again is reflected by the spectrum of G which must contain the spectra of these front divisors (Section 4.5, Theorem 4.7). Beside these symmetries there are other structural properties a (multi-di-)graph may have which also imply or are equivalent to the existence of some special front divisor; many of these properties are, in a sense, "similar" to symmetry properties expressed in terms of the automorphism group.

Consider, for example, the set $\{D_1, D_2, \ldots, D_6\}$ of all strongly connected multidigraphs on two vertices having all front valencies equal to 3 (Fig. 4.6) and let G denote a cubic multigraph: what does $D_i \mid_f G$ mean? $D_i \mid_f G$ is equivalent to the existence of a D_i-feasible coloration of G. Starting from such a coloration it is easy to check the proposition now to be formulated.

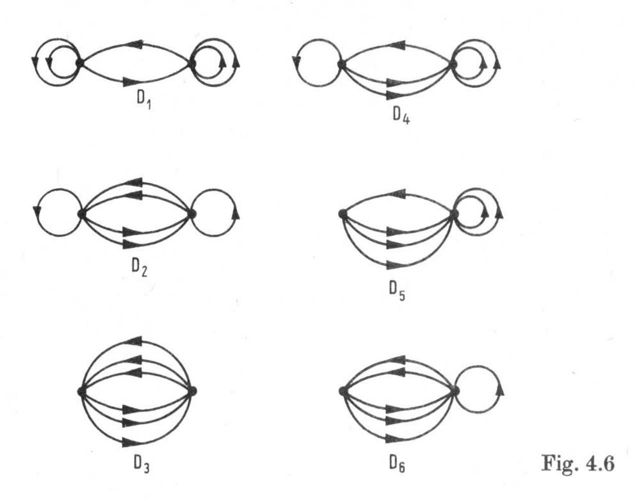

Fig. 4.6

Let $E_1, E_2, \ldots, E_6$ be the following properties:

E_1: G decomposes into a linear factor† L and a quadratic factor Q the circuits (components) of which can be coloured in two colours in such a way that each edge of L joins distinctly coloured circuits of Q.

E_2: G decomposes into a quadratic factor Q and a linear factor L the edges of which can be coloured in two colours in such a way that each edge of Q is adjacent with two distinctly coloured edges of L.

E_3: G is bipartite.

E_4, E_5: G contains a set of disjoint copies of the tree T_4 or T_5, respectively, depicted in Fig. 4.7 which together cover all vertices of G.

E_6: G has $n = 10k$ vertices which can be partitioned into two classes $\mathscr{A}$ and $\mathscr{B}$ with $|\mathscr{A}| = 4k$ and $|\mathscr{B}| = 6k$ where $\mathscr{B}$ spans in G a regular subgraph of degree one.

Fig. 4.7

† Recall that a linear or quadratic factor of a regular graph G is a partial graph of G which is regular of degree 1 or 2, respectively.

Proposition 4.5: *For $i = 1, 2, \ldots, 6$, $D_i \mid_f G$ if and only if G has property $\boldsymbol{E}_i$.*

The spectra of the D_i are

$$\boldsymbol{Sp}_P(D_i) = \{3, \varrho_i\}$$

with $\varrho_1 = 1$, $\varrho_2 = -1$, $\varrho_3 = -3$, $\varrho_4 = 0$, $\varrho_5 = -1$, $\varrho_6 = -2$ so that we conclude: *If G has property $\boldsymbol{E}_i$, then it has ϱ_i as one of its eigenvalues* ($i = 1, 2, \ldots, 6$).

A bit more generally: Let T, T', T'' denote trees; we shall say that *a graph G has a T-cover* (*or a (T', T'')-cover*) if G contains a set $\mathscr{S}$ of disjoint copies of T (or of T', T'') which together cover all vertices of G (and where, in the case of a (T', T'')-cover, $\mathscr{S}$ is the union of two disjoint sets $\mathscr{S}'$, $\mathscr{S}''$ all trees of $\mathscr{S}'$ or $\mathscr{S}''$ being isomorphic with T' or T'', respectively, and where each edge of G not covered by one of the trees joins a tree of $\mathscr{S}'$ with a tree of $\mathscr{S}''$).

Clearly, a linear factor of a regular graph is a K_2-cover, so, in a way, the T-cover concept generalizes the concept of a linear factor.

In this terminology, the above listed properties $\boldsymbol{E}_2, \ldots, \boldsymbol{E}_6$ simply mean:

$\boldsymbol{E}_2$: G has a (K_2, K_2)-cover,

$\boldsymbol{E}_3$: G has a (K_1, K_1)-cover,

$\boldsymbol{E}_4$: G has a T_4-cover,

$\boldsymbol{E}_5$: G has a T_5-cover,

$\boldsymbol{E}_6$: G has a (K_1, K_2)-cover.

Definition. A *symmetric tree $T = T^q_{r,m}$ of degree r ($q = 1$ or 2; $r \geqq 3$, $m \geqq 0$)* is a tree with the following properties:

(i) each vertex of T has either valency 1 or valency r,

(ii) T has a central element c which is a vertex if $q = 1$, and an edge if $q = 2$; the distance between c and each of the vertices of valency 1 is equal to m (see Figs. 4.7 and 4.8).

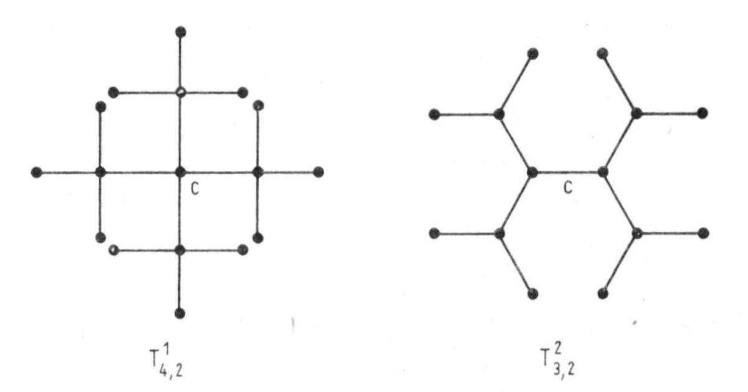

Fig. 4.8

In [FiSa], T-covers and (T', T'')-covers of regular graphs of degree r are investigated where T or T', T'', respectively, are symmetric trees of degree r: it turns out that the property of having such a cover is equivalent with the existence of a special front divisor; thus a necessary condition for G to have a T- or (T', T'')-cover can be formulated in terms of the spectrum of G. In particular, in the case of a $T^2_{r,m}$-cover the spectrum of the corresponding front divisor can explicitly be calculated; the result is the following:

Assume that the connected graph G which is regular of degree r has a $T^2_{r,m}$-cover: then each of the numbers

$$2\sqrt{r-1}\cos\frac{\mu\pi}{m+1} \qquad (\mu = 1, 2, \ldots, m)$$

(which are all distinct) is contained in $\mathbf{Sp}_P(G)$.

It sometimes happens that G can be covered by symmetric trees (with fixed parameters) in more than one way: this indicates that the eigenvalues ($\neq r$) of the corresponding divisor will appear in the spectrum of G with a multiplicity greater than 1 — but we shall not go into further detail.

Remark: As noted above, a linear factor is a K_2-cover, i.e., a $T^2_{r,0}$-cover. But in this case we obtain no statement — as could have been expected.

4.8. Miscellaneous results and problems

1. As an application of the divisor concept to coding theory† we shall outline an elementary proof of *Lloyd's Theorem* due to D. M. Cvetković and J. H. van Lint [CvLi].

We need some preparations. Consider a set $\mathcal{F}$ of b distinct symbols which we call the *alphabet*. The elements of $\mathcal{F}^n$ will be called *words of length n*. In $\mathcal{F}^n$ the *Hamming distance d* is defined by

$$d(x, y) = |\{i \mid x_i \neq y_i, 1 \leq i \leq n\}|.$$

A subset $\mathcal{C}$ of $\mathcal{F}^n$ is called a *perfect e-code* if the spheres

$$\mathcal{S}_e(c) := \{x \in \mathcal{F}^n \mid d(x, c) \leq e\},$$

where c runs through $\mathcal{C}$, form a partition of $\mathcal{F}^n$.

In 1957 S. P. Lloyd [Lloy] proved a strong necessary condition for the existence of a binary (i.e. $b = 2$) perfect e-code; since 1972 several authors (see [Bass], [Del1], [Lens]; see also [Big3], [Sm,D2]) have proved that the theorem which is always referred to as Lloyd's Theorem holds for all b:

Lloyd's Theorem: *If a perfect e-code of length n over an alphabet of b symbols exists, then the e zeros x_j of the polynomial*

$$\psi_{enb}(x) := \sum_{i=0}^{e} (-1)^i (b-1)^{e-i} \binom{n-x}{e-i} \binom{x-1}{i}$$

are distinct positive integers $\leq n$.

Sketch of a proof. From other parts of the theory it is well known that the zeros x_j of $\psi_{enb}(x)$ are all distinct and $\neq 0$; all that matters is to show that $x_j \in \{1, 2, \ldots, n\}$.

Assume that $\mathcal{C}$ is a perfect e-code; define the distance $d(x, \mathcal{C})$ of x from $\mathcal{C}$ by

$$d(x, \mathcal{C}) := \min\{d(x, c) \mid c \in \mathcal{C}\}$$

and denote by $\mathcal{C}_i$ the set

$$\mathcal{C}_i := \{x \in \mathcal{F}^n \mid d(x, \mathcal{C}) = i\} \qquad (i = 1, 2, \ldots, e).$$

Clearly, the sets $\mathcal{C}_i$ form a partition of the space $\mathcal{F}^n$.

† For the general concepts of coding theory consult, e.g., [Lint] (Lloyd's theorem on p. 111).

Consider now the sum G of n copies of the graph K_b: according to no. 10 of Section 2.6, the distinct eigenvalues of G are the numbers

$$\lambda_j = b(n-j) - n = bn - n - bj \qquad (j = 0, 1, \ldots, n).$$

The vertices of G can be understood as the elements of $\mathscr{F}^n$: then the partition $\mathscr{C}_0, \mathscr{C}_1, \ldots, \mathscr{C}_e$ defines a front divisor H of G with the adjacency matrix

$$\boldsymbol{B} = \begin{pmatrix} 0 & n(b-1) & & & & \\ 1 & b-2 & (n-1)(b-1) & & & 0 \\ & 2 & 2(b-2) & (n-2)(b-1) & & \\ & & \cdot\ \cdot\ \cdot & \cdot\ \cdot\ \cdot & \cdot\ \cdot\ \cdot & \\ & 0 & & e-1 & (e-1)(b-2) & (n-e+1)(b-1) \\ & & & & e & n(b-1)-e \end{pmatrix}$$

Since the spectrum of H is contained in the spectrum of G, each of the $e+1$ eigenvalues μ_i of H, i.e. each of the zeros of the polynomial $P_H(\mu) = |\mu \boldsymbol{I} - \boldsymbol{B}|$, is equal to one of the numbers λ_j:

$$\mu_i = bn - n - bj_i, \qquad j_i \in \{0, 1, \ldots, n\}.$$

This means that each root x_i of the equation

$$P_H(bn - n - bx) = 0$$

is equal to one of the numbers $0, 1, \ldots, n$. By developing the determinant $|(bn - n - bx)\,\boldsymbol{I} - \boldsymbol{B}|$ (see [Big 3] or [CvLi]) it can be shown that

$$P_H(bn - n - bx) = e! \cdot x \cdot \psi_{enb}(x),$$

and so we conclude that $\psi_{enb}(x)$ has e distinct zeros x_j where $x_j \in \{1, 2, \ldots, n\}$.

This proves the theorem.

2. Show that the main part of the spectrum of a graph G (see Section 1.8) is contained in the spectrum of any front divisor of G.

(D. M. Cvetković [Cve 18])

3. Assume that the regular graph G of degree r with n vertices has two disjoint independent sets $\mathscr{S}_1, \mathscr{S}_2$ each of m vertices such that

(i) every vertex of $\mathscr{S}_i$ is adjacent to exactly k vertices of $\mathscr{S}_j$ $(i, j = 1, 2;\ i \neq j)$,

(ii) every vertex not in $\mathscr{S}_i \cup \mathscr{S}_j$ is adjacent to exactly s vertices of each of $\mathscr{S}_1, \mathscr{S}_2$.

(G is then said to have the *double-m-property*.) Show that $(n - 2m)\,s = m(r - k)$ and that G has the eigenvalues $k - 2s$ and $-k$.

(J. J. Seidel, private communication)

4. Let $\boldsymbol{E}_{p,q} = \boldsymbol{E}_{q,p}$ be the following property which a multigraph G can have.

$\boldsymbol{E}_{p,q}$: The vertex set $\mathscr{X}$ of G can be partitioned into two disjoint non-void sets $\mathscr{X}_p, \mathscr{X}_q$ which induce subgraphs G_p, G_q of G that are regular of degrees p, q, respectively.

Clearly, G has property $\boldsymbol{E}_{0,0}$ if and only if G is bipartite: so, in a way, $\boldsymbol{E}_{p,q}$ generalizes bipartiteness.

Now assume that G is a connected regular graph of degree r. If for $r = 3$ all possible combinations of p, q $(p, q = 0, 1, 2)$ are taken into consideration the six resulting properties are precisely the properties $\boldsymbol{E}_1, \boldsymbol{E}_2, \ldots, \boldsymbol{E}_6$ discussed in Section 4.7.

The property $\boldsymbol{E}_{p,q}$ is equivalent to the property of G having a special front divisor, namely a multi-digraph with adjacency matrix $\begin{pmatrix} p & r-p \\ r-q & q \end{pmatrix}$. Thus $\boldsymbol{E}_{p,q}$ implies $p + q - r \in \mathbf{Sp}_P(G)$.

(H.-J. Finck, H. Sachs [FiSa])

5. Let G be a connected regular graph of degree r. If G has a

$(T^1_{r,0}, T^1_{r,1})$-cover, then $\{-r, 0\} \subset \boldsymbol{Sp}_P(G)$,

$\cdots (T^1_{r,1}, T^1_{r,2})$-cover, then $\left\{-r, -\sqrt{r}, 0, \sqrt{r}\right\} \subset \boldsymbol{Sp}_P(G)$,

$\cdots (T^2_{r,0}, T^2_{r,1})$-cover, then $\{1 - r, 1\} \subset \boldsymbol{Sp}_P(G)$,

$\cdots (T^1_{r,0}, T^2_{r,1})$-cover, then $\left\{\frac{1}{2}\left(1 - r - \sqrt{r^2 + 2r - 3}\right), \frac{1}{2}\left(1 - r + \sqrt{r^2 + 2r - 3}\right)\right\}$
$\subset \boldsymbol{Sp}_P(G)$,

$\cdots (T^1_{r,0}, T^2_{r,2})$-cover, then $\left\{-r, -\sqrt{r-1}, \sqrt{r-1}\right\} \subset \boldsymbol{Sp}_P(G)$.

In the 1st, 2nd, and 5th case G is bipartite (indicated by $-r \in \boldsymbol{Sp}_P(G)$).

(H.-J. Finck, H. Sachs [FiSa])

6. Let G be an arbitrary connected multigraph. The following four statements are equivalent:

(i) G is bipartite,

(ii) $K_2 \mid G$,

(iii) $-1 \in \boldsymbol{Sp}_Q(G)$,

(iv) $\boldsymbol{Sp}_Q(G)$ is symmetric with respect to the zero point of the real axis.

7. Find the front divisors which, within the set of all connected multigraphs, characterize the set of all multigraphs which are a) regular of degree r or b) semiregular of degrees r, s, respectively ($r, s > 0$); use these front divisors to factor a) K_{r+1}, b) $K_{r,s}$ by means of the Rempel-Schwolow algorithm; c) perform the corresponding procedure for the complete k-partite graph $K_{n_1,n_2,\ldots,n_k}$.

8. There are graphs with no non-trivial automorphism which have a non-trivial front divisor: Let, for example, G be any cubic graph which has no automorphism except for the identity and let G' denote its "subdivision graph" obtained from G by subdividing each of its edges by an additional vertex of valency 2; clearly, G', too, has no non-trivial automorphism. Nevertheless, G' which is semiregular of degrees 2 and 3 has a non-trivial front divisor, namely the multi-digraph with adjacency matrix $\begin{pmatrix} 0 & 2 \\ 3 & 0 \end{pmatrix}$.

9. Let G be a connected multigraph with n vertices. The product $G \times K_2$ (see Section 2.5) is bipartite. Let C denote a component of $G \times K_2$: then C is isomorphic with G if G is bipartite, and $C = G \times K_2$ otherwise. In each case, $G \mid C$ and

$$P_{G \times K_2}(\lambda) = (-1)^n P_G(\lambda) \cdot P_G(-\lambda).$$

A bit more generally: Let G and H be connected multigraphs, not isolated vertices. If at least one of G, H is non-bipartite, then $G \times H$ is connected, and if G and H are both bipartite, then $G \times H$ decomposes into exactly two components which are also bipartite. Let C denote a component of $G \times H$: then $G \mid C$, $H \mid C$ and, according to Theorem 4.8 (Section 4.5),

$$Q_G(\lambda) \mid Q_{G \times H}(\lambda) \text{ and } Q_H(\lambda) \mid Q_{G \times H}(\lambda);$$

if G and H are both bipartite, then

$$Q_G^2(\lambda) \mid Q_{G \times H}(\lambda) \quad \text{and} \quad Q_H^2(\lambda) \mid Q_{G \times H}(\lambda).$$

5. The Spectrum and the Group of Automorphisms

The results obtained and the methods used in relating the spectrum of a (multi-)(di-)graph to its symmetries expressed in terms of its automorphism group Γ are many and varied so that not all details can be reproduced here; we shall give the basic facts and outline the trends along which the general theory develops.

We shall start with some straightforward investigations (mainly concerning simple eigenvalues) and then turn to the powerful methods of the theory of linear representations of finite groups (we assume that the reader is familiar with the basic concepts of representation theory). We shall see that it makes a great difference whether we regard Γ as a concrete group of permutations or merely as an abstract group: we just mention that, given an arbitrary family of finite abstract groups $\gamma_1, \gamma_2, \ldots, \gamma_k$, there are always k pairwise non-isomorphic cospectral graphs $G_1, G_2, \ldots, G_k$ with G_i having γ_i as its automorphism group ($i = 1, 2, \ldots, k$) (L. Babai; see Section 5.4).

We shall, however, not pursue the question how the results can be used for the construction of families of non-isomorphic cospectral graphs in this chapter (except for some remarks in Sections 5.4 and 5.5) as this problem is to be dealt with in Chapter 6.

For a general survey, the reader is referred to N. Biggs' excellent book [Big5].

5.1. Symmetry and simple eigenvalues

In this section we shall establish by elementary methods some remarkable relations between the spectrum $\boldsymbol{Sp}_P(G)$ of a (multi-)graph G and its automorphism group $\Gamma = \Gamma(G)$.

The group Γ is realized by the set of all permutation matrices $\boldsymbol{P}$ which commute with the adjacency matrix $\boldsymbol{A}$ of G (see Section 4.4):

$$\boldsymbol{P} \in \Gamma \Leftrightarrow \boldsymbol{PA} = \boldsymbol{AP}. \tag{5.1}$$

We shall use the letter $\boldsymbol{P}$ for permutation matrices as well as for the underlying permutations; if the automorphism group is to be regarded as an abstract group, this will be indicated by small letters γ, p, etc.

For an arbitrary multigraph G with non-trivial automorphism group $\boldsymbol{\Gamma}$, let $\boldsymbol{x}$ be an eigenvector of $\boldsymbol{A}$ with corresponding eigenvalue λ and let $\boldsymbol{P} \in \boldsymbol{\Gamma}$, then

$$\boldsymbol{A}\boldsymbol{x} = \lambda\boldsymbol{x}$$

implies

$$\boldsymbol{A} \cdot \boldsymbol{P}\boldsymbol{x} = \boldsymbol{P}\boldsymbol{A}\boldsymbol{x} = \boldsymbol{P}\lambda\boldsymbol{x} = \lambda \cdot \boldsymbol{P}\boldsymbol{x};$$

this means that together with $\boldsymbol{x}$ all vectors $\boldsymbol{P}\boldsymbol{x}$ (for $\boldsymbol{P} \in \boldsymbol{\Gamma}$) are eigenvectors of G. If the vectors $\boldsymbol{x}$ and $\boldsymbol{P}\boldsymbol{x}$ prove linearly independent (and that is, in a sense, "in general" the case), then λ must have a multiplicity $m > 1$. For what follows this simple observation is of basic significance.

(**A**) Let G be a multigraph and assume that λ is a simple eigenvalue of G with corresponding real eigenvector $\boldsymbol{x} = (x_1, x_2, \ldots, x_n)^\mathsf{T}$. Then for any $\boldsymbol{P} \in \boldsymbol{\Gamma}(G)$ the eigenvectors $\boldsymbol{x}$ and $\boldsymbol{P}\boldsymbol{x}$ are linearly dependent, i.e. there is a real number μ such that $\boldsymbol{P}\boldsymbol{x} = \mu\boldsymbol{x}$. Let $\boldsymbol{C}$ be any cycle of $\boldsymbol{P}$; assume, without loss of generality, that $\boldsymbol{C} = (1, 2, \ldots, t)$ and that the partial vector $\boldsymbol{x}' = (x_1, x_2, \ldots, x_t)^\mathsf{T}$ of $\boldsymbol{x}$ is non-zero. $\boldsymbol{P}\boldsymbol{x} = \mu\boldsymbol{x}$ implies

$$\boldsymbol{C}\boldsymbol{x}' = \mu\boldsymbol{x}', \tag{5.2}$$

whence $\boldsymbol{C}^t\boldsymbol{x}' = \boldsymbol{x}' = \mu^t\boldsymbol{x}'$. This means $\mu^t = 1$, and we conclude

$$\boldsymbol{P}\boldsymbol{x} = \mu\boldsymbol{x}, \text{ where } \begin{cases} \mu = 1 & \text{if } t \text{ is odd,} \\ \mu = \pm 1 & \text{if } t \text{ is even.} \end{cases} \tag{5.3}$$

From (5.2) we deduce:

(i) If t is odd, then $x_1 = x_2 = \cdots = x_{t-1} = x_t$,

if t is even, then $x_1 = x_2 = \cdots = x_{t-1} = x_t$

or $x_1 = -x_2 = \cdots = x_{t-1} = -x_t$.

(**B**) If, in particular, G has all distinct eigenvalues then

$$\boldsymbol{P}\boldsymbol{x}^i = \mu_i\boldsymbol{x}^i, \qquad \mu_i = \pm 1$$

and consequently

$$\boldsymbol{P}^2\boldsymbol{x}^i = \boldsymbol{x}^i$$

for all $\boldsymbol{P} \in \boldsymbol{\Gamma}$ and all eigenvectors $\boldsymbol{x}^i$ of $\boldsymbol{A}$. This means that the whole space spanned by the eigenvectors is pointwise invariant under $\boldsymbol{P}^2$, hence $\boldsymbol{P}^2 = \boldsymbol{I}$. Thus we have proved a first simple statement:

Theorem 5.1 (A. Mowshowitz [Mow3]; M. Petersdorf, H. Sachs [PeS2]): *If a multigraph has all distinct eigenvalues then all of its non-trivial automorphisms* $\boldsymbol{P}$ *are involutions, i.e.*

$$\boldsymbol{P} \in \boldsymbol{\Gamma} \Rightarrow \boldsymbol{P}^2 = \boldsymbol{I}.$$

This, in particular, implies that $\boldsymbol{\Gamma}$ *is abelian.*

(**C**) We return to point (A).

Regard Γ as a group of permutations acting on the vertex set of G. For each pair of vertices i, j contained in the same orbit of Γ there is a $\boldsymbol{P} \in \Gamma$ which has a cycle containing both i and j. So we deduce from (i)

(ii) For any orbit Ω of Γ and for each pair of vertices $i, j \in \Omega$ the corresponding components x_i, x_j of $\boldsymbol{x}$ satisfy $x_i = x_j$ or $x_i = -x_j$ (possibly $x_i = x_j = 0$).

Clearly, all vertices of an orbit have the same valency.

(**D**) Assume that G is connected and Γ is transitive; then G is regular of a certain degree r, and r is a simple eigenvalue of G with corresponding eigenvector $\boldsymbol{j} = (1, 1, \ldots, 1)^\mathsf{T}$. Suppose that $\lambda \neq r$. From (ii) ensues

$$|x_1| = |x_2| = \cdots = |x_n| > 0; \tag{5.4}$$

we can assume $x_i = \pm 1$.

Since $\boldsymbol{x}$ and $\boldsymbol{j}$ are perpendicular, we obtain

$$x_1 + x_2 + \cdots + x_n = 0.$$

Because of (5.4), this is possible only if n is even and if the number of positive components of $\boldsymbol{x}$ equals the number of negative components.

If $n = 2m$, then there are exactly m vertices i with $x_i = 1$ and exactly m vertices k with $x_k = -1$ ($i, k \in \{1, 2, \ldots, 2m\}$). Among the r neighbours of a vertex i with $x_i = 1$ there is a certain number of vertices i', say q, with $x_{i'} = 1$, and for the remaining $r - q$ neighbours k' of i we have $x_{k'} = -1$. From the equation $\lambda x_i = \sum_{j \cdot i} x_j$ (equivalent with $\lambda \boldsymbol{x} = \boldsymbol{A}\boldsymbol{x}$) we obtain $\lambda \cdot 1 = q \cdot 1 + (r - q)(-1)$, i.e. $\lambda = 2q - r$. Thus we have proved

Theorem 5.2 (M. Petersdorf, H. Sachs [PeS2]): *Let G be a connected regular multigraph of degree r on n vertices which has a transitive automorphism group, and let λ be a simple eigenvalue of G. Then*

$$\lambda = r \textit{ if } n \textit{ is odd, and}$$

$$\lambda = 2q - r \textit{ with } q \in \{0, 1, \ldots, r\} \textit{ if } n \textit{ is even.}$$

(**E**) For the case $n = 2m$ the argument can be carried a bit further:

1. Put $\mathscr{X}^+ = \{i \mid x_i = +1\}$, $\mathscr{X}^- = \{k \mid x_k = -1\}$ and consider the induced subgraphs G^+, G^-, and the partial graph G^* of G spanned by $\mathscr{X}^+$, $\mathscr{X}^-$, and by the set of edges connecting $\mathscr{X}^+$ with $\mathscr{X}^-$, respectively. Clearly, G^+ and G^- are disjoint subgraphs, each on m vertices and each regular of degree $r^+ = r^- = q$, whereas G^* is a regular bipartite graph on $2m$ vertices of degree $r^* = r - q$. If l denotes the number of edges of G^+, then by counting the half-edges we obtain $2l = m \cdot q$, hence if m is odd, q must be even, $q = 2h$, say, and we conclude that λ is of the form $2q - r = 4h - r$.

2. Suppose that there is, besides λ, still another simple eigenvalue $\tilde{\lambda} \neq r$. Let the eigenvector corresponding to $\tilde{\lambda}$ be $\tilde{\boldsymbol{x}} = (\tilde{x}_1, \tilde{x}_2, \ldots, \tilde{x}_n)^\mathsf{T}$ with $\tilde{x}_i = \pm 1$. In the same way as we did for λ and $\boldsymbol{x}$ we define now for $\tilde{\lambda}$ and $\tilde{\boldsymbol{x}}$ the corresponding sets $\widetilde{\mathscr{X}}^+$, $\widetilde{\mathscr{X}}^-$ and graphs $\tilde{G}^+$, $\tilde{G}^-$, $\tilde{G}^*$ with degrees $\tilde{r}^+ = \tilde{r}^- = \tilde{q}$, $\tilde{r}^* = r - \tilde{q}$, respectively. By $\boldsymbol{x}$, $\tilde{\boldsymbol{x}}$

the vertex set $\mathscr{X}$ of G is partitioned into four disjoint sets

$$\mathscr{X}^{++} := \mathscr{X}^{+} \cap \widetilde{\mathscr{X}}^{+}, \quad \mathscr{X}^{+-} := \mathscr{X}^{+} \cap \widetilde{\mathscr{X}}^{-}, \quad \mathscr{X}^{-+} := \mathscr{X}^{-} \cap \widetilde{\mathscr{X}}^{+}, \quad \mathscr{X}^{--} := \mathscr{X}^{-} \cap \widetilde{\mathscr{X}}^{-};$$

none of these sets is empty since otherwise $\widetilde{\mathscr{X}}^{+} = \mathscr{X}^{+}, \widetilde{\mathscr{X}}^{-} = \mathscr{X}^{-}$ or $\widetilde{\mathscr{X}}^{+} = \mathscr{X}^{-}, \widetilde{\mathscr{X}}^{-} = \mathscr{X}^{+}$, i.e., $\tilde{\boldsymbol{x}} = \boldsymbol{x}$ or $\tilde{\boldsymbol{x}} = -\boldsymbol{x}$, which would imply $\tilde{\lambda} = \lambda$.

For $\boldsymbol{P} \in \boldsymbol{\Gamma}(G)$, according to (5.3), $\boldsymbol{P}\boldsymbol{x} = \pm\boldsymbol{x}$: this means that the permutation $\boldsymbol{P}$ either maps $\mathscr{X}^{+}$ onto $\mathscr{X}^{+}$ and $\mathscr{X}^{-}$ onto $\mathscr{X}^{-}$ or $\mathscr{X}^{+}$ onto $\mathscr{X}^{-}$ and $\mathscr{X}^{-}$ onto $\mathscr{X}^{+}$, and the analogue holds for $\widetilde{\mathscr{X}}^{+}, \widetilde{\mathscr{X}}^{-}$.

Let $i \in \mathscr{X}^{++}$, $j \in \mathscr{X}^{+-}$. Since $\boldsymbol{\Gamma}$ is transitive, there is a permutation $\boldsymbol{P}'$ with $\boldsymbol{P}'(i) = j$; clearly, $\boldsymbol{P}'$ maps $\mathscr{X}^{+}$ onto itself and interchanges $\widetilde{\mathscr{X}}^{+}$ and $\widetilde{\mathscr{X}}^{-}$, so it interchanges $\mathscr{X}^{++}$ and $\mathscr{X}^{+-}$. This implies that $|\mathscr{X}^{++}| = |\mathscr{X}^{+-}|$, hence $m = |\mathscr{X}^{+}| = |\mathscr{X}^{++} \cup \mathscr{X}^{+-}| = |\mathscr{X}^{++}| + |\mathscr{X}^{+-}|$ must be even.

3. Under the restriction that only graphs (without loops and multiple edges) are taken into consideration we obtain still another condition a simple eigenvalue $\lambda \neq r$ must satisfy: if G is a graph, then so are G^{+}, G^{-}, and G^{*} and we conclude that their degrees satisfy the relations

$$r^{+} = r^{-} = q \leqq m - 1, \qquad r^{*} = r - q \leqq m.$$

Hence

$$r - n = r - 2m \leqq 2q - r = \lambda \leqq 2m - 2 - r = n - r - 2.^{\dagger}$$

Because of Theorem 5.2, $-r \leqq \lambda \leqq r - 2$, therefore

$$\max(-r, r - n) \leqq \lambda \leqq \min(r - 2, n - r - 2).$$

Combining these results we obtain

Theorem 5.3 (H. Sachs, M. Stiebitz [SaSt]): *Let G be a connected regular multigraph of degree r with $n = 2m$ vertices which has a transitive automorphism group.*

(a) *If m is odd, then G has at most two simple eigenvalues, namely $\lambda_1 = r$ and, possibly, one other which has the form $\lambda = 4h - r$ with $h \in \left\{0, 1, 2, \ldots, \left[\frac{r-1}{2}\right]\right\}$.*††

---

† Actually, the inequality $r - n \leqq \lambda \leqq n - r - 2$ holds for every eigenvalue $\lambda \neq r$ of any graph G which is regular of degree r (see Section 2.7, no. 3).

†† Only recently H. Sachs and M. Stiebitz [SaSt] obtained the following more general result: *Let G be a connected multigraph with n vertices which has a transitive automorphism group $\boldsymbol{\Gamma}$ and let $n = 2^q m$ where m is an odd integer ($q \in \{0, 1, 2, \ldots\}$). Then G has at most 2^q simple eigenvalues. The bound is attained by multigraphs G_q defined as follows: G_0 consists of an isolated vertex; if G_q has adjacency matrix $\boldsymbol{A}_q$ then the adjacency matrix of G_{q+1} is*

$$\boldsymbol{A}_{q+1} = \begin{pmatrix} 2\boldsymbol{A}_q & \boldsymbol{I} \\ \boldsymbol{I} & 2\boldsymbol{A}_q \end{pmatrix}.$$

Note that G_q is a bipartite multigraph with 2^q vertices which is regular of degree $r_q = 2^q - 1$: G_q reduces to the q-dimensional cube graph if all multiple edges are replaced by single ones. *G_q has the equidistant spectrum* $[r_q, r_q - 2, r_q - 4, \ldots, -r_q]$.

In [SaSt] there is also given another (better) upper bound for the number of simple eigenvalues which depends on $\boldsymbol{\Gamma}$, and which is, at the same time, a lower bound for the number of rational eigenvalues (counting their multiplicities).

(b) *If G is a graph and λ is a simple eigenvalue of G, then $\lambda = r$ or*

$$\max(-r, r - n) \leqq \lambda \leqq \min(r, n - r) - 2.$$

Note that the last inequality is more restrictive than the inequality $-r \leqq \lambda \leqq r - 2$ which ensues from Theorem 5.2 if and only if $r > m$.

For more results on the spectra of graphs with transitive automorphism group and, especially, of Cayley graphs expressed in terms of group characters, see nos. 5 and 6 of Section 5.5. See also no. 10 of Section 5.5.

For multigraphs which admit a cyclic permutation we note without proof

Theorem 5.4 (H. Sachs [Sac 1]): *Let G be a connected regular multigraph of degree r with n vertices whose automorphism group contains a cyclic permutation. Then*

$$P_G(\lambda) = \prod_{d|n} \left(f_d(\lambda)\right)^{v_d}$$

where $f_d(\lambda) = \lambda^{u_d} + \cdots$ is a polynomial with integral coefficients which is irreducible over the rational number field. The degrees u_d of the $f_d(\lambda)$ and the exponents v_d satisfy $u_d \cdot v_d = \varphi(d)$ where $\varphi(N)$ is Euler's function (= number of prime residues modulo *N). Furthermore, $f_1(\lambda) = \lambda - r$, $v_1 = 1$ and, if n is even, also $v_2 = 1$; for $d > 2$ the number v_d is an even integer. The zeros of $f_d(\lambda)$ are real sums of r d-th roots of unity.* (For the last statement see also [Djok], [Lová].)

For an application of Theorem 5.4 see Section 7.1, Theorem 7.2.

(**F**) We shall call a graph $G = (\mathcal{X}, \mathcal{U})$ *weakly symmetric* if its automorphism group $\boldsymbol{\Gamma}$ acts transitively not only on the set of vertices but also on the set of edges of G, i.e., if for any pair of edges (i, j), $(k, l) \in U$ the group $\boldsymbol{\Gamma}$ contains a permutation $\boldsymbol{P}'$ with $\boldsymbol{P}'(i) = k$, $\boldsymbol{P}'(j) = l$ or a permutation $\boldsymbol{P}''$ with $\boldsymbol{P}''(i) = l$, $\boldsymbol{P}''(j) = k$; G is called *symmetric* if the word "or" can in fact be replaced by "and".

Theorem 5.5 (see N. Biggs [Big 5]†; see also J. H. Smith [Sm, J]: *Let G be a connected weakly symmetric graph of degree r and let λ be a simple eigenvalue of G. Then $\lambda = r$ if G is non-bipartite, and $\lambda = r$ or $\lambda = -r$ if G is bipartite.*

For the proof we continue with the above notation (see point (E)). By Theorem 5.2, the assertion holds if n is odd. Let $n = 2m$, suppose $\lambda \neq r$, and consider the vertex sets $\mathcal{X}^+$ and $\mathcal{X}^-$. Because of $\lambda \neq r$ we have $q < r$; therefore the degree $r^* = r - q$ of G^* is positive. Suppose $q > 0$, then there are three distinct vertices $i_1, i_2 \in \mathcal{X}^+$ and $k \in \mathcal{X}^-$ such that both edges (i_1, i_2) and (i_1, k) are in G. By our hypothesis there is a permutation $\boldsymbol{P} \in \boldsymbol{\Gamma}$ with

$$\text{(a)}\ \boldsymbol{P}(i_1) = i_1,\quad \boldsymbol{P}(i_2) = k \quad \text{or} \quad \text{(b)}\ \boldsymbol{P}(i_1) = k,\quad \boldsymbol{P}(i_2) = i_1;$$

† In [Big 5] it is assumed that G is symmetric but that does not make much difference for the proof. — It can easily be proved that a connected regular graph of *odd* degree which is weakly symmetric is also symmetric (see W. T. Tutte, Connectivity in Graphs. University of Toronto Press, London/Oxford University Press, 1966; pp. 59–60). According to an oral communication by N. Biggs (August, 1978), D. Holt (Oxford) found a connected regular graph of degree 4 with 27 vertices which is weakly symmetric without being symmetric (see: J. Graph Theory 5 (1981), 201–204).

according to (5.3), $\boldsymbol{Px} = \mu \boldsymbol{x}$ with $\mu = \pm 1$, and consequently

in case (a): $x_{i_1} = \mu x_{i_1}$ and $x_{i_2} = \mu x_k$,

in case (b): $x_{i_1} = \mu x_k$ and $x_{i_2} = \mu x_{i_1}$.

In both cases $x_{i_1} x_{i_2} = x_{i_1} x_k$, which is a contradiction since $x_{i_1} = x_{i_2} = 1$, $x_k = -1$.

Thus we conclude that $q = 0$, this means that $G = G^*$, i.e., G is bipartite, and $\lambda = 2q - r = -r$.

This proves the theorem.

For a theorem on digraphs generalizing Theorem 5.5 see [SaSt].

(G) We now turn our attention to the general case (with no restrictions on Γ but always assuming that λ is a simple eigenvalue) and resume the investigations of point (C). Let Γ have orbits $\Omega_1, \Omega_2, \ldots, \Omega_s$. According to (ii), to each Ω_σ there corresponds a non-negative number y_σ such that $x_i = y_\sigma$ or $x_i = -y_\sigma$ whenever the vertex i belongs to Ω_σ. Let r_σ be the common valency of all vertices of Ω_σ; we will call the numbers $r_1, r_2, \ldots, r_s$ the orbit valencies of G. From each of the Ω_σ select an arbitrary vertex $i = i_\sigma$ and suppose that, among the neighbours j of i_σ which lie in Ω_τ, there are exactly $a_{\sigma\tau}$ vertices with $x_j = y_\tau$ and, if $y_\tau \neq 0$, exactly $b_{\sigma\tau}$ vertices with $x_j = -y_\tau$; if $y_\tau = 0$, put $b_{\sigma\tau} = 0$ $(\sigma, \tau = 1, 2, \ldots, s)$. Then $x_{i_\sigma} = \varepsilon_\sigma y_\sigma$ with $\varepsilon_\sigma = \pm 1$, hence

$$\lambda \varepsilon_\sigma y_\sigma = \lambda x_{i_\sigma} = \sum_{j \cdot i_\sigma} x_j = \sum_{\tau=1}^{s} (a_{\sigma\tau} - b_{\sigma\tau})\, y_\tau \qquad (\sigma = 1, 2, \ldots, s), \tag{5.5}$$

where

$$\sum_{\tau=1}^{s} (a_{\sigma\tau} + b_{\sigma\tau}) = r_\sigma, \qquad a_{\sigma\tau} \geqq 0, \qquad b_{\sigma\tau} \geqq 0 \qquad (\sigma, \tau = 1, 2, \ldots, s). \tag{5.6}$$

With $d_{\sigma\tau} := a_{\sigma\tau} - b_{\sigma\tau}$, (5.5) can be rewritten as

$$\lambda y_\sigma = \sum_{\tau=1}^{s} \varepsilon_\sigma d_{\sigma\tau} y_\tau \qquad (\sigma = 1, 2, \ldots, s). \tag{5.5$'$}$$

The $a_{\sigma\tau}$, $b_{\sigma\tau}$, $d_{\sigma\tau}$ satisfy

$$\sum_{\tau=1}^{s} (a'_{\sigma\tau} + b'_{\sigma\tau}) = r_\sigma, \qquad d'_{\sigma\tau} = a'_{\sigma\tau} - b'_{\sigma\tau}, \qquad a'_{\sigma\tau} \geqq 0, \qquad b'_{\sigma\tau} \geqq 0 \tag{5.7}$$

$$(\sigma, \tau = 1, 2, \ldots, s).$$

Clearly, this system has only a finite number of solutions which can easily be listed, and together with the solution $(a_{\sigma\tau}, b_{\sigma\tau}, d_{\sigma\tau})$ there is also a solution with $d'_{\sigma\tau} = -d_{\sigma\tau}$ (σ fixed, $\tau = 1, 2, \ldots, s$). So we can drop the factor ε_σ in (5.5)$'$ and claim that the y_σ, λ satisfy a system of equations with integral coefficients

$$\lambda y_\sigma = \sum_{\tau=1}^{s} d_{\sigma\tau} y_\tau \tag{5.8}$$

or briefly, if we put $\boldsymbol{D} := (d_{\sigma\tau})$,

$$\lambda \boldsymbol{y} = \boldsymbol{D} \boldsymbol{y}; \tag{5.8$'$}$$

therefore, $|\lambda \boldsymbol{I} - \boldsymbol{D}| = 0$. So we have proved

Theorem 5.6 (M. PETERSDORF, H. SACHS [PeS2]): *Let G be a multigraph with orbit valencies $r_1, r_2, \ldots, r_s$ and let λ be a simple eigenvalue of G. Then λ satisfies an equation*

$$|\lambda \boldsymbol{I} - \boldsymbol{D}| = 0$$

of order s with integral coefficients where $\boldsymbol{D} = (d_{\sigma\tau})$ is a solution[†] *of* (5.7). *The set of all solutions of* (5.7) *is finite and depends only on the orbit valencies of G.*

For given positive integers $r_1, r_2, \ldots, r_s$, denote the set of all eigenvalues of all matrices $\boldsymbol{D}$ which are solutions of (5.7) by $\mathcal{S}_{r_1,r_2,\ldots,r_s}$. This finite set contains the set of all simple eigenvalues a multigraph G with orbit valencies $r_1, r_2, \ldots, r_s$ can have.

The sets $\mathcal{S}_{r_1,r_2,\ldots,r_s}$ were investigated by L. L. KRAUS and D. M. CVETKOVIĆ [KrC2] who established — among other results — that all integers k with $|k| \leq \max(r_1, r_2, \ldots, r_s)$ are contained in $\mathcal{S}_{r_1,r_2,\ldots,r_s}$, that if $\lambda \in \mathcal{S}_{r_1,r_2,\ldots,r_s}$ then also $-\lambda \in \mathcal{S}_{r_1,r_2,\ldots,r_s}$, and that for $(r_1, r_2) = (1, 2), (1, 3), (2, 2), (2, 3), (3, 3)$ to each number α contained in $\mathcal{S}_{r_1,r_2}$ there exists a graph (possibly with loops)[††] with orbit valencies r_1, r_2 which has α as a simple eigenvalue. Furthermore, in [KrC2] all sets $\mathcal{S}_{r_1,r_2}$ with $1 \leq r_1, r_2 \leq 5$ are listed, the largest one being $\mathcal{S}_{5,4} = \mathcal{S}_{4,5}$ with 203 numbers; see Appendix, Table 7. As an example,

$$\mathcal{S}_{3,3} = \left\{0, \pm 1, \pm 2, \pm 3, \pm\sqrt{3}, \pm\sqrt{5}, \pm 1 \pm\sqrt{2}, \pm\frac{1}{2} \pm\frac{1}{2}\sqrt{17}\right\}$$

(see also [Pete], [PeS2]).

(**H**) The next theorem will be proved in Section 5.2 (see p. 145).

Theorem 5.7 (M. PETERSDORF, H. SACHS [PeS2]): *Let G be a multigraph, let $\boldsymbol{P} \in \Gamma(G)$, the cycle decomposition of $\boldsymbol{P}$ consisting of exactly $\alpha(\boldsymbol{P})$ odd and $\beta(\boldsymbol{P})$ even cycles. Then G has at most $\alpha(\boldsymbol{P}) + 2\beta(\boldsymbol{P})$ simple eigenvalues.*

Remark 1. If G has n vertices and if all eigenvalues are distinct, then Theorem 5.7 implies $\alpha(\boldsymbol{P}) + 2\beta(\boldsymbol{P}) \geq n$ for each $\boldsymbol{P} \in \Gamma(G)$: this is possible only if each odd cycle has length 1 and each even cycle has length 2, i.e., only if $\boldsymbol{P}^2 = \boldsymbol{I}$. That is precisely the content of Theorem 5.1.

Remark 2. The results of this section show already that, generally speaking, a multigraph G with a rich automorphism group Γ has only a small number of distinct eigenvalues. One should expect that also the converse is true; indeed, if G has only two distinct eigenvalues (the case of only one eigenvalue is trivial), then the group Γ is the richest possible, namely the symmetric group. However, among strongly regular graphs which have exactly three distinct eigenvalues (see Theorem 3.32, Section 3.4) we find surprising examples: there are strongly regular graphs with trivial automorphism group (e.g., A. J. H. PAULUS [Paul] found an example on 26 vertices).

---

† We say that $(d_{\sigma\tau})$ is a *solution of* (5.7) if there are integers $a_{\sigma\tau}, b_{\sigma\tau}$ such that the $3s^2$ numbers $a_{\sigma\tau}, b_{\sigma\tau}, d_{\sigma\tau}$ satisfy (5.7).

†† Only recently M. SCHULZ established that for each $\alpha \in \mathcal{S}_{3,3}$ there exists a *schlicht* graph with orbit valencies $r_1 = r_2 = 3$ which has α as a simple eigenvalue; see [103] of the additional list of references.

5.2. The spectrum and representations of the automorphism group

We shall now use the powerful tools provided by the theory of linear representations of finite groups to establish more relations between the spectrum of a multi-(di-)graph G and its automorphism group.

We assume that the reader is familiar with the basic concepts and theorems of representation theory. There exist several excellent text-books and monographs on representation theory and its applications that mathematicians as well as users can refer to; we just mention [Boer], [CuRe], [Ha,L], [Scho], [Wign]; [БеФо], [Вигн], [КэРа], [Найм].

We recall that the direct sum (block diagonal sum) $\boldsymbol{M}_1 \dotplus \boldsymbol{M}_2 \dotplus \cdots \dotplus \boldsymbol{M}_s$ of square matrices $\boldsymbol{M}_1, \boldsymbol{M}_2, \ldots, \boldsymbol{M}_s$ is defined as the square matrix

$$\begin{pmatrix} \boldsymbol{M}_1 & & \boldsymbol{O} \\ & \boldsymbol{M}_2 & \\ \boldsymbol{O} & & \ddots \ \boldsymbol{M}_s \end{pmatrix}.$$

If $\boldsymbol{M}_1 = \boldsymbol{M}_2 = \cdots = \boldsymbol{M}_{a_1} = \boldsymbol{N}_1$, $\boldsymbol{M}_{a_1+1} = \boldsymbol{M}_{a_1+2} = \cdots = \boldsymbol{M}_{a_1+a_2} = \boldsymbol{N}_2, \ldots, \boldsymbol{M}_{s-a_t+1} = \boldsymbol{M}_{s-a_t+2} = \cdots = \boldsymbol{M}_s = \boldsymbol{N}_t$, we shall write

$$\boldsymbol{M}_1 \dotplus \boldsymbol{M}_2 \dotplus \cdots \dotplus \boldsymbol{M}_s = a_1 \circ \boldsymbol{N}_1 \dotplus a_2 \circ \boldsymbol{N}_2 \dotplus \cdots \dotplus a_t \circ \boldsymbol{N}_t.$$

Note that in this notation $a \circ \boldsymbol{N} = \boldsymbol{I}_a \otimes \boldsymbol{N}$ whereas $a\boldsymbol{N}$ is something quite different.

If the family $\mathscr{R}$ of square matrices is a representation of the abstract group γ, then the matrix $\boldsymbol{R} \in \mathscr{R}$ representing the group element $p \in \gamma$ is denoted by $\boldsymbol{R}(p)$; if $\gamma = \{p_1, p_2, \ldots, p_g\}$, then $\mathscr{R} = \{\boldsymbol{R}_1, \boldsymbol{R}_2, \ldots, \boldsymbol{R}_g\}$ where $\boldsymbol{R}_i = \boldsymbol{R}(p_i)$.

There are several matrices in block diagonal form which are similar† to the matrices to be considered and which play a role in the subsequent investigations; namely:

1. Let $\boldsymbol{J}_m(\lambda)$ denote the *Jordan matrix* of order m and parameter λ, that is a square matrix (h_{ij}) with $h_{11} = h_{22} = \cdots = h_{mm} = \lambda$, $h_{21} = h_{32} = \cdots = h_{m,m-1} = 1$, and $h_{ij} = 0$ if $i \neq j$, $j + 1$. Any square matrix of order n can be transformed into its *canonical form* (or *Jordan normal form*)

$$\boldsymbol{J} = \boldsymbol{J}(\boldsymbol{A}) = \boldsymbol{J}_{m_1}(\lambda_1^*) \dotplus \boldsymbol{J}_{m_2}(\lambda_2^*) \dotplus \cdots \dotplus \boldsymbol{J}_{m_q}(\lambda_q^*), \tag{5.9}$$

where $m_1 + m_2 + \cdots + m_q = n$ and where the λ_i^* are (not necessarily distinct) eigenvalues of $\boldsymbol{A}$.

2. If, in particular, $\boldsymbol{A}$ is symmetric (for example if $\boldsymbol{A}$ is the adjacency matrix of a multigraph) or a permutation matrix, then it is similar to a diagonal matrix

$$\boldsymbol{A}^{\boldsymbol{T}} = \lambda^{(1)}\boldsymbol{I}_{m_1} \dotplus \lambda^{(2)}\boldsymbol{I}_{m_2} \dotplus \cdots \dotplus \lambda^{(s)}\boldsymbol{I}_{m_s} \tag{5.10}$$

where the $\lambda^{(\sigma)}$ are the distinct eigenvalues of $\boldsymbol{A}$, $\lambda^{(\sigma)}$ having multiplicity m_σ.

---

† The square matrix $\boldsymbol{B}$ is called *similar* to the square matrix $\boldsymbol{A}$ if there is a non-singular square matrix $\boldsymbol{T}$ which transforms $\boldsymbol{A}$ into $\boldsymbol{B}$, i.e. such that $\boldsymbol{T}^{-1}\boldsymbol{A}\boldsymbol{T} = \boldsymbol{B}$, or $\boldsymbol{A}\boldsymbol{T} = \boldsymbol{T}\boldsymbol{B}$. — We sometimes write $\boldsymbol{A}^{\boldsymbol{T}} := \boldsymbol{T}^{-1}\boldsymbol{A}\boldsymbol{T}$, not to be confused with the transpose $\boldsymbol{A}^{\mathsf{T}}$ of $\boldsymbol{A}$.

3. If $\mathscr{R}$ is a representation of the group γ by matrices of order n, then there is a non-singular matrix $\boldsymbol{U}$ which simultaneously transforms all matrices $\boldsymbol{R}(p)$ $(p \in \gamma)$ into the block diagonal form

$$\boldsymbol{R}^{\boldsymbol{U}}(p) = a_1 \circ \boldsymbol{R}^{(1)}(p) \dotplus a_2 \circ \boldsymbol{R}^{(2)}(p) \dotplus \cdots \dotplus a_k \circ \boldsymbol{R}^{(k)}(p) \tag{5.11}$$

where the families $\mathscr{R}^{(\varkappa)} = \{\boldsymbol{R}^{(\varkappa)}(p) \mid p \in \gamma\}$ are the k non-equivalent irreducible representations of γ; we then say that $\mathscr{R}^{(\varkappa)}$ is an *irreducible component of $\mathscr{R}$ with multiplicity $a_\varkappa$* (note that $a_\varkappa = 0$ simply means that $\mathscr{R}$ has no irreducible component equivalent with $\mathscr{R}^{(\varkappa)}$, so in (5.11) the terms $a_\varkappa \circ \boldsymbol{R}^{(\varkappa)}(p)$ with $a_\varkappa = 0$ can be cancelled). Instead of (5.11) we also write

$$\mathscr{R}^{\boldsymbol{U}} = a_1 \circ \mathscr{R}^{(1)} \dotplus a_2 \circ \mathscr{R}^{(2)} \dotplus \cdots \dotplus a_k \circ \mathscr{R}^{(k)} \tag{5.11$'$}$$

and call this decomposition the *totally reduced form of $\mathscr{R}$*. Here k is the number of classes of conjugates in γ, the order $n_\varkappa$ of $\boldsymbol{R}^{(\varkappa)}(p)$ is equal to the degree of $\mathscr{R}^{(\varkappa)}$ (= dimension of the corresponding invariant subspace of $\mathbf{S}^n$) and we have

$$a_1 n_1 + a_2 n_2 + \cdots + a_k n_k = n;$$

furthermore:

$$n_1^2 + n_2^2 + \cdots + n_k^2 = g,$$

where g is the order of γ.

4. If Γ is a representation of γ by permutation matrices $\boldsymbol{P}$, then the rows and columns of all $\boldsymbol{P}$ can be arranged in such a way that to each orbit of Γ there corresponds a block whose order is equal to the size of the orbit. We shall then say that Γ *is in an orbit separated form.*

Now let G be a multigraph with adjacency matrix $\boldsymbol{A}$ and with automorphism group Γ;† of course, Γ is a faithful representation $\mathscr{R}$ of the corresponding abstract group γ. Then there are two non-singular matrices $\boldsymbol{T}$, $\boldsymbol{U}$ such that $\boldsymbol{T}$ transforms $\boldsymbol{A}$ into its diagonal form (5.10) and $\boldsymbol{U}$ transforms $\Gamma = \mathscr{R}$ into its totally reduced form (5.11)$'$. What happens if, conversely, (i) $\boldsymbol{T}$ is applied to $\mathscr{R}$ and (ii) $\boldsymbol{U}$ is applied to $\boldsymbol{A}$?

(i) Put $\boldsymbol{T}^{-1}\boldsymbol{R}(p)\,\boldsymbol{T} = \boldsymbol{R}^{\boldsymbol{T}}(p)$ and $\{\boldsymbol{R}^{\boldsymbol{T}}(p) \mid p \in \gamma\} = \mathscr{R}^{\boldsymbol{T}}$. Every matrix $\boldsymbol{R}^{\boldsymbol{T}}(p)$ commutes with $\boldsymbol{T}^{-1}\boldsymbol{A}\boldsymbol{T} = \boldsymbol{A}^{\boldsymbol{T}} = \lambda^{(1)}\boldsymbol{I}_{m_1} \dotplus \lambda^{(2)}\boldsymbol{I}_{m_2} \dotplus \cdots \dotplus \lambda^{(s)}\boldsymbol{I}_{m_s}$ and a simple calculation shows that $\boldsymbol{R}^{\boldsymbol{T}}(p)$ has also block diagonal form, namely

$$\boldsymbol{R}^{\boldsymbol{T}}(p) = \tilde{\boldsymbol{R}}_1(p) \dotplus \tilde{\boldsymbol{R}}_2(p) \dotplus \cdots \dotplus \tilde{\boldsymbol{R}}_s(p) \tag{5.12}$$

where $\tilde{\boldsymbol{R}}_\sigma(p)$ is a square matrix of order m_σ: so

$$\mathscr{R}^{\boldsymbol{T}} = \widetilde{\mathscr{R}}_1 \dotplus \widetilde{\mathscr{R}}_2 \dotplus \cdots \dotplus \widetilde{\mathscr{R}}_s \tag{5.12$'$}$$

where $\widetilde{\mathscr{R}}_\sigma = \{\tilde{\boldsymbol{R}}_\sigma(p) \mid p \in \gamma\}$ is a representation of γ of degree m_σ.

† Note that the considerations and results of this section remain valid if Γ is replaced by any one of its subgroups.

(ii) Put $U^{-1}AU = A^U$. Every matrix $U^{-1}R(p)\,U = R^U(p) = a_1 \circ R^{(1)}(p) \dotplus a_2 \circ R^{(2)}(p) \dotplus \cdots \dotplus a_k \circ R^{(k)}(p)$ commutes with A^U:

$$R^U(p) \cdot A^U = A^U \cdot R^U(p).$$

Writing A^U in block form according to the block form of $R^U(p)$ and multiplying block by block shows that A^U has also block diagonal form, as a simple consequence of Schur's Lemma (see, for example, [Boer], pp. 20-21). This time we find k blocks $A_\varkappa$ of size $a_\varkappa n_\varkappa$ ($\varkappa = 1, 2, \ldots, k$) or fewer, if blocks of size zero (which actually do not appear) are not counted:

$$A^U = A_1 \dotplus A_2 \dotplus \cdots \dotplus A_k. \tag{5.13}$$

In pursuing theoretical questions the mathematician will often assume that he knows the eigenvalues: he is interested in obtaining information about structural properties, in particular about the automorphisms of the graph, from the spectrum whereas the chemist usually knows the molecule and its symmetries (i.e., the graph and its automorphisms): he is interested in information about the possible energy levels (i.e., the eigenvalues). So the mathematician finds himself in a situation to which (i) rather applies than (ii), and conversely for the chemist (it would be difficult, however, to give this statement a precise meaning).

Let us start with case (i). For each $\sigma \in \{1, 2, \ldots, s\}$ there is a non-singular matrix V_σ of order m_σ which transforms $\widetilde{\mathscr{R}}_\sigma$ into its totally reduced form:

$$\widetilde{\mathscr{R}}_\sigma^{V_\sigma} = a_{\sigma 1} \circ \mathscr{R}^{(1)} \dotplus a_{\sigma 2} \circ \mathscr{R}^{(2)} \dotplus \cdots \dotplus a_{\sigma k} \circ \mathscr{R}^{(k)}.$$

Then $V = V_1 \dotplus V_2 \dotplus \cdots \dotplus V_s$ transforms $\mathscr{R}^T$ into its totally reduced form, namely

$$\mathscr{R}^{TV} = \widetilde{\mathscr{R}}_1^{V_1} \dotplus \widetilde{\mathscr{R}}_2^{V_2} \dotplus \cdots \dotplus \widetilde{\mathscr{R}}_s^{V_s}.$$

Applying V to A^T changes nothing:

$$A^{TV} = \lambda^{(1)} I_{m_1} \dotplus \lambda^{(2)} I_{m_2} \dotplus \cdots \dotplus \lambda^{(s)} I_{m_s}.$$

Geometrically this means that the irreducible invariant subspaces of $\mathbf{S}^n$ are subspaces of the eigenspaces of A.

We draw two conclusions:

a) The eigenvalues can be arranged in $a_1 + a_2 + \cdots + a_k$ families $\mathscr{E}_{\varkappa\alpha}$ ($\varkappa = 1, 2, \ldots, k$; $\alpha = 1, 2, \ldots, a_\varkappa$) (there is no $\mathscr{E}_{\varkappa\alpha}$ if $a_\varkappa = 0$) corresponding to the irreducible representations where $\mathscr{E}_{\varkappa\alpha}$ contains exactly $n_\varkappa$ eigenvalues which are all equal (this is information also wanted by the chemist!). So *there cannot be more than* $a_1 + a_2 + \cdots + a_k$ *distinct eigenvalues*. The determination of the $a_\varkappa$ (which depend on Γ, not on γ) is a standard problem of representation theory, which can be solved by means of the character of Γ (see below).

b) The eigenspaces of A of dimensions $m_1, m_2, \ldots, m_s$ are left invariant by the members of $\mathscr{R}^{TV}$. Since $\mathscr{R}^{TV}$ is a faithful representation of γ, this means that γ is (isomorphic to) a subgroup of $\omega(m_1) \oplus \omega(m_2) \oplus \cdots \oplus \omega(m_s)$, where $\omega(m)$ denotes the abstract real orthogonal group of dimension m.

Thus we have proved

Theorem 5.8 (L. Babai [Bab 2][†]): *Let G be a multigraph having s distinct eigenvalues with respective multiplicities $m_1, m_2, \ldots, m_s$. Then the abstract automorphism group γ of G is a subgroup of*

$$\omega(m_1) \oplus \omega(m_2) \oplus \cdots \oplus \omega(m_s),$$

where $\omega(m)$ denotes the real orthogonal group of dimension m.

Theorem 5.1 is now a simple consequence of Theorem 5.8: To show this, let ζ_q denote the cyclic group of order q; note that $\omega(1) = \zeta_2$. Under the hypothesis of Theorem 5.1 all eigenvalues are distinct, i.e., $s = n$, $m_1 = m_2 = \cdots = m_n = 1$, and we conclude that

$$\gamma \text{ is a subgroup of } \zeta_2^n,$$

so $p^2 = 1$ for each $p \in \gamma$. This proves the assertion. (See also Section 5.5, no. 3.)

Up to this point we have considered only undirected multigraphs, but all considerations of this section remain valid also for any multi-digraph whose adjacency matrix is similar to a diagonal matrix; of course, in Theorem 5.8 the real orthogonal groups $\omega(m_\sigma)$ are then to be replaced by the unitary groups $\boldsymbol{u}(m_\sigma)$ of dimension m_σ[††], and Theorem 5.1 does not hold for digraphs (compare Theorem 5.9). If we want to extend the results to the case of an arbitrary multi-digraph, we shall have to deal with the Jordan normal form of the adjacency matrix which, in general, is no longer a diagonal matrix — a fact which, in some cases, makes the proofs much more complicated.

We shall now formulate some results for (multi-)digraphs.

Assume that G is any multi-digraph that has all distinct eigenvalues. Then the adjacency matrix $\boldsymbol{A}$ of G is similar to the diagonal matrix $(\lambda_1) \dotplus (\lambda_2) \dotplus \cdots \dotplus (\lambda_n)$, and, since $m_1 = m_2 = \cdots = m_n = 1$, we conclude that all irreducible components of $\Gamma(G)$ are of degree 1. Clearly, this means that $\Gamma(G)$ is abelian. Thus we have proved

Theorem 5.9 (C.-Y. Chao [Chao]): *If the eigenvalues of an arbitrary multi-digraph G are all distinct, then $\Gamma(G)$, the automorphism group of G, is abelian.*

This result has been generalized by A. Mowshowitz and again by L. Babai:

Theorem 5.10 (A. Mowshowitz [Mow 4]): *If the adjacency matrix $\boldsymbol{A}$ of a digraph G is non-derogatory (i.e., if its minimal and characteristic polynomials are identical) over a field $\mathcal{F}$, then $\Gamma(G)$ is abelian.*

We quote the original proof given by A. Mowshowitz: Since $\boldsymbol{A}$ is non-derogatory,

---

† A related result has only recently been obtained by C. D. Godsil [Gods].

†† Here we refer to the well-known fact that every finite subgroup of $\mathfrak{GL}(n, \mathbf{C})$ (the group of all linear mappings of the n-dimensional vector space over the complex number field $\mathbf{C}$ onto itself) is equivalent to a subgroup of the unitary group $\boldsymbol{u}(n)$; see, for example, the book of J.-P. Serre, Représentations linéaires des groupes finis, Hermann, Paris 1967 (Lineare Darstellung endlicher Gruppen, Akademie-Verlag, Berlin 1972), § 1.3.

the centralizer of $\boldsymbol{A}$ is just the ring of polynomials in $\boldsymbol{A}$ over $\mathscr{F}$.† Thus, every $\boldsymbol{P} \in \Gamma(G)$ is a polynomial in $\boldsymbol{A}$, from which the result follows.

Theorem 5.11 (L. Babai [Bab2]): *If any Jordan matrix occurs at most once in the canonical form of the adjacency matrix of the multi-digraph G, then Γ(G) is abelian.*

This theorem is an immediate consequence of the following result of L. Babai generalizing Theorem 5.8, which we quote without proof.

Theorem 5.12 (L. Babai [Bab2]): *Let the canonical form of the adjacency matrix* $\boldsymbol{A}$ *of the multi-digraph G be*

$$\boldsymbol{J}(\boldsymbol{A}) = k_1 \circ \boldsymbol{J}_{m_1}(\lambda_1^*) \dotplus k_2 \circ \boldsymbol{J}_{m_2}(\lambda_2^*) \dotplus \cdots \dotplus k_p \circ \boldsymbol{J}_{m_p}(\lambda_p^*)$$

($k_1 m_1 + k_2 m_2 + \cdots + k_p m_p = n$; the pairs (m_j, λ_j^*) are different for different values of j, but, possibly, $m_i = m_j$ or $\lambda_i^* = \lambda_j^*$ with $i \neq j$.) *Then $\Gamma(G)$ is isomorphic to a subgroup of*

$$\boldsymbol{u}(k_1) \oplus \boldsymbol{u}(k_2) \oplus \cdots \oplus \boldsymbol{u}(k_p),$$

where $\boldsymbol{u}(k)$ denotes the k-dimensional unitary group.

For the *proof of Theorem* 5.11 just notice that, by hypothesis, $k_1 = k_2 = \cdots = k_p = 1$ and that $\boldsymbol{u}(1)$ is abelian.

We now return to multigraphs and give the missing

Proof of Theorem 5.7. Let $\boldsymbol{P} \in \Gamma(G)$ and let the permutation $\boldsymbol{P}$ have exactly α odd and β even cycles. Consider $\boldsymbol{P}$ as the adjacency matrix of a digraph P^*. Clearly, the characteristic polynomial of $\boldsymbol{P}$ is the product of the characteristic polynomials of the cycles of P^*. Since the characteristic polynomial of a cycle of length l is $\lambda^l - 1$, we see that the only real eigenvalues of $\boldsymbol{P}$ are 1 with multiplicity $\alpha + \beta$ and, if $\beta > 0$, -1 with multiplicity β: thus $\boldsymbol{P}$ has exactly $\alpha + 2\beta$ real eigenvalues. Now let $\boldsymbol{T}$ be a real orthogonal matrix which transforms $\boldsymbol{A}$ into its diagonal form (5.10). Then $\boldsymbol{T}$ transforms $\Gamma(G)$, regarded as a representation $\mathscr{R}$ of γ, into the block diagonal form (5.12)′; in particular, since $\boldsymbol{P} = \boldsymbol{R}(p)$, according to (5.12)

$$\boldsymbol{T}^{-1}\boldsymbol{P}\boldsymbol{T} = \tilde{\boldsymbol{R}}_1(p) \dotplus \tilde{\boldsymbol{R}}_2(p) \dotplus \cdots \dotplus \tilde{\boldsymbol{R}}_s(p),$$

where $\tilde{\boldsymbol{R}}_\sigma(p)$ is a real matrix of order m_σ. Suppose that G has exactly t simple eigenvalues: then exactly t of the numbers m_σ are equal to 1, consequently, exactly t of the matrices $\tilde{\boldsymbol{R}}_\sigma(p)$ have order 1. So $\boldsymbol{T}^{-1}\boldsymbol{P}\boldsymbol{T}$ has at least t real eigenvalues, and since $\boldsymbol{P}$ and, together with $\boldsymbol{P}$, also $\boldsymbol{T}^{-1}\boldsymbol{P}\boldsymbol{T}$ has exactly $\alpha + 2\beta$ real eigenvalues, we conclude that $t \leqq \alpha + 2\beta$. This proves the theorem.

We shall now turn our attention to case (ii) (p. 143). Let G be a multigraph. Formula (5.13) indicates that if we know how to obtain the totally reduced form (5.11)′ of $\Gamma(G)$ (regarded as a faithful representation $\mathscr{R}$ of γ), i.e., if we know the transformation $\boldsymbol{U}$, then we shall be able to reduce the task of finding the eigenvalues

† See, for example, Д. А. Супруненко, Р. И. Тышкевич, Перестановочные Матрицы. Наука и Техника, Минск 1966; D. A. Suprunenko and R. I. Tyshkevich, Commutative Matrices, Academic Press, New York 1968.

of $\boldsymbol{A}$ to several similar tasks but each of smaller size (i.e., we shall be able to partially factor the characteristic polynomial of G).

We shall confine ourselves to briefly describing a procedure frequently being used by chemists and physicists.[†]

Let G be an arbitrary multigraph on n vertices with the adjacency matrix $\boldsymbol{A}$ and abstract automorphism group $\gamma = \{p_1, p_2, \ldots, p_g\}$ faithfully represented by $\mathscr{R} = \Gamma(G) = \{\boldsymbol{P}_1, \boldsymbol{P}_2, \ldots, \boldsymbol{P}_g\}$; the $\boldsymbol{P}_i$ are permutation matrices which commute with $\boldsymbol{A}$. All these matrices represent operators acting in the n-dimensional space $\mathbf{S}^n$ with its natural basis consisting of the unit vectors $\boldsymbol{e}_1, \boldsymbol{e}_2, \ldots, \boldsymbol{e}_n$. The problem is to find a basis in which $\mathscr{R}$ has its totally reduced form (5.11)'; for the same basis $\boldsymbol{A}$ will then have the block diagonal form

$$\boldsymbol{A}^U = \boldsymbol{A}_1 \dotplus \boldsymbol{A}_2 \dotplus \cdots \dotplus \boldsymbol{A}_k, \tag{5.13}$$

where the block $\boldsymbol{A}_\varkappa$ is a square matrix of order $a_\varkappa n_\varkappa$, $n_\varkappa$ being the degree of the irreducible representation $\mathscr{R}^{(\varkappa)}$ of γ and $a_\varkappa$ being the multiplicity of $\mathscr{R}^{(\varkappa)}$ as an irreducible component of $\mathscr{R}$ (see (5.11)').

Let $\mathscr{R}' = \{\boldsymbol{R}'(p) \mid p \in \gamma\}$ be any representation of γ. The *character* $\chi_i' = \chi(p_i : \mathscr{R}')$ *of* p_i *with respect to* $\mathscr{R}'$ is defined as the trace of $\boldsymbol{R}'(p_i)$,

$$\chi_i' = \operatorname{tr} \boldsymbol{R}'(p_i) = \sum_{j=1}^{n'} \big(\boldsymbol{R}'(p_i)\big)_{jj},$$

and the *character* $\boldsymbol{\chi}' = \boldsymbol{\chi}(\mathscr{R}')$ *of* $\mathscr{R}'$ is defined as the vector

$$\boldsymbol{\chi}' = (\chi_1', \chi_2', \ldots, \chi_g').$$

The character of an irreducible representation $\mathscr{R}^{(\varkappa)}$ of γ is

$$\boldsymbol{\chi}(\mathscr{R}^{(\varkappa)}) = \boldsymbol{\chi}^{(\varkappa)} = (\chi_1^{(\varkappa)}, \chi_2^{(\varkappa)}, \ldots, \chi_g^{(\varkappa)});$$

p_1 is usually taken to be the unity element of γ and $\mathscr{R}^{(1)}$ denotes the trivial (or totally symmetric) irreducible representation with $\boldsymbol{R}^{(1)}(p) = (1)$ for each $p \in \gamma$, so $\chi_1^{(\varkappa)}$ is simply the degree of $\mathscr{R}^{(\varkappa)}$,

$$\chi_1^{(\varkappa)} = n_\varkappa \qquad (\varkappa = 1, 2, \ldots, k),$$

and

$$\chi_i^{(1)} = 1 \qquad (i = 1, 2, \ldots, g).$$

For many groups the characters of their irreducible representations have been tabulated.[††] Clearly, any two elements of γ that belong to the same conjugacy class

---

[†] Chemists speak about the *group of symmetries of a molecule* rather than about the automorphism group of the corresponding graph. U. Wild, J. Keller, and Hs. H. Günthard [WiKG] explained that, in fact, the automorphism group is in the background of the procedure.

[††] For practical purposes it is a great handicap that the procedure can only be applied if character tables are available. Therefore, E. Heilbronner [Hei3] and B. J. McClelland [McC2] developed methods for factoring the characteristic polynomial which avoid the use of character tables, but these procedures are applicable only to multigraphs with very special symmetry properties; see Section 4.6, Remark 1. — Using group theoretical means (representations, characters) E. Heilbronner [Hei2] developed expressions for the eigenvalues of some graphs of particular types.

have the same character (with respect to any fixed representation), since the matrices representing these elements are similar: so the character tables usually include only one representative from each of the k conjugacy classes. Assuming that $\{p_1 = 1, p_2, \ldots, p_k\}$ is such a set of representatives, the table of characters is a square scheme of the form

γ	p_1	p_2	$\cdots$	p_k
$\mathscr{R}^{(1)}$	$\chi_1^{(1)}$	$\chi_2^{(1)}$	$\cdots$	$\chi_k^{(1)}$
$\mathscr{R}^{(2)}$	$\chi_1^{(2)}$	$\chi_2^{(2)}$	$\cdots$	$\chi_k^{(2)}$
.	.	.	$\cdots$	.
.	.	.	$\cdots$	.
.	.	.	$\cdots$	.
$\mathscr{R}^{(k)}$	$\chi_1^{(k)}$	$\chi_2^{(k)}$	$\cdots$	$\chi_k^{(k)}$

For our ends, however, it is a bit more convenient to use the full character table with g columns headed $p_1, p_2, \ldots, p_g$. Note that, if γ is abelian, then $k = g$; hence for abelian groups the character table is always the full table.

The rows of the (full) table are orthogonal, namely

$$\sum_{i=1}^{g} \bar{\chi}_i^{(\varkappa)} \chi_i^{(\mu)} = g \cdot \delta_{\varkappa\mu} \qquad (\varkappa, \mu = 1, 2, \ldots, k). \tag{5.14}$$

With respect to the new basis we are looking for, $\mathscr{R}$ has its totally reduced form

$$\mathscr{R}^{\boldsymbol{U}} = a_1 \circ \mathscr{R}^{(1)} \dotplus a_2 \circ \mathscr{R}^{(2)} \dotplus \cdots \dotplus a_k \circ \mathscr{R}^{(k)}. \tag{5.11$'$}$$

$\chi(\mathscr{R})$ does not depend on the special basis, so

$$\chi(\mathscr{R}) = \chi(\mathscr{R}^{\boldsymbol{U}}) = (\chi_1, \chi_2, \ldots, \chi_g). \tag{5.15}$$

The numbers χ_i are, on the one hand, determined by G and can easily be found by simple inspection:

$$\chi_i = \operatorname{tr} \boldsymbol{P}_i$$

is the number of vertices remaining fixed under the automorphism $\boldsymbol{P}_i$; on the other hand, (5.11)′ and (5.15) imply

$$\chi_i = \sum_{\varkappa=1}^{k} a_\varkappa \chi_i^{(\varkappa)} \qquad (i = 1, 2, \ldots, g). \tag{5.16}$$

By virtue of (5.14) we obtain from (5.16)

$$a_\varkappa = \frac{1}{g} \sum_{i=1}^{g} \bar{\chi}_i^{(\varkappa)} \cdot \chi_i \qquad (\varkappa = 1, 2, \ldots, k). \tag{5.17}$$

Note that

$$a_1 = \frac{1}{g} \sum_{i=1}^{g} \chi_i \tag{5.18}$$

is always positive since $\chi_i \geqq 0$ and not all χ_i can be zero (for example, $\chi_1 = n$), i.e., in (5.11)′ the trivial irreducible representation $\mathcal{R}^{(1)}$ is always present. In fact, a_1 has a simple meaning: according to a theorem of BURNSIDE (see, for example, [Big 2], p. 5), a_1 is equal to the number of orbits of Γ.

The new basis can be obtained in the following way. For any irreducible representation $\mathcal{R}^{(\varkappa)}$ with $a_\varkappa \neq 0$, form the vectors

$$\boldsymbol{v}_i^{(\varkappa)} = \sum_{j=1}^{g} \bar{\chi}_j^{(\varkappa)} \boldsymbol{P}_j \boldsymbol{e}_i \qquad (i = 1, 2, \ldots, n). \tag{5.19}$$

Among these vectors specify $a_\varkappa n_\varkappa$ independent ones (which are always present). Perform this process for each $\mathcal{R}^{(\varkappa)}$ with $a_\varkappa \neq 0$. The set of all vectors so obtained is the new basis (which is, in general, not uniquely determined).

The eigenvectors of $\boldsymbol{A}$ belonging to the block $\boldsymbol{A}_\varkappa$ in (5.13) ($a_\varkappa \neq 0$) must be in the space spanned by the vectors $\boldsymbol{v}_i^{(\varkappa)}$ ($i = 1, 2, \ldots, n$). Therefore, these eigenvectors are of the form

$$\sum_{i=1}^{n} c_i \boldsymbol{v}_i^{(\varkappa)} = \sum_{i=1}^{n} c_i \sum_{j=1}^{g} \bar{\chi}_j^{(\varkappa)} \boldsymbol{P}_j \boldsymbol{e}_i \tag{5.20}$$

with suitably chosen numbers c_i. Inserting this expression into the system of equations for the eigenvectors of G will lead to a reduction of this system as shall now be demonstrated by a simple

Example. Consider the graph $G =$ 1—2—3. Its automorphism group γ is the (cyclic) group of order 2:

$$\gamma = \{p_1, p_2\} \text{ with } p_2^2 = p_1 = 1.$$

Since γ is abelian, $k = g = 2$, $n_1 = n_2 = 1$. The (full) table of characters of γ is

γ	p_1	p_2
$\mathcal{R}^{(1)}$	1	1
$\mathcal{R}^{(2)}$	1	-1

By (5.17), $a_1 = \frac{1}{2}(1 \cdot 3 + 1 \cdot 1) = 2$, $a_2 = \frac{1}{2}\big(1 \cdot 3 + (-1) \cdot 1\big) = 1$. (5.19) yields

$$\begin{aligned}
\varkappa = 1:\quad & \boldsymbol{v}_1^{(1)} = 1 \cdot \boldsymbol{e}_1 + 1 \cdot \boldsymbol{e}_3 = \boldsymbol{e}_1 + \boldsymbol{e}_3\\
& \boldsymbol{v}_2^{(1)} = 1 \cdot \boldsymbol{e}_2 + 1 \cdot \boldsymbol{e}_2 = 2\boldsymbol{e}_2\\
& \boldsymbol{v}_3^{(1)} = 1 \cdot \boldsymbol{e}_3 + 1 \cdot \boldsymbol{e}_1 = \boldsymbol{e}_1 + \boldsymbol{e}_3 = \boldsymbol{v}_1^{(1)},\\
\varkappa = 2:\quad & \boldsymbol{v}_1^{(2)} = 1 \cdot \boldsymbol{e}_1 + (-1) \cdot \boldsymbol{e}_3 = \boldsymbol{e}_1 - \boldsymbol{e}_3\\
& \boldsymbol{v}_2^{(2)} = 1 \cdot \boldsymbol{e}_2 + (-1) \cdot \boldsymbol{e}_2 = \boldsymbol{o}\\
& \boldsymbol{v}_3^{(2)} = 1 \cdot \boldsymbol{e}_3 + (-1) \cdot \boldsymbol{e}_1 = -\boldsymbol{e}_1 + \boldsymbol{e}_3 = -\boldsymbol{v}_1^{(2)}.
\end{aligned}$$

So, for $\varkappa = 1$ we obtain $a_1 n_1 = 2$ independent vectors, namely, $\boldsymbol{v}_1^{(1)}$ and $\boldsymbol{v}_2^{(1)}$. Hence, according to (5.20), two of the eigenvectors of G are of the form

$$c_1(\boldsymbol{e}_1 + \boldsymbol{e}_3) + c_2\boldsymbol{e}_2 = c_1\boldsymbol{e}_1 + c_2\boldsymbol{e}_2 + c_1\boldsymbol{e}_3. \tag{5.21}$$

For $\varkappa = 2$ we obtain only $a_2 n_2 = 1$ independent vector, namely, $\boldsymbol{v}_1^{(2)} = \boldsymbol{e}_1 - \boldsymbol{e}_3$. Hence $\boldsymbol{e}_1 - \boldsymbol{e}_3$ is itself an eigenvector of G.

Case $\varkappa = 1$. Inserting (5.21) into our eigenvalue problem $\boldsymbol{A}\boldsymbol{x} = \lambda\boldsymbol{x}$ we obtain

$$\begin{pmatrix} 0 & 1 & 0 \\ 1 & 0 & 1 \\ 0 & 1 & 0 \end{pmatrix} \begin{pmatrix} c_1 \\ c_2 \\ c_1 \end{pmatrix} = \lambda \begin{pmatrix} c_1 \\ c_2 \\ c_1 \end{pmatrix}, \qquad \begin{pmatrix} c_2 \\ 2c_1 \\ c_2 \end{pmatrix} = \begin{pmatrix} \lambda c_1 \\ \lambda c_2 \\ \lambda c_1 \end{pmatrix}.$$

So there remains a system of only two independent equations

$$c_2 = \lambda c_1, \qquad 2c_1 = \lambda c_2$$

or, equivalently,

$$\begin{pmatrix} 0 & 1 \\ 2 & 0 \end{pmatrix} \begin{pmatrix} c_1 \\ c_2 \end{pmatrix} = \lambda \begin{pmatrix} c_1 \\ c_2 \end{pmatrix}, \tag{5.22}$$

yielding the eigenvalues $\lambda = \pm\sqrt{2}$.

Case $\varkappa = 2$. Here the problem reduces to only one independent equation, namely, $\lambda = 0$.

Remark: Note that vertices 1 and 3 are in the same orbit of Γ: so, if we apply the procedure described in the proof of Theorem 4.7 (Section 4.5) to the matrix $\boldsymbol{A}$, i.e., if we add the third column to the first column and subtract the first row from the third row, we shall obtain a matrix $\begin{pmatrix} \boldsymbol{F} & \boldsymbol{B} \\ \boldsymbol{O} & \boldsymbol{C} \end{pmatrix}$ where the matrix $\boldsymbol{F}$ of order 2 is the adjacency matrix of the front divisor of G induced by the orbits of Γ (the orbits being regarded as colour classes). Indeed, we obtain $\boldsymbol{F} = \begin{pmatrix} 0 & 1 \\ 2 & 0 \end{pmatrix}$. That is the same matrix as in (5.22), and it can be shown that also in the general case the matrix $\boldsymbol{A}_1$ in (5.13) is similar to the adjacency matrix of the front divisor of G induced by the orbits of Γ (see next section).

5.3. The front divisor induced by a subgroup of the automorphism group

Let G be any strongly connected† multi-digraph with non-trivial automorphism group Γ, and let Σ be any non-trivial subgroup of Γ (possibly $\Sigma = \Gamma$). We shall now relate the results of Sections 4.4 (Theorem 4.3) and 4.5 (proof of Theorem 4.7) to the decomposition of Σ, regarded as a faithful representation $\mathcal{R}$ of the subgroup σ of the abstract automorphism group γ, into irreducible components.

† The strong connectedness of G is, of course, no essential restriction.

Let Σ have orbits $\Omega_1, \Omega_2, \ldots, \Omega_m$. By Theorem 4.3, Σ determines a front divisor $F = D_\Sigma$ of G whose vertices correspond to the orbits; we shall call F the *front divisor induced by* (*the orbits of*) Σ.

We recall from the proof of Theorem 4.7 how the adjacency matrices of F and of its codivisors C with respect to G can be obtained from the adjacency matrix of G:

From each orbit Ω_i we select a vertex in an arbitrary way, call it V_i, and number the remaining vertices V of G from $m + 1$ through n in such a way that $i < k \leqq m$ and $V_p \in \Omega_i \setminus \{V_i\}$, $V_q \in \Omega_k \setminus \{V_k\}$ imply $p < q$. The adjacency matrix of G corresponding to this labelling is denoted $\boldsymbol{G} = (g_{pq})$. For $k = 1, 2, \ldots, m$ we add to the k-th column all other columns belonging to vertices of Ω_k, and for $i = 1, 2, \ldots, m$ we subtract the i-th row from all other rows which belong to vertices of Ω_i. Eventually we obtain a matrix

$$\boldsymbol{M} = \begin{pmatrix} \boldsymbol{F} & \boldsymbol{B} \\ \boldsymbol{O} & \boldsymbol{C} \end{pmatrix}, \tag{5.23}$$

where $\boldsymbol{F}$ is the adjacency matrix of F and $\boldsymbol{C}$ is the adjacency matrix of the codivisor of F which corresponds to the special choice of the representatives $V_1, V_2, \ldots, V_m$.

We shall show that the procedure of adding columns and subtracting rows can be given the form of a transformation of $\boldsymbol{G}$ by a matrix $\boldsymbol{Q}$ so that

$$\boldsymbol{M} = \boldsymbol{G}^{\boldsymbol{Q}} = \boldsymbol{Q}^{-1}\boldsymbol{G}\boldsymbol{Q}. \tag{5.24}$$

To this end we define square matrices

$$\boldsymbol{E}_{pq} = (\delta_{ip}\delta_{jq})_{i,j=1,2,\ldots,n} \qquad (p, q = 1, 2, \ldots, n;\ p \neq q)$$

and note that

$$(\boldsymbol{I} + \boldsymbol{E}_{pq})^{-1} = \boldsymbol{I} - \boldsymbol{E}_{pq}.$$

Right-hand multiplication of any matrix with n columns by $\boldsymbol{I} + \boldsymbol{E}_{pq}$ is equivalent to adding the p-th column to the q-th column, and left-hand multiplication of any matrix with n rows by $\boldsymbol{I} - \boldsymbol{E}_{pq} = (\boldsymbol{I} + \boldsymbol{E}_{pq})^{-1}$ is equivalent to subtracting the q-th row from the p-th one. Note that $\boldsymbol{I} \pm \boldsymbol{E}_{pq}$ and $\boldsymbol{I} \pm \boldsymbol{E}_{rs}$ commute if $r \neq q$ and $s \neq p$.

Thus, in the transition from $\boldsymbol{G}$ to $\boldsymbol{M}$, the adding of columns is equivalent to the right-hand multiplication of $\boldsymbol{G}$ by the matrix

$$\boldsymbol{Q} = \prod_{i=1}^{m} \prod_{V_p \in \Omega_i \setminus \{V_i\}} (\boldsymbol{I} + \boldsymbol{E}_{pi})$$

(note that all factors commute since the subscripts p and i satisfy $i \leqq m < p$), and, analogously, the subtracting of rows is equivalent to the left-hand multiplication of $\boldsymbol{G}\boldsymbol{Q}$ by $\boldsymbol{Q}^{-1}$, which establishes (5.24).

The matrix $\boldsymbol{Q}$ transforms every $\boldsymbol{P} = \boldsymbol{R}(p) \in \Sigma$ into a matrix of the form

$$\boldsymbol{R}^{\boldsymbol{Q}}(p) = \boldsymbol{Q}^{-1}\boldsymbol{R}(p)\,\boldsymbol{Q} = \begin{pmatrix} \boldsymbol{I}_m & \boldsymbol{Y}(p) \\ \boldsymbol{O} & \boldsymbol{R}'(p) \end{pmatrix} \tag{5.25}$$

with some matrices $\boldsymbol{Y}(p)$, $\boldsymbol{R}'(p)$. This will immediately become evident if the rows and columns are arranged so that all permutation matrices of Σ appear in orbit separated form

$$\boldsymbol{P}' = \boldsymbol{P}^{(1)} \dotplus \boldsymbol{P}^{(2)} \dotplus \cdots \dotplus \boldsymbol{P}^{(m)};$$

the adding of columns then means that in each block $\boldsymbol{P}^{(i)}$ all columns different from the first one are added to the first one which results in a first column with all entries equal to 1, and the subsequent subtraction of rows means that in each block the first row is subtracted from all the others so that in the resulting blocks all entries in the first column — except for the first one — are equal to zero while the first one is 1. Rearranging the rows and columns according to the numbering adopted above results in a matrix of the form (5.25).

Note that $\mathscr{R}^Q = \{\boldsymbol{R}^Q(p) \mid p \in \sigma\}$ is a faithful representation of σ.

According to a theorem of H. MASCHKE (see, for example, [Boer], pp. 44-45) there is a matrix

$$\boldsymbol{T} = \begin{pmatrix} \boldsymbol{I}_m & \boldsymbol{Z} \\ \boldsymbol{O} & \boldsymbol{I}_{n-m} \end{pmatrix}$$

which simultaneously transforms all matrices $\boldsymbol{R}^Q(p)$ into the block diagonal form

$$\boldsymbol{R}^{QT}(p) = \boldsymbol{T}^{-1}\boldsymbol{R}^Q(p)\,\boldsymbol{T} = \begin{pmatrix} \boldsymbol{I}_m & \boldsymbol{O} \\ \boldsymbol{O} & \boldsymbol{R}'(p) \end{pmatrix} = \boldsymbol{I}_m \dotplus \boldsymbol{R}'(p). \tag{5.26}$$

Since the number m of orbits of Γ is equal to the multiplicity a_1 of the trivial irreducible representation $\mathscr{R}^{(1)}$ as a component of $\mathscr{R}$ (see the remark following (5.18)), we have succeeded in separating the trivial components from the representation $\mathscr{R}$:

$$\mathscr{R}^{QT} = m \circ \mathscr{R}^{(1)} \dotplus \mathscr{R}',$$

where $\mathscr{R}' = \{\boldsymbol{R}'(p) \mid p \in \sigma\}$ is a faithful representation of σ without trivial components. The matrices $\boldsymbol{R}'(p)$ are real and can — without knowing $\boldsymbol{T}$ — be calculated from Σ by simple additions and subtractions (evidently, all resulting entries are 0, +1, or −1).

We are interested in the effect of the transformation $\boldsymbol{T}$ if applied to $\boldsymbol{M}$. Note that $\boldsymbol{T}^{-1} = \begin{pmatrix} \boldsymbol{I}_m & -\boldsymbol{Z} \\ \boldsymbol{O} & \boldsymbol{I}_{n-m} \end{pmatrix}$. Recalling (5.23) and (5.24) we obtain

$$\boldsymbol{G}^{QT} = \boldsymbol{T}^{-1}\boldsymbol{M}\boldsymbol{T} = \begin{pmatrix} \boldsymbol{F} & \boldsymbol{H} \\ \boldsymbol{O} & \boldsymbol{C} \end{pmatrix}$$

with $\boldsymbol{H} = \boldsymbol{F}\boldsymbol{Z} + \boldsymbol{B} - \boldsymbol{Z}\boldsymbol{C}$. We shall show that $\boldsymbol{H}$ is, in fact, a zero matrix.

All matrices $\boldsymbol{P} \in \Sigma$ commute with $\boldsymbol{G}$, hence all matrices $\boldsymbol{P}^{QT} = \boldsymbol{R}^{QT}(p) = \begin{pmatrix} \boldsymbol{I}_m & \boldsymbol{O} \\ \boldsymbol{O} & \boldsymbol{R}'(p) \end{pmatrix}$ commute with $\boldsymbol{G}^{QT} = \begin{pmatrix} \boldsymbol{F} & \boldsymbol{H} \\ \boldsymbol{O} & \boldsymbol{C} \end{pmatrix}$:

$$\begin{pmatrix} \boldsymbol{I}_m & \boldsymbol{O} \\ \boldsymbol{O} & \boldsymbol{R}'(p) \end{pmatrix} \begin{pmatrix} \boldsymbol{F} & \boldsymbol{H} \\ \boldsymbol{O} & \boldsymbol{C} \end{pmatrix} = \begin{pmatrix} \boldsymbol{F} & \boldsymbol{H} \\ \boldsymbol{O} & \boldsymbol{C} \end{pmatrix} \begin{pmatrix} \boldsymbol{I}_m & \boldsymbol{O} \\ \boldsymbol{O} & \boldsymbol{R}'(p) \end{pmatrix} \text{ for all } p \in \sigma.$$

Multiplying block by block and equating we obtain

$$\boldsymbol{H} = \boldsymbol{H} \cdot \boldsymbol{R}'(p) \text{ for all } p \in \sigma, \tag{5.27}$$

$$\boldsymbol{R}'(p) \cdot \boldsymbol{C} = \boldsymbol{C} \cdot \boldsymbol{R}'(p) \text{ for all } p \in \sigma. \tag{5.28}$$

We keep (5.28) in mind and exploit (5.27).

Put $\boldsymbol{X} := (\boldsymbol{O}_m, \boldsymbol{H}) \cdot \boldsymbol{Q}^{-1}$ where $\boldsymbol{O}_m$ is the square zero matrix of order m ($(\boldsymbol{O}_m, \boldsymbol{H})$ is a matrix with m rows and n columns). Then

$$\begin{aligned} \boldsymbol{X} \cdot \boldsymbol{P} &= (\boldsymbol{O}_m, \boldsymbol{H}) \cdot \boldsymbol{Q}^{-1}\boldsymbol{P}\boldsymbol{Q} \cdot \boldsymbol{Q}^{-1} = (\boldsymbol{O}_m, \boldsymbol{H}) \cdot \boldsymbol{R}^{\boldsymbol{Q}}(p) \cdot \boldsymbol{Q}^{-1} \\ &= (\boldsymbol{O}_m, \boldsymbol{H}) \cdot \begin{pmatrix} \boldsymbol{I}_m & \boldsymbol{Y}(p) \\ \boldsymbol{O} & \boldsymbol{R}'(p) \end{pmatrix} \cdot \boldsymbol{Q}^{-1} \\ &= \big(\boldsymbol{O}_m, \boldsymbol{H}\boldsymbol{R}'(p)\big) \cdot \boldsymbol{Q}^{-1} = (\boldsymbol{O}_m, \boldsymbol{H}) \cdot \boldsymbol{Q}^{-1} = \boldsymbol{X}, \end{aligned}$$

i.e.

$$\boldsymbol{X}\boldsymbol{P} = \boldsymbol{X} \text{ for all } \boldsymbol{P} \in \Sigma.$$

Therefore, any two columns of $\boldsymbol{X}$ which belong to the same orbit of Σ are equal. That means that $\boldsymbol{X}$ has the form

$$\boldsymbol{X} = (\boldsymbol{x}^1\boldsymbol{x}^2 \cdots \boldsymbol{x}^m, \quad \boldsymbol{x}^1\boldsymbol{x}^1 \cdots \boldsymbol{x}^1 \quad \boldsymbol{x}^2\boldsymbol{x}^2 \cdots \boldsymbol{x}^2 \quad \cdots \quad \boldsymbol{x}^m\boldsymbol{x}^m \cdots \boldsymbol{x}^m),$$

where the $\boldsymbol{x}^i$ are columns vectors. $(\boldsymbol{O}_m, \boldsymbol{H}) = \boldsymbol{X}\boldsymbol{Q}$ is retrieved from $\boldsymbol{X}$ by adding, for $k = 1, 2, \ldots, m$, to the k-th column of $\boldsymbol{X}$ all other columns belonging to the orbit Ω_k; so we obtain

$$(\boldsymbol{O}_m, \boldsymbol{H}) = (t_1\boldsymbol{x}^1 t_2\boldsymbol{x}^2 \cdots t_m\boldsymbol{x}^m, \quad \boldsymbol{x}^1\boldsymbol{x}^1 \cdots \boldsymbol{x}^1 \quad \boldsymbol{x}^2\boldsymbol{x}^2 \cdots \boldsymbol{x}^2 \quad \cdots \quad \boldsymbol{x}^m\boldsymbol{x}^m \cdots \boldsymbol{x}^m)$$

with $t_k = |\Omega_k| \geqq 1$ $(k = 1, 2, \ldots, m)$. We conclude that $\boldsymbol{H}$ is a zero matrix.

That was our assertion.

Summary of results: *Let G be a strongly connected multi-digraph with non-trivial automorphism group Γ, and let Σ be any non-trivial subgroup of Γ (possibly $\Sigma = \Gamma$). Then there is a matrix $\boldsymbol{W} = \boldsymbol{Q}\boldsymbol{T}$ which simultaneously transforms*

(a) *the representation $\mathscr{R} = \Sigma$ of the abstract group σ into the form*

$$\mathscr{R}^{\boldsymbol{W}} = m \circ \mathscr{R}^{(1)} \dotplus \mathscr{R}',$$

where $\mathscr{R}'$ is a faithful representation of σ without trivial components, and

(b) *the adjacency matrix $\boldsymbol{G}$ of G into the corresponding form*

$$\boldsymbol{G}^{\boldsymbol{W}} = \boldsymbol{F} \dotplus \boldsymbol{C},$$

where $\boldsymbol{F}$ is the adjacency matrix of the front divisor of G induced by Σ and $\boldsymbol{C}$ is the adjacency matrix of one of its codivisors.

The matrices $\boldsymbol{R}'(p) \in \mathscr{R}'$ which have only entries 0, $+1$, and -1 can easily (without calculating $\boldsymbol{W}$) be obtained from the $\boldsymbol{R}(p) = \boldsymbol{P} \in \Sigma$; all of them commute with $\boldsymbol{C}$ (see (5.28)).

5.4. Cospectral graphs with prescribed (distinct) automorphism groups

Having in mind the results of the preceding sections one might be inclined to conjecture that the (abstract) automorphism group of a graph is determined by its spectrum, but that is not at all true. On the contrary, L. BABAI showed that, in a sense, the spectrum and the abstract automorphism group of a graph have "almost nothing" to do with one another: by an ingenious construction making use of an idea of A. J. SCHWENK [Schw1] (see Section 6.1, Corollary to Theorem 6.3) he established the following theorem which we quote without proof.

Theorem 5.13 (L. BABAI [Bab2], [Bab3]): *Given a finite family of (not necessarily different) finite (abstract) groups $\gamma_1, \gamma_2, \ldots, \gamma_N$, there are pairwise non-isomorphic finite graphs $G_1, G_2, \ldots, G_N$ such that for $i = 1, 2, \ldots, N$ the automorphism group of G_i is isomorphic to γ_i, and all graphs G_i have the same spectrum.*

This statement can even be generalized to endomorphism monoids [Bab2].

The main result of [Bab2] asserts that *any category of finite algebraic systems of a given finite type is isomorphic to a full subcategory of the category of finite graphs such that the graphs corresponding to algebraic systems having common underlying set are cospectral.*

For some more results on cospectral graphs and corresponding automorphism groups see nos. 7, 8, 9 of Section 5.5.

5.5. Miscellaneous results and problems

1. If the characteristic polynomial of a graph G is irreducible over the rational number field, then $\Gamma(G)$ is trivial. (For the proof use Theorems 4.3, 4.5.)

(A. MOWSHOWITZ [Mow 4])

2. Let G be a graph with at least three vertices, which has all distinct eigenvalues. Then $\Gamma(G)$ is non-transitive. (For the proof use Theorem **5.2.**)

(A. MOWSHOWITZ [Mow1], see also [ElTu])

3. Let ψ denote the group of quaternions.

a) If the abstract automorphism group γ of the multigraph G contains ψ as a subgroup then G has an eigenvalue of multiplicity $\geqq 4$.

b) There is a graph G with the following properties:
(i) all of its eigenvalues have multiplicities $\leqq 4$,
(ii) $\gamma = \psi$.

(L. BABAI, private communication)

4. Let G be a digraph on n vertices having a circulant adjacency matrix with non-repeated eigenvalues. Then $\Gamma(G)$ is isomorphic to ζ_n, the cyclic group of order n.

(B. ELSPAS, J. TURNER [ElTu])

5. L. LOVÁSZ showed that the determination of the spectrum of a graph which has a transitive automorphism group can be reduced to the same task for some Cayley graph† and found

†For the definition of Cayley graphs see no. 6 of this Section.

a formula for certain power sums of the eigenvalues from which he derived the following rule for calculating the eigenvalues:

Let G be a graph on n vertices with transitive automorphism group; denote this group, or any transitive subgroup of it, by $\boldsymbol{\Gamma} = \{\boldsymbol{P}_1, \boldsymbol{P}_2, \ldots, \boldsymbol{P}_g\}$, faithfully representing the abstract group $\gamma = \{p_1, p_2, \ldots, p_g\}$. Let $n_\varkappa$ and $\boldsymbol{\chi}^{(\varkappa)} = (\chi_1^{(\varkappa)}, \chi_2^{(\varkappa)}, \ldots, \chi_g^{(\varkappa)})$ $(\varkappa = 1, 2, \ldots, k)$, respectively, be the degrees and the characters of the irreducible representations $\mathscr{R}^{(\varkappa)}$ of γ.

Denote by w_{im} the number of walks of length m connecting a vertex to its image under $\boldsymbol{P}_i \in \boldsymbol{\Gamma}$, and put

$$f_{\varkappa m} = \frac{1}{g} \sum_{i=1}^{g} w_{im} \chi_i^{(\varkappa)},$$

$$F_\varkappa(x) = \begin{vmatrix} 1 & f_{\varkappa 1} & \cdots & f_{\varkappa, n_\varkappa - 1} & f_{\varkappa n_\varkappa} \\ 0 & 1 & \cdots & f_{\varkappa, n_\varkappa - 2} & f_{\varkappa, n_\varkappa - 1} \\ \cdot & \cdot & \ddots & \cdot & \cdot \\ 0 & 0 & \cdots & 1 & f_{\varkappa 1} \\ \dfrac{1}{n_\varkappa!} & \dfrac{x}{(n_\varkappa - 1)!} & \cdots & \dfrac{x^{n_\varkappa - 1}}{1!} & x^{n_\varkappa} \end{vmatrix}$$

$(\varkappa = 1, 2, \ldots, k)$. Write down each root of the equation

$$F_\varkappa(x) = 0$$

$n_\varkappa$ times for $\varkappa = 1, 2, \ldots, k$. Remove $g - n$ zeros from this sequence. The remaining numbers are the eigenvalues of G.

If γ is abelian, then $k = g = n$ and all $n_\varkappa = 1$, therefore the above rule simplifies considerably: We obtain one eigenvalue associated with each irreducible representation $\mathscr{R}^{(\varkappa)}$, namely

$$\lambda^{(\varkappa)} = \sum_{\boldsymbol{P}_i(j) \cdot j} \chi_i^{(\varkappa)} \qquad (\varkappa = 1, 2, \ldots, n),$$

where j is any fixed vertex of G (the summation runs through all i for which the image of j under $\boldsymbol{P}_i$ is adjacent to j).

Note that for abelian groups the $\chi_i^{(\varkappa)}$ are roots of unity and compare the result with Theorem 5.4 (Section 5.1).

(L. Lovász [Lová])

6. L. Babai succeeded in simplifying Lovász' formula for the power sums of the eigenvalues of Cayley graphs mentioned in no. 5.

Let γ be a finite group with generating set δ satisfying $\delta = \delta^{-1}$, $1 \notin \delta$. The *Cayley graph* $G = G(\gamma, \delta)$ with vertex set γ is defined as follows: The vertices p and q are connected by an edge if and only if $p^{-1}q \in \delta$ (note that together with $p^{-1}q \in \delta$ also $q^{-1}p \in \delta$)†. Clearly, the left regular permutation representation of γ (which is transitive) is a subgroup of the automorphism group of G.

We continue with the notation of no. 5. The character of $p \in \gamma$ with respect to $\mathscr{R}^{(\varkappa)}$ is denoted by $\chi^{(\varkappa)}(p)$. L. Babai proved:

The eigenvalues of the Cayley graph $G = G(\gamma, \delta)$ can be arranged in families $\boldsymbol{\Phi}_{\varkappa\nu}$ $(\nu = 1, 2, \ldots, n_\varkappa;\ \varkappa = 1, 2, \ldots, k)$ such that

† The condition $\delta = \delta^{-1}$ guarantees that $G(\gamma, \delta)$ is an undirected graph; it is no longer needed if Cayley digraphs are admitted: In a Cayley digraph $G = G(\gamma, \delta)$ there is an arc directed from vertex p to vertex q if and only if $q^{-1}p \in \delta$.
All the other statements of no. 6 remain unchanged.

(i) $\Phi_{\varkappa\nu}$ contains exactly $n_\varkappa$ eigenvalues which are all equal; if the common value of these eigenvalues is denoted by $\lambda_{\varkappa\nu}$, then

(ii) $\lambda_{\varkappa 1}^t + \lambda_{\varkappa 2}^t + \cdots + \lambda_{\varkappa n_\varkappa}^t = \sum_{p_{i_1}, p_{i_2}, \ldots, p_{i_t} \in \delta} \chi^{(\varkappa)}(p_{i_1} p_{i_2} \cdots p_{i_t})$

($\varkappa = 1, 2, \ldots, k$) for every natural number t.

Using this formula for $t = 1, 2, \ldots, n_\varkappa$ one can obtain a polynomial of degree $n_\varkappa$ with roots $\lambda_{\varkappa 1}, \lambda_{\varkappa 2}, \ldots, \lambda_{\varkappa n_\varkappa}$.

If γ is abelian, then all $n_\varkappa = 1$ and (ii) reduces to

$$\lambda^{(\varkappa)} = \sum_{p \in \delta} \chi^{(\varkappa)}(p) \qquad (\varkappa = 1, 2, \ldots, n)$$

which is equivalent to the formula of no. 5.

(L. Babai [Bab 1])

7. For any positive integer k and for any of the following types of connectedness there are k non-isomorphic cospectral digraphs with given automorphism group:

(i) Weak but not unilateral,
(ii) Unilateral but not strong,
(iii) Strong but not symmetric,
(iv) Symmetric.

(V. Krishnamoorthy, K. R. Parthasarathy [KrP2])

8. Let p denote a prime and k any positive integer. For $p > 64k$ there exist k non-isomorphic cospectral Cayley graphs of the dihedral group $\boldsymbol{D}_p$.

(L. Babai [Bab 1])

9. Let G be a graph on n vertices with adjacency matrix $\boldsymbol{A}$ and automorphism group Γ and let $\sigma(\boldsymbol{A})$ denote the set of distinct eigenvalues of $\boldsymbol{A}$. If $\lambda \in \sigma(\boldsymbol{A})$ has multiplicity $m(\lambda)$, let $\{z_i \mid i = 1, 2, \ldots, m(\lambda)\}$ be an orthonormal basis of the eigenspace of $\boldsymbol{A}$, associated with the eigenvalue λ. Let $\boldsymbol{Z}_\lambda$ be the $n \times m(\lambda)$ matrix with i-th column z_i. Set $\boldsymbol{A}_\lambda = \boldsymbol{Z}_\lambda \boldsymbol{Z}_\lambda^{\mathsf{T}}$: then (as is well known)

$$\boldsymbol{A} = \sum_{\lambda \in \sigma(\boldsymbol{A})} \lambda \boldsymbol{A}_\lambda,$$

$$\boldsymbol{A}_\lambda^2 = \boldsymbol{A}_\lambda, \quad \boldsymbol{A}_\lambda \boldsymbol{A}_\mu = \boldsymbol{A}_\mu \boldsymbol{A}_\lambda = \boldsymbol{O} \quad \text{for} \quad \lambda, \mu \in \sigma(\boldsymbol{A}), \quad \lambda \neq \mu.$$

C. D. Godsil denotes the i-th row of $\boldsymbol{Z}_\lambda$ by $\boldsymbol{w}_\lambda(i)$ and calls it the *weight vector on λ of the vertex i of G*. Since $\boldsymbol{A}\boldsymbol{Z}_\lambda = \lambda \boldsymbol{Z}_\lambda$,

$$\lambda \boldsymbol{w}_\lambda(i) = \sum_{j \cdot i} \boldsymbol{w}_\lambda(j).$$

Theorem: Let G be a regular graph of degree r with Γ acting transitively on both its vertices and its edges. Let λ be an eigenvalue such that all the weight vectors of G on λ are distinct.
(i) If G contains a clique on c vertices, then $m(\lambda) \geqq c - 1$ and if $m(\lambda) = c - 1$, $\lambda = -r/(c - 1)$.
(ii) If G is connected and $m(\lambda) = 2$, then $r = 2$, i.e., G is a circuit.

(C. D. Godsil [Gods])

6. Characterization of Graphs by Means of Spectra

In this chapter various forms of the following problem are considered:

The spectrum or some spectral characteristics of a graph are given. Determine all graphs from a given class of graphs having that spectrum or those spectral characteristics.

Thus, the possibility of graph identification is investigated. In the preceding chapters in the majority of cases, in contrast to the subject of this chapter, procedures enabling the determination of structural details of a graph on the basis of its spectrum have been described.

Since a graph is in general not characterized by its spectrum, some families of non-isomorphic cospectral graphs are represented in the first section. Further, several cases in which a spectral characterization is possible are described.

6.1. Some families of non-isomorphic cospectral graphs

At one time it was conjectured that non-isomorphic graphs have different spectra [GüPr], [Har2], but in [CoSi1] L. Collatz and U. Sinogowitz showed that the two graphs in Fig. 6.1 are cospectral (= isospectral). A *P*air of *I*sospectral *N*on-isomorphic *G*raphs is sometimes called a *PING*.

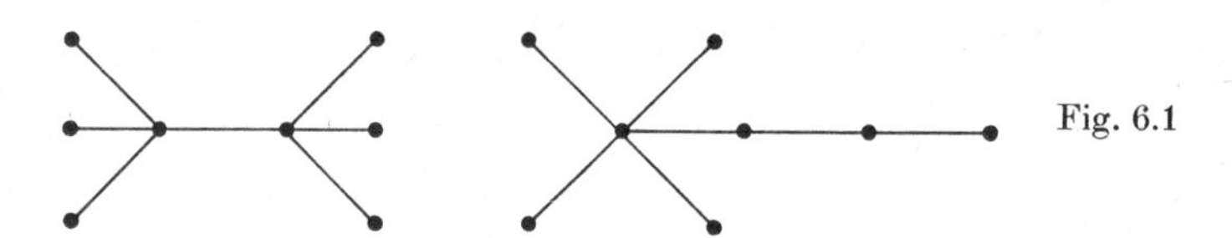

Fig. 6.1

In this section the known constructions of non-isomorphic cospectral graphs will be surveyed. These are important for the following reasons:

(i) From families of non-isomorphic cospectral graphs it is possible, by observing the properties that are not constant throughout the family to determine properties that are not dependent on the spectrum. From Fig. 6.1, for example, it is seen that graphs with different degree sequences can have the same spectrum.

(ii) By deriving properties that must be constant for all graphs in a cospectral family, it is possible to construct families of graphs where no two graphs have the same spectrum by letting these properties vary. The degree of a regular graph or, indeed, regularity itself is such a property.

Among all graphs with four or fewer vertices, no pair has the same spectrum. For graphs with five vertices, however, there is a pair of cospectral graphs as is shown in Fig. 6.2. From this example we see that the connectedness of a graph is not, in general, determined by the spectrum of a graph.

Fig. 6.2

A pair of connected cospectral graphs with six vertices exists and is given in Fig. 6.3 [Bak2]. Their common characteristic polynomial is

$$\lambda^6 - 7\lambda^4 - 4\lambda^3 + 7\lambda^2 + 4\lambda - 1.$$

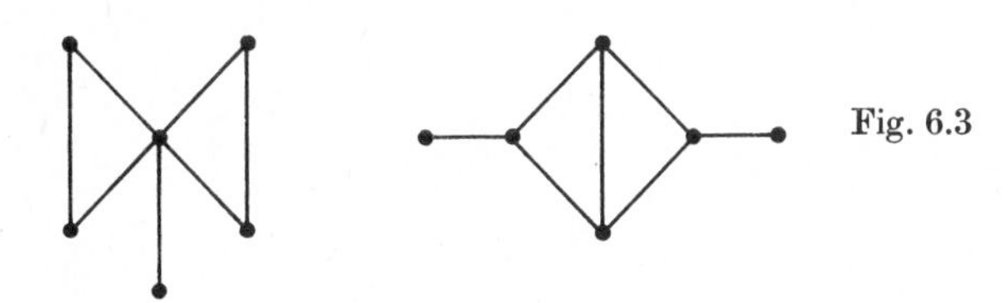

Fig. 6.3

A. J. Hoffman and D. K. Ray-Chaudhuri [HoR2] have given a pair of non-isomorphic, regular, connected cospectral graphs with twelve vertices, and they are displayed in Fig. 6.4. Note that the first of these graphs is planar and the second is non-planar. The common spectrum of these graphs is:

$$[4, 2, 2, 2, 0, 0, 0, -2, -2, -2, -2, -2].$$

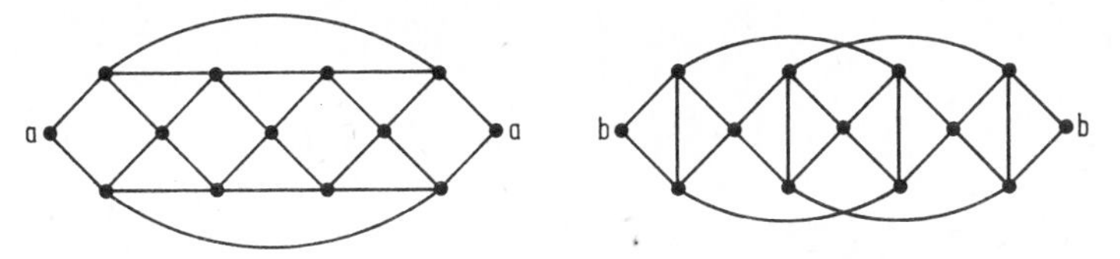

Vertices with the same label are identified

Fig. 6.4

There are two cospectral pairs with sixteen vertices [Hof3], [Shr2], a cospectral quadruple with twenty-eight vertices [Sei2], and a set of ninety-one cospectral graphs with thirty-five vertices [BuS2]. All of these are regular and connected.

There have also been constructed seven regular graphs with twenty-five vertices such that each is cospectral with, but not isomorphic to, its complement. All fourteen graphs are cospectral; they have 12, 2, and -3 as eigenvalues with respective multiplicities of 1, 12, and 12 [Paul]. Pairs of non-regular, complementary, cospectral, but non-isomorphic, graphs have also been found [GoM2].

In addition there are three cospectral pairs of graphs among all cubic graphs with 14 vertices and a pair of cospectral cubic graphs on 20 vertices [BuČCS], [BuCv].

We can use the examples in Fig. 6.1 and Fig. 6.2 to form more general cospectral families. Let G be any graph, and let G_k be the direct sum of k copies of the first graph in Fig. 6.1, $n - k$ copies of the second graph in Fig. 6.1, and G. Then the family of graphs $\{G_k \mid k = 1, 2, \ldots, n\}$ is a cospectral one, and each graph in this family contains a copy of G. This family may be made arbitrarily large.

The complete bipartite graph $K_{m,n}$ has $\pm \sqrt{mn}$ as simple eigenvalues and 0 as an eigenvalue with multiplicity $m + n - 2$. Let r be any positive integer and let d be a divisor of r. Define G_d to be the graph which consists of the direct sum of $K_{d,\frac{r}{d}}$ and $r + 1 - d - \frac{r}{d}$ isolated vertices. Then the family $\{G_d \mid d \text{ divides } r\}$ is a cospectral one.

In fact arbitrarily large families of regular connected graphs also exist as the following argument shows.

Theorem 6.1 (A. J. Hoffman [Mow5]): *For any positive integer n there is an integer N such that for any $m \geqq N$ there exist n non-isomorphic cospectral regular connected graphs with m vertices each.*

Proof: Let $N = 12n + 4$, G_1 and G_2 be the two graphs in Fig. 6.4, and G_t be a regular graph of degree 4 with t vertices, $t = 5, 6, \ldots, 16$. Now if $m \geqq N$, then $m = 12k + t$ where $k \geqq n - 1$ and $5 \leqq t \leqq 16$. Let H_i be the direct sum of i copies of G_1, $k - i$ copies of G_2, and G_t. The H_i are regular, not connected, and cospectral for $i = 0, \ldots, k$. By Theorem 2.6, if $\overline{H}_i$ is the complement of H_i, then $\{\overline{H}_i \mid i = 1, \ldots, k\}$ forms a regular connected cospectral family.

In fact Theorem 6.1 can be expanded so that each member of the cospectral family contains a given graph.

Theorem 6.2: *Let G be a given graph. Then for any integer n there is an integer $\overline{N}$ such that for any $m \geqq \overline{N}$ there exist n non-isomorphic, cospectral, regular, connected graphs each of which has m vertices and contains G (as an induced subgraph).*

Proof. Consider the graph which is the complement of G. It can be imbedded in a regular graph H of degree d if d is greater than the degree of any vertex in the graph [Hof7]. The graphs $\overline{H}_i$ constructed in the proof of Theorem 6.1 are of degree $m - 5$. Choose $\overline{N}$ large enough such that $\overline{N} - 5$ is greater than both the maximum degree of the complement of G and $N - 5$ where N is the bound from Theorem 6.1. Then define G_i as the direct sum of $\overline{H}_i$ and H. Finally, it is clear that the $\overline{G}_i$ are cospectral, non-isomorphic, regular, connected, and each $\overline{G}_i$ contains G. Hence the proof is complete.

Suppose we have two graphs G_1 and G_2 with $v_1 \in \mathscr{V}(G_1)$ and $v_2 \in \mathscr{V}(G_2)$; the *coalescence* of G_1 and G_2 with respect to v_1 and v_2 is formed by identifying v_1 and v_2

and is denoted by $G_1 \cdot G_2$. In other words, $\mathscr{V}(G_1 \cdot G_2) = \mathscr{V}(G_1) \cup \mathscr{V}(G_2) \cup \{v^*\} - \{v_1, v_2\}$, with two vertices in $G_1 \cdot G_2$ adjacent if they are adjacent in G_1 or G_2, or if one is v^* and the other is adjacent to v_1 or v_2 in G_1 or G_2. In terms of the adjacency matrix (denoted by $\boldsymbol{A}(G)$) this implies

$$\boldsymbol{A}(G_1 \cdot G_2) = \begin{pmatrix} 0 & \boldsymbol{r} & \boldsymbol{s} \\ \boldsymbol{r}^{\mathsf{T}} & \boldsymbol{A}(G_1 - v_1) & \boldsymbol{O} \\ \boldsymbol{s}^{\mathsf{T}} & \boldsymbol{O} & \boldsymbol{A}(G_2 - v_2) \end{pmatrix}$$

where $\boldsymbol{r}$ is the vector of off-diagonal terms in $\boldsymbol{A}(G_1)$ in the row corresponding to v_1 and $\boldsymbol{s}$ is the vector of off-diagonal terms in $\boldsymbol{A}(G_2)$ in the row corresponding to v_2. If, for a graph G with n vertices, $\Phi(G) := (-1)^n P_G(\lambda)$ and $\boldsymbol{B} = \boldsymbol{A}(G_1 \cdot G_2) - \lambda \boldsymbol{I}$, then

$$\Phi(G_1 \cdot G_2) = \det \boldsymbol{B} = \sum_{\sigma \in \sigma_n} (-1)^{\operatorname{sgn} \sigma} B_{1,\sigma(1)} \cdots B_{n,\sigma(n)},$$

where σ_n is the symmetric group of order n.

Consider permutations of the following types: (i) $\sigma(1) = 1$ or $\sigma(1) = j$ where j is a column that corresponds to $\boldsymbol{A}(G_1 - v_1)$, and (ii) $\sigma(1) = 1$ or $\sigma(1) = j$ where j is a column that corresponds to $\boldsymbol{A}(G_2 - v_2)$. Since the four upper left-hand blocks are merely $\boldsymbol{A}(G_1)$, the permutations in case (i) contribute $\Phi(G_1)\,\Phi(G_2 - v_2)$ to $\Phi(G_1 \cdot G_2)$. Similarly the permutations of case (ii) contribute $\Phi(G_1 - v_1)\,\Phi(G_2)$ to $\Phi(G_1 \cdot G_2)$. The permutations such that $\sigma(1) = 1$ have been counted twice and contribute $-\lambda\Phi(G_1 - v_1)\,\Phi(G_2 - v_2)$. Hence

$$\Phi(G_1 \cdot G_2) = \Phi(G_1)\,\Phi(G_2 - v_2) + \Phi(G_1 - v_1)\,\Phi(G_2) + \lambda\Phi(G_1 - v_1)\,\Phi(G_2 - v_2).$$

Thus the following lemma is clear.

Lemma 6.1: *If G_2 and G_2' have the same spectrum, and $G_2 - v_2$ and $G_2' - v_2'$ also have the same spectrum, then $G_1 \cdot G_2$ and $G_1 \cdot G_2'$ have the same spectrum.*

A. J. Schwenk [Schw1] has used Lemma 6.1 to construct large families of cospectral trees. He notes that the two trees in Fig. 6.5 satisfy the hypothesis of Lemma 6.1. In particular $G_2 \cong G_2'$ and

$$\Phi(G_2 - v_2) = \Phi(G_2' - v_2') = \lambda^2(\lambda^2 - 2)\,(\lambda^4 - 4\lambda^2 + 2).$$

Fig. 6.5

Now let G_1 be any given graph with $v_1 \in \mathscr{V}(G_1)$, and take n copies of G_2. Form a new graph by identifying v_1 and all the v_2 in the copies of G_1 and G_2. The idea is to remove one copy of G_2 attached by coalescence at v_2 and replace it by G_2' attached at v_2'. By Lemma 6.1 the new graph is cospectral with the old one. By iterating the replacement of G_2 with G_2', a cospectral family with $n + 1$ members arises.

Theorem 6.3: *Given a graph G and an integer n, there exists a cospectral family $\{G_i \mid i = 1, \ldots, n\}$ such that each G_i contains a copy of G and all the circuits in G_i are contained in G.*

Corollary (A. J. Schwenk [Schw1]): *There exist arbitrarily large families of cospectral trees.*

A. J. Schwenk has also shown that as the number of vertices gets large, the probability that a tree contains a subgraph isomorphic to G_2 which may be replaced by G_2' approaches 1. Hence the probability that a tree is characterized by its spectrum approaches zero as the number of vertices gets large.

There are other methods for constructing cospectral graphs. If G_1, G_2, G_3, and G_4 are arbitrary graphs then, according to [ŽiTR], the graphs in Fig. 6.6 are cospectral. The proof is based on Theorem 2.12.

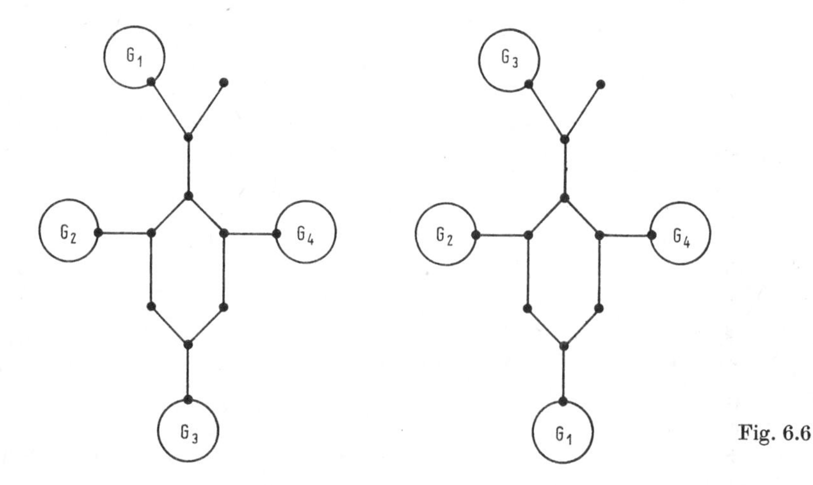

Fig. 6.6

A. Mowshowitz [Mow5] has shown that for any non-negative integer k the trees of Fig. 6.7 are cospectral. In the case of $k = 0$ this reduces to the original cospectral pair of graphs of L. Collatz and U. Sinogowitz in Fig. 6.1.

Other cospectral graphs can be formed by using the sum of graphs. If G_1 and G_2 are cospectral, and G is any given graph, then let H_k be the sum formed by taking one summand of G, k summands of G_1, and $n - k$ summands of G_2. Then the $\{H_k \mid k = 0, 1, \ldots, n\}$ is a cospectral family, and G is contained in each H_k. Cospectral families with other special constraints can be formed using the sum of graphs [Doo7]. Cospectral graphs can be constructed similarly using the NEPS (see Theorem 2.23).

Since the chromatic numbers of the graphs of Fig. 6.12 are not equal, we deduce that the chromatic number of a graph is in general not determined by its spectrum. Moreover, A. J. Hoffman [Hof11] has constructed PINGs in which one graph has chromatic number equal to 3 and the other one has chromatic number equal to $2^k + 1$, k being a given non-negative integer.

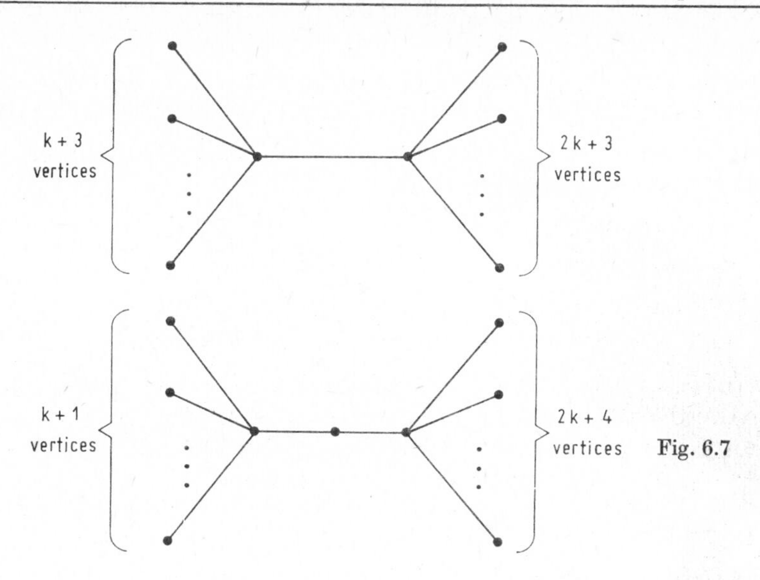

Fig. 6.7

In [CvG3] a procedure is given by which all graphs, cospectral with a given graph G, can be determined if the greatest eigenvalue of G is not greater than 2.

Results on families of cospectral graphs with distinct automorphism groups have already been reported about in Section 5.4.

Further data about cospectral graphs can be found in [Bab1], [Bab2], [Bab3], [BaHa], [BeJa], [Bruc], [BuCS], [Cha2], [Cook], [Cve7], [Cve9], [Djok], [Doo6], [Doo10], [EdGG], [Fish], [Gut8], [GoM2], [Har2], [HaKMR], [Her1], [Her2], [HeEl], [Hof1], [KrP1], [KrP2], [LiSe], [Meye], [RaTŽ], [Pons], [Sei1], [Sei3], [Sei 4], [Sei6], [StMa], [Turn2], [Кел5].

See also Sections 6.2, 6.5, 8.4, Chapter 9, and Appendix.

6.2. The characterization of a graph by its spectrum

In the last section several families of cospectral graphs were constructed. This might give the impression that very few if any graphs are characterized by their spectra, but it will be shown that this is not the case. Indeed, sometimes only meager information concerning the spectrum of a graph is sufficient to characterize it. The first type of graph to be considered will be those with a small number of distinct eigenvalues. For these the following lemma is useful:

Lemma 6.2: *Let G be a regular connected graph with n vertices whose spectrum is $\mathcal{S}$ and whose set of distinct eigenvalues is $\mathcal{T}$. Suppose $|\mathcal{T}| \leqq 4$. Then the following are equivalent:*

(i) *H is cospectral with G.*

(ii) *H is regular, connected, has n vertices and has $\mathcal{T}$ for its set of distinct eigenvalues.*

Proof.

(i) $\Rightarrow$ (ii): By Theorems 3.22 and 3.23, the regularity of G implies the regularity of H, and since the maximum eigenvalue is simple, H is connected.

(ii) $\Rightarrow$ (i): Let $\lambda_1 > \lambda_2 > \lambda_3 > \lambda_4$ be the distinct eigenvalues with respective multiplicities m_1, m_2, m_3, and m_4. Now

(a) $m_1 = 1$, since the graph is connected.

(b) $m_1 + m_2 + m_3 + m_4 = n$, the number of vertices.

(c) $m_1\lambda_1 + m_2\lambda_2 + m_3\lambda_3 + m_4\lambda_4 = 0$.

(d) $m_1\lambda_1^2 + m_2\lambda_2^2 + m_3\lambda_3^2 + m_4\lambda_4^2 = n\lambda_1$, since the graph is regular.

The equations (a)—(d) determine m_1, m_2, m_3, and m_4 uniquely and hence the spectrum is known. A virtually identical argument is used for $|\mathcal{T}| < 4$.

Suppose a graph G with adjacency matrix $\boldsymbol{A}$ has one distinct eigenvalue λ with multiplicity m. Then since $\operatorname{tr} \boldsymbol{A} = 0$, it must be that $\lambda = 0$. Since the minimal polynomial is $m(x) = x - \lambda$, it must be that $\boldsymbol{A} = \boldsymbol{O}$, and G consists of m isolated vertices. If G has two distinct eigenvalues $\lambda_1 > \lambda_2$ with multiplicities m_1 and m_2, then the minimal polynomial is $m(x) = (x - \lambda_1)(x - \lambda_2)$, and $\boldsymbol{A}^2 - (\lambda_1 + \lambda_2)\boldsymbol{A} + \lambda_1\lambda_2\boldsymbol{I} = \boldsymbol{O}$. Since† $a_{kk}^{(2)} = -\lambda_1\lambda_2$ for all k, G is regular with degree $-\lambda_1\lambda_2$. Since the degree is the greatest eigenvalue, it must be that $\lambda_2 = -1$. Further, if two vertices are non-adjacent, then they are not joined by a path of length 2. Hence G is a direct sum of m_1 complete graphs of order $\lambda_1 + 1$.

Theorem 6.4 (M. Doob [Doo3]): *G has one eigenvalue if and only if G is a totally disconnected graph. G has two distinct eigenvalues $\lambda_1 > \lambda_2$ with multiplicities m_1 and m_2 if and only if G is the direct sum of m_1 complete graphs of order $\lambda_1 + 1$. In this case $\lambda_2 = -1$ and $m_2 = m_1\lambda_1$.*

In particular, complete graphs and graphs with no edges are characterized by their spectra.

With three distinct eigenvalues the situation becomes a bit more complex. Suppose G is bipartite, with $\lambda_1 > \lambda_2 > \lambda_3$ being the distinct eigenvalues with respective multiplicities m_1, m_2, and m_3. Then by Theorem 3.11, $\lambda_1 = -\lambda_3$, $m_1 = m_3$, and $\lambda_2 = 0$ and by Theorem 3.13 each connected component has diameter 2 or less. Thus each connected component is a complete bipartite graph or an isolated vertex.

Theorem 6.5 (M. Doob [Doo3]): *Let G be a bipartite graph with eigenvalues $\lambda_1 > \lambda_2 > \lambda_3$ with respective multiplicities m_1, m_2, and m_3. Then $\lambda_3 = -\lambda_1$, $\lambda_2 = 0$, $m_3 = m_1$, and G is the direct sum of m_1 complete bipartite graphs K_{r_i,s_i} where $r_is_i = \lambda_1^2$, $i = 1, \ldots, m_1$, and $m_2 - \sum_{i=1}^{m_1} (r_i + s_i - 2)$ isolated vertices.*

Corollary: *If G is as in Theorem 6.5 and $\lambda_1^2 = p$, a prime integer, then G is characterized by its spectrum.*

---

† We resume our notation from Section 1.8, Theorem 1.9:

$$\boldsymbol{A} = (a_{ij}), \quad \boldsymbol{A}^q = (a_{ij}^{(q)}), \quad \text{etc.}$$

Since the spectrum of the direct sum of two graphs is the union of their spectra, a graph with three distinct eigenvalues is either a direct sum of complete graphs of two different orders or the direct sum of graphs with three distinct eigenvalues. Thus with no essential loss of generality it may be assumed that the graph is connected.

By Theorem 3.22, any graph that is cospectral with a regular graph is itself regular. Thus if we are considering a family of regular graphs, we need only look among other regular graphs to determine whether or not a cospectral mate exists. A regular graph that has $r > \lambda_2 > \lambda_3$ as distinct eigenvalues by Theorem 3.25 satisfies the equation

$$\boldsymbol{A}^2 - (\lambda_2 + \lambda_3)\,\boldsymbol{A} + \lambda_2\lambda_3\boldsymbol{I} = \frac{(r - \lambda_2)\,(r - \lambda_3)}{n}\,\boldsymbol{J}$$

where r is the degree of the graph and n is the number of vertices. The reader will recall from Chapter 3 that such graphs are strongly regular.

By Theorem 6.4, G is neither a complete graph nor the complement of a complete graph, so that there exist two adjacent vertices and two non-adjacent vertices. Now $a_{kk}^{(2)} = r$ and $a_{kk} = 0$, so

$$\boldsymbol{A}^2 - (\lambda_2 + \lambda_3)\,\boldsymbol{A} + \lambda_2\lambda_3\boldsymbol{I} = (r + \lambda_2\lambda_3)\,\boldsymbol{J}.$$

Hence for two non-adjacent vertices $a_{ij}^{(2)} = r + \lambda_2\lambda_3$ and for two adjacent vertices $a_{ij}^{(2)} = r + \lambda_2\lambda_3 + \lambda_2 + \lambda_3$. Hence $\lambda_2\lambda_3$ is an integer and $\lambda_2 + \lambda_3$ is an integer. Finally since $1 \leqq a_{ij}^{(2)} \leqq r$ for a pair of non-adjacent vertices, and $0 \leqq a_{ij}^{(2)} \leqq r - 1$ for a pair of adjacent vertices, it follows that

$$-(r - 1) \leqq \lambda_2\lambda_3 \leqq 0 \quad \text{and} \quad -r \leqq \lambda_2 + \lambda_3 \leqq r - 2.$$

Can these bounds be attained? Consider first the case of a graph G where $\lambda_2\lambda_3 = 0$. Then the distinct eigenvalues are r, 0, and λ_3. Since $\lambda_2 + \lambda_3$ is an integer, let $\lambda_3 = -m$. Hence the number of vertices is $r + m$, and by Theorem 2.6 the eigenvalues of the complement of G are $m - 1$ and -1. Hence by Theorem 6.4, G is the complement of $\frac{r}{m} + 1$ complete graphs, and G is a complete multipartite graph.

Theorem 6.6 (M. Doob [Doo3]): *A regular graph G has eigenvalues r, 0, and λ_3 if and only if the complement of G is the direct sum of $-\frac{r}{\lambda_3} + 1$ complete graphs of order $-\lambda_3$.*

Corollary (H.-J. Finck [Finc]): *Regular complete multipartite graphs are characterized by their spectra.*

A. K. Kel'mans [Кел 3], [Кел 5] proved the more general theorem that *any complete multipartite graph is characterized by its C-spectrum.*

J. H. Smith [Sm,J] has generalized Theorem 6.6 to the following theorem.

Theorem 6.7 (J. H. Smith [Sm,J]): *A graph has exactly one positive eigenvalue if and only if its non-isolated vertices form a complete multipartite graph.*

Proof. We may ignore isolated vertices. If a graph G is not complete multipartite, it contains an induced subgraph as given in Fig. 6.8a, since the non-adjacency relation of the vertices of G must be non-transitive. But x is not an isolated vertex·

in G. In this way we conclude that G contains at least one of graphs b), c), d) from Fig. 6.8 as an induced subgraph. All of these graphs have two positive eigenvalues and, according to Theorem 0.10, H has at least two. The characteristic polynomial of the complete multipartite graph is given by (2.51). Having in view the signs of its coefficients, we see that this polynomial has exactly one positive eigenvalue. This completes the proof of the Theorem.

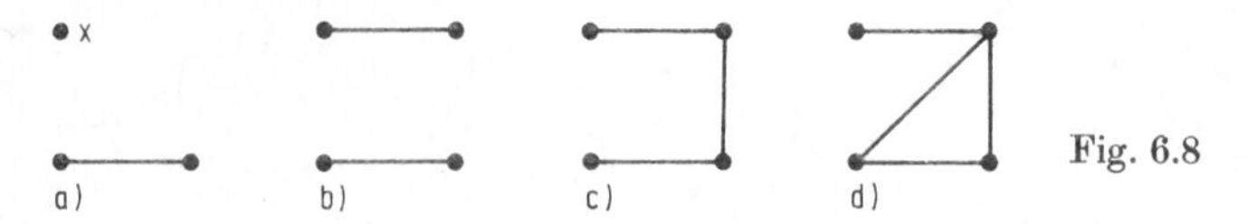

Fig. 6.8

D. T. Malbaški noticed that if a graph G has exactly one positive eigenvalue, then the non-isolated vertices of G form a complete k-partite graph with $k = 1 + p_-$, where p_- is the number of negative eigenvalues of G.

Complete multipartite graphs are in general not characterized by their spectra. For example, graphs $K_{18,3,3}$ and $K_{9,9,2} \dot{+} 4K_1$ represent a PING. There exist infinitely many PINGs of a similar type.

We continue with the analysis of spectral properties of regular graphs having three distinct eigenvalues.

Now suppose $\lambda_2 + \lambda_3 = -r$. Then by Theorem 3.25 for any pair of adjacent vertices $a_{ij}^{(2)} + r = r + \lambda_2\lambda_3$ so that $\lambda_2\lambda_3 \geqq 0$. Hence by the bounds on the product of the eigenvalues, $\lambda_2\lambda_3 = 0$, and by Theorem 6.6, G is a complete multipartite graph, and since $\lambda_3 = -r$, G is a complete bipartite graph.

If $\lambda_2 + \lambda_3 = r - 2$, then the number of vertices is $\dfrac{r}{r + \lambda_1\lambda_2} + 1$, and since $\lambda_1\lambda_2 \geqq -(r-1)$, it must be that $r \geqq n - 1$. Hence the graph is a complete graph; but since a complete graph has only two distinct eigenvalues, no graph attains the bound. Thus $\lambda_2 + \lambda_3 \leqq r - 3$, and this bound is attained by a circuit of length 5.

A regular graph of degree r and diameter d can have at most $1 + r + r(r-1) + \cdots + r(r-1)^d$ vertices. A graph that attains the bound is called a *Moore graph of diameter d*.

Returning to spectral properties, suppose $\lambda_2\lambda_3 = -(r-1)$. From Theorem 3.25, if two vertices are non-adjacent, there is exactly one path of length 2 joining them. This implies that the number of vertices is $1 + r + r(r-1) = 1 + r^2$ so that the graph must be a Moore graph of diameter 2, and, again by Theorem 3.25, it must be that $\lambda_2\lambda_3 = -1$. Thus $\boldsymbol{A}^2 + \boldsymbol{A} - (r-1)\boldsymbol{I} = \boldsymbol{J}$, $P(x) = x^2 + x - (r-1)$ is the polynomial of the graph, and the eigenvalues of the graph are r and the two roots of $P(x)$. Hence $\lambda_2 + \lambda_3 = -1$, $\lambda_2 - \lambda_3 = (4r-3)^{1/2}$, and the respective multiplicities satisfy $m_2 + m_3 = r^2$ and $m_2\lambda_2 + m_3\lambda_3 = -r$; thus $r(r-2) = (m_2 - m_3)(4r-3)^{1/2}$. If $(4r-3)^{1/2}$ is not an integer, then $m_2 = m_3$ and $r = 0$ or $r = 2$. If $(4r-3)^{1/2}$ is an integer, then $(4r-3)^{1/2} = 2t + 1$, and $2t + 1$ divides

$$\begin{aligned} 16r(r-2) &= 4r(4r-8) = [(2t+1)^2 + 3] \cdot [(2t+1)^2 - 5] \\ &= (2t+1)[(2t+1)^3 + 8(2t+1)] - 15. \end{aligned}$$

Hence $2t + 1$ divides 15 so that $t = 0, 1, 2$ or 7 and $r = 1, 3, 7$ or 57. Now $r = 0$ and $r = 1$ are impossible, so the only possible degrees are $r = 2, 3, 7$ or 57. If $r = 2$, the only graph with five vertices is the cycle. If $r = 3$, the unique graph with ten vertices such that $\boldsymbol{A}^2 + \boldsymbol{A} - 2\boldsymbol{I} = \boldsymbol{J}$ is the Petersen graph. If $r = 7$, there is a unique graph of this type which is displayed in [HoSi] and is called the *Hoffman-Singleton graph*.† For the final case, $r = 57$, it is unknown whether or not such graphs exist. Graphs satisfying $\boldsymbol{A}^2 + \boldsymbol{A} - (r - 1)\,\boldsymbol{I} = \boldsymbol{J}$ are precisely the Moore graphs of diameter 2; they were first investigated by A. J. Hoffman and R. R. Singleton [HoSi] and despite eighteen years effort the final case of $r = 57$ is yet unsolved. The following table indicates the eigenvalues in each case:

r	λ_2	λ_3	m_1	m_2	m_3
2	$\frac{1}{2}\left(-1 + \sqrt{5}\right)$	$\frac{1}{2}\left(-1 - \sqrt{5}\right)$	1	2	2
3	1	-2	1	5	4
7	2	-3	1	28	21
57	7	-8	1	1729	1520

Various other properties related to the case $r = 57$ and others have appeared in the literature [Dam1], [BaIt], [BeLo], [Free], [BoDo].

Theorem 6.8 (A. J. Hoffman, R. R. Singleton [HoSi], generalized by M. Doob): *If G is a regular graph with distinct eigenvalues $r > \lambda_2 > \lambda_3$ and $\lambda_2\lambda_3 = -(r - 1)$, then G is a Moore graph of diameter 2. Such graphs exist for $r = 2, 3, 7$ and possibly for $r = 57$.*

The only connected bipartite graph with three distinct eigenvalues is the complete bipartite graph. There are, however, many bipartite graphs with four distinct eigenvalues and, unlike the case of three eigenvalues, not all of these graphs are characterized by their spectra. To construct some of these, a *balanced incomplete block design*, denoted *BIBD*, will now be defined. A BIBD consists of v elements

---

† J. J. Seidel has given a geometric realization of the Hoffmann-Singleton graph HS by means of the projective geometry $\mathbb{PG}\,(3, 2)$. This geometry can be defined in the following way: $\mathbb{PG}\,(3, 2)$ has as its points the 15 four-tuples $X = (x_1, x_2, x_3, x_4)$ where x_i is 0 or 1 and not all the x_i are zero. Considering $\{0, 1\}$ as the two element field, addition of four-tuples is defined in the usual coordinatewise manner. For any two points X_1 and X_2, the line through X_1 and X_2 consists of the three points X_1, X_2, and $X_1 + X_2$: thus there are 35 lines and each point is incident with exactly 7 lines. Given three non-collinear points X_1, X_2, and X_3, the plane through X_1, X_2, and X_3 consists of the seven points X_1, X_2, X_3, $X_1 + X_2$, $X_1 + X_3$, $X_2 + X_3$, and $X_1 + X_2 + X_3$. We proceed to the Seidel realization of HS: The $7^2 + 1 = 50$ vertices of HS correspond to the points and lines of $\mathbb{PG}\,(3, 2)$. Now it is possible to identify the 35 lines of $\mathbb{PG}\,(3, 2)$ with the 35 triples from a set of seven elements such that two lines intersect in $\mathbb{PG}\,(3, 2)$ if and only if as triples they intersect in exactly one element. With this construction we define two vertices in the graph to be adjacent if they correspond to two lines which are disjoint as triples, or if they correspond to a point and a line which are incident.

and b subsets of these elements called *blocks* such that (i) each element is contained in r blocks, (ii) each block contains k elements, and (iii) each pair of elements is simultaneously contained in λ^* blocks. The integers (v, b, r, k, λ^*) are called the *parameters* of the design. (Usually, the parameter λ^* of a BIBD is simply denoted by λ, but we shall write λ^* in order to avoid any confusion with the eigenvalues.)

In the particular case $r = k$ the design is called *symmetric*. There is a rich and extensive literature concerning the theory and construction of block designs. A survey may be found in [Ha,M]. Given a BIBD, the graph of the design is formed in the following way: the $b + v$ vertices of the graph correspond to the blocks and elements of the design with two vertices adjacent if and only if one corresponds to a block and the other corresponds to an element contained in that block. Clearly the graph is bipartite with each vertex of degree r or k depending on whether it corresponds to an element or a block. It is not hard to determine the spectrum of this graph once the incidence matrix of the design is defined. It is a (0, 1) matrix of order $v \times b$ whose rows correspond to the elements and whose columns correspond to the blocks of the design. An entry is 1 if and only if the element corresponding to the row is contained in the block corresponding to the column. Directly from the definitions of a design it follows that the incidence matrix $\boldsymbol{B}$ satisfies $\boldsymbol{B}\boldsymbol{B}^\mathsf{T} = (r - \lambda^*)\,\boldsymbol{I} + \lambda^*\boldsymbol{J}$ and $\boldsymbol{J}\boldsymbol{B} = k\boldsymbol{J}$. From these equations there follow the two fundamental relationships between the parameters of a design which are $vr = bk$ and $\lambda^*(v - 1) = r(k - 1)$. The first follows from $vr\boldsymbol{J}_{v\times b} = \boldsymbol{J}_v(\boldsymbol{B}\boldsymbol{J}_b) = (\boldsymbol{J}_v\boldsymbol{B})\,\boldsymbol{J}_b = bk\boldsymbol{J}_{v\times b}$, and the second follows from a similar evaluation of $\boldsymbol{J}\boldsymbol{B}\boldsymbol{B}^\mathsf{T}\boldsymbol{J}$. Now observe that the adjacency matrix of the graph of a BIBD has the form

$$\boldsymbol{A}(G) = \begin{pmatrix} \boldsymbol{O} & \boldsymbol{B} \\ \boldsymbol{B}^\mathsf{T} & \boldsymbol{O} \end{pmatrix},$$

where $\boldsymbol{B}$ is the incidence matrix of the design. Hence $\boldsymbol{A}^2(G) = \boldsymbol{B}\boldsymbol{B}^\mathsf{T} \dotplus \boldsymbol{B}^\mathsf{T}\boldsymbol{B}$.

Since $\boldsymbol{B}\boldsymbol{B}^\mathsf{T} = (r - \lambda^*)\,\boldsymbol{I} + \lambda^*\boldsymbol{J}$ and $\boldsymbol{B}^\mathsf{T}\boldsymbol{B}$ have the same non-zero eigenvalues, the eigenvalues of $\boldsymbol{A}^2(G)$ are rk with multiplicity 2, $r - \lambda^*$ with multiplicity $2(v - 1)$, and 0 with multiplicity $b - v$. Thus the eigenvalues of the bipartite graph G are $\pm(rk)^{\frac{1}{2}}$, $\pm(r - \lambda^*)^{\frac{1}{2}}$, and 0 with multiplicities of 1, $v - 1$, and $b - v$.

It is interesting to note in passing that the parameters of the design completely determine the spectrum of its graph. Hence two non-isomorphic designs with the same parameters determine two cospectral non-isomorphic graphs, and there is a cospectral family of non-isomorphic graphs with 125 vertices, which contains 163, 929, 929, 318, 400 members [Wi,RM]. Because of this phenomenon, a graph of a BIBD is said to be characterized by its spectrum if a graph with the same spectrum is the graph of a BIBD with the same parameters.

It has been seen that the parameters of the design determine the spectrum of the graph. Under what circumstances is the converse true, i.e., given a graph with the eigenvalues of a BIBD, is it in fact the graph of a BIBD with appropriate parameters? This question can be answered affirmatively for symmetric designs, for suppose for positive integers k and λ^* there is a graph with $\pm k$ as simple eigenvalues and $\pm(k - \lambda^*)^{\frac{1}{2}}$ as eigenvalues with multiplicity $v - 1$ each. From Theorem 3.22 and Theorem 3.11, the graph is regular and bipartite. Hence $A^2(G) = B^{\mathsf{T}}B \dotplus BB^{\mathsf{T}}$ where BB^{T} has k^2 and $k - \lambda^*$ as eigenvalues of multiplicities 1 and $v - 1$. BB^{T} and J commute and consequently have a common set of eigenvectors from which follows that

$$(BB^{\mathsf{T}} - \lambda^* J)\, x = (k - \lambda^*)\, x \qquad \text{for all vectors } x.$$

Thus $BB^{\mathsf{T}} = \lambda^* J + (k - \lambda^*)\, I$ and B is the incidence matrix of an appropriate BIBD.

Theorem 6.9: *The graph of a symmetric design is characterized by its spectrum (in the above defined sense).*

There are other graphs that are characterized by their spectra. Any regular graph with n vertices of degree 0, 1, 2, $n - 3$, $n - 2$, or $n - 1$ is so characterized. By Theorem 2.6 and Theorem 6.4, the cases of graphs with degree 0, 1, $n - 2$, and $n - 1$ are easy. If a graph is regular of degree 2, then it is the direct sum of circuits. Now, the next to largest eigenvalue of a circuit with n vertices is $2 \cos \frac{2\pi}{n}$. Hence from the next to largest eigenvalue of G the number of vertices in the largest circuit can be determined. Deleting that connected component from the graph, the order of the remaining circuits of the graph may be determined by iteration of the above argument. Hence graphs of degree 2 are determined by their spectra and by Theorem 2.6 graphs of degree $n - 3$ are also determined by their spectra.

Several other families of graphs that can be characterized by their spectra have appeared in the literature. The techniques used tend to be somewhat involved; some of these will be indicated in the next section.

We shall now give a spectral characterization, of a somewhat different type, of certain digraphs. The proof is taken from the original paper.

Theorem 6.10 (B. Elspas, J. Turner [ElTu]): *Let $\mathscr{H}$ be the set of all digraphs (or graphs) with n (n being a prime number) vertices and with a cyclic adjacency matrix (such digraphs are polynomials of a cycle;* see Section 2.1). *Then non-isomorphic digraphs from $\mathscr{H}$ have different spectra.*

Proof. Let A and B be the adjacency matrices of two digraphs from $\mathscr{H}$. We shall prove that, if A and B have the same eigenvalues, then there exists a permutation matrix P of the form $P = (\delta_{qi,j})_1^n$ such that $B = P^{-1}AP$, where δ_{kj} is the Kronecker δ-symbol, q is an integer with $1 \leq q \leq n - 1$, and qi has to be reduced modulo n. Notice that the proof is valid if A and B are arbitrary cyclic [illegible]trices of prime order, $n > 2$, with rational entries.

Let $a_0, a_1, \ldots, a_{n-1}$ and $b_0, b_1, \ldots, b_{n-1}$ be the entries of the first rows of matrices A and B. Let

$$f_a(x) = \sum_{i=0}^{n-1} a_i x^i, \qquad f_b(x) = \sum_{i=0}^{n-1} b_i s^i.$$

Then it is known that the eigenvalues of A and B, respectively, are given by

$$\alpha_k = f_a(\omega^k), \qquad k = 0, 1, \ldots, n - 1,$$
$$\beta_k = f_b(\omega^k), \qquad k = 0, 1, \ldots, n - 1,$$

where ω is any primitive n-th root of 1. By assumption, the set of α_k's is the same as the set of β_k's. In particular, $\alpha_1 = \beta_q$ for some integer $q = 0, 1, \ldots, n-1$. We may, in fact, assume that $q \neq 0$, for, if $n - 1 > 1$ (as we assume), then replacement of the arbitrary primitive root ω by any one of the other $n - 2$ primitive roots $\omega^2, \omega^3, \ldots, \omega^{n-1}$ will result in a permutation of the α_i, $i = 1, \ldots, n-1$, so that α_1 is replaced by an α_c that matches an eigenvalue $\beta_{q'}$ with $q' \neq 0$. So, without loss of generality, $q \neq 0$.

Now, the equation $\alpha_1 = \beta_q$ states that

$$F(\omega) = 0,$$

where $F(x) = \sum a_i x^i - \sum b_i x^{qi(\mathrm{mod}\, n)}$ is a rational polynomial of degree $< n$. Hence, $F(x)$ is divisible by the minimum polynomial (minimum function) of ω over the rational numbers. This minimum polynomial is the cyclotomic polynomial of order n, which for $n =$ prime is known to be

$$\varphi_n(x) = x^{n-1} + \cdots + x + 1.$$

But $\varphi_n(\omega^k) = 0$ for $k = 1, \ldots, n-1$, so that also $F(\omega^k) = 0$ for $k = 1, \ldots, n-1$. That is, we have proved that

$$\alpha_k = \sum a_i \omega^{ki} = \sum b_i \omega^{kqi} = \beta_{kq} \qquad \text{for } k = 1, \ldots, n-1$$

(with subscripts taken modulo n throughout).

Now, since q is relatively prime to n, the index kq runs through the non-zero residues mod n as k runs through $1, \ldots, n-1$. Since the complete sets of n eigenvalues are identical, and the α_k are paired off with the β_k for $k = 1, \ldots, n-1$ by the above relation, it follows that also $\alpha_0 = \beta_0$. In other words, $\alpha_k = \beta_{qk}$ for $k = 0, 1, \ldots, n-1$.

The entries a_j of the circulant matrix $\boldsymbol{A}$ may be computed from the eigenvalues

$$a_j = \frac{1}{n} \sum_{k=0}^{n-1} \alpha_k \omega^{-kj}.$$

Putting $j = qi$, we have

$$a_{qi} = \frac{1}{n} \sum \alpha_k \omega^{-kqi} = \frac{1}{n} \sum \beta_{qk} \omega^{-kqi} = \frac{1}{n} \sum_{s=0}^{n-1} \beta_s \omega^{-si} = b_i.$$

The relation $b_i = a_{qi}$ is equivalent to the matrix relation $\boldsymbol{B} = \boldsymbol{P}\boldsymbol{A}\boldsymbol{P}^{\mathsf{T}} = \boldsymbol{P}\boldsymbol{A}\boldsymbol{P}^{-1}$, where $\boldsymbol{P}$ is the permutation matrix whose i, j-th element is $\delta_{qi,j}$. This completes the proof.

The special case of this theorem, related to graphs, was proved by J. TURNER [Turn 1].

6.3. The characterization and other spectral properties of line graphs

Line graphs have been considered in Section 2.4. These graphs have been particularly interesting in terms of their spectral properties. One such property is the lower bound on the eigenvalues as was first noted by A. J. HOFFMAN [Hof 6]. If $\boldsymbol{R}$ is the incidence matrix of a graph G, then $\boldsymbol{R}^{\mathsf{T}}\boldsymbol{R} = 2\boldsymbol{I} + \boldsymbol{A}(L(G))$, and since $\boldsymbol{R}^{\mathsf{T}}\boldsymbol{R}$ is positive semidefinite, the eigenvalues of $\boldsymbol{A}(L(G))$ are all bounded below by -2. If the least eigenvalue of a graph is denoted by $\lambda(G)$, then $\lambda(L(G)) \geqq -2$. When is this bound attained? If $\boldsymbol{A}(L(G))$ acts on a vector $\boldsymbol{x}$, the coordinates correspond to the vertices of $L(G)$ which in turn correspond to the edges of G. Note that for any vertex in G the sum of the coordinates corresponding to the edges incident to that vertex is zero if and only if $\boldsymbol{Rx} = \boldsymbol{o}$, and that this implies that $\boldsymbol{A}(L(G))\,\boldsymbol{x} = -2\boldsymbol{x}$. Define $\boldsymbol{x}$ to be in the kernel of $\boldsymbol{R}$ if $\boldsymbol{Rx} = \boldsymbol{o}$. Thus a sufficient (and, as will be shown, necessary) condition for $\lambda(L(G)) = -2$ is that the kernel of $\boldsymbol{R}$ contains a non-zero vector. If G contains a closed walk of even length whose edges in order are $e_1, e_2, \ldots, e_{2m}$, a vector $\boldsymbol{x}$ can be constructed by letting the value of the coordinate corresponding to e_i be $(-1)^i$ and letting all other coordinates be zero. (If an edge appears more than once in the walk, then the coordinate corresponding to that edge is the sum of the values of the e_i that are equal to that edge.) The vector so constructed is in the kernel of $\boldsymbol{R}$, and, if it is non-zero, $\lambda(L(G)) = -2$. Further, if one connected component contains two odd circuits, then an even closed walk may be found by considering the edges in order around the first circuit, along a path, if necessary, to the second circuit, around the second circuit, and finally back along the path to the first circuit. Then once again a non-zero vector can be constructed in the kernel of $\boldsymbol{R}$. Thus if a graph G contains an even circuit or if a connected component contains two odd circuits, then $\lambda(L(G)) = -2$. This condition is necessary as well as sufficient, as can be seen in two steps: first, $\boldsymbol{x}$ is in the kernel of $\boldsymbol{R}$ is equivalent to $\boldsymbol{R}^\mathsf{T}\boldsymbol{Rx} = \boldsymbol{o}$, and second, there is a non-zero vector in the kernel of $\boldsymbol{R}$ if and only if G contains an even circuit or two odd circuits in one connected component. For the first step, suppose $\boldsymbol{R}^\mathsf{T}\boldsymbol{Rx} = \boldsymbol{o}$. Then we have $\boldsymbol{x}^\mathsf{T}\boldsymbol{R}^\mathsf{T}\boldsymbol{Rx} = \boldsymbol{o}$ so that $\boldsymbol{Rx} = \boldsymbol{o}$. For the second step, let $\boldsymbol{x}$ be a non-zero vector in the kernel of $\boldsymbol{R}$. Consider the subgraph determined by the edges corresponding to non-zero coordinates, and let $\overline{G}$ be one of the connected components of that subgraph. Define $\overline{\boldsymbol{x}}$ to be the restriction of $\boldsymbol{x}$ to the coordinates corresponding to edges of $\overline{G}$, and let $\overline{\boldsymbol{R}}$ be the incidence matrix of $\overline{G}$. Then $\overline{\boldsymbol{R}}\overline{\boldsymbol{x}} = \boldsymbol{o}$ and the sum of the coordinates corresponding to edges incident to any particular vertex of $\overline{G}$ is zero. Since $\overline{\boldsymbol{x}}$ is non-zero in every coordinate, $\overline{G}$ has no vertices of degree one and, in particular, is not a tree. If $\overline{G}$ contains an even circuit or two odd circuits, then the desired conclusion is satisfied. If $\overline{G}$ contains precisely one circuit and that circuit is odd, then, since there are no vertices of degree one, $\overline{G}$ is an odd circuit. If $e_1, e_2, \ldots, e_{2n+1}$ are the edges of the circuit in order and $\overline{x}_1, \ldots, \overline{x}_{2m+1}$ are the corresponding coordinates of $\overline{\boldsymbol{x}}$, then $\overline{x}_i = -\overline{x}_{i+1}$, $i = 1, \ldots, 2m$, and $\overline{x}_1 = -\overline{x}_{2m+1}$. This implies that $\overline{\boldsymbol{x}} = \boldsymbol{o}$, which is a contradiction. Hence the following theorem has been proved.

Theorem 6.11 (M. Doob [Doo2], [Doo8]): *For any graph G, $\lambda(L(G)) \geqq -2$. Equality holds if and only if G contains an even circuit or two odd circuits in the same connected component.*

In fact the circuit structure of G can be used to determine a basis of eigenvectors for the eigenspace corresponding to the eigenvalue -2. In particular the multiplicity of that eigenvalue is the number of independent even closed walks of G. Further details concerning this construction can be found in [Doo8].

For any graph G, it has been shown that $\lambda(L(G)) \geqq -2$. Is this a characterization of line graphs? The answer is no, for $\lambda(K_{1,3}) = -\sqrt{3}$ and $\lambda(K_{1,4}) = -2$. What if only regular graphs are considered? The answer is still no, for the Petersen graph, the graph cospectral and not isomorphic to $L(K_{4,4})$, the three graphs cospectral but not isomorphic to the triangular graph $T(8)$, and one of the graphs in Fig. 6.4 are not line graphs but all have least eigenvalue equal to -2 (for definitions and some details see p. 174, in particular Theorems 6.18 and 6.16). If we define $H(n)$ to be the regular graph with $2n$ vertices and degree equal to $2n - 2$, then, since $H(n)$ is a complete multipartite graph, by Theorem 6.6 the distinct eigenvalues are $2n - 2$, 0, and -2. Since two non-adjacent vertices can be mutually adjacent to at most four other vertices in a line graph, $H(n)$ is not a line graph for $n > 3$.

The following important theorem asserts that if the degree is large enough then any regular graph with least eigenvalue not larger than -2 is either a line graph or has a connected component isomorphic to $H(n)$.

Theorem 6.12 (A. J. Hoffman, D. K. Ray-Chaudhuri [HoR3]): *Let H be a regular connected graph with degree greater than* 16 *and* $\lambda(H) \geqq -2$. *Then either* $H = H(n)$ *for some n or* $H = L(G)$ *for some G.*

This bound on the degree is the best possible since the Schläfli graph (for the definition see p. 184; see also Theorem 6.29, p. 185) is not a line graph, has least eigenvalue equal to -2, and is regular and connected with degree equal to 16.

A similar result for non-regular graphs has been given in the following theorem by D. K. Ray-Chaudhuri [Ray1]. Define $d(G)$ to be the minimum degree of the vertices of the graph, and $\alpha(u, v)$ to be the number of vertices adjacent to u and not adjacent to v.

Theorem 6.13 (D. K. Ray-Chaudhuri [Ray1]): *If G is a graph such that* (i) $d(G) > 45$, (ii) $\alpha(u, v) > 2$ *for any pair of adjacent vertices, and* (iii) $\lambda(G) \geqq -2$, *then G is a line graph.*

The proofs of Theorem 6.12 and Theorem 6.13 are rather involved. The idea is to characterize first a line graph by certain subgraphs relating to the clique structure and then to show that different subgraphs from these cannot occur when the least eigenvalue is equal to -2. A partial proof will be given here to indicate the spirit of these theorems. The reader desiring an understanding in greater depth is referred to the papers of A. J. Hoffman and D. K. Ray-Chaudhuri.

First it should be noted that whether or not a graph is a line graph can be determined from local properties. (Indeed, L. W. Beineke [Bein] has shown that a graph is not a line graph, if and only if at least one of nine graphs with at most six vertices appears as an induced subgraph.) If v is a vertex of the graph G, define G_v to be the subgraph induced by the vertices of G adjacent to v. Notice that v itself is not a vertex in G_v. Say that v is contained in a *pair of bridged cliques* $C_1(v)$ and $C_2(v)$ if the two cliques (i.e., complete subgraphs) $C_1(v)$ and $C_2(v)$ satisfy

(i) $C_1(v) \cap C_2(v) = \{v\}$;

(ii) the vertices of G_v are precisely those of $C_1(v) \cup C_2(v) - \{v\}$;

(iii) the edges of G_v can be partitioned into the edges in $C_1(v)$, the edges in $C_2(v)$, and a set of independent (mutually non-adjacent) edges.

If $G = L(H)$ for some H, then every vertex v of G is contained in a pair of bridged cliques, namely, the vertices that correspond to edges of H joined to one or the other of the end points of v viewed as an edge of H. Conversely, under the hypothesis of Theorem 6.12 or Theorem 6.13, if every vertex in a graph G is contained in a pair of bridged cliques, then $G = L(H)$ for some graph H. This follows since the set of all bridged cliques possesses these properties:

(i) $|C_i(v) \cap C_j(u)| \leqq 1$, or $C_i(v) = C_j(u)$, $i, j = 1, 2$;

(ii) every vertex v is contained in exactly two of the bridged cliques, i.e., $C_1(v)$ and $C_2(v)$;

(iii) the vertices adjacent to v are precisely

$$C_1(v) \cup C_2(v) - \{v\}.$$

The vertices of H then correspond to the bridged cliques with two vertices adjacent if and only if they have non-empty intersection. Thus the proof of Theorem 6.12 or Theorem 6.13 can be done locally, i.e., it need only be shown that every vertex is contained in a pair of bridged cliques.

To proceed, it is necessary to show that certain subgraphs cannot appear in a graph with least eigenvalue equal to -2. For this the following lemma is needed.

Lemma 6.3 (A. J. Hoffman, D. K. Ray-Chaudhuri [HoR2]): *If $\overline{G}$ is an induced subgraph of G, $\lambda(\overline{G}) = \lambda(G)$, and $\overline{\boldsymbol{x}}$ is an eigenvector of $\overline{G}$ with corresponding eigenvalue $\lambda(\overline{G})$ and $\boldsymbol{x}$ is formed from $\overline{\boldsymbol{x}}$ by adding coordinates equal to zero, then $\boldsymbol{x}$ is an eigenvector of G with corresponding eigenvalue equal to $\lambda(G)$.*

Proof. Since $\boldsymbol{A}(G)$ is a real symmetric matrix, any non-zero vector $\boldsymbol{z}$ can be written as a sum of eigenvectors. Thus

$$\frac{\boldsymbol{z}^{\mathsf{T}} \boldsymbol{A}(G)\, \boldsymbol{z}}{\|\boldsymbol{z}\|^2} \geqq \lambda(G)$$

with equality if and only if $\boldsymbol{z}$ is an eigenvector corresponding to the eigenvalue $\lambda(G)$.

Consider the vectors $\boldsymbol{x}$ and $\overline{\boldsymbol{x}}$ from the hypothesis of the lemma. Since $\|\boldsymbol{x}\| = \|\overline{\boldsymbol{x}}\|$ and $\overline{\boldsymbol{x}}^{\mathsf{T}} \boldsymbol{A}(\overline{G})\, \overline{\boldsymbol{x}} = \boldsymbol{x}^{\mathsf{T}} A(G)\, \boldsymbol{x}$, it must be that $\boldsymbol{x}$ is an eigenvector of $A(G)$ with eigenvalue $\lambda(G)$. In addition, if G is regular, then the vector $(1, 1, \ldots, 1)^{\mathsf{T}}$ is an eigenvector orthogonal to $\boldsymbol{x}$ and hence the sum of the coordinates of $\overline{\boldsymbol{x}}$ must be equal to zero. Thus the following theorem has been proved:

Theorem 6.14 (A. J. Hoffman, D. K. Ray-Chaudhuri [HoR2]): *If $\overline{G}$ is an induced subgraph of a regular graph G, $\lambda(\overline{G}) = \lambda(G)$, and $\overline{\boldsymbol{x}}$ is an eigenvector of $\overline{G}$ with corresponding eigenvalue equal to $\lambda(\overline{G})$, then*

(i) *the sum of the coordinates of $\overline{\boldsymbol{x}}$ is zero and*

(ii) *if $v \in \mathscr{V}(G) - \mathscr{V}(\overline{G})$, then the sum of the coordinates of $\overline{\boldsymbol{x}}$ corresponding to vertices in $\overline{G}$ adjacent to v is zero.*

For the following discussion a numeric label of a vertex will indicate the coordinate of $\bar{x}$ corresponding to that vertex. A vertex with no label will be the vertex v of part (ii) of the conclusion. Because of part (i) the graphs in Fig. 6.9 cannot appear as an induced subgraph of a regular graph with least eigenvalue equal to -2. Because of part (ii) the graphs in Fig. 6.10 cannot appear as an induced subgraph of a graph with least eigenvalue equal to -2.

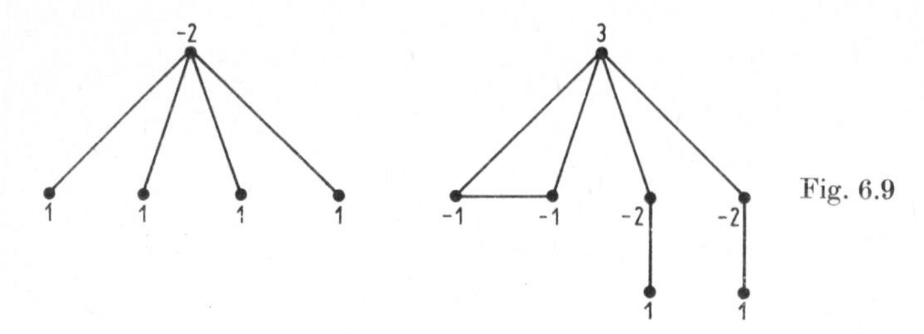

Fig. 6.9

From Fig. 6.9 it is seen that $K_{1,4}$ cannot be an induced subgraph of a regular graph with least eigenvalue equal to -2. What if one additional edge is added to $K_{1,4}$; that is, can the graph in Fig. 6.11 be an induced subgraph of a regular graph with least eigenvalue equal to -2? Since this graph has least eigenvalue strictly

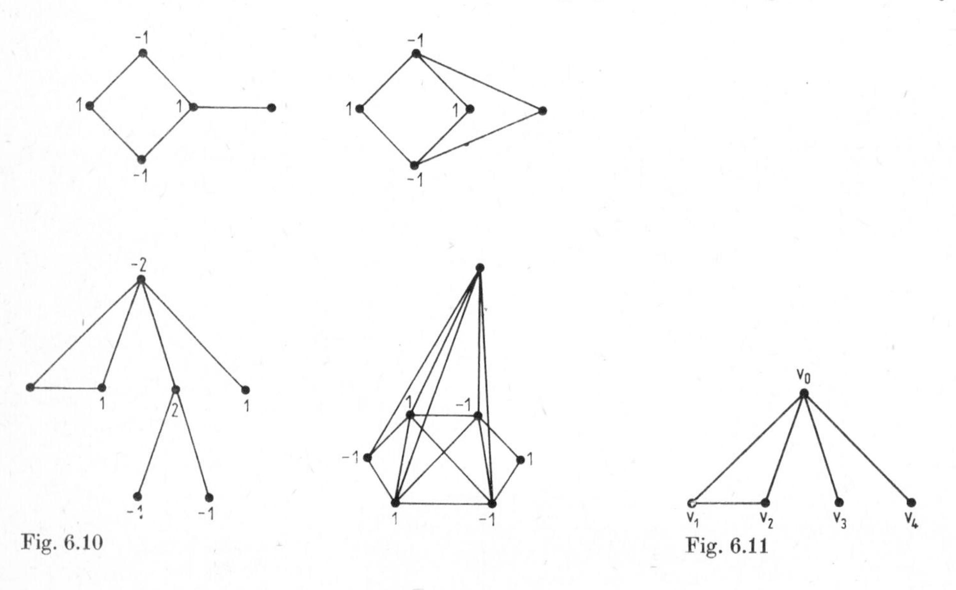

Fig. 6.10

Fig. 6.11

greater than -2, on the face of it Theorem 6.14 would not seem applicable. However, let v_0 be the vertex of degree 4, v_1 and v_2 be vertices of degree 2, and v_3 and v_4 be the vertices of degree 1 as in Fig. 6.11. Since the graph is regular and v_1, v_2, and v_4 are adjacent to v_0 but not v_3, there must be three vertices v_5, v_6, and v_7 adjacent to v_3 but not to v_0. If v_5 is adjacent to exactly one of v_1, v_2, and v_4, or to both v_1 and v_2, then a subgraph isomorphic to the first graph in Fig. 6.9 is induced, and this is impossible. If v_5 is adjacent to both v_2 and v_4, a subgraph isomorphic to the second graph in Fig. 6.10 is induced. Hence v_5 is not adjacent to v_0, v_1, v_2, or v_4. By symmetry the same can be said of v_6 and v_7. Now, if two of the vertices v_5, v_6, v_7 are not adjacent, then a subgraph isomorphic to the third graph in Fig 6.10 is induced, and so v_5, v_6, v_7 induce a clique. By the symmetric argument there are three vertices v_8, v_9, and v_{10} adjacent to v_4 but not v_0. These three vertices then also induce a clique, and none are adjacent to v_0, v_1, v_2, or v_3. Suppose one of v_5, v_6, v_7 is not adjacent to one of v_8, v_9, v_{10}. Then these two vertices along with v_0, v_1, v_2, v_3, and v_4 form a subgraph isomorphic to the second graph in Fig. 6.9. Hence it must be that v_5, v_6, v_7, v_8, v_9, and v_{10} induce a clique. But then v_3, v_4, v_5, v_6, v_7, v_8, and v_9 induce a subgraph isomorphic to the fourth graph in Fig. 6.10. Hence the original graph in Fig. 6.11 is impossible.

What if $K_{1,3}$ is considered as a possible subgraph? $K_{1,3}$ does indeed occur as an induced subgraph in the Petersen graph, the graph cospectral with but not isomorphic to $L(K_{4,4})^\dagger$, the three graphs cospectral with but not isomorphic to the triangular graph $T(8)^\dagger$, and one of the graphs in Fig. 6.4. However, the hypothesis of Theorem 6.12 does not allow graphs of small degree, and under this condition it is impossible to have a subgraph isomorphic to $K_{1,3}$. By eliminating possible subgraphs with $\lambda(G) > -2$ in this manner, it is possible to force each vertex to be contained in a pair of bridged cliques; and this is precisely the method used by D. K. RAY-CHAUDHURI to complete the proof of Theorem 6.13.

Now suppose G is regular and connected and $G = L(H)$. With no essential loss of generality it may be assumed that H has no isolated vertices and thus is connected. Since G is regular, the sum of the degrees of two adjacent vertices is constant in H. Thus if a vertex u_1 of H has degree r_1 and is adjacent to a vertex of degree r_2, then every vertex whose distance from u_1 is even has degree r_1 and every vertex whose distance from u_1 is odd has degree r_2. If $r_1 = r_2$, then the graph is regular. If $r_1 \neq r_2$, then the graph has a bipartition of the vertices into two subsets such that two vertices in the same subset have the same degree and every edge joins vertices in different subsets. Such a graph where the two subsets have cardinalities n_1 and n_2 and the degree of a vertex in the respective subsets is r_1 and r_2 is called semiregular with parameters (n_1, n_2, r_1, r_2). In particular the graph of a BIBD is semiregular with parameters (v, b, r, k).

Theorem 6.15 (D. K. RAY-CHAUDHURI [Ray1]): *Let H be a regular connected graph with degree $r > 9$, and suppose G and $L(H)$ have the same spectrum. Then $G = L(H')$ where either*

(i) *H' is regular with degree r, or*

(ii) *H' is semiregular with parameters (n_1, n_2, r_1, r_2) where $r_1 + r_2 = 2n$.*

Proof. Since H is regular, $L(H)$ has $2r - 2$ as its dominant eigenvalue, and $\lambda(L(H)) \geq -2$. Hence Theorem 6.12 is applicable and G is either a line graph or $H(n)$ for some n. But by the Corollary to Theorem 6.6 the latter case is impossible so that $G = L(H')$ for some H'. Since $L(H)$ is regular, G must have the same degree and the proof is complete.

† See p. 174, Theorems 6.18 and 6.16, respectively.

Note that (i) and (ii) in the conclusion are not mutually exclusive. But by considering the multiplicity of -2 in $L(H)$ the following corollary results.

Corollary: *If H and H' are regular and $L(H)$ and $L(H')$ are cospectral, then H and H' are cospectral. In particular H is bipartite if and only if H' is bipartite.*

The triangular graph $T(n)$ may be defined by $T(n) \cong L(K_n)$. The eigenvalues of the triangular graph are $2n - 4$, $n - 4$, and -2. Thus if $n > 10$ and G has the same spectrum as $T(n)$, then by Theorem 6.12, $G = L(H)$. Now by Theorem 3.32 two non-adjacent vertices of G are mutually adjacent to four other vertices and hence H is a clique and $T(n)$ is characterized by its spectrum for $n > 10$. If $n = 1, 2, 3$, or 4, it is easy to see that $T(n)$ is characterized by its spectrum. For the cases $n = 5, 6, 7$, 9, and 10, $T(n)$ is also characterized by its spectrum as is shown by A. J. Hoffman [Hof2] in a proof for all $n \neq 8$ that is independent of Theorem 6.12. L. C. Chang [Cha2] and E. Seiden [Seid] have shown that there are exactly three exceptional graphs cospectral with $T(8)$. These will be constructed in Section 6.5. Hence the following has been proved.

Theorem 6.16: *The triangular graph $T(n)$ is characterized by its spectrum if $n \neq 8$. If $n = 8$, there are exactly three exceptional graphs.*

In light of Theorem 3.32, the following theorem is equivalent to Theorem 6.16.

Theorem 6.17: *If $n \neq 8$, $G \cong L(K_n)$ if and only if*

(i) *G has $\binom{n}{2}$ vertices;*

(ii) *G is regular of degree $2(n - 2)$;*

(iii) *every pair of non-adjacent vertices is mutually adjacent to four other vertices;*

(iv) *every pair of adjacent vertices is mutually adjacent to $n - 2$ other vertices.*

Now, consider the line graph of the regular complete bipartite graph. Since $L(K_{n,n}) = K_n + K_n$, the eigenvalues of this graph are $2n - 2$, $n - 2$, and -2. Suppose G and $L(K_{n,n})$ are cospectral with $n > 9$. Then by Theorem 6.12, $G = L(H)$, and by Theorem 6.15, H is regular or semiregular. In the former case by Corollary of Theorem 6.15, H and $K_{n,n}$ are cospectral, and by Theorem 6.5, $H \cong K_{n,n}$. In the latter case, by Theorem 3.32 every pair of non-adjacent vertices is mutually adjacent to two vertices so that H is a complete bipartite graph. If H is not regular, then $L(H)$ has four distinct eigenvalues. Hence H is regular and once again $H \cong K_{n,n}$. Hence $L(K_{n,n})$ is characterized by its spectrum if $n > 9$. S. S. Shrikhande [Shr2] has given a completely independent proof that includes $n = 5, 6, 7, 8, 9$, and $L(K_{n,n})$ is characterized by its spectrum in these cases. For $n = 4$ there is exactly one counterexample. For $n = 1, 2, 3$, it is easy to see that $L(K_{n,n})$ is characterized by its spectrum.

Theorem 6.18 (S. S. Shrikhande [Shr2]): *Unless $n = 4$, $L(K_{n,n})$ is characterized by its spectrum. If $n = 4$, there is exactly one exceptional graph.*

This theorem will be proved independently in Section 6.4.

Since $L(K_{n,n})$ is strongly regular, Theorem 6.18 can be restated to form a theorem similar to Theorem 6.17. In fact a more general theorem is given.

Theorem 6.19 (A. J. HOFFMAN [Hof4], J. W. MOON [Moo1]): *Unless* $m = n = 4$, $G \cong L(K_{m,n})$ *if and only if*

(i) *G has $m \cdot n$ vertices;*

(ii) *G is regular of degree $m + n - 2$;*

(iii) *every pair of non-adjacent vertices is mutually adjacent to two other vertices;*

(iv) *every pair of adjacent vertices is mutually adjacent to $m - 2$ or $n - 2$ other vertices.*

An *Hadamard matrix of order n* is a matrix with entries equal to 1 or -1 such that any pair of rows is orthogonal. The order of an Hadamard matrix is always 1, 2, or a multiple of 4. A symmetric Hadamard matrix with constant diagonal is always of order 1, 2, or $4t^2$ for some integer t. Considerable effort has been expended in constructing Hadamard matrices. A survey of results can be found in W. D. WALLIS, A. P. STREET, and J. S. WALLIS [WaSW].

The following generalization of Theorem 6.19 appears independently in papers of M. DOOB [Doo4] and D. M. CVETKOVIĆ [Cve9].

Theorem 6.20† (M. DOOB [Doo4]; D. M. CVETKOVIĆ [Cve9]): *Let* $G \cong L(K_{m,n})$, $m + n > 18$. *Then G is characterized by its spectrum unless* $\{m, n\} = \{2t^2 + t, 2t^2 - t\}$ *and there exists a symmetric Hadamard matrix with constant diagonal of order* $4t^2$.

The idea of the proof of Theorem 6.20 is as follows: since $L(K_{m,n}) \cong K_m + K_n$, the distinct eigenvalues of that graph are $m + n - 2$, $m - 2$, $n - 2$, and -2. Thus Theorem 6.12 implies that if G is cospectral with $L(K_{m,n})$ then $G = L(H)$ for some graph H. If H is not regular, then H is semiregular with parameters (n_1, n_2, r_1, r_2) with $r_2 < r_1$. If $\boldsymbol{R}$ is the incidence matrix of H, then r_1 is an eigenvalue of $\boldsymbol{R}\boldsymbol{R}^\mathsf{T}$ so that $r_1 - 2$ is an eigenvalue of $L(H)$. Hence it follows that $\{r_1, r_2\} = \{m, n\}$ and $H \cong K_{m,n}$. If, on the other hand, H is regular, then the eigenvalues of H are $\frac{1}{2}(m+n)$, $\pm\frac{1}{2}(m - n)$ and, if H is bipartite, $-\frac{1}{2}(m + n)$. If H is indeed bipartite, the multiplicities of $\frac{1}{2}(m - n)$ and $-\frac{1}{2}(m - n)$ which are $m - 2$ and $n - 2$ must be equal. Hence the eigenvalues of H are $\pm m$ and 0, and by Theorem 6.5, $H = K_{m,m}$. Thus the only case in which a new cospectral graph can arise is if H is regular and not bipartite with $\frac{1}{2}(m+n)$ and $\pm\frac{1}{2}(m - n)$ as eigenvalues. Let $t = \frac{1}{2}(m + n)$ be the degree of H. It then follows that $m = 2t^2 + t$, $n = 2t^2 - t$ and that the number of vertices of H is $4t^2 - 1$. Define the matrix $\boldsymbol{C}$ by $\boldsymbol{C} = \boldsymbol{J} - 2\boldsymbol{A}(H)$ and form $\bar{\boldsymbol{C}}$ by adding one extra row and column with every entry equal to one. $\bar{\boldsymbol{C}}$ is then the desired Hadamard matrix. In addition, if we start with a symmetric Hadamard matrix with constant diagonal, then by multiplying appropriate rows and columns

† A more recent result has allowed the restriction $m + n > 18$ to be replaced by $\{m, n\} \neq \{6, 3\}$. In the latter case one cospectral mate exists. See the remarks on root systems at the end of the section.

by -1, an Hadamard matrix with last row and column having every entry equal to1 results. By reversing the above argument, a graph which is non-isomorphic to but cospectral with $L(K_{m,n})$ results, and hence the proof of the theorem is complete.

It is known [Doo4] that if $m = 2$, $n = 2$, or $|m - n| = 1$, then $L(K_{m,n})$ is characterized by its spectrum.

The line graph of a BIBD was defined in Section 6.2. For a symmetric design with parameters (v, k, λ^*), $k > 9$, the line graph of the design has $2k - 2$, $k - 2 \pm (k - \lambda^*)^{1/2}$, and -2 as eigenvalues, and hence if G is cospectral with this graph, then $G \cong L(H)$ for some H. If H is regular, then, by Corollary of Theorem 6.15 and Theorem 6.9, H is the graph of a BIBD with parameters (v, k, λ^*). If H is not regular, then H is semiregular with parameters (n_1, n_2, r_1, r_2), $r_1 < r_2$. This implies that r_1 is an eigenvalue of $L(H)$, and since $r_1 + r_2 = 2k$, the eigenvalues of $L(H)$ are $r_1 + r_2 - 2$, $r_1 - 2$, $r_2 - 2$, and -2. If $\boldsymbol{B}$ is the incidence matrix of H, then $(\boldsymbol{B}\boldsymbol{B}^\mathsf{T} - r_1\boldsymbol{I}) \dotplus (\boldsymbol{B}^\mathsf{T}\boldsymbol{B} - r_2\boldsymbol{I}) = \boldsymbol{A}(H)^2$. Hence $\boldsymbol{A}(H)$ has $\pm\sqrt{r_1 r_2}$ and 0 as distinct eigenvalues, and since H is connected, it must be that H is a complete bipartite graph. The number of vertices in $L(K_{r_1,r_2})$ is $r_1 r_2$ which is not equal to vk. Thus for $k > 9$ the line graph of a symmetric BIBD is characterized by its spectrum. A. J. Hoffman and D. K. Ray-Chaudhuri [HoR 2] have shown independently that the same is true for $k \leqq 9$ with the sole exception of $(v, k, \lambda^*) = (4, 3, 2)$. In this case there is one exception which is illustrated in Fig. 6.4.

Theorem 6.21: *Unless* $(v, k, \lambda^*) = (4, 3, 2)$, *the line graph of a symmetric* BIBD *is characterized by its spectrum. If* $(v, k, \lambda^*) = (4, 3, 2)$, *there is exactly one exceptional graph.*

(Note again that in general there exist BIBDs with non-isomorphic but cospectral line graphs.)

Corollary (A. J. Hoffman [Hof5]): *The line graph of a projective plane is characterized by its spectrum.*

What if the line graph of an asymmetric BIBD is considered? Since $K_{m,n}$ is the graph of a trivial BIBD with parameters (m, n, n, m, n), Theorem 6.20 is applicable. A more interesting result occurs when $\lambda^* = 1$.

Theorem 6.22† (M. Doob [Doo9]): *If* $r + k > 18$, *then the line graph of a* BIBD *with parameters* $(v, b, r, k, 1)$ *is characterized by its spectrum.*

As an immediate corollary the line graph of a finite affine plane of order n is characterized by its spectrum if $n > 8$. The smaller values of n have also been resolved in an independent proof of A. J. Hoffman and D. K. Ray-Chaudhuri [HoR 1].

Corollary: *The line graph of a finite affine plane is always characterized by its spectrum.*

---

† A more recent result allows the restriction $r + k > 18$ to be deleted. See the remarks on root systems at the end of the section.

It is not known to what extent Theorem 6.21 and Theorem 6.22 can be extended to other BIBDs. It is true that the line graph of the Petersen graph and the graph of the BIBD with parameters $(v, b, r, k; \lambda^{\dagger}) = (6, 10, 5, 3, 2)$ have cospectral line graphs. There is reason to conjecture that if the line graph of a BIBD is not characterized by its spectrum, then the parameters of the design must be of the form $(v, b, r, k, \lambda^*) = \big((n+1)(2n+1), (n+1)(2n+3), m(2n+3), m(2n+1), 2m^2 + m(m-1)/n\big)$, where m and n are positive integers with $m \leqq n+1$. Note that the example mentioned above is the case where $m = n = 1$ and the exceptional graphs for Theorem 6.20 lie in the case where $t = m = n + 1$. Note also that the validity of this conjecture implies that for any $\lambda_0^* > 2$ there is at most one set of parameters $(v, b, r, k, \lambda_0^*)$ whose line graph is not characterized by its spectrum.

After this section was written, a new geometric method for showing spectral characterizations for graphs with $\lambda(G) \geqq -2$ was developed by P. J. Cameron, J. M. Goethals, J. J. Seidel, and E. E. Shult [CaGSS]. The main idea is as follows: if $\boldsymbol{A}$ is the adjacency matrix of a graph G with $\lambda(G) \geqq -2$, then $\boldsymbol{A} + 2\boldsymbol{I}$ is positive semidefinite so that $\boldsymbol{A} + 2\boldsymbol{I} = \boldsymbol{M}\boldsymbol{M}^{\mathsf{T}}$ for some matrix $\boldsymbol{M}$. Viewing the rows of $\boldsymbol{M}$ as vectors, each has length $\sqrt{2}$, and the inner product between any two of them is 0 or 1. Hence lines joining the rows of $\boldsymbol{M}$ (now viewed as points in $\mathbb{R}^n$) and the origin must mutually intersect at angles of 60° or 90°. Conversely, given lines in $\mathbb{R}^n$ that mutually intersect at the origin with angles of 60° or 90°, if vectors are formed along these lines with length $\sqrt{2}$ and inner products equal to 0 or 1, then a graph with $\lambda(G) \geqq -2$ results by letting the vectors be the rows of $\boldsymbol{M}$ and $\boldsymbol{A}(G) = \boldsymbol{M}\boldsymbol{M}^{\mathsf{T}} - 2\boldsymbol{I}$. Such lines can arise in only a limited number of ways, as described by the theory of root systems ([Cart], [Wybo]). The root systems $\mathscr{A}_{n-1}$, $\mathscr{D}_n$, and $\mathscr{E}_8$, which in fact represent maximal systems of lines at 60° or 90° in the corresponding Euclidean spaces, are described as follows; let $\{\boldsymbol{e}_1, \ldots, \boldsymbol{e}_n\}$ be an orthonormal basis of $\mathbb{R}^n$. Then $\mathscr{A}_{n-1}$ is the set of $\frac{1}{2} n(n-1)$ lines that pass through the origin and $\boldsymbol{e}_i - \boldsymbol{e}_j$, $i \neq j$, $\{i, j\} \subseteqq \{1, \ldots, n\}$, and $\mathscr{D}_n$ is the set of $n(n-1)$ lines that pass through the origin and $\pm\boldsymbol{e}_i \pm \boldsymbol{e}_j$, $i \neq j$, $\{i, j\} \subseteqq \{1, \ldots, n\}$. $\mathscr{E}_8$ consists of the 84 lines in $\mathbb{R}^9$ that pass through the origin and $\frac{1}{3} \sum_{h=1}^{9} \boldsymbol{e}_h - (\boldsymbol{e}_i + \boldsymbol{e}_j + \boldsymbol{e}_k)$, $\{i, j, k\} \subseteqq \{1, \ldots, 9\}$ plus the 36 lines in $\mathscr{A}_8$. All graphs with $\lambda(G) \geqq -2$ can be constructed by taking a subset of these root systems and picking vectors along these lines with length $\sqrt{2}$ and mutual inner products of 0 or 1. As it turns out, graphs arising from the root system $\mathscr{A}_{n-1}$ are precisely the line graphs of bipartite graphs, while graphs arising from $\mathscr{D}_n$ are precisely the generalized line graphs.† All other graphs with $\lambda(G) \geqq -2$

† A regular graph on $2k$ vertices of the degree $2k - 2$ is called a *cocktail party graph* and is denoted by $CP(k)$. Let G be a graph on the vertices $1, \ldots, n$ and let $a_1, \ldots, a_n$ be non-negative integers. The *generalized line graph* $L(G; a_1, \ldots, a_n)$ *of* G is the graph obtained from the graphs $L(G), CP(a_1), \ldots, CP(a_n)$ if, for every i, all the vertices of $CP(a_i)$ are joined by edges with every vertex in $L(G)$ which corresponds to an edge of G incident to the vertex i. If all a_i's are zero, the generalized line graph is reduced to a line graph.

must arise from $\mathscr{E}_8$, a set of 120 lines, and hence P. J. CAMERON, J. M. GOETHALS, J. J. SEIDEL, and E. E. SHULT have reduced the problem of describing these graphs to a finite although very large problem. They have also shown that a connected graph that is not a generalized line graph and satisfies $\lambda(G) \geqq -2$ can have at most 36 vertices, while any such graph that is regular can have at most 28 vertices. All regular graphs of this type are known [BuCS]; there are exactly 187 of them which are not line graphs or cocktail party graphs† of which 68 are cospectral with line graphs.

From this work on root systems, it can be seen why Dynkin diagrams, an object studied concerning root systems in Lie algebras, include those graphs whose maximum eigenvalue is at most 2 (these graphs are described in Section 2.7). Such graphs also arise in the study of discrete groups generated by reflections of points by hyperplanes in $\mathbb{R}^n$ [Coxe].

6.4. Metrically regular graphs

An *m-class association scheme* consists of a set of n *objects* and m symmetric relations called *associations* such that

(i) each pair of elements satisfies exactly one of the relations;

(ii) for each object x the number $n_i(x)$ of objects that are in the i-th association with x is independent of the choice of x;

(iii) if x and y are i-th associates, the number $p^i_{j,k}(x, y)$ of objects that are simultaneously j-th associates of x and k-th associates of y is independent of the choice of i-th associates x and y.

Association schemes were first introduced by R. C. BOSE and T. SHIMAMOTO [BoSh] and a survey of their classification can be found in D. RAGHAVARAO [Ragh] or P. DEMBOWSKI [Demb]. The numbers v, $n_i(x)$, $p^i_{j,k}(x, y)$, $i, j, k = 1, \ldots, n$, are called the *parameters* of the association scheme.

Given a graph of diameter D, we may define D association classes by letting x and y be k-th associates if $d(x, y) = k$ where $d(x, y)$ is the distance from vertex x to vertex y. If these association classes give rise to an association scheme, the graph is called *metrically regular*.

Some examples of metrically regular graphs are as follows:

(1) *Lattice graphs of characteristic* n: The vertices consist of integral m-tuples $(x_1, \ldots, x_m)$, $1 \leqq x_i \leqq n$, with two vertices joined if as m-tuples they differ in exactly one coordinate. Thus two vertices are k-th associates if they differ in exactly k coordinates. The n-dimensional cube is a special case of a lattice graph.

(2) *Binomial coefficient graphs of characteristic* n: The vertices consist of the $\binom{n}{m}$ subsets of size m of a set of size n. Two vertices are joined if their intersection

† See the preceding footnote.

has cardinality $m - 1$. Thus two vertices are k-th associates if their intersection has cardinality $m - k$.

(3) *Strongly regular graphs*: By definition these graphs yield two-class association schemes. Since strongly regular graphs have interesting spectral properties, it is natural to investigate the spectral properties of metrically regular graphs.

First define the (0, 1) matrices $\boldsymbol{A}_0, \ldots, \boldsymbol{A}_m$ by $\boldsymbol{A}_0 = \boldsymbol{I}$ and $(\boldsymbol{A}_i)_{j,k} = 1$ if and only if the vertices j and k are i-th associates. Extend the definitions of the parameters in the natural way so that $p^i_{j,0}(x, y) = \delta_{ij}$ and $p^0_{j,k}(x, y) = n_j(x)\,\delta_{jk}$ where δ_{ij} is the Kronecker δ-symbol. It then follows by direct computation that $\boldsymbol{A}_j\boldsymbol{A}_k = \sum_{i=0}^{m} p^i_{j,k}\boldsymbol{A}_i$, and hence any pair of matrices in the linear span of the $\boldsymbol{A}_i$'s commute. This includes the $\boldsymbol{A}_i$'s themselves and $\boldsymbol{J} = \sum_{i=0}^{m} \boldsymbol{A}_i$. Further, since the $\boldsymbol{A}_i$'s are linearly independent, they form a basis for the commutative algebra generated by the $\boldsymbol{A}_i$'s. Let $\boldsymbol{P}_k$ be defined by $\boldsymbol{P}_k = (p^i_{j,k})$, $0 \leqq i, j, k \leqq m$. The eigenvalues of a strongly regular graph are determined by its parameters. The next theorem shows that the eigenvalues of a metrically regular graph are also determined by its parameters.

Theorem 6.23 (R. C. Bose, D. M. Messner [BoMe]): *The distinct eigenvalues of* $\sum_{i=1}^{m} r_i\boldsymbol{A}_i$ *and* $\sum_{i=1}^{m} r_i\boldsymbol{P}_i$ *are the same. In particular the distinct eigenvalues of a metrically regular graph are the same as those of* $\boldsymbol{P}_1$.

Proof. Count the number of sequences of vertices (x, u, v, y) such that $d(x, u) = k$, $d(u, v) = j$, $d(v, y) = i$, and $d(x, y) = s$. First let $r = d(x, v)$; then the number of such sequences is $\sum_{r=0}^{m} p^r_{j,k}\, p^s_{r,i}$. Now let $r = d(u, y)$; then the number of such sequences is $\sum_{r=0}^{m} p^r_{i,j} p^s_{k,r}$. Using the equality of these two sums, it follows by direct computation that $\boldsymbol{P}_j\boldsymbol{P}_k = \sum_{i=0}^{m} p^i_{j,k}\boldsymbol{P}_i$. Since this is precisely the formula for $\boldsymbol{A}_j\boldsymbol{A}_k$, it follows that if $\boldsymbol{A} = \sum_{i=1}^{m} r_i\boldsymbol{A}_i$, $\boldsymbol{P} = \sum_{i=1}^{m} r_i\boldsymbol{P}_i$, and $q(x)$ is a polynomial, then $q(\boldsymbol{A}) = \sum_{i=0}^{m} s_i\boldsymbol{A}_i$ implies that $q(\boldsymbol{P}) = \sum_{i=0}^{m} s_i\boldsymbol{P}_i$. Thus if $m(x)$ is the minimal polynomial for $\boldsymbol{A}$, then $m(\boldsymbol{P}) = \boldsymbol{O}$, and each distinct eigenvalue of $\boldsymbol{P}$ is an eigenvalue of $\boldsymbol{A}$. By the symmetric reasoning, every distinct eigenvalue of $\boldsymbol{A}$ is also an eigenvalue of $\boldsymbol{P}$.

Since the eigenvalues of a metrically regular graph are determined by its parameters, a graph that has the same parameters as a metrically regular graph will have the same eigenvalues. For graphs of diameter greater than 3 the converse is not true, for there is a graph cospectral with the m-dimensional cube, Q_m, that is not metrically regular for all $m \geqq 4$. If the diameter is less than 3, however, it is easy to see that a graph with the eigenvalues of a metrically regular graph is metrically regular with the same parameters. If a graph has the same distinct eigenvalues as a metrically regular graph of diameter 3, then, by Lemma 6.2 (p. 161), it is cospectral with it. It is not known whether this implies that the graph is metrically regular, but it would be reasonable to conjecture that it does.

The lattice graph of characteristic n and diameter 3 and the binomial coefficient graph of characteristic n and diameter 3 have been of particular interest insofar as their spectral characterizations are concerned. The former graph is called the *cubic lattice graph* while the latter graph is called the *tetrahedral graph*. The lattice graph will now be looked at in some detail.

The cubic lattice graph has an interesting spectral characterization by virtue of a characterization of that graph by R. LASKAR [Las1] and M. AIGNER [Aign]. Define $\Delta(x, y)$ to be the number of vertices adjacent to both x and y. They have shown that for $n \neq 4$ a graph is a cubic lattice graph with characteristic n, if and only if the following properties hold:

($\boldsymbol{b}_1$) the number of vertices is n^3;

($\boldsymbol{b}_2$) the graph is connected and regular of degree $3(n-1)$;

($\boldsymbol{b}_3$) if $d(x, y) = 1$, then $\Delta(x, y) = n - 2$;

($\boldsymbol{b}_4$) if $d(x, y) = 2$, then $\Delta(x, y) = 2$;

($\boldsymbol{b}_5$) if $d(x, y) = 2$, then there exist $n - 1$ vertices z such that $d(x, z) = 1$ and $d(y, z) = 3$.

Further, if $n = 4$, M. AIGNER has shown that there is exactly one exceptional graph.

It is easy to determine the spectrum of the cubic lattice graph of characteristic n since it is isomorphic to $K_n + K_n + K_n$. The eigenvalues are thus $3n - 3$, $2n - 3$, $n - 3$, and -3 with respective multiplicities 1, $3(n-1)$, $3(n-1)^2$, and $(n-1)^3$. Hence by Lemma 6.2 a graph has the spectrum of the cubic lattice graph if and only if it is regular, connected, has n^3 vertices, and has $3n - 3$, $2n - 3$, $n - 3$, and -3 as distinct eigenvalues. This gives rise to the following theorem.

Theorem 6.24 (R. LASKAR [Las2]; D. M. CVETKOVIĆ [Cve7]): *For all $n \neq 4$, G is a cubic lattice graph of characteristic n if and only if G satisfies the following properties*:

($\mathbf{P}_1$) *the number of vertices is n^3;*

($\mathbf{P}_2$) *G is regular and connected;*

($\mathbf{P}_3$) *$n_2(x) = 3(n-1)^2$ for all x;*

($\mathbf{P}_4$) *the distinct eigenvalues are $3n - 3$, $2n - 3$, $n - 3$, and -3.*

Theorem 6.24 will be proved in two steps. As the first part of Lemma 6.4, it will be shown that if G satisfies ($\mathbf{P}_1$), ($\mathbf{P}_2$), and ($\mathbf{P}_4$), then ($\boldsymbol{b}_1$), ($\boldsymbol{b}_2$), ($\boldsymbol{b}_3$), and ($\boldsymbol{b}_4$) imply ($\boldsymbol{b}_5$). As the second part of Lemma 6.4, it will be shown that if G satisfies ($\mathbf{P}_1$), ($\mathbf{P}_2$), and ($\mathbf{P}_4$), then ($\mathbf{P}_3$) is true if and only if ($\boldsymbol{b}_3$) and ($\boldsymbol{b}_4$) are true. This will complete the proof.

Lemma 6.4: *Let G be a graph satisfying* ($\mathbf{P}_1$), ($\mathbf{P}_2$), *and* ($\mathbf{P}_4$) *then*

(i) *if G satisfies* ($\boldsymbol{b}_3$) *and* ($\boldsymbol{b}_4$), *then G satisfies* ($\boldsymbol{b}_5$) *and*

(ii) *G satisfies* ($\boldsymbol{b}_3$) *and* ($\boldsymbol{b}_4$) *if and only if G satisfies* ($\mathbf{P}_3$).

Proofs. (i) From the polynomial of the graph G, it follows that $\boldsymbol{A}^3 - 3(n-3)\,\boldsymbol{A}^2 + (2n^2 - 18n + 27)\,\boldsymbol{A} + (6n^2 - 27n + 27)\,\boldsymbol{I} = 6\boldsymbol{J}$. Hence if $d(x, y) = 2$, then $a_{xy}^{(2)} = 2$, and from the above equation $a_{xy}^{(3)} = 6(n-2)$. Hence there are $6(n-2)$ paths of the form (x, v, w, y). For each v such that v and y are joined, there are

$n - 2$ choices for w by the assumption of $(\boldsymbol{b}_3)$, and since x and y are not joined, there are two such v. Hence this accounts for $2(n - 2)$ paths and the remaining $4(n - 2)$ paths have v and y non-adjacent. Now each v adjacent to x satisfying $d(v, y) = 2$ determines two paths of this sort by $(\boldsymbol{b}_4)$, so there are $2(n - 2)$ vertices adjacent to x such that $d(v, y) = 2$. By $(\boldsymbol{b}_4)$ there are two vertices adjacent to x such that $d(v, y) = 1$. Hence the remaining vertices adjacent to x satisfy the condition of $(\boldsymbol{b}_5)$ and there are $3(n - 1) - 2(n - 2) - 2 = n - 1$ of them.

(ii) First assume that G satisfies $(\mathbf{P}_1)$, $(\mathbf{P}_2)$, $(\mathbf{P}_4)$, $(\boldsymbol{b}_3)$, and $(\boldsymbol{b}_4)$. For a given vertex x, consider all paths (x, w, y) of length 2. The total number of such paths is $d^2 = 9(n - 1)^2$. If $d(x, y) = 0$, i.e., $x = y$, then there are $d = 3(n - 1)$ such paths. If $d(x, y) = 1$, then (x, w, y, x) is a closed walk of length 3. Looking at the diagonal elements in the matrix equation that arises from the polynomial of the graph, it follows that the number of such closed walks is $3(n - 1)(n - 2)$. Hence for the remaining $9(n - 1)^2 - 3(n - 1) - 3(n - 1)(n - 2) = 6(n - 1)^2$ paths, $d(x, y) = 2$. Now for each such y, by $(\boldsymbol{b}_4)$ there are two paths (x, w, y) so that the total number of such y's, i.e., $n_2(x)$, is $3(n - 1)^2$.

Conversely, suppose G satisfies $(\mathbf{P}_1)$, $(\mathbf{P}_2)$, $(\mathbf{P}_3)$, and $(\mathbf{P}_4)$. If x and y are adjacent, $\Delta(x, y)$ is precisely the number of closed walks (x, u, y, x) of length 3. Thus $\sum_{y:d(x,y)=1} \Delta(x,y) = a_{xx}^{(3)} = 3(n - 1)(n - 2)$. Hence the average over all y adjacent to x is $\bar{\Delta}_1 = n - 2$. As before there are $6(n - 1)^2$ paths (x, w, y) where $d(x, y) = 2$ and hence from the property $(\mathbf{P}_3)$ the average is $\bar{\Delta}_2 = 2$. Define $r(i)$ and $s(i)$ to be the respective cardinalities of the sets $\{y \mid d(x, y) = 2, \Delta(x, y) = i\}$ and $\{y \mid d(x, y) = 1, \Delta(x, y) = i\}$. Then $\sum r(i)(i - 2) = 0 = \sum s(i)(i - n + 2)$, and $r(1) = \sum_{i \neq 1} r(i)(i - 2)$ and $s(n - 3) = \sum_{i \neq n-3} s(i)(i - n + 2)$. Now define the auxiliary matrix $\boldsymbol{C} = (c_{ij})$ by

$$\boldsymbol{C} = \frac{1}{2}\left(\boldsymbol{A}^2 - (n - 4)\boldsymbol{A} - 3(n - 1)\boldsymbol{I}\right).$$

From the polynomial of the graph it follows that $a_{ii}^{(3)} = 3(n - 1)(n - 2)$, and multiplying that matrix equation by $\boldsymbol{A}$, it follows that $a_{ii}^{(4)} = 3(n - 1)(n^2 + 3n - 3)$. By direct computation $(\boldsymbol{C}^2)_{ii} = 3n(n - 1)$, so that $\sum_{j=1}^{n^3} c_{ij}^2 = 3n(n - 1)$. In addition $\sum_{i=1}^{n^3} c_{ij} = (\boldsymbol{CJ})_{ii} = 3n(n - 1)$. Hence $\boldsymbol{C}$ has the property that the sum of the entries in any row is the sum of the squares of those entries. Now since

$$c_{xy} = \begin{cases} \frac{1}{2}\Delta(x, y) & \text{if } d(x, y) = 2, \\ \frac{1}{2}\left(\Delta(x, y) - n + 4\right) & \text{if } d(x, y) = 1, \\ 0 & \text{otherwise,} \end{cases}$$

it follows that

$$\sum r(i)\frac{i}{2} + \sum s(i)\frac{i - n + 4}{2} = \sum_y c_{xy} = \sum_y c_{xy}^2 = \sum r(i)\left(\frac{i}{2}\right)^2 + \sum s(i)\left(\frac{i - n + 4}{2}\right)^2.$$

Using this equality plus the previously derived equalities involving $r(i)$ and $s(i)$, after some simplification it follows that

$$\sum_{i \neq 1} r(i)\,(i-1)\,(i-2) + \sum_{i \neq n-3} s(i)\,(i-n+2)\,(i-n+3) = 0.$$

Since each term is non-negative, $r(i) = 0$ for $i \neq 1, 2$ and $s(i) = 0$ for $i \neq n-2$, $n-3$. Thus $r(1) = r(2)\,(2-2) = 0$ and $s(n-3) = s(n-2)\,(n-2-n+2) = 0$.

Hence $\Delta(x, y) = 2$ if $d(x, y) = 2$ and $\Delta(x, y) = n-2$ if $d(x, y) = 1$, and the proof is complete.

It has been seen that the cubic lattice graph is almost characterized by its spectrum when $n \neq 4$. That is, if G has the spectrum of the cubic lattice graph plus $(\mathbf{P}_3)$ or $(\boldsymbol{b}_3)$ and $(\boldsymbol{b}_4)$, then G is isomorphic to the cubic lattice graph. Can these properties be weakened? They can, as the following Theorem indicates.

Theorem 6.25: *If $n \neq 4$ and G satisfies* $(\mathbf{P}_1)$, $(\mathbf{P}_2)$, $(\mathbf{P}_3)$, *and*

$(\boldsymbol{b}_4')$ *if $d(x, y) = 2$, then $\Delta(x, y) \neq 1$,*

then G is isomorphic to the cubic lattice graph of characteristic n.

Proof. Proceeding in the spirit of Cvetković [Cve 7], let

$$m = \min\,\{\Delta(x, y) : d(x, y) = 1\},$$

and let i and j be a pair of vertices that attain that minimum. Then $a_{ij}^{(3)} = \sum_k a_{ik}^{(2)} a_{kj}$. Breaking up this expression into three summands that correspond to vertices k such that $d(i, k)$ is 0, 1, or 2, it follows that $a_{ij}^{(3)} \geqq 3(n-1) + m^2 + 2(3n-4-m)$. On the other hand, $a_{ij}^{(3)}$ can be evaluated since $\boldsymbol{A}^3 - 3(n-3)\,\boldsymbol{A}^2 + (2n^2 - 18n + 27)\,\boldsymbol{A} + (6n^2 - 27n + 27)\,\boldsymbol{I} = 6\boldsymbol{J}$. The resulting inequality implies that $m \geqq n-2$. Thus the matrix $\boldsymbol{C}$ defined in the proof of Lemma 6.4 has entries such that $c_{xy} = 0$ if $d(x, y) = 0$ or $d(x, y) = 3$, and $c_{xy} \geqq 1$ otherwise. Hence $c_{xy} \leqq c_{xy}^2$ and, as was seen before, $\sum c_{xy} = \sum c_{xy}^2$. Thus $\boldsymbol{C}$ is a $(0, 1)$ matrix and $c_{xy} = 1$ if $d(x, y) = 1$ or $d(x, y) = 2$; this is equivalent to $(\boldsymbol{b}_3)$ and $(\boldsymbol{b}_4)$ and so G is isomorphic to the cubic lattice graph of characteristic n.

A closer examination into the sum of graphs can prove interesting for elucidating the spectral properties of the cubic lattice graph. A graph G is called *prime* if $G \cong G_1 + G_2$ implies that either G_1 or G_2 is a graph consisting of a single vertex. The cubic lattice graph is not prime, of cource, since it is isomorphic to $K_n + \big(L(K_{n,n})\big)$.

Theorem 6.26: *Suppose $n \neq 4$, G satisfies* $(\mathbf{P}_1)$, $(\mathbf{P}_2)$, $(\mathbf{P}_4)$ *and G is not prime. Then G is isomorphic to the cubic lattice graph of characteristic n.*

Proof. Let $G \cong G_1 + G_2$. Since G is regular, it must be that G_1 and G_2 are also regular. If G_1 and G_2 are not complete graphs, then $\lambda(G) = \lambda(G_1) + \lambda(G_2) \leqq -\big(1 + \sqrt{5}\big) < -3$ (see Section 6.6, no. 7) which is impossible. Without loss of generality let $G_1 \cong K_m$. Then $\lambda(G_2) = -2$. Let d be the degree of G_2 and let λ be an eigenvalue, $\lambda \neq d$, $\lambda \neq -2$. Then $d + m - 1 > d - 1 > \lambda - 1 > -2 - 1$ are eigenvalues of $G_1 + G_2$ and hence respectively equal to $3n-3$, $2n-3$, $n-3$, and -3. This implies that $m = n$, $d = 2n - 2$, and $\lambda = n - 2$. Hence by Theorem 6.18, $G_2 \cong L(K_{n,n})$ and $G \cong K_n + L(K_{n,n})$.

An independent proof of Theorem 6.18 follows from Theorem 6.24.

Lemma 6.5: *Suppose G and $L(K_{n,n})$ are cospectral. Then $G + K_n$ is cospectral with the cubic lattice graph of characteristic n and has the property $n_2(x) = 3(n - 1)^2$.*

Proof. It is obvious that the two graphs are cospectral. If (v_1, v_2) and (u_1, u_2) are two vertices in $G_1 + G_2$, then the distance between them is the sum of the distance between u_1 and v_1 in G_1 and the distance between u_2 and v_2 in G_2. Since G and $L(K_{n,n})$ are cospectral, $n_1(x) = 2n - 2$ and $n_2(x) = (n - 1)^2$. Hence for the graph $G + K_n$, $n_2(x) = (n - 1)^2 + (2n - 2)(n - 1) = 3(n - 1)^2$.

Theorem 6.27 (C. R. Cook [Cook]): *If $n \neq 4$, $L(K_{n,n})$ is characterized by its spectrum For $n = 4$ the exceptional graph in the characterization of Aigner is $G + K_4$ where G is the graph cospectral with $L(K_{4,4})$.*

An analysis of the tetrahedral graph that is identical to that of the cubic lattice graph produces the following theorem.

Theorem 6.28 (R. C. Bose, R. Laskar [BoLa]): *The graph G is isomorphic to the tetrahedral graph if $n > 16$ and G satisfies the following properties:*

(P_1) *the number of vertices is* $\binom{n}{3}$;

(P_2) *G is regular and connected;*

(P_3) $n_2(x) = \frac{3}{2}(n - 3)(n - 4)$ *for all* x;

(P_4) *the distinct eigenvalues of G are* $3n - 9$, $2n - 9$, $n - 7$, *and* -3.

Hence there exist theorems that characterize the cubic lattice graph and the tetrahedral graph by their spectra assuming that one additional property concerning $n_2(x)$ is true. It is a reasonable but unproven conjecture that the condition concerning $n_2(x)$ can be dropped.†

At present relatively little is known about metricly regular graphs with diameter larger than 3. P. Delsarte [Del4] has investigated association schemes from the graph theoretic point of view. A special case of metrically regular graphs, namely distance transitive graphs, has received some attention (see Section 7.2).

6.5. The $(-1, 1, 0)$-adjacency matrix and Seidel switching

In Chapter 1 the $(-1, 1, 0)$-adjacency matrix or Seidel matrix has been defined (see 1.2, (1.10)–(1.14)). If $\boldsymbol{S}$ is the $(-1, 1, 0)$-adjacency matrix and $\boldsymbol{A}$ is the usual adjacency matrix, then $\boldsymbol{S} = \boldsymbol{J} - 2\boldsymbol{A} - \boldsymbol{I}$. For a regular graph $\boldsymbol{A}$, $\boldsymbol{S}$, and $\boldsymbol{J}$ commute

† Added in proof: According to a private communication, J. M. Goethals and J. J. Seidel have proved this conjecture (unpublished). See also P. T. Rolland (no. [99] of the supplementary bibliography).

and hence can be simultaneously diagonalized and have a common set of eigenvectors, yielding the following lemma.

Lemma 6.6: *Let G be a regular connected graph with n vertices and eigenvalues $\lambda_1 = r > \lambda_2 \geqq \lambda_3 \geqq \cdots \geqq \lambda_n$. Then the eigenvalues of $\boldsymbol{S}$ are* $n - 2r - 1$, $-2\lambda_2 - 1$, $-2\lambda_3 - 1, \ldots, -2\lambda_n - 1$. (See Section 1.3, (1.34).)

Corollary: *Two regular connected graphs of the same degree are cospectral if and only if their* $(-1, 1, 0)$*-adjacency matrices are cospectral.*

Now if $\boldsymbol{U}$ is a diagonal matrix with entries of ± 1 on the diagonal, then $\boldsymbol{USU}$ is similar to $\boldsymbol{S}$ and hence cospectral. If $\mathscr{V}_1$ is the set of vertices of G corresponding to those diagonal entries equal to -1 and $\mathscr{V}_2$ is the complementary set of vertices, then $\boldsymbol{USU}$ is the $(-1, 1, 0)$-adjacency matrix of the graph where two vertices in $\mathscr{V}_1$ or two vertices in $\mathscr{V}_2$ are adjacent if and only if they were adjacent in G while a vertex in $\mathscr{V}_1$ and a vertex in $\mathscr{V}_2$ are adjacent if and only if they were not adjacent in G. In other words, the adjacencies between vertices in $\mathscr{V}_1$ and those in $\mathscr{V}_2$ are switched while those within $\mathscr{V}_1$ and within $\mathscr{V}_2$ remain unchanged. This process of forming cospectral graphs with respect to the $(-1, 1, 0)$-adjacency matrix was introduced by J. J. Seidel and is called *Seidel switching* (see [LiSe]). By Corollary of Lemma 6.6, switching a regular connected graph into another regular connected graph of the same degree results in a pair of (not necessarily non-isomorphic) cospectral graphs. The following examples which are due to J. J. Seidel [Sei2] are relevant to the exceptional graphs discussed in Section 6.3.

Example 1: $L(K_{n,n})$ is characterized by its spectrum unless $n = 4$. If $n = 4$, let $G = L(K_{4,4})$ and let $\mathscr{V}_1$ be eight vertices which induce a circuit of length 8. Switching with respect to $\mathscr{V}_1$ produces a cospectral graph. As was seen in Lemma 6.5 and Theorem 6.27, this also yields the exceptional graph for the cubic lattice graph.

Example 2: The triangular graph $T(n)$ is characterized by its spectrum unless $n = 8$. Let $\mathscr{V}_1$ be four vertices, no two of which are joined, $\mathscr{V}_1'$ be eight vertices that induce a circuit of length 5 and a circuit of length 3, and $\mathscr{V}_1''$ be eight vertices that induce a circuit of length 8. Then switching the triangular graph $T(8)$ with respect to $\mathscr{V}_1$, $\mathscr{V}_1'$, and $\mathscr{V}_1''$ produces the three exceptional graphs cospectral with $T(8)$.

The Seidel switching generates, in a natural way, an equivalence relation in the set of graphs, where two graphs G_1, G_2 are equivalent if and only if G_1 can be switched into G_2 with respect to some subset of its vertices. Equivalence classes under Seidel switching of graphs up to 7 vertices are given in the Appendix.

Theorem 6.12 states that if $\lambda(G) = -2$, and G is regular and connected with degree greater than 16, then either G is a line graph or $G = H(n)$. J. J. Seidel [Sei2] has extended this result for graphs with three distinct eigenvalues by analysis of the $(-1, 1, 0)$-adjacency matrix.

The *Clebsch graph* may be defined as the graph whose vertices are 16 lines on the Clebsch quartic surface, any pair or lines being adjacent if and only if they do not intersect. The vertices of the *Schläfli graph* correspond to 27 lines on a general cubic surface, two vertices being adjacent if and only if the corresponding lines do not intersect.

Theorem 6.29: *The only graphs G with three distinct eigenvalues and $\lambda(G) = -2$ are $H(n)$, $L(K_{n,n})$ and the exceptional graph when $n = 4$, $T(n)$ and the three exceptional graphs when $n = 8$, the Petersen graph, the Clebsch graph, and the Schläfli graph.*

Theorem 6.29 is also interesting from a group theoretic point of view. A transitive rank three permutation group has the property that the stabilizer of any point x has three orbits which may be denoted by $\{x\}$, $\Gamma(x)$, and $\Delta(x)$. If the group acts on a set with an even number of vertices, let $\Delta(x)$ always denote the unique orbit of even cardinality. A graph may then be defined by letting the vertex set be the set on which the group is acting and by letting x be adjacent to y whenever $y \in \Gamma(x)$. The resulting graph is called a *rank three graph* and is strongly regular. Such graphs with least eigenvalue equal to -2 have been characterized [Sim 3]. It is interesting to note that M. Aschbacher [Asch] has shown that the possible Moore graph with $r = 57$ could never be a rank three graph, this in spite of the fact that the Moore graphs with $r = 2, 3, 7$ are all rank three. Further results pertaining to rank three graphs have appeared in [HeHi], [Hig 4], [Hig 5], [Hig 6], [Sim 1], [Sim 2].

6.6. Miscellaneous results and problems

1. (a) Use the graphs $K_{d,\frac{r}{d}}$ to show that there is a family of 9 cospectral graphs with 245 vertices.

(b) Use the cospectral pair in Fig. 6.2 to form a family of 50 cospectral graphs with 245 vertices.

(c) From the 91 cospectral strongly regular graphs with 35 vertices constructed by F. C. Bussemaker and J. J. Seidel [BuS 2], show that if $n = 35t$, then there exists a family of $\binom{90+t}{90}$ cospectral graphs. Use this to construct 10^{10} cospectral graphs with 245 vertices each.

2. From the equation for $\Phi(G_1 \cdot G_2)$, p. 159, show that the characteristic polynomial of both graphs in Fig. 6.7 is $\lambda^{3k+4}(\lambda^4 - (3k+7)\lambda^2 + (2k+3)(k+3))$.

3. A regular graph with distinct eigenvalues $\lambda_1 > \lambda_2 > \lambda_3$ satisfies

$$0 \geqq (\lambda_2 + 1)(\lambda_3 + 1) \geqq -\lambda_1 + 1.$$

What graphs attain these bounds?

4. If G has $\lambda_1 > \lambda_2 > \lambda_3$ as distinct eigenvalues and $\lambda_2 + \lambda_3 = r - 3$, then $G \cong C_5$.

5. (a) A regular graph G has ± 1 and $\pm r$ as distinct eigenvalues if and only if each connected component is isomorphic to $K_{r+1,r+1}$ with a 1-factor deleted.

(M. Doob [Doo 3])

(b) G is regular and bipartite with four distinct eigenvalues if and only if G is the graph of a symmetric BIBD.

(M. Doob)

6. Let $\boldsymbol{R}$ be the incidence matrix of a graph G. Show that $\boldsymbol{Rx} = \boldsymbol{o}$ if and only if the sum of the coordinates of $\boldsymbol{x}$ corresponding to edges incident to any given vertex is 0. $\boldsymbol{R}^{\mathsf{T}}\boldsymbol{Rx} = \boldsymbol{o}$ if and only if the corresponding sums for any pair of adjacent vertices add to 0.

7. If G is regular and not a complete graph, then $\lambda(G) \leqq -\frac{1}{2}\left(1 + \sqrt{5}\right)$.

8. A regular graph has G_v equal to a pair of disjoint cliques for every v if and only if $G = L(H)$ where H contains no circuits of length three or G is isomorphic to K_3. What about non-regular graphs?

9. Under the hypothesis of Theorem 6.12 or Theorem 6.13, if each vertex of G is contained in a unique pair of bridged cliques, then G is a line graph.

(D. K. RAY-CHAUDHURI [Ray 1])

10. If v_0 is adjacent to v_1, v_2, and v_3, then each vertex adjacent to v_0 is adjacent to none, one, two, or three of v_1, v_2, and v_3. Using this fact, show that a graph satisfying the hypothesis of Theorem 6.12 cannot contain $K_{1,3}$ as a subgraph.

11. If G is cospectral with $L(K_{4,4})$, then either G_v is the direct sum of two copies of K_3 and $G \cong L(K_{4,4})$ or $G_v \cong C_6$ and G is the unique exceptional graph. Apply this result to Example 1 of the Seidel switching technique.

12. In the original formulation of Theorem 6.19, (iv) was replaced by (iv'): $n\binom{m}{2}$ pairs of adjacent vertices are mutually adjacent to $m - 2$ other vertices while $m\binom{n}{2}$ vertices are mutually adjacent to $n - 2$ other vertices. Use the matrix equation $(m - n)\boldsymbol{B} = \boldsymbol{A}^2 - (n - 4)\boldsymbol{A} - (m + n - 4)\boldsymbol{I} - 2\boldsymbol{J}$ to show that (iv) and (iv') are equivalent.

13. $L(K_{3,5})$ is characterized by its spectrum.

(M. DOOB [Doo 1])

14. If a semiregular graph H has $n_2 > n_1$, then $n_1 - 2$ is an eigenvalue of $L(H)$.

(D. M. CVETKOVIĆ [Cve 9])

15. Give an example of two semiregular graphs with different parameters whose line graphs have the same distinct eigenvalues, but also show that if the line graphs of two semiregular graphs are cospectral, then the two semiregular graphs have the same parameters.

16. If H is semiregular, r_1 and r_2 have different parity, and $L(H)$ and $L(H')$ are cospectral, then H' is semiregular.

17. A metrically regular graph with diameter m has $m + 1$ distinct eigenvalues.

18. For any graph G, $G + K_4 + K_4$ is not characterized by its spectrum.

19. Use Theorem 6.18 to show that $L(K_{3,3})$ is self-complementary.

20. Let Q_m be the graph of the m-dimensional unit cube (for its spectrum see Section 2.6). For $m = 1, 2, 3$ the graph Q_m is characterized by its spectrum. For $m \geqq 4$ there exists always a graph which is non-isomorphic but cospectral to Q_m.

(A. J. HOFFMAN [Hof 3], M. DOOB, unpublished)

21. For any positive integer n let G_n be the graph whose vertices are all ordered pairs (i, j) of residues mod n and whose edges go from (i, j) to $(i, j + 1)$ and $(i + 1, j)$ for all i, j. Let $P_n(x)$ be the polynomial of G_n (see Theorem 3.7). If $n = 2, 4$, or an odd prime, and if H is a graph with n^2 vertices that belongs to $P_n(x)$, then H is isomorphic to G_n.

(A. J. HOFFMAN, M. H. MCANDREW [HoMc])

22. If a regular connected graph has three distinct eigenvalues, then the number of vertices can be determined from these eigenvalues, but the same is not true for graphs with four distinct eigenvalues.

(M. DOOB [Doo 3])

23. The three graphs of Fig. 6.12 are mutually non-isomorphic but have the same characteristic polynomial $\lambda^7 - 11\lambda^5 - 10\lambda^4 + 16\lambda^3 + 16\lambda^2$.

(F. HARARY, C. KING, A. MOWSHOWITZ, R. C. READ [HaKMR])

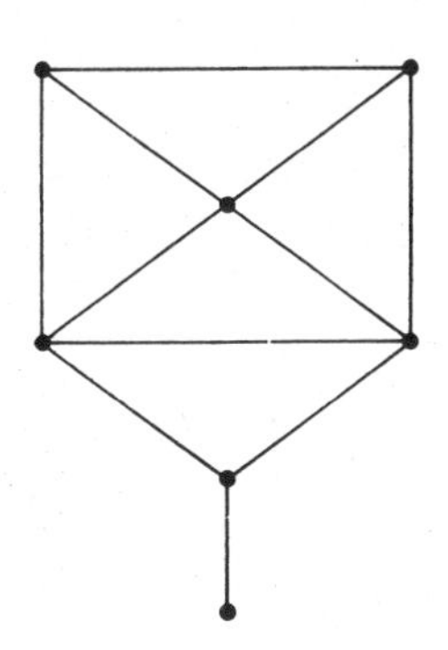

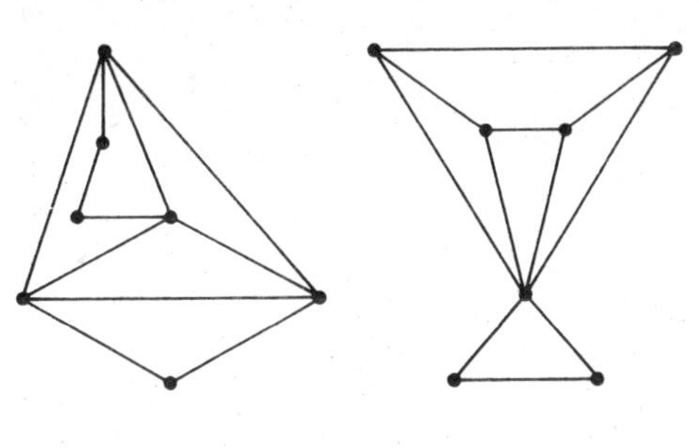

Fig. 6.12

24. Let T_1 and T_2 be trees. If T_1 and T_2 are cospectral, then $L(T_1)$ and $L(T_2)$ have the same numbers of vertices and edges.

(A. MOWSHOWITZ [Mow5])

25. Show that the number of sets of three mutually non-adjacent vertices of a tree T is determined by the spectrum of T.

26. If G is a graph of diameter D, then $-2 \leqq \lambda(L(G)) \leqq -2\cos\dfrac{\pi}{D+1}$, where $\lambda(L(G))$ is the least eigenvalue of $L(G)$. Further, these bounds are the best possible.

(M. DOOB [Doo2], [Doo8])

27. The graph $C_m[\overline{K}_n]$ (the definition of the lexicographic product is in Section 2.5) is characterized by its spectrum for odd m and for even $m \leqq 8$. It is conjectured that this characterization holds for the remaining values of m also.

(M. C. HEYDEMANN [Heyd])

28. Let $\boldsymbol{S}$ be the Seidel adjacency matrix of a graph G with an even number of vertices. Then $\dfrac{1}{2}(\boldsymbol{S}^2 - \boldsymbol{I})$ is an integral matrix and $\boldsymbol{S}$ is non-singular. If G has $2k + 1$ vertices, then the rank of $\boldsymbol{S}$ is at least $2k$.

(R. A. GIBBS [Gib2])

29. Let G be a regular connected graph with n vertices and degree r. If the distinct eigenvalues are $\lambda_1 = r, \lambda_2, \ldots, \lambda_t$, then $(r - \lambda_2)(r - \lambda_3) \cdots (r - \lambda_t)$ is an integer divisible by n.

(D. M. CVETKOVIĆ [Cve16])

30. A regular graph has $\pm r$ and $\pm\mu$ as distinct eigenvalues if and only if each connected component is isomorphic to the graph of a symmetric BIBD with parameters (v, k, λ^*) $= \left(\dfrac{r^2 - \mu^2}{r - \mu^2}, r, r - \mu^2\right)$.

31. Show that the two graphs in Fig. 6.4 are switching equivalent.

32. For all graphs in a switching class of graphs with an even number of vertices, the dissection of this number into the numbers of vertices with even and odd degrees is the same. Both the graphs of Fig. 6.13 have the Seidel spectrum $\pm\sqrt{17}, \pm\sqrt{5}, \pm\sqrt{5}, \pm 1$, but they are not switching equivalent, since their dissections are different.

(J. J. SEIDEL [Sei 6])

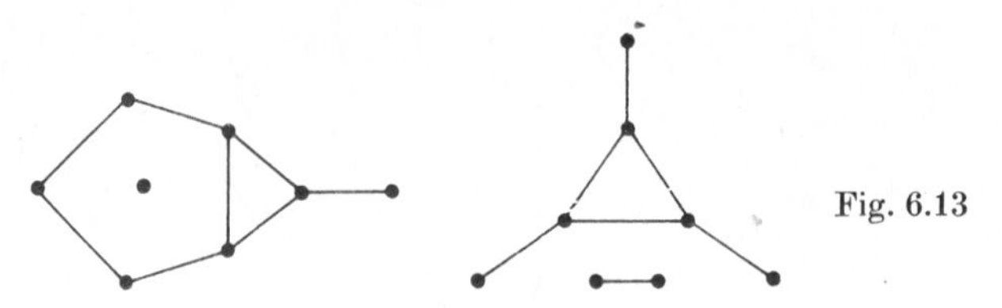

Fig. 6.13

33. If G is a regular graph of degree ≥ 13 and if the least eigenvalue of G is -2, then G does not contain $K_{1,3}$ as an induced subgraph. This result is best possible in the sense that the number 13 cannot be reduced.

(A. J. HOFFMAN, D. K. RAY-CHAUDHURI [HoR 3])

34. Find the smallest integer n for which there exist two cospectral, non-isomorphic, regular graphs with n vertices.

7. Spectral Techniques in Graph Theory and Combinatorics

In Chapter 1 and 3—6 several connections between spectral and structural properties of graphs have been described. Now we shall use these results to prove some theorems having non-spectral character. In the proofs of such theorems we always have the following scheme: by assumption $\boldsymbol{A}$ (not containing any statement on spectra) of a theorem a class of graphs is determined and some facts $\boldsymbol{B}$ (also having nothing to do with the spectrum) for this class have to be proved. By $\boldsymbol{A}$ we determine the spectrum or some spectral properties of these graphs. The facts $\boldsymbol{B}$ are then deduced from the spectrum. The material from the different sections in this chapter is united by this common scheme which can be called *spectral techniques in graph theory.*

7.1. The existence and the non-existence of certain combinatorial objects

The existence of combinatorial objects of different kinds can be investigated by means of graph spectra, at least in principle. This can be done along the following lines. Suppose that the considered object exists. Associate a graph to this object in a certain way (sometimes even in the case when the object is a graph itself). From the desired properties of the combinatorial object find the spectrum of the associated graph. Finally find graphs having this spectrum or disprove the existence of such graphs, which in fact answers the question about the existence of the original object.

If we want to prove the *existence* of a certain object in this way, we then have the difficult task of deciding whether a graph with the given spectrum exists. Unfortunately, we do not know how to solve this problem in the general case apart from considering all graphs with the given number of vertices and checking whether the spectrum of any one of these graphs coincides with the given spectrum. Of course, in special cases one can expect the solution. For example, in [CvG 3] the problem is solved for spectra consisting of numbers not greater than 2. For some particular graphs one can use theorems on graph spectra to learn as much as possible about the graph structure, and then continue the construction. by non-spectral means. A few such examples can be found in [BuCv].

The problem of deciding whether a graph with the given spectrum exists, can be considered as the basic one in the theory of graph spectra. But it is very difficult and, in fact, represents not only one problem but a great area of investigation. For example, one of the basic problems in the theory of block designs, i.e., the problem of deciding whether a block design with the given parameters exists, can be reduced to our problem using some theorems from Chapter 6 providing spectral characterizations of block designs. Theorem 6.9 can be reformulated in the following way:

A symmetric BIBD *with the parameters* v, k, λ *exists if and only if there exists a graph with the spectrum consisting of eigenvalues* k, $(k-\lambda)^{\frac{1}{2}}$, $-(k-\lambda)^{\frac{1}{2}}$, $-k$ *with the multiplicities* 1, $v-1$, $v-1$, 1 *respectively.*

Theorem 6.21 can also be stated in a similar form [Ray 2].

On the other hand, many different means have been used for proving the existence of block designs; the proofs are often very hard, even in quite special cases.

Proving the existence of a graph with the given spectrum means, in fact, to construct it. An idea how to do this using the theory of graph spectra could be to start with graphs with known spectra and perform some graph operations (such as described in Chapter 2) on them in order to obtain a graph with the desired spectrum. Although one must not expect too much from this idea, in special cases the construction can be attained. We shall describe a few examples of this kind in the next sections.

If we want to prove the *non-existence* of a certain object, we have much more chance to do it by spectral methods, providing, of course, that the object really does not exist. We do not need the knowledge of the whole object as in the case of the existence; it is sufficient to derive from the spectrum some structural detail which contradicts the assumed properties of the object wanted or leads to a logical contradiction. For example, the non-existence of cubic graphs, having certain spectra consisting of integers, has been proved by calculating the number D_4, of circuits of length 4 (using Theorem 3.27), and by establishing that the obtained values for D_4 are negative [BuCv].

A standard device in this field is to start with the set of distinct eigenvalues and to calculate the corresponding multiplicities using the trace of the adjacency matrix. Sometimes the obtained multiplicities are not non-negative integers and this, of course, means that the graph does not exist (see [Cve 16] and the next section).

There are many nice non-existence proofs of such kind in graph theory and combinatorics. We have already described some of them in earlier chapters. An excellent example of this type is the investigation of the existence of Moore graphs, described in Chapter 6. In Chapter 4 we have proved the non-existence of certain perfect codes (Lloyd's Theorem) and the non-existence of certain coverings of graphs by trees using the notion of divisor. In the next section the investigation of the existence of strongly regular graphs will be described with many details. Now we shall give a few other examples of non-existence proofs.

The proof of the next theorem follows that of the original paper.

Theorem 7.1 (W. G. Brown [Brow]): *There are no regular graphs of degree* r *having* r^2+2 *vertices and girth* 5.

Remark. H. Sachs [Sac 13] showed the existence of regular graphs of degree r and girth g for all $r, g \geqq 2$. Let $f(r, g)$ denote the smallest integer n for which there exists a regular graph of degree r and girth g. P. ERDÖS and H. SACHS [ErSa] proved that

$$r^2 + 1 \leqq f(r, 5) \leqq 4(r-1)(r^2 - r + 1).$$

Equality holds in the first inequality only for $r = 2, 3, 7$, and possibly 57 and the bound is attained by Moore graphs which have been described in Chapter 6. Theorem 7.1 says that $f(r, 5) \neq r^2 + 2$.

Proof of the theorem. Suppose G is regular of valency r. Consider all paths of length 2 from some vertex v; their end points, exclusive of v, are $r(r-1)$ in number, the end points of paths of length 1 from v are r in number. No two of these r^2 vertices can coincide if G is to have girth 5. We assume henceforth that G has $n = r^2 + 2$ vertices; thus there exists in G exactly one vertex which cannot be reached from v along a path of length less than 3; this vertex we shall denote by v^* for any given v. Clearly $(v^*)^* = v$. We thus obtain the matrix equation

$$\boldsymbol{A}^2 + \boldsymbol{A} - (r-1)\boldsymbol{I} = \boldsymbol{J} - \boldsymbol{B},$$

where $\boldsymbol{J}$ is the $n \times n$ matrix all of whose entries are 1, and $\boldsymbol{B}$ is a symmetric permutation matrix with zeros on the main diagonal. By a suitable relabelling of G we can arrange that $\boldsymbol{B}$ be a direct sum of matrices $\begin{pmatrix} 0 & 1 \\ 1 & 0 \end{pmatrix}$. It follows incidentally that n is even and so $r \equiv 0 \pmod 2$. $\boldsymbol{J} - \boldsymbol{B} - \boldsymbol{I}$ is the adjacency matrix of a regular graph of degree $n - 2$ with n vertices and the corresponding spectrum has been determined in Section 2.6. Hence $\boldsymbol{J} - \boldsymbol{B}$ has the eigenvalues $n - 1, 1, -1$, of multiplicities 1, $n/2$ and $n/2 - 1$, respectively. Any eigenvector of $\boldsymbol{A}$ having eigenvalue k must also be an eigenvector of $\boldsymbol{J} - \boldsymbol{B}$ with eigenvalue $k^2 + k - (r-1)$. As $\boldsymbol{A}$ is real and symmetric, it must have $n/2$ eigenvalues k satisfying

$$k^2 + k - (r-1) = 1, \quad \text{i.e., } k = (-1 \pm s)/2 \text{ where } s = \sqrt{4r+1}$$

and $n/2 - 1$ eigenvalues k satisfying

$$k^2 + k - (r-1) = -1, \quad \text{i.e., } k = (-1 \pm t)/2 \text{ where } t = \sqrt{4r-7}.$$

We shall impose on r the condition that the trace of $\boldsymbol{A}$ be zero; four cases must be considered.

Case 1: *s and t both rational,* hence both integral. The only two odd positive integers whose squares differ by 8 are 1 and 9, so $r = 2$. But G would then be a hexagon, whose girth is 6, not 5.

Case 2: *s and t both irrational.* First suppose s and t are linearly dependent over the rationals. Then s^2 and t^2 must have the same square-free part α, which must divide their difference, 8. By hypothesis, $\alpha > 1$; but α cannot be even, since s^2 and t^2 are odd. Thus s and t must be linearly independent. But this implies that the eigenvalues $(-1 \pm s)/2$ occur in pairs; so also do the eigenvalues $(-1 \pm t)/2$. This is impossible, since one of $n/2$, $n/2 - 1$ is odd.

Case 3: *s irrational, t rational.* Here t is an odd integer, so $-1 \pm t$ is even. Thus the sum of the eigenvalues $(-1 \pm s)/2$ is an integer. These eigenvalues occur in pairs, and so their sum is $-n/4$. But $4 \mid n$ implies that $r^2 \equiv 2 \pmod 4$ which is impossible.

Case 4: *s rational, t irrational.* The eigenvalues $(-1 \pm t)/2$ must occur in pairs, so their sum is $(-1/2)(n/2-1)$. Suppose the multiplicity of the eigenvalue $(-1+s)/2$ is m. Then the trace of $\boldsymbol{A}$ is

$$0 = r + m(-1+s)/2 + (n/2-m)(-1-s)/2 + (-1/2)(n/2-1).$$

Because $n = r^2 + 2$ and $r = (s^2-1)/4$, this yields a quintic equation for s:

$$s^5 + 2s^4 - 2s^3 - 20s^2 + (33-64m)\,s + 50 = 0.$$

Any positive rational solutions s must be among the integral divisors of 50, viz. 1, 2, 5, 10, 25, 50. Of these, only three yield solutions, namely

$$\begin{array}{lll} s = 1 & s = 5 & s = 25 \\ m = 1 & m = 12 & m = 6565 \\ r = 0 & r = 6 & r = 156. \end{array}$$

The case $s = 1$ is obviously of no interest: it yields a graph of infinite girth.

We eliminate the cases $s = 5$, $s = 25$ by the following argument. As G has girth 5, it cannot contain any triangles. Hence $\boldsymbol{A}^3$ has only zeros on its main diagonal. The eigenvalues of $\boldsymbol{A}^3$ are the cubes of the eigenvalues of $\boldsymbol{A}$. Hence we can impose on the eigenvalues of $\boldsymbol{A}$ the additional condition that the sum of their cubes be zero. In neither of the remaining cases this condition is satisfied.

This completes the proof.

Theorem 7.2 (H. Sachs [Sac 1]): *Let p and q be primes, $p \equiv q \equiv 3 \pmod 4$, and let s be a positive integer. There are no cyclic self-complementary graphs with $n = p^{2s}$ or with $n = pq$ vertices.*

The idea of the proof is the following. Suppose that a graph G from the theorem exists. The corresponding adjacency matrix can be taken as cyclic and G can be understood as a polynomial of a cycle $\vec{C}_n$ (see Section 2.1). The eigenvalues of G are then sums of the n-th roots of unity. Considering primitive roots of unity in the case $n = p^{2s}$ we can conclude that the characteristic polynomial of G can be factored in the form

$$P_G(\lambda) = \left(\lambda - \frac{1}{2}(n-1)\right) \prod_{\sigma=1}^{2s} \left(g_{p^\sigma}(\lambda)\right)^{v_{p^\sigma}},$$

where $g_d(\lambda)$ is an irreducible monic polynomial of some degree u_d, where $u_d v_d = \varphi(d)$, $\varphi(d)$ being the number of primitive d-th roots of unity (cf. Section 5.1, Theorem 5.4). One can show that this polynomial is not of the form required by Theorem 3.28.

Similar considerations hold also in the case $n = pq$.

There are many other non-existence proofs in the literature. We shall mention a few of them here.

R. Singleton [Sin 1], [Sin 2] investigates the existence of regular graphs of degree r, diameter d, girth $g = 2d$ and $n = 2\big((r-1)^d - 1\big)/(r-2)$ vertices. For $d \neq 2, 3, 4, 6$

no graphs exist. For $d = 2$ there is only one graph for each r. For $d = 3$ and $d = 4$ there is one graph for each finite projective geometry with r points on a line, of dimension 2 and 3, respectively.

A. Gewirtz [Gew 1], [Gew 2] considers the existence of connected graphs of diameter d (> 1), girth $g = 2d$ and for any pair of vertices which are at distance d there exist t distinct paths of length d connecting them. It is proved that such graphs are regular. If $d = 2$ and $t \neq 2, 4, 6$, then there exists at most a finite number of graphs with a particular value t.

W. Feit, G. Higman [FeHi] investigated the existence of certain geometric structures called *generalized polygons*. In terms of graphs, a *non-degenerate generalized polygon* is a connected bipartite graph G of diameter d and girth $g = 2d$. Let G be semiregular with vertex degrees r_1 and r_2. If $r_1 = r_2$, we have $d = 2, 3, 4, 6$. If $r_1 \neq r_2$, the additional values $d = 8, 12$ are possible. Some similar problems have been treated by S. E. Payne [Payn]. A regular bipartite graph with $n = 2v$ vertices of degree $r = s + 1$, with the girth $g = 2k$ satisfies the inequality $v \geqq v_0 = 1 + s + s^2 + \cdots + s^{k-1}$, with equality if and only if the graph is a non-degenerate generalized polygon. For $k \geqq 3$ and $s \geqq 3$ we have $v \neq v_0 + 1$.

H. S. Wilf [Wil 1] gave a proof by eigenvalues of the so-called *friendship theorem* which reads: *in a party of n people, suppose that every pair of people has exactly one common friend. Then there is a person* (leader) *in the party who knows everyone else.* Consider a geometric structure in which the points are the people of the party and the lines are the sets $l(x)$, $l(x)$ being the sets of friends of x. Under the assumption that the leader does not exist one can prove that this geometric structure is a projective plane (for the definition and elementary properties see, for example, [Ha,M]). Consider now a graph in which the vertices are the people, two of them being adjacent if they know each other. It follows that this graph is strongly regular of degree $m + 1$ (for some m, i.e. the order of the plane) that it has $m^2 + m + 1$ vertices and that each pair of vertices has exactly one common adjacent vertex. The non-existence of this graph has been proved along the lines which will be described in the next section.

H. L. Skala [Skal] has relaxed the conditions of the friendship theorem so that two friends need not have a common friend. The friendship set is now a graph in which each pair of vertices has at most one common adjacent vertex. If such a graph is regular we again have a Moore graph. Some necessary conditions for the existence of the friendship sets in the non-regular case have been given in [Skal] in terms of the irreducibility of certain polynomials. See also [Pars].

7.2. Strongly regular graphs and distance-transitive graphs

Strongly regular graphs have already been defined in Section 3.4. The theory of strongly regular graphs was introduced by R. C. Bose [Bos 1] in connection with so called partial geometries and association schemes with two classes (see Section 6.4). Almost at the same time D. G. Higman [Hig 1] also found these graphs when studying permutation groups of rank 3.

In fact strongly regular graphs are equivalent with association schemes with two classes. Let G be a strongly regular graph of degree r with each adjacent pair of vertices having e common adjacent vertices and each non-adjacent pair of vertices having f common neighbours. We can now associate a 2-class association scheme to G taking the vertices of G as treatments and defining that two vertices (treatments) are first or second associates according to whether (or not) they are adjacent.

The quantities r, e, f will be called *parameters of a strongly regular graph.*

The main problem in the field of strongly regular graphs is the *problem of their existence.* We shall describe only those parts of the existence problem, which are related to the eigenvalues. For a general survey on strongly regular graphs see [Huba]. Note that strongly regular graphs became a very interesting object for group theorists when it was discovered that some interesting groups (especially simple groups) occur as the automorphism groups of these graphs [Huba], [HiSi], [Hest].

Let $d(x, y)$ denote the distance between vertices x and y. A connected graph G is said to be *distance-transitive,* if for any vertices x, y, u, v satisfying $d(x, y) = d(u, v)$ there exists an automorphism of G which takes x to u and y to v. Of course, a distance-transitive graph is regular. Note that a distance-transitive graph of diameter 2 is strongly regular. There are strongly regular graphs which are not distance-transitive. For example A. J. H. Paulus [Paul] found a strongly regular graph on 26 vertices whose automorphism group is trivial.

A distance-transitive graph is a special case of metrically regular graphs, i.e., it is equivalent to an association scheme. In fact, a distance-transitive graph G of diameter d carries an association scheme with d classes which can be constructed by taking two vertices of G to be i-th associates if their distance is i.

Distance-transitive graphs have been investigated by N. L. Biggs and others [Big 1]–[Big 7], [BiSm], [Sm,D 1], [Sm,D 2]. Although the underlying theory is contained in the work of R. C. Bose about *association schemes* and D. G. Higman about *coherent configurations* (a generalization of the association schemes) [BoMe], [Hig 1], [Hig 2], [Hig 3], the structure developed by N. L. Biggs is very nice in application, especially because many existence theorems mentioned in the previous section can be proved in this way. We shall explain shortly this topic after describing procedures for the investigation of the existence of strongly regular graphs.

As proved in Theorem 3.32, strongly regular graphs have exactly three distinct eigenvalues, and parameters r, e, f and eigenvalues $\lambda^{(1)} > \lambda^{(2)} > \lambda^{(3)}$ are connected by

$$r = \lambda^{(1)}, \quad e = \lambda^{(1)} + \lambda^{(2)} + \lambda^{(3)} + \lambda^{(2)}\lambda^{(3)}, \quad f = \lambda^{(1)} + \lambda^{(2)}\lambda^{(3)}$$

or

$$\lambda^{(1)} = r, \quad \lambda^{(2,3)} = \frac{1}{2}\left(e - f \pm \sqrt{(e - f)^2 - 4(f - r)}\right).$$

Distinct eigenvalues are determined by the parameters, and vice versa, and therefore eigenvalues can be also used as parameters of the graph. So the eigenvalues can be used for classification of strongly regular graphs.

Note that the complement of a strongly regular graph is again strongly regular.

If the parameters r, e, f are given, the number n of vertices and multiplicities p_i

of eigenvalues $\lambda^{(i)}$ $(i = 1, 2, 3)$ can be determined. Except for the direct sum of complete graphs a strongly regular graph is connected and we have $p_1 = 1$. Now we can write

$$1 + p_2 + p_3 = n, \tag{7.1}$$

$$r + p_2\lambda^{(2)} + p_3\lambda^{(3)} = 0, \tag{7.2}$$

$$r^2 + p_2(\lambda^{(2)})^2 + p_3(\lambda^{(3)})^2 = nr. \tag{7.3}$$

The last condition says that the graph is regular (see Theorem 3.22). Quantities n, p_2, and p_3 can easily be calculated.

Expressing $\lambda^{(2)}$ and $\lambda^{(3)}$ in (7.2) as functions of e and f we get

$$(p_2 - p_3)\sqrt{(e-f)^2 - 4(f-r)} + 2r + (e-f)(p_2 + p_3) = 0. \tag{7.4}$$

Further we have two possibilities,

1. $(e-f)^2 - 4(f-r)$ is not the square of an integer. Then $p_2 = p_3$, and from (7.4) we get $p_2 = p_3 = \dfrac{r}{f-e}$. Since p_2, p_3 are positive integers, we have $f - e > 0$ and $(f-e) \mid r$. We shall prove that $f - e = 1$. From (7.1) we have $n = 1 + 2\dfrac{r}{f-e}$ and then from (7.3) the relation $r = 1 - (f-e) + \dfrac{2f}{f-e}$ can be obtained. If $f - e \geqq 2$, then $r \leqq 1 - 2 + f = f - 1$ which is obviously impossible. Hence $f - e = 1$, $p_2 = p_3 = r$ and $n = 1 + 2r$. Further $r = 2f$ and $n = 4f + 1$.
2. $(e-f)^2 - 4(f-r)$ is the square of an integer, say s. Then p_2 and p_3 need not be equal. Eliminating n and p_3 from equations (7.1)–(7.3) we get

$$p_2 = \frac{r}{2fs}\big((r - 1 + f - e)(s + f - e) - 2f\big).$$

In this way we have proved the following theorem.

Theorem 7.3 (W. S. Connor, W. H. Clatworthy [CoCl], D. G. Higman [Hig1]): *If a strongly regular graph G with the parameters r, e, f exists, then*

1^0 *$r = 2f$, $e = f - 1$ (and G has $4f + 1$ vertices), or*

2^0 *$(e-f)^2 - 4(f-r) = s^2$ for some positive integer s, and the expression*

$$\frac{r}{2fs}\big((r - 1 + f - e)(s + f - e) - 2f\big)$$

is a positive integer.

Note that sometimes a strongly regular graph can satisfy both 1^0 and 2^0 from Theorem 7.3 (for example, $n = 25$, $r = 12$, $e = 5$, $f = 6$, [Paul]). In the case 2^0 the eigenvalues are integers.

Strongly regular graphs can be described also in terms of the $(-1, 1, 0)$-adjacency matrix. Having in mind the relations between eigenvalues of the $(0, 1)$- and $(-1, 1, 0)$-

adjacency matrix, the last matrix has eigenvalues $\varrho_0 = n - 1 - 2r$, $\varrho_1 = -1 - 2\lambda^{(3)}$, $\varrho_2 = -1 - 2\lambda^{(2)}$. One can easily see that in the case 1° of the last theorem the eigenvalues are 0 and $\pm\sqrt{n}$. In the case 2°, ϱ_1 and ϱ_2 are odd integers. But sometimes ϱ_0 coincides with ϱ_1 or ϱ_2. Therefore, speaking in terms of $(-1, 1, 0)$-adjacency matrix, a strongly regular graph has two or three distinct eigenvalues. But now the complete and totaly disconnected and graphs also have two distinct eigenvalues.

Coincidence of ϱ_0 with one of ϱ_1, ϱ_2 means that the strongly regular graph has certain additional regularities. The proper explanation will be given in the next section. For the moment notice that if $\boldsymbol{S}$ has two distinct eigenvalues ϱ_1 and ϱ_2 then $(\boldsymbol{S} - \varrho_1\boldsymbol{I}) \times (\boldsymbol{S} - \varrho_2\boldsymbol{I}) = \boldsymbol{O}$. In the case of three distinct eigenvalues, according to Theorem 3.25 we would have $(\boldsymbol{S} - \varrho_1\boldsymbol{I})(\boldsymbol{S} - \varrho_2\boldsymbol{I}) = (n - 1 + \varrho_1\varrho_2)\boldsymbol{J}$ with $n - 1 + \varrho_1\varrho_2 \neq 0$. The first equation remains satisfied if we replace $\boldsymbol{S}$ by $\boldsymbol{DSD}$ ($\boldsymbol{D}$ diagonal matrix with 1's and -1's on the diagonal) and the second does not. That means that we can switch a strongly regular graph into another strongly regular graph only in the case of two distinct eigenvalues.

Let $p(x, y)$ be the number of vertices which are adjacent to x and non-adjacent to y. A graph is called *strong* if it is not complete and not totally disconnected and if $p(x, y) + p(y, x)$ depends only on whether x and y are adjacent or non-adjacent. The notion of strong graph is a generalization of the notion of strongly regular graph since each strongly regular graph is strong. Clearly, a graph is strongly regular if and only if it is strong and regular. Strong graphs have been introduced by J. J. Seidel [Sei2].

Theorem 7.4: *A graph G with the Seidel adjacency matrix $\boldsymbol{S}$ is strong if and only if one of the following holds:*

1° *$\boldsymbol{S}$ has three distinct eigenvalues (ϱ_0 being simple) and G is (strongly) regular;*

2° *$\boldsymbol{S}$ has two distinct eigenvalues and G is not complete and not totally disconnected.*

In both cases the corresponding eigenvalues ϱ_1, ϱ_2 and values α, β of $p(x, y) + p(y, x)$ for adjacent and non-adjacent pairs of vertices, respectively, are related by $2\alpha = -(\varrho_1 - 1)(\varrho_2 - 1)$, $2\beta = -(\varrho_1 + 1)(\varrho_2 + 1)$.

The proof of this theorem is based on the fact that the elements $s_{ij}^{(2)}$ of $\boldsymbol{S}^2$ satisfy

$$s_{ij}^{(2)} = \begin{cases} n - 1 & \text{if} \quad i = j, \\ n - 2 - 2\big(p(x_i, x_j) + p(x_j, x_i)\big) & \text{if} \quad i \neq j \end{cases}$$

(x_i, x_j representing vertices of the graph) as can be seen by direct computation. If G is a strong graph, we have

$$\boldsymbol{S}^2 = (n - 1)\boldsymbol{I} + (n - 2)\boldsymbol{J} - \alpha(\boldsymbol{J} - \boldsymbol{S}) - \beta(\boldsymbol{J} + \boldsymbol{S})$$

and

$$\boldsymbol{S}^2 - (\alpha - \beta)\boldsymbol{S} - (n - 1)\boldsymbol{I} = (n - 2 - \alpha - \beta)\boldsymbol{J}.$$

Thus $\boldsymbol{S}$ has two or three distinct eigenvalues depending on whether $n - 2 - \alpha - \beta = 0$ or $\neq 0$. In the second case G must be regular (and, of course, strongly regular) because the vector with all components equal to 1 is an eigenvector of $\boldsymbol{S}$. Relations between α, β and ϱ_1, ϱ_2 can be obtained using the fact that ϱ_1 and ϱ_2 are the roots of the equation $\varrho^2 - (\alpha - \beta)\,\varrho - (n-1) = 0$.

Returning to strongly regular graphs, let us mention that the necessary conditions for the existence from Theorem 7.3 are very restrictive. The paper [ApJIP] contains a list of possible parameter sets of strongly regular graphs on 300 vertices and less, which was obtained by computer search based on restrictions of Theorem 7.3. M. D. Hestenes (unpublished) computed such parameter sets for graphs up to 900 vertices.

All strongly regular graphs up to 28 vertices are known [Poзe], [ApJIP]. The circuits C_4 and C_5 and the Petersen graph belong to the group of strongly regular graphs with a small number of vertices. For some parameter sets there are many non-isomorphic (but, of course, cospectral) strongly regular graphs (see Section 6.1). Moreover, sometimes such strongly regular graphs are not mutually switching equivalent but belong to different switching classes. F. C. Bussemaker and J. J. Seidel [BuS2] found 91 classes of strongly regular cospectral graphs on 35 vertices.

There are infinite series of strongly regular graphs. Trivial examples are complete bipartite graphs $K_{n,n}$, their line graphs $L(K_{n,n})$, line graphs of complete graphs $L(K_n)$, and regular graphs $H(n) = CP(n)$ on $2n$ vertices of degree $2n-2$, sometimes called "*coctail party graphs*" (see p. 177).

Non-trivial examples of strongly regular graphs have been constructed in the majority of cases by non-spectral means. As a starting point for construction, several combinatorial objects such as projective planes and projective geometries, block designs, Steiner triple systems, Latin squares, Hadamard matrices etc. have been used. A description of these construction diverges from the purposes of this book and only a few examples will be reviewed in Section 7.8.

We shall now describe a spectral construction. The Hoffman-Singleton graph G mentioned in Section 6.2, is strongly regular with the eigenvalues 7, 2, -3 of multiplicities 1, 28, 21, respectively. According to Theorem 2.15, its line graph $L(G)$ has eigenvalues 12, 7, 2, -2 with the multiplicities 1, 28, 21, 125, respectively. Let $\boldsymbol{B}$ be the adjacency matrix of $L(G)$. Since G has girth 5, exactly one path of length 2 connects any two vertices of distance 2 in $L(G)$. Therefore, $\boldsymbol{A} = \boldsymbol{B}^2 - 5\boldsymbol{B} - 12\boldsymbol{I}$ is a (0, 1) matrix. Eigenvalues of $\boldsymbol{A}$ are 72, 2, -18 with the multiplicities 1, 153, 21, respectively. Hence, $\boldsymbol{A}$ is the adjacency matrix of a strongly regular graph on 175 vertices.

See also Section 7.8, no. 15.

Further facts about strongly regular graphs can be found in the books [Big5], [CaLi] and expository papers [Huba], [Sei3], [Sei4], [Hig4], [Hig7].

Let us now turn back to distance-transitive graphs. We can again find some necessary conditions for the existence of such graphs in terms of the multiplicities of eigenvalues.

Instead of a set of parameters we shall consider the intersection matrix $\boldsymbol{B}$ defined by $\boldsymbol{B} = \boldsymbol{P}_1^{\mathrm{T}}$ where $\boldsymbol{P}_1$ is defined in Section 6.4 and is related to the corresponding association scheme. The (i, j)-entry $(i, j = 0, 1, \ldots, d)$ of $\boldsymbol{B}$ is equal to the number of vertices which are at distance i from x and adjacent to y provided that x and y

are at distance j. Hence $\boldsymbol{B}$ is tridiagonal of the form

$$\boldsymbol{B} = \begin{pmatrix} 0 & 1 & & & 0 \\ k_1 & a_1 & b_2 & & \\ & c_1 & a_2 & \ddots & \\ & & c_2 & \ddots & \ddots \\ & & & \ddots & \ddots & b_d \\ 0 & & & & c_{d-1} & a_d \end{pmatrix}$$

where d is the diameter of the graph. The entries of $\boldsymbol{B}$ are, of course, non-negative integers, and those above and below the main diagonal are positive since the graph is connected. All column sums are the same since the graph is regular. The transpose of the intersection matrix in fact defines a front divisor of the graph (see Chapter 4).

In order to save the space the matrix $\boldsymbol{B}$ can be conventionally written as a $3 \times (d+1)$ matrix:

$$\begin{pmatrix} * & 1 & b_2 & \cdots & b_{d-1} & b_d \\ 0 & a_1 & a_2 & \cdots & a_{d-1} & a_d \\ k_1 & c_1 & c_2 & \cdots & c_{d-1} & * \end{pmatrix}$$

Matrix $\boldsymbol{B}$ written in this form is called the *intersection array of the graph*. If G is a distance-transitive graph with the intersection matrix $\boldsymbol{B}$, then the whole spectrum can be calculated from $\boldsymbol{B}$. One part of this statement follows from Theorem 6.23. A general method for calculating the multiplicities of eigenvalues in association schemes has been given by R. C. Bose and D. M. Messner [BoMe]. But in special case of distance-transitive graphs we have the following nice theorem.

Theorem 7.5 (N. L. Biggs [Big1], [Big2]): *Let $\boldsymbol{B}$ be the intersection matrix of a distance-transitive graph G of diameter d and let $\lambda_0 = r, \lambda_1, \ldots, \lambda_d$ be eigenvalues of $\boldsymbol{B}$ with the eigenvectors $u_0 = (1, k_1, k_2, \ldots, k_d)^{\mathsf{T}}$ and $u_i = (u_{i0}, \ldots, u_{id})$, $i = 1, \ldots, d$, respectively. Then the spectrum of G consists of eigenvalues $\lambda_0 = r, \lambda_1, \ldots, \lambda_d$ with the multiplicities*

$$p_i = \left(\sum_{j=0}^{d} k_j\right) \left(\sum_{j=0}^{d} \frac{u_{ij}^2}{k_j}\right)^{-1} \qquad (k_0 = 1).$$

If G exists, then according to Theorem 7.5 quantities $k_1, \ldots, k_d$ and $p_1, \ldots, p_d$ must be positive integers. If a given tridiagonal matrix $\boldsymbol{B}$ satisfies all mentioned necessary conditions for the existence of G (and in particular the last two), it is called *feasible*.

Of course, if a matrix is feasible, this does not mean that the corresponding distance-transitive graph exists. In any particular case further investigations are necessary.

Using this technique, N. L. Biggs and D. H. Smith [BiSm] proved that there are exactly 12 regular connected graphs of degree 3 which are distance-transitive.

7.3. Equiangular lines and two-graphs

We turn now to some applications of graph spectra in geometry. In Chapter 6 we have seen how a set of vectors, angles between which have only two possible values, can be associated with a graph. This has been done by interpreting the matrix $\boldsymbol{A} - \lambda_{\min}\boldsymbol{I}$ ($\boldsymbol{A}$ adjacency matrix of the graph, $\lambda_{\min}$ its least eigenvalue) as the Gramian matrix of a set of vectors in the Euclidean space of the corresponding dimension. If we do the same thing with the Seidel matrix $\boldsymbol{S}$ of a graph, the corresponding set of lines will have all angles between lines the same. As a contrast to Chapter 6, where geometrical means have been used to prove theorems about eigenvalues of graphs, we shall use now eigenvalues in investigating geometrical objects.

J. J. Seidel and others [LiSe], [LeS2] studied systems of equiangular lines. The original problem was to find the maximal number of equiangular lines in a Euclidian space. It turned out that sets of equiangular lines are equivalent to equivalence classes of graphs under switching (see Section 6.5) and also to some homogeneous hypergraphs called *two-graphs*. A set of lines in Euclidean r-dimensional space $\mathbb{R}^r$ is a *set of equiangular lines* whenever each piar of lines has the same angle.

For example, the six lines connecting the antipodal pairs of vertices of a regular icosahedron are equiangular. The value of common angle is $\arccos 1/\sqrt{5}$.

Next a few pages have been written mainly according to [Sei4].

Lemma 7.1: *To any set of non-orthogonal equiangular lines there is associated a switching class of graphs.*

Proof. Without loss of generality we consider n equiangular lines through the origin of $\mathbb{R}^r$, with the angle $0 \leqq \varphi < \frac{1}{2}\pi$. Each line is represented by a spanning unit vector $\boldsymbol{p}_i$, in either of the two directions, $i = 1, 2, \ldots, n$. The inner product of any such vectors $\boldsymbol{p}_i$ and $\boldsymbol{p}_j$, $i \neq j$, satisfies

$$\langle \boldsymbol{p}_i, \boldsymbol{p}_j \rangle = \pm\cos\varphi, \qquad 0 < \cos\varphi \leqq 1,$$

the sign depending on whether the angle between $\boldsymbol{p}_i$ and $\boldsymbol{p}_j$ is acute or obtuse. Consider the following matrices of the order n:

$$\boldsymbol{P} = (\langle \boldsymbol{p}_i, \boldsymbol{p}_j \rangle) = \begin{pmatrix} 1 & & \pm\cos\varphi \\ & \ddots & \\ \pm\cos\varphi & & 1 \end{pmatrix}, \qquad \boldsymbol{S} = \frac{1}{\cos\varphi}(\boldsymbol{P} - \boldsymbol{I}) = \begin{pmatrix} 0 & & \pm 1 \\ & \ddots & \\ \pm 1 & & 0 \end{pmatrix}.$$

The matrix $\boldsymbol{S}$ is taken to be the Seidel adjacency matrix of a graph on n vertices. If any $\boldsymbol{p}_i$ is replaced by its opposite $-\boldsymbol{p}_i$, then in $\boldsymbol{S}$ the i-th row and the i-th column are multiplied by -1, hence the graph is switched with respect to the i-th vertex. Since replacing any subset of $\boldsymbol{p}_1, \ldots, \boldsymbol{p}_n$ by the opposite vectors yields any graph in the switching class, the assertion is proved.

A set of lines is said to be a *set of dependent lines*, if the corresponding spanning unit vectors are dependent.

Theorem 7.6 (J. H. VAN LINT, J. J. SEIDEL [LiSe]): *There is a one-to-one correspondence between he switching classes of graphs on n vertices and the dependent sets of n equiangular lines.*

Proof. According to Lemma 7.1, to any set of n equiangular lines in $\mathbb{R}^r$, $n > r$, there is associated a switching class of graphs on n vertices. The class of the totally disconnected graph belongs to the set of n coinciding lines.

Conversely, let $\boldsymbol{S}$ be the adjacency matrix of any graph on $n > 1$ vertices. Let $-\alpha$ be the smallest eigenvalue of $\boldsymbol{S}$. By Theorem 0.10 we have $\alpha \geqq 1$. The matrix $\boldsymbol{S} + \alpha \boldsymbol{I}$ has the smallest eigenvalue 0. Hence $\boldsymbol{P} := \boldsymbol{I} + \boldsymbol{S}\alpha^{-1}$ is singular, symmetric, positive semi-definite, with diagonal $\boldsymbol{I}$. Therefore, $\boldsymbol{P}$ is the Gramian matrix of a dependent set of n unit vectors whose inner products equal $\pm\alpha^{-1}$. These yield n dependent equiangular lines, which coincide for $\alpha = 1$. The associated switching class contains the graph from which we started.

This completes the proof.

Let us turn now to two-graphs. Two-graphs have been introduced by G. HIGMAN (unpublished) and investigated in [Tay1], [Tay2].

A set of cardinality k will be called a *k-set.*

Let $\mathcal{X}$ denote a finite set, and let $\mathcal{X}^{(3)}$ denote the set of all 3-subsets of $\mathcal{X}$. We shall consider sets $\Delta \subset \mathcal{X}^{(3)}$ of 3-subsets of $\mathcal{X}$ whose elements will be called *triples.*

Definition 7.1. A *two-graph* $(\mathcal{X}, \Delta)$ is a pair of a vertex set $\mathcal{X}$ and a triple set $\Delta \subset \mathcal{X}^{(3)}$ such that each 4-subset of $\mathcal{X}$ contains an even number of triples of Δ.

For any $x \in \mathcal{X}$, the triple set Δ of any two-graph $(\mathcal{X}, \Delta)$ is determined by its triples containing x. Indeed, $\{x_1, x_2, x_3\} \in \Delta$ whenever an odd number of the remaining 3-subsets of $\{x, x_1, x_2, x_3\}$ belongs to Δ.

Theorem 7.7: *Given n, there is a one-to-one correspondence between the two-graphs and the switching classes of graphs on n vertices.*

Proof. Let $(\mathcal{X}, \mathcal{E})$ be any graph. It can easily be seen that the set Δ of the 3-subsets of $\mathcal{X}$ which carry an odd number of edges is invariant under switching of $(\mathcal{X}, \mathcal{E})$. By considering all possible graphs on four vertices (see Appendix) one can see that in these graphs the number of induced subgraphs on three vertices, having an odd number of edges, is even; it follows that $(\mathcal{X}, \Delta)$ is a two-graph. Conversely, let $(\mathcal{X}, \Delta)$ be any two-graph. Select any $x \in \mathcal{X}$ and partition $\mathcal{X} \setminus \{x\}$ into any two disjoint sets $\mathcal{X}_1$ and $\mathcal{X}_2$. Let $\mathcal{E}$ consist of the following pairs:

$\{x, x_1\}$, for all $x_1 \in \mathcal{X}_1$;

$\{x_1, x_1'\}$, for all $x_1, x_1' \in \mathcal{X}_1$ with $\{x, x_1, x_1'\} \in \Delta$;

$\{x_2, x_2'\}$, for all $x_2, x_2' \in \mathcal{X}_2$ with $\{x, x_2, x_2'\} \in \Delta$;

$\{x_1, x_2\}$, for all $x_1 \in \mathcal{X}_1, x_2 \in \mathcal{X}_2$ with $\{x, x_1, x_2\} \notin \Delta$.

Thus, we associate to $(\mathcal{X}, \Delta)$ a class of graphs $(\mathcal{X}, \mathcal{E})$. By construction, the set of 3-subsets of $\mathcal{X}$ carrying an odd number of edges from $(\mathcal{X}, \mathcal{E})$ is again Δ. It can easily be seen that the class of graphs $(\mathscr{X}, \mathscr{E})$ is an equivalence class of graphs under switching. Also distinct switching classes yield distinct two-graphs. This completes the proof of the theorem.

According to the last two theorems, there is a one-to-one correspondence between the two-graphs on n vertices and the dependent sets of n equiangular lines.

C. L. MALLOWS and N. J. SLOANE [MaSl] established that the number of switching classes of graphs on n vertices (or the number of two-graphs on n vertices) is equal to the number of Eulerian graphs (all vertex degrees even) on n vertices. However, there is no one-to-one correspondence between the set of Eulerian graphs on n vertices and the set of switching classes of graphs on n vertices if n is even. Namely, some switching classes contain more than one Euler graph and others contain no Euler graphs. For odd n this correspondence exists, as proved by J. J. SEIDEL [Sei6]. In particular this means that we can find all graphs on an odd number of vertices just by starting from Eulerian graphs and switching them in all possible ways.

The spectrum of a two-graph can be defined as the common spectrum of graphs in the corresponding switching class of graphs.

***Definition* 7.2.** A two-graph $(\mathscr{X}, \Delta)$ is *regular* whenever each pair of elements of $\mathscr{X}$ is contained in the same number a of triples of Δ.

Theorem 7.8: *A two-graph is regular if and only if it has exactly two distinct eigenvalues. The parameters and the eigenvalues are related by*

$$n = 1 - \varrho_1\varrho_2, \qquad a = -\frac{1}{2}(\varrho_1 + 1)(\varrho_2 + 1).$$

Proof. Having in mind Theorem 7.7, the regularity condition is equivalent to

$$a = p(u, v) + p(v, u) = n - 2 - p(x, y) - p(y, x),$$

for all adjacent $\{x, y\}$ and all non-adjacent $\{u, v\}$ of any graph in the corresponding switching class. In terms of the eigenvalues this reads, by Theorem 7.4,

$$a = -\frac{1}{2}(\varrho_1 + 1)(\varrho_2 + 1) = n - 2 + \frac{1}{2}(\varrho_1 - 1)(\varrho_2 - 1),$$

whence $n - 1 + \varrho_1\varrho_2 = 0$, proving the theorem.

Theorem 7.9: *Let ϱ be an eigenvalue of a two-graph on n vertices with the multiplicity $n - p$. Then $\varrho^2 \leqq p(n-1)/(n-p)$, and equality holds if and only if the two-graph is regular.*

Proof. For the $(-1, 1, 0)$-adjacency matrix $\boldsymbol{S}$ of a graph with n vertices we have $\operatorname{tr} \boldsymbol{S} = 0$ and $\operatorname{tr} \boldsymbol{S}^2 = n(n-1)$. Let now $\boldsymbol{S}$ be the adjacency matrix of any graph in the corresponding switching class and let $\varrho_1, \varrho_2, \ldots, \varrho_p$ be its eigenvalues different from ϱ. Then $\sum_{i=1}^{p} \varrho_i = -(n-p)\,\varrho$, $\sum_{i=1}^{p} \varrho_i^2 = n(n-1) - (n-p)\,\varrho^2$, which implies $\sum_{i,j=1}^{p} (\varrho_i - \varrho_j)^2 = 2n\big(p(n-1) - \varrho^2(n-p)\big) \geqq 0$. If equality holds, all ϱ_i's must be equal and the two-graph has exactly two distinct eigenvalues; hence, it is regular.

For a very nice survey of two-graphs and related topics the reader is referred to [Sei4].

We now turn back to the original problem of determining the maximal number of equiangular lines in a Euclidean space.

The sets of n equiangular lines in Euclidean space $\mathbb{R}^r$ of dimension r may be interpreted as the sets of n equidistant points in elliptic space $\mathbb{E}^{r-1}$ of dimension $r - 1$. In fact, this was the original setting for the present problems (see [LiSe]).

The Gramian matrix $\boldsymbol{P}$ of the inner product of the vectors $\boldsymbol{p}_1, \ldots, \boldsymbol{p}_n$ in $\mathbb{R}^r$ has order n, is symmetric and positive semi-definite of rank $\leq r$, and hence has smallest eigenvalue 0 of multiplicity $\geqq n - r$. The matrix $\boldsymbol{S}$ has the smallest eigenvalue $-1/\cos\varphi$ of multiplicity $\geqq n - r$. Therefore, the number n of equiangular lines in $\mathbb{R}^r$ is large if a graph may be found whose adjacency matrix $\boldsymbol{S}$ has its smallest eigenvalue of a large multiplicity (see [LiSe]). We refer to [LeS2] for the state of affairs concerning the maximum number of equiangular lines in $\mathbb{R}^r$.

We shall give here only a few results concerning this problem.

Theorem 7.10 (P. M. Neumann [LeS2]): *If $\mathbb{R}^r$ contains v equiangular lines with the angle* $\arccos\alpha$, *and if $v > 2r$, then $\frac{1}{\alpha}$ is an odd integer.*

Proof. Let G be the Gramian matrix of a set of v unit vectors in $\mathbb{R}^r$ with mutual inner products $\pm\alpha$. Then $\boldsymbol{S} = \frac{1}{\alpha}(\boldsymbol{G} - \boldsymbol{I})$ has the smallest eigenvalue $-\frac{1}{\alpha}$ with multiplicity m where $m \geqq v - r$. Since $\boldsymbol{S}$ is an integral matrix, $-\frac{1}{\alpha}$ is an algebraic integer, and every algebraic conjugate of $-\frac{1}{\alpha}$ is also an eigenvalue of $\boldsymbol{S}$ with multiplicity m. If $v > 2r$, then $m > \frac{1}{2}v$ and $\boldsymbol{S}$, being a $v \times v$ matrix, cannot have more than one eigenvalue of multiplicity m. Therefore $-\frac{1}{a}$ is rational, hence an integer. Also $-\frac{1}{\alpha}$ cannot be an even integer, since in that case the (0, 1)-adjacency matrix of the corresponding graph would have a rational non-integer eigenvalue which is impossible.

This completes the proof.

Let $n_\alpha(r)$ be the maximum number of lines in Euclidean r-dimensional space $\mathbb{R}^r$ such that the angle of each pair of lines equals $\arccos\alpha$ $(\alpha > 0)$ and let $n(r) = \max_\alpha n_\alpha(r)$.

Theorem 7.9 can be used for obtaining an upper bound for $n_\alpha(r)$. Put $\varrho = -\frac{1}{\alpha}$ in Theorem 7.9. Then p is equal to the dimension r of the corresponding space $\mathbb{R}^r$. Solving the inequality of Theorem 7.9 in n, we get the following theorem.

Theorem 7.11 (J. H. van Lint, J. J. Seidel [LiSe]): *If $r < \frac{1}{\alpha^2}$, then $n_\alpha(r) \leqq \frac{r(1-\alpha^2)}{1 - r\alpha^2}$;*

if equality holds, then the two-graph defined by the equiangular set of lines is regular (i.e., graphs in the corresponding switching class are strong).

This theorem shows why strongly regular graphs (and especially those whose Seidel adjacency matrix has only two eigenvalues) are important in studying equiangular sets of lines.

According to [LeS2] we have $n(3) = n(4) = 6$, $n(5) = 10$, $n(6) = 16$, $n(7) = n(8) = \cdots = n(13) = 28$, $n(15) = 36$ and so on. The value of $n(14)$ is not known. For example, ten equiangular lines in $\mathbb{R}^5$ can be obtained from the Petersen graph whose Seidel adjacency matrix has eigenvalues 3 and -3, the multiplicities in both cases being equal to 5.

The problem of equiangular lines is very difficult and is far from being solved in general cases.

7.4. Connectedness and bipartiteness of certain graph products

P. M. Weichsel [Weic] has observed the interesting fact that the product of two connected graphs can be a disconnected graph. Such an example is given in Fig. 7.1.

Similar situations appear in a NEPS (whose special case is the product), in Boolean functions and in polynomials of graphs. (For the definition and spectral properties of these operations see Chapter 2.)

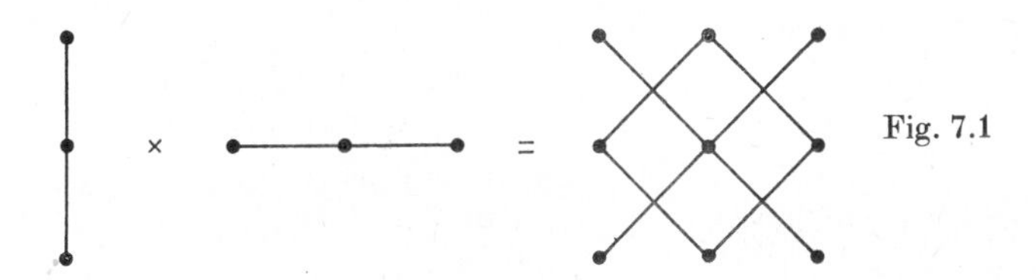

Fig. 7.1

In this section we shall investigate conditions under which the results of some operations on graphs are connected graphs. Since in a NEPS this question is related in a natural way to the question whether or not the resulting graph is bipartite, we shall treat the problem of bipartiteness of a NEPS, too.

We shall investigate the connectedness of these operations on graphs by using Theorems 3.34 and 3.35.

Consider the NEPS of graphs $G_1, \ldots, G_n$ each containing at least one edge. The indices $r_1, \ldots, r_n$ of the graphs are then positive. By analysis of expression (2.44) we see that the index of the NEPS is obtainable from (2.44) if we put $i_1 = i_2 = \cdots = i_n = 1$, i.e. according to the accepted convention $\lambda_{ji_j} = \lambda_{j1} = r_j$ $(j = 1, \ldots, n)$. Thus, for the index r of the NEPS we have

$$r = \Lambda_{1,\ldots,1} = \sum_{\beta \in \mathscr{B}} r_1^{\beta_1} \cdots r_n^{\beta_n} \qquad (> 0). \tag{7.5}$$

We shall consider only the NEPS with the basis $\mathscr{B}$, for which there exists in $\mathscr{B}$ for every j $(j = 1, \ldots, n)$ at least one n-tuple $(\beta_1, \ldots, \beta_n)$ with $\beta_j = 1$. We will denote this condition by ($\boldsymbol{D}$). This condition implies that the index r of a NEPS effectively depends on each r_j.

If $G_1, \ldots, G_n$ are connected graphs, positive eigenvectors $\boldsymbol{x}_1, \ldots, \boldsymbol{x}_n$ belong to indices $r_1, \ldots, r_n$. It can be easily verified (see also Theorems 2.23 and 2.24) that the eigenvector $\boldsymbol{x} = \boldsymbol{x}_1 \otimes \cdots \otimes \boldsymbol{x}_n$ belongs to the index r of the NEPS. Thus, $\boldsymbol{x}$ is a positive vector, too.

According to Theorem 3.35, we then have that the number of components of the NEPS is equal to the multiplicity of index r. Hence, it is necessary to investigate whether or not $\Lambda_{i_1,\ldots,i_n}$ is equal to r for some n-tuple $(i_1, \ldots, i_n)$ different from $(1, \ldots, 1)$. Thus, it is necessary for this that at least for one j ($j = 1, \ldots, n$) the relation $|\lambda_{ji_j}| = |\lambda_{j1}| = r_j$ holds. Since $G_1, \ldots, G_n$ are connected graphs, their indices are simple eigenvalues and the above equality can be satisfied only if $\lambda_{ji_j} = -r_j$. According to Theorem 3.4 we then have that G_j is a bipartite graph.

Hence, the possible disconnectedness of the NEPS of connected graphs, each containing at least one edge, appears as a consequence of the bipartiteness of these graphs. However, the bipartiteness does not always cause disconnectedness. The structure of the considered NEPS has a certain influence, too.

By further analysis we see that the desired n-tuple of indices $i_1, \ldots, i_n$ must be such that for every $i_j \neq 1$ the graph G_j is bipartite, i.e., that $\lambda_{ji_j} = -r_j$, and that every summand in (2.44) contains an even number of quantities λ_{ji_j} ($i_j \neq 1$).

In order to formulate the theorem with precise conditions for connectedness (and later for bipartiteness) of the NEPS we introduce the following definition.

Definition 7.3. A function in several variables is called *even* (*odd*) *with respect to a given non-empty subset of variables* if the function does not change its value (changes only its sign) when all the variables from that subset are simultaneously changed in sign. The function is *even* (*odd*) if at least one non-empty subset of variables exists with respect to which the function is even (odd).

According to the previous results we get the following theorem.

Theorem 7.12 (D. M. Cvetković [Cve9]): *Let $G_1, \ldots, G_n$ be connected graphs each containing at least one edge. Suppose also that $G_{i_1}, \ldots, G_{i_s}$ ($\{i_1, \ldots, i_s\} \subset \{1, \ldots, n\}$) are bipartite. Then the* NEPS *with the basis $\mathscr{B}$ satisfying condition* ($\boldsymbol{D}$), *of graphs $G_1, \ldots, G_n$ is a connected graph if and only if the function*

$$\sum_{\beta \in \mathscr{B}} x_1^{\beta_1} \cdots x_n^{\beta_n} \tag{7.6}$$

is never even w.r.t. any non-empty subset of the set $\{x_{i_1}, \ldots, x_{i_s}\}$. In the case of disconnectedness the number of components is equal to the multiplicity of the index of the NEPS.

The following theorem is a specialization of the preceding one.

Theorem 7.13 (D. M. Cvetković [Cve2], [Cve9]): *Let $G_1, \ldots, G_n$ be connected graphs each containing at least two vertices. Then the p-sum of these graphs is a connected graph if and only if one of the following conditions holds:*

1⁰ *p is equal to n and at most one of the graphs is bipartite;*

2^0 *p is odd and less than n;*

3^0 *p is even and less than n, where at least one of the graphs $G_1, \ldots, G_n$ is not bipartite. If p is equal to n and exactly l ($l > 1$) of the graphs $G_1, \ldots, G_n$ are bipartite, the p-sum has 2^{l-1} components. If p is even and less than n, and if all the graphs $G_1, \ldots, G_n$ are bipartite, then the p-sum has two components.*

It is proved in [Weic] that the product of two connected graphs is a connected graph if and only if at least one of the graphs (= factors) is non-bipartite. (See also Section 4.8, no. 9.)

In [McAn] it is proved that among other things the product of the connected graphs $G_1, \ldots, G_n$ has exactly

$$q = \begin{cases} 1 & \text{if } s \geqq n-1, \\ 2^{n-s-1} & \text{if } s \leqq n-1 \end{cases}$$

components, where exactly s of graphs $G_1, \ldots, G_n$ are non-bipartite.

It is noticed in [Sabi] that the sum of connected graphs is a connected graph.

Theorem 7.13 represents an amalgamation and a generalization of these particular results.

Papers [McAn], [HaTr], [HaWi], [Brua], [Mill], [Aber], [Reid], [Визи] deal also with connectedness of several binary operations on graphs. In these papers the connectedness is investigated directly, by proving the existence of a path between two arbitrary vertices of the considered graph.

We proceed to the investigation of the connectedness of the Boolean function. We shall first prove a lemma.

Lemma 7.2: *The arbitrary Boolean function $G = f(G_1, \ldots, G_n)$ of regular graphs $G_1, \ldots, G_n$ is a regular graph.*

Proof. According to (2.47) the index r of G is given by

$$r = \Lambda_{1,\ldots,1} = \sum_{\beta \in \mathscr{F}} \lambda_{11}^{[\beta_1]} \cdots \lambda_{n1}^{[\beta_n]} = \sum_{\beta \in \mathscr{F}} r_1^{[\beta_1]} \cdots r_n^{[\beta_n]}, \tag{7.7}$$

with the convention $r_j^{[1]} = r_j$, $r_j^{[0]} = \bar{r}_j = m_j - 1 - r_j$ $(j = 1, \ldots, n)$, $\bar{r}_j$ being the index of $\overline{G}_j$. The eigenvector $\boldsymbol{v} = \boldsymbol{v}_1 \otimes \cdots \otimes \boldsymbol{v}_n$, where $\boldsymbol{v}_1, \ldots, \boldsymbol{v}_n$ are eigenvectors of the indices $r_1, \ldots, r_n$ of the graphs $G_1, \ldots, G_n$, corresponds to r. Since $G_1, \ldots, G_n$ are regular graphs, the vectors $\boldsymbol{v}_1, \ldots, \boldsymbol{v}_n$ have all components equal to 1, and thus the vector $\boldsymbol{v}$ also has this property. According to Theorem 3.33 we get the statement of the lemma.

For regular graphs the multiplicity of the index is equal to the number of components of the graph. Thus, the Boolean function will be a connected graph if and only if the eigenvalue (7.7) is simple. Hence, as in Theorem 7.12, it is necessary to investigate whether or not the expression (2.47) gives the value (7.7) for some n-tuple $(i_1, \ldots, i_n) \neq (1, \ldots, 1)$.

We consider primarily the case where for all $j = 1, \ldots, n$ there exists among the n-tuples of $\mathscr{F}$ one for which $\beta_j = 1$ and another for which $\beta_\cdot = 0$. Denote this condition by ($\boldsymbol{E}$). Then, there exist in (2.47) one term containing λ_{ji_j} and another term containing $\bar{\lambda}_{ji_j}$.

We will also introduce the restriction that $G_1, \ldots, G_n$ are connected graphs, that they contain at least one edge, and that they are not complete graphs.

In order to obtain (7.7) from (2.47) for the n-tuple $(i_1, \ldots, i_n)$, it is necessary that indices $i_1, \ldots, i_n$ can be taken so that $|\lambda_{ji_j}| = |\lambda_{j1}|$ $(= r_j)$ and $|\bar{\lambda}_{ji_j}| = |\bar{\lambda}_{j1}|$ $(= m_j - 1 - r_j)$ hold for every $j = 1, \ldots, n$. Since G_j is a connected graph, the realization of the first condition is possible only if we take $\lambda_{ji_j} = -r_j$. According to Theorem 3.4 the eigenvalue $-r_j$ is to be found in the spectrum of the graph G_j if and only if G_j is bipartite. Further, if $\lambda_{ji_j} = -r_j$, according to Theorem 2.6 we have $\bar{\lambda}_{ji_j} = r_j - 1$, and the second of the above conditions holds if $r_j - 1 = m_j - 1 - r_j$, i.e., if $m_j = 2r_j$. It can be easily seen that in that case G_j is a complete bipartite graph.

Thus for disconnectedness of the Boolean function under the considered conditions, it is necessary that some of graphs $G_1, \ldots, G_n$ be bicomplete. (G is called *bicomplete* if G is a complete bipartite graph). However, bicompleteness does not always cause disconnectedness. The structure of the considered Boolean function has a certain influence.

Let the graphs $G_{j_1}, \ldots, G_{j_s}$ $(\{j_1, \ldots, j_s\} \subset \{1, \ldots, n\})$ be bicomplete. In this case the eigenvalue (7.7) will be a multiple root if there exists a non-empty subset of the set $\{\lambda_{j_1 i_{j_1}}, \ldots, \lambda_{j_s i_{j_s}}\}$ such that every summand of (2.47) contains an even number of quantities from that subset. We see now that quantities $\bar{\lambda}_{ji_j}$ play no role at all. Hence, instead of (2.47) we consider the function

$$\sum_{(\beta_1, \ldots, \beta_n) \in \mathcal{F}} x_1^{\beta_1} \cdots x_n^{\beta_n}. \tag{7.8}$$

According to the above, we have the following theorem.

Theorem 7.14 (D. M. Cvetković [Cve5], [Cve9]): *Let $G_1, \ldots, G_n$ be regular connected graphs each containing at least one edge. Suppose also that none of $G_1, \ldots, G_n$ is complete. Further, let graphs $G_{j_1}, \ldots, G_{j_s}$ $(\{j_1, \ldots, j_s\} \subset \{1, \ldots, n\})$ be bicomplete. The Boolean function $G = f(G_1, \ldots, G_n)$, satisfying condition* ($\boldsymbol{E}$), *is a connected graph if and only if there exists no non-empty subset of variables $x_{j_1}, \ldots, x_{j_s}$ with respect to which the function* (7.8) *is even.*

Example 1. For the disjunction of two graphs we have $\mathcal{F} = \{(1, 0), (0, 1), (1, 1)\}$, and the function (7.8) has the form $x_1 + x_2 + x_1 x_2$. The function satisfies condition ($\boldsymbol{E}$) and is not even. Therefore, the disjunction of regular connected graphs which are not complete and which contain each at least one edge is a connected graph.

Example 2. Consider the negation of the exclusive disjunction. Here $\mathcal{F} = \{(1, 1), (0, 0)\}$ and (7.8) becomes $x_1 x_2 + 1$. This function is even w.r.t. $\{x_1, x_2\}$ and $f(G_1, G_2)$ is a disconnected graph if G_1 and G_2 are bicomplete. The following fact is of interest. The product of bicomplete (in general, bipartite) graphs is a disconnected graph (see p. 205), $f(G_1, G_2)$ contains all the edges of the product $G_1 \times G_2$ and some others, but nevertheless it remains a disconnected graph. Note that $G_1 \times G_2$ is the NEPS corresponding to $f(G_1, G_2)$ (see Section 2.5, Definition 3 on p. 67).

We see that in general the connectedness of the NEPS corresponding to a Boolean function depends on the evenness of the function (7.8) in the same way as the connectedness of the Boolean function. So we have the following theorem.

Theorem 7.15 (D. M. Cvetković [Cve9]): *Under conditions of Theorem* 7.14 *the Boolean function graph and its corresponding* NEPS *are either both connected or both disconnected.*

The connectedness of the Boolean function in a more general case is described in the following theorem in terms of a set $\mathcal{H}$ which is formed from $\mathcal{F}$ by the following three operations:

1⁰ If for some j ($j = 1, \ldots, n$) all n-tuples from $\mathcal{F}$ have ones in the j-th coordinate and the graph G_j is connected and bipartite, these ones remain also in the n-tuples of $\mathcal{H}$. If the graph does not have this property, then these 1's are changed to 0's.

2⁰ If for some j ($j = 1, \ldots, n$) all n-tuples from $\mathcal{F}$ have 0's in the j-th coordinate and the complement $\overline{G}_j$ of the graph G_j is connected and bipartite, then all such 0's are changed to 1's. If G_j does not have the above property, zeros remain on the j-th places in all n-tuples of $\mathcal{H}$.

3⁰ If for some j ($j = 1, \ldots, n$) at least one n-tuple from $\mathcal{F}$ has 1 in the j-th coordinate and if at least one n-tuple has 0 on the same coordinate, then: *a*) in the corresponding n-tuples of $\mathcal{H}$ nothing is to be changed in the j-th coordinates if G_j is a bicomplete graph; *b*) 1's change into 0's and vice versa, if G_j has two components with the same number of vertices, each representing a complete graph; *c*) in all n-tuples in the set $\mathcal{H}$ we put zeros if neither of the preceding two cases occurs.

Theorem 7.16 (D. M. Cvetković [Cve9]): *Let $G_1, \ldots, G_n$ be regular non-complete graphs each containing at least one edge. The Boolean function $f(G_1, \ldots, G_n)$ is a connected graph if and only if the following conditions hold:*

1⁰ *If for some j ($j = 1, \ldots, n$) all n-tuples from $\mathcal{F}$ contain* 1's *in the j-th coordinate, then the graph G_j is connected;*

2⁰ *if for some j ($j = 1, \ldots, n$) all n-tuples from $\mathcal{F}$ contain* 0's *in the j-th coordinate, then the graph G_j is ∇-prime;*

3⁰ *the function $\sum_{\beta \in \mathcal{H}} x_1^{\beta_1} \cdots x_n^{\beta_n}$ is not even.*

Note that all properties of $G_1, \ldots, G_n$, which affect the connectedness of the Boolean function, can be determined by means of their spectra (see Chapter 3).

We shall now investigate the question of connectedness of a polynomial of a connected graph. Let us consider the square of a graph, defined in Section 2.1, as an illustrative example.

Let G be an undirected connected graph without loops or multiple edges and let $\lambda_1, \ldots, \lambda_n$ ($\lambda_1 \geqq \lambda_2 \geqq \cdots \geqq \lambda_n$) be its eigenvalues. Also let G contain at least one edge. Then the index λ_1 is greater than 0 and a positive eigenvector $\boldsymbol{x}$ belongs to it.

The graph G^2 has eigenvalues $\lambda_1^2, \ldots, \lambda_n^2$. The vector x is an eigenvector belonging to λ_1^2, too. According to Theorem 3.35, G^2 will be a disconnected graph if and only if the multiplicity of the eigenvalue λ_1^2 is greater than 1. This will appear only if $\lambda_1^2 = \lambda_n^2$, i.e., $\lambda_n = -\lambda_1$. On the basis of Theorem 3.4, G is then a bipartite graph.

In this way the following theorem is arrived at:

Theorem 7.17: *The graph G^2 of an undirected connected graph G without loops or multiple edges and with at least two vertices is a connected graph if and only if G is not bipartite. If G is bipartite, G^2 contains exactly two components.*

Note that this theorem still holds even when the definition of the graph square is modified in the following way:

The "*square*" $G^{\{2\}}$ of the above described graph G is an undirected graph without loops or multiple edges with the same vertex set as G and in which different vertices are adjacent if and only if there is in G at least one path of length 2 between them.

We shall now deduce the conditions under which the NEPS of connected graphs is a bipartite graph.

All components of a NEPS of connected graphs have the same index r. Thus, the number of components of such a NEPS is equal to the multiplicity of its index. A NEPS is bipartite if, naturally, all its components are bipartite. According to Theorem 3.4, each component must then contain the number $-r$ in the spectrum. Since no component contains in the spectrum the number $-r$ with multiplicity greater than 1, it follows that a necessary and sufficient condition for the bipartiteness of a NEPS is that the numbers r and $-r$ have the same multiplicity in the spectrum of the NEPS.

We see from (2.44) that the number $-r$ can exist in the spectrum of NEPS only if some of the graphs are bipartite and if there exist subsets of variables $x_1, \ldots, x_n$ w.r.t. which the function (7.6) is odd. According to the foregoing facts we get the following theorem.

Theorem 7.18 (D. M. Cvetković [Cve9]): *Let $G_1, \ldots, G_n$ be connected graphs, each containing at least one edge. Suppose also that $G_{i_1}, \ldots, G_{i_s}$ ($\{i_1, \ldots, i_s\} \subset \{1, \ldots, n\}$) are bipartite. Then the* NEPS *with the basis $\mathscr{B}$ satisfying condition* (**D**), *of graphs $G_1, \ldots, G_n$ is bipartite if and only if the number of non-empty subsets of the set $\mathscr{L} = \{x_{i_1}, \ldots, x_{i_s}\}$, w.r.t. which the function* (7.6) *is even, is smaller by* 1 *than the number of such subsets w.r.t. which it is odd.*

This theorem represents the basis for proving the following theorem which gives the precise conditions under which the p-sum is a bipartite graph.

Theorem 7.19 (D. M. Cvetković [Cve2], [Cve9]): *Let $G_1, \ldots, G_n$ be connected graphs each containing at least two vertices. Then the p-sum of these graphs is a bipartite graph if and only if one of the following conditions holds:*

1⁰ *p is equal to n and at least one of the graphs $G_1, \ldots, G_n$ is bipartite;*

2⁰ *p is odd and less than n and all the graphs $G_1, \ldots, G_n$ are bipartite.*

Proof. Let $p = n$. The function (7.6) is then of the form $x_1 \cdots x_n$. If $\mathscr{L} = \emptyset$ (see Theorem 7.18), the p-sum is not bipartite. Let $\mathscr{L}$ contain l ($l \geqq 1$) elements. Then the function $x_1 \cdots x_n$ is even w.r.t. exactly $\binom{l}{2} + \binom{l}{4} + \cdots = 2^{l-1} - 1$ nonempty subsets of $\mathscr{L}$ and is odd w.r.t. exactly $\binom{l}{1} + \binom{l}{3} + \cdots = 2^{l-1}$ such subsets. According to Theorem 7.18, the p-sum is then bipartite.

Let p be odd and less than n. The function (7.6) is then not even (Theorem 7.13). It is odd only w.r.t. the set of all variables because if it were odd w.r.t. a proper subset of variables, then one summand among the summands of (7.6) would exist, containing an even number of variables from the same subset, which is in contradiction with the assumption of the oddness of (7.6). Thus, for the bipartiteness of p-sum in this case it is necessary (and sufficient) that all the graphs $G_1, \ldots, G_n$ are bipartite.

Finally, if p is even and less than n, using similar reasoning, we see that (7.6) cannot be odd.

This completes the proof of the theorem.

7.5. Determination of the number of walks

It has been proved in Section 1.8 that the number of walks in a graph is related to the spectrum of the graph. The generating function $H_G(t)$ of the numbers of walks is determined in terms of the characteristic polynomials of the graph and of its complement.

We shall now derive the generating function for graphs obtained by the use of some operations on graphs.

G' is the graph obtained from G when one (simply counted) loop is added to each of the vertices of G.

Theorem 7.20 (D. M. Cvetković [Cve8]): *For the generating function $H_G(t)$ for the numbers of walks of the graph G the following formulas hold:*

$$H_{G'}(t) = \frac{1}{1-t} H_G\left(\frac{t}{1-t}\right), \tag{7.9}$$

$$H_{\bar{G}}(t) = \frac{H_G\left(-\frac{t}{t+1}\right)}{t+1-tH_G\left(-\frac{t}{t+1}\right)}, \tag{7.10}$$

$$H_{G_1 \dot{+} G_2}(t) = H_{G_1}(t) + H_{G_2}(t), \tag{7.11}$$

$$H_{G_1 \nabla G_2}(t) = \frac{H_{G_1}(t) + H_{G_2}(t) + 2tH_{G_1}(t)\, H_{G_2}(t)}{1 - t^2 H_{G_1}(t)\, H_{G_2}(t)}. \tag{7.12}$$

Proof. According to (1.59) we have

$$H_{G'}(t) = \frac{1}{t}\left[(-1)^n \frac{P_{J-A-I}\left(-\frac{1}{t}\right)}{P_{A+I}\left(\frac{1}{t}\right)} - 1\right]$$

$$= \frac{1}{t}\left[(-1)^n \frac{P_{J-A}\left(-\frac{1}{t}+1\right)}{P_A\left(\frac{1}{t}-1\right)} - 1\right]$$

$$= \frac{1}{1-t}\,\frac{1}{\frac{t}{1-t}}\left[(-1)^n \frac{P_{J-A}\left(-\frac{1}{\frac{t}{1-t}}\right)}{P_A\left(\frac{1}{\frac{t}{1-t}}\right)} - 1\right] = \frac{1}{1-t}\,H_G\left(\frac{t}{1-t}\right),$$

which proves (7.9). Formula (7.10) can be directly verified if one uses (1.59) and the following formula analogous to (1.59):

$$H_{\overline{G}}(t) = \frac{1}{t}\left[(-1)^n \frac{P_G\left(-\frac{t+1}{t}\right)}{P_{\overline{G}}\left(\frac{1}{t}\right)} - 1\right].$$

The relation (7.11) is obvious. For the proof of the relation (7.12) one uses the relations (2.1), (7.10), and (7.11):

$$H_{G_1 \nabla G_2}(t) = H_{\overline{\overline{G}_1 \dot{+} \overline{G}_2}}(t) = \frac{H_{\overline{G}_1 \dot{+} \overline{G}_2}\left(-\frac{t}{t+1}\right)}{t+1-tH_{\overline{G}_1 \dot{+} \overline{G}_2}\left(-\frac{t}{t+1}\right)}$$

$$= \frac{H_{\overline{G}_1}\left(-\frac{t}{t+1}\right) + H_{\overline{G}_2}\left(-\frac{t}{t+1}\right)}{t+1-t\left(H_{\overline{G}_1}\left(-\frac{t}{t+1}\right) + H_{\overline{G}_2}\left(-\frac{t}{t+1}\right)\right)}. \tag{7.13}$$

From (7.10) one obtains

$$H_{\overline{G}}\left(-\frac{t}{t+1}\right) = \frac{(t+1)\,H_G(t)}{1+tH_G(t)}. \tag{7.14}$$

Using (7.14) from (7.13) one obtains (7.12). This completes the proof.

Hence, according to this theorem, we can determine the generating function for any graph if we know the generating functions for elementary graphs (see Section 2.2).

Example 1. A regular graph of the degree r, with n vertices, has obviously $N_k = nr^k$ walks of length k, and therefore

$$H_G(t) = \sum_{k=0}^{+\infty} nr^k t^k = \frac{n}{1 - rt} \qquad \left(|t| < \frac{1}{r}\right).$$

For $r = n - 1$ one obtains the complete graphs and for $r = 0$ the graph which contains only isolated vertices. The graph which has only one vertex and no edges or loops has the generating function of the form $H_G(t) = 1$.

Example 2. The bicomplete graph K_{n_1,n_2} can be represented as the ∇-product of graphs G_1 and G_2, both of which contain only isolated vertices. We then have $H_{G_1}(t) = n_1$ and $H_{G_2}(t) = n_2$, and according to (7.12), for the bicomplete graph we have

$$H_{K_{n_1,n_2}}(t) = \frac{n_1 + n_2 + 2n_1n_2t}{1 - n_1n_2t^2}.$$

Example 3. Consider the k-complete graph $K_{n_1,\ldots,n_k}$. First we have

$$H_{\overline{K}_{n_1,\ldots,n_k}}(t) = \sum_{i=1}^{k} \frac{n_i}{1 - (n_i - 1)\,t},$$

because $\overline{K}_{n_1,\ldots,n_k}$ is a direct sum of complete graphs with $n_1, \ldots, n_k$ vertices, respectively. According to (7.10) we have

$$H_{K_{n_1,\ldots,n_k}}(t) = \frac{\displaystyle\sum_{i=1}^{k} \frac{n_i}{1 + n_it}}{1 - t \displaystyle\sum_{i=1}^{k} \frac{n_i}{1 + n_it}}.$$

We shall consider now the non-complete extended p-sum of graphs.

The adjacency matrix $\mathbf{A}$ for the NEPS with the basis $\mathcal{B}$ of graphs $G_1, \ldots, G_n$, whose adjacency matrices are $\boldsymbol{A}_1, \ldots, \boldsymbol{A}_n$, respectively, is given by (2.41). Then also the relation (2.45) holds.

From the definition of the Kronecker product of matrices the following relation can be easily proved:

$$\operatorname{sum}(\boldsymbol{X} \otimes \boldsymbol{Y}) = \operatorname{sum} \boldsymbol{X} \operatorname{sum} \boldsymbol{Y}, \tag{7.15}$$

where $\boldsymbol{X}$ and $\boldsymbol{Y}$ are arbitrary two matrices and sum $\boldsymbol{X}$ denotes the sum of all entries of $\boldsymbol{X}$.

According to (2.45) and (7.15) we have

$$\operatorname{sum} \mathbf{A}^k = \sum_{s_1,\ldots,s_q} \frac{k!}{s_1! \cdots s_q!} \operatorname{sum} \boldsymbol{A}_1^{l_1} \cdots \operatorname{sum} \boldsymbol{A}_n^{l_n},$$

i.e.,

$$N_k = \sum_{s_1,\ldots,s_q} \frac{k!}{s_1! \cdots s_q!} N_{l_1}^1 \cdots N_{l_n}^n, \tag{7.16}$$

where N_k denotes the number of walks of length k in the NEPS and N_k^i $(i = 1, \ldots, n)$ the number of walks of length k in the graph G_i.

The number of walks in a graph can always be represented in the form of (1.58). Let the formula

$$N_k^j = \sum_{i_j} C_{ji_j} \lambda_{ji_j}^k \qquad (j = 1, \ldots, n), \tag{7.17}$$

hold for G_j. Then (7.17) takes the form

$$\begin{aligned} N_k &= \sum_{s_1,\ldots,s_q} \frac{k!}{s_1! \cdots s_q!} \sum_{i_1} C_{1i_1} \lambda_{1i_1}^{l_1} \cdots \sum_{i_n} C_{ni_n} \lambda_{ni_n}^{l_n} \\ &= \sum_{i_1,\ldots,i_n} C_{1i_1} \cdots C_{ni_n} \sum_{s_1,\ldots,s_q} \frac{k!}{s_1! \cdots s_q!} \lambda_{1i_1}^{l_1} \cdots \lambda_{ni_n}^{l_n} \\ &= \sum_{i_1,\ldots,i_n} C_{1i_1} \cdots C_{ni_n} \sum_{s_1,\ldots,s_q} \frac{k!}{s_1! \cdots s_q!} R_1^{s_1} \cdots R_q^{s_q} \\ &= \sum_{i_1,\ldots,i_n} C_{1i_1} \cdots C_{ni_n} (R_1 + \cdots + R_q)^k, \end{aligned}$$

where $R_1, \ldots, R_q$ have the same meaning as in the proof of Theorem 2.23. We there fore arrive at the following theorem.

Theorem 7.21 (D. M. Cvetković [Cve4], [Cve9]): *Let* $N_k^j = \sum_{i_j} C_{ji_j} \lambda_{ji_j}^k$ $(j = 1, \ldots, n)$ *denote the number of walks of length* k *for* G_j. *Then the* NEPS *with the basis* $\mathcal{B}$ *of graphs* $G_1, \ldots, G_n$ *contains*

$$N_k = \sum_{i_1,\ldots,i_n} C_{1i_1} \cdots C_{ni_n} \left(\sum_{\beta \in \mathcal{B}} \lambda_{1i_1}^{\beta_1} \cdots \lambda_{ni_n}^{\beta_n} \right)^k$$

walks of length k.

A somewhat better result is proved in [Cve14].

We describe now some combinatorial problems related to the number of walks in a graph. See also Section 8.3.

1^0 The variation of the k-th class with repetitions of the set $\mathcal{X} = \{x_1, \ldots, x_n\}$ is every ordered k-tuple $(x_{i_1}, \ldots, x_{i_k})$ where $i_j \in \{1, \ldots, n\}$ $(j = 1, \ldots, k)$. The number of such variations is $\overline{V}_n^k = n^k$.

During the formation of variations with repetitions it is possible to impose certain restrictions. We shall consider variations with restrictions of the following type.

Let $\mathcal{X}_i$ $(i = 1, \ldots, n)$ be a family of subsets of $\mathcal{X}$. A pair (x_i, x_j) is a *permitted pair* if and only if $x_j \in \mathcal{X}_i$. The square matrix $\boldsymbol{A} = (a_{ij})_1^n$, where $a_{ij} = 1$ if (x_i, x_j) is a permitted pair and $a_{ij} = 0$ otherwise, is called the *matrix of permitted pairs*. The *matrix of restrictions* $\bar{\boldsymbol{A}}$ is obtained from $\boldsymbol{A}$ by interchanging 0 and 1.

A k-tuple $(x_{i_1}, \ldots, x_{i_k})$ where $x_{i_j} \in \mathcal{X}_{i_{j-1}}$ $(j = 2, \ldots, k)$ is called an *$\boldsymbol{A}$-variation with repetitions of the k-th class*.

In [Cve8] the number $\overline{V}_n^k(\boldsymbol{A})$ of $\boldsymbol{A}$-variations with repetitions of the k-th class of a set with n elements is determined.

We shall connect this problem with the problem of determining the number of walks in a graph.

If $\boldsymbol{A}$ is interpreted as the adjacency matrix of digraph G with vertices $x_1, \ldots, x_n$, it can easily be seen that the number $\overline{V}_n^k(\boldsymbol{A})$ is equal to the number of walks of length $k-1$ in G.

Theorem 7.22 (D. M. Cvetković [Cve8]): *The function*

$$G(t) = (-1)^n \frac{P_{\bar{A}}\left(-\frac{1}{t}\right)}{P_A\left(\frac{1}{t}\right)}$$

is the generating function for the numbers $\overline{V}_n^k(\boldsymbol{A})$ *of* $\boldsymbol{A}$*-variations of the k-th class with repetitions where the relation* $\overline{V}_n^k(\boldsymbol{A}) = \frac{1}{k} G^{(k)}(0)$ $(k = 1, 2, \ldots)$ *holds. For* $k = 0$ *one obtains* $\overline{V}_n^0(\boldsymbol{A}) = 1$.

Proof. We obviously have

$$G(t) = 1 + tH_G(t), \tag{7.18}$$

where $H_G(t)$ is the generating function for the numbers of walks for the digraph G whose adjacency matrix is $\boldsymbol{A}$.

Similarly, as in the proof of Theorem 1.11, we obtain

$$H_G(t) = \frac{\operatorname{sum}\{\boldsymbol{I} - t\boldsymbol{A}\}}{|\boldsymbol{I} - t\boldsymbol{A}|}, \tag{7.19}$$

$$\operatorname{sum}\{\boldsymbol{I} - t\boldsymbol{A}\} = \frac{1}{t}\left(|\boldsymbol{I} + t(\boldsymbol{J} - \boldsymbol{A})| - |\boldsymbol{I} - t\boldsymbol{A}|\right). \tag{7.20}$$

Combining (7.19) and (7.20) we get

$$H_G(t) = \frac{1}{t}\left[(-1)^n \frac{P_{\bar{A}}\left(-\frac{1}{t}\right)}{P_A\left(\frac{1}{t}\right)} - 1\right], \tag{7.21}$$

where, naturally, $\bar{\boldsymbol{A}} = \boldsymbol{J} - \boldsymbol{A}$. The theorem then follows from (7.18) and (7.21).

2° We shall determine the number of walks of length k in the path P_n with n vertices. The adjacency matrix of P_n is of the form

$$\begin{pmatrix} 0 & 1 & & & 0 \\ 1 & 0 & 1 & & \\ & \ddots & \ddots & \ddots & \\ & & 1 & 0 & 1 \\ 0 & & & 1 & 0 \end{pmatrix}.$$

In Section 2.6 we deduced that the eigenvalues of this matrix are $\lambda_i = 2\cos\frac{i\pi}{n+1}$ $(i = 1, \ldots, n)$. It can be easily verified that $u_{ij} = \sqrt{\frac{2}{n+1}} \sin\frac{ij\pi}{n+1}$ $(j = 1, \ldots, n)$ are the coordinates of the normalized eigenvector u_i belonging to λ_i (see [CoSi 1]). By the use of (1.58) we obtain for the number N_{kn} of walks of length k in P_n the expression

$$N_{kn} = \frac{2^{k+1}}{n+1} \sum_{l=1}^{\left[\frac{n+1}{2}\right]} \cot^2 \frac{2l-1}{n+1}\frac{\pi}{2} \cos^k \frac{2l-1}{n+1}\pi. \tag{7.22}$$

This result is related to the following three problems treated in literature.

(i) In [Доче] the following problem is solved.

Determine the number N_{kn} of all possible lines in the plane consisting of segments of length $\sqrt{2}$ and with direction coefficients ± 1. The lines start from one of the points $(0, 0), (0, 1), \ldots, (0, k-1)$ and without leaving the rectangle $0 \leq x \leq n$, $0 \leq y \leq k-1$, terminate in the points $(n, 0), (n, 1), \ldots, (n, k-1)$.

This task is of interest in some problems of the theory of functional spaces.

The solution is given by (7.22).

(ii) A particular result from [Cve 6] reads:

The number N_{kn} of ways in which a king (chess-man) *can make a series of k moves on a one-dimensional chess board* (see Fig. 7.2) *is given by* (7.22).

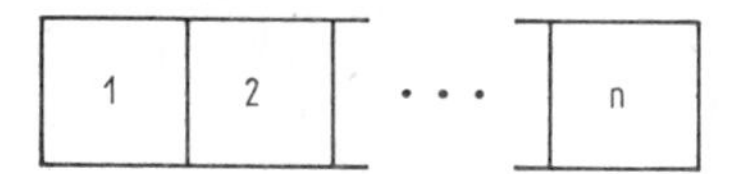

Fig. 7.2

This is obvious if we use the concept of a graph corresponding to a chess-man on a given chess-board. The vertices of this graph correspond to the squares of the chess-board and two vertices are adjacent if and only if the chess-man can go from one square to the other in one move.

In the considered case the corresponding graph is equal to the path P_n.

For some chess-men, the graphs of their moving on two- or higher-dimensional chess-boards can be expressed in form of a NEPS (or by the use of some similar operations) of graphs of its moving on one-dimensional boards. So it is mentioned in [Aber], that the graph of the rook's move on a two-dimensional square board is equal to sum of graphs of the rook's move on a one-dimensional board with itself. (This is mentioned in [Aber] for the case when the rook moves only to an adjacent square, but it is obvious that this holds also in the general case.)

In [Cve 6] the king's moving on a two-dimensional square board was considered. From this consideration it follows that everything that has been said for the rook is valid also for the king, only the sum of graphs is to be interchanged with the strong product (NEPS with the basis containing all possible n-tuples) of graphs.

If the rook moves only to an adjacent square the corresponding graph of its moving on a one-dimensional board is not different from the graph of the king's move on the same board. The number of walks of length k is given by (7.22). Then according to Theorem 7.21 we have for the number of ways that a rook (if it moves along to an adjacent square), i.e., a king makes a series of k moves on an s-dimensional chess-board of the type $n_1 \times \cdots \times n_s$:

$$N_k = \sum_{i_1,\ldots,i_s} C_{1i_1} \cdots C_{si_s}(\lambda_{1i_1} + \cdots + \lambda_{si_s})^k, \tag{7.23}$$

$$N_k = \sum_{i_1,\ldots,i_s} \left(\prod_{j=1}^{s} C_{ji_j}\right)\left(\prod_{j=1}^{s} (\lambda_{ji_j} + 1) - 1\right)^k, \qquad i_j = 1, \ldots, n_j \quad (j = 1, \ldots, s), \tag{7.24}$$

where

$$C_{ji_j} = \frac{2}{n_j + 1} \cot^2 \frac{2i_j - 1}{n_j + 1} \frac{\pi}{2}, \qquad \lambda_{ji_j} = 2 \cos \frac{2i_j - 1}{n_j + 1} \pi.$$

In [Cve6] a special case of the result (7.24) was obtained. This is the case $s = 2$, $n_1 = n_2 = n$, which corresponds to the king's moving on the $n \times n$ board.

(iii) The number of n-tuples $(a_1, \ldots, a_n)$ of non-negative integers satisfying $|a_{i+1} - a_i| = 1$ $(i = 1, \ldots, n - 1)$ and some other additional conditions were determined in [Carl]. These results can be related to the above problem (see [CvS1]).

3° In [MoAb] the combinations for the set $\{1, \ldots, n\}$ with restricted differences and span were considered. For the combination of the p-th class $\{x_1, \ldots, x_p\}$, $1 \leqq x_1 < \cdots < x_p \leqq n$, differences are defined by $d_j = x_{j+1} - x_j$, $j = 1, \ldots, p - 1$, and span by $d = x_p - x_1$. The number of combinations, satisfying restrictions $k \leqq d_j \leqq k'$, $j = 1, \ldots, p - 1$, $l \leqq n - d \leqq l'$, is determined. Special cases of these combinations with a generalization in another direction can be dealt with by the use of the above proposed procedure. Consider combinations of the p-th class with repetitions for which we have $d_j \in \mathcal{M}$ $(j = 1, \ldots, p)$, $\mathcal{M} \subset \{1, \ldots, n\}$ and for which the greatest element m from $\mathcal{M}$ satisfies the condition $pm < 2n$. The number of such combinations (we shall call them $\mathcal{M}$-combinations) can be determined by considering the graph G, having vertices $x_1, \ldots, x_n$ and in which, for every pair of indices (i, j), a (directed) edge leads from the vertex x_i to the vertex x_j if and only if $(j - i) \pmod{n} \in \mathcal{M}$. Besides, G has a loop at every vertex. Exactly p closed walks of length p correspond to each of the considered combinations. Only those closed walks which, due to the existence of loops, do not leave the vertex from which they start correspond to none of the combinations. The number of such closed walks is equal to n. According to Theorem 1.9 the number of closed walks of length k is equal to the trace of the k-th power of the adjacency matrix. If $\{\lambda_1, \ldots, \lambda_n\}$ is the spectrum of the graph G, the number of considered combinations, except for some trivial cases, is equal to $\frac{1}{p}\left(\sum_{i=1}^{n} \lambda_i^p - n\right)$.

M. Tuero [Tuer] determines the spectra of some graphs of the described form. The following theorem is due to L. L. Kraus.

Let $\mathcal{M} = \{m_1, \ldots, m_\mu\}$. The above mentioned digraph G can be represented in the form $G = P(\vec{C}_n)$, where $P(x) = 1 + \sum_{k=1}^{\mu} x^{m_k}$. The eigenvalues of G are (see Section 2.1) $\lambda_l = P\left(e^{\frac{2\pi l}{n} i}\right)$, $l = 1, \ldots, n$. Further we have

$$\sum_{l=1}^{n} \lambda_l^p = \sum_{l=1}^{n} \sum_{\mathcal{J}} \frac{p!}{j_0! j_1! \cdots j_\mu!} \prod_{k=1}^{\mu} \exp \frac{2\pi l b m_k j_k}{n} i$$

$$= \sum_{\mathcal{J}} \frac{p!}{j_0! j_1! \cdots j_\mu!} \sum_{l=1}^{n} \exp\left(i \frac{2\pi}{n} \sum_{k=1}^{\mu} m_k j_k\right) l,$$

where $\sum_{\mathcal{J}}$ denotes summation over all elements of the set

$$\mathcal{J} = \left\{(j_0, j_1, \ldots, j_\mu) \,\middle|\, \sum_{k=0}^{\mu} j_k = p; \quad j_0, j_1, \ldots, j_\mu \text{ are non-negative integers}\right\}.$$

We see that

$$\sum_{l=1}^{n} \exp\left(i \frac{2\pi}{n} \sum_{k=1}^{\mu} m_k j_k\right) l = \begin{cases} n \text{ if } \sum_{k=1}^{\mu} m_k j_k \text{ is divisible by } n, \\ 0 \text{ otherwise.} \end{cases}$$

Thus we conclude the following theorem.

Theorem 7.23: *The number of $\mathcal{M}$-combinations is equal to*

$$\frac{n}{p}\left(\sum_{\mathcal{J}'} \frac{p!}{j_0! j_1! \cdots j_\mu!} - 1\right),$$

where

$$\mathcal{J}' = \left\{(j_0, j_1, \ldots, j_\mu) \,\middle|\, \sum_{k=0}^{\mu} j_k = p, \quad \sum_{k=1}^{\mu} m_k j_k \text{ is divisible by } n\right\}.$$

4⁰ Asymptotical enumeration of certain types of permutations of a finite set can be reduced to the determination of the number of walks in a suitably chosen graph. Let us consider the permutations π of the set $\{1, 2, \ldots, n\}$ in which the position of elements after the permutation is restricted by the following inequalities:

$$|\pi(i) - i| \leqq k, \qquad i = 1, 2, \ldots, n.$$

Such restrictions will be denoted by $\mathbf{R}^{(k)}$. $N(\mathbf{R}^{(k)}, n)$ is the number of permutations of this type.

We shall explain a method of enumeration of permutations under restrictions $\mathbf{R}^{(k)}$ according to [Lehm]. The method was suggested by D. Blackwell.

We may assume that n is much larger than k since n will tend to infinity. Consider a partially built permutation

$$\begin{pmatrix} 1 & 2 & \cdots & h-1 \\ \pi(1) & \pi(2) & \cdots & \pi(h-1) \end{pmatrix}, \tag{7.25}$$

where $k < h < n$. Now we have to select a value for $\pi(h)$. The choice depends on $\mathbf{R}^{(k)}$ and on the already selected values for $\pi(h-1), \pi(h-2), \ldots, \pi(h-k)$. In some cases the choice of $\pi(h)$ may be forced by the fact that certain values have not yet been chosen. Hence there are a number of states of incompletion (or partial completion) of the permutation (7.25). These states can be taken as vertices of a digraph G. There exists an oriented edge from i to j in G if and only if there is a value $\pi(h)$ which, beginning at an incomplete permutation in state i, will produce a state j. The value of

$$\lim_{n\to+\infty} N(\mathbf{R}^{(k)}, n)^{\frac{1}{n}} = \mu(\mathbf{R}^{(k)})$$

is equal to the greatest eigenvalue of G, i.e., to the dynamic mean of the valencies of the vertices of G (see Section 1.8), since each walk of length m from G corresponds to a manner of continuing an incomplete permutation of length h until the one of length $h + m$. The quantity $\mu(\mathbf{R}^{(k)})$ is called *index of mobility of* $\mathbf{R}^{(k)}$ in [Lehm].

Consider, as an example, restrictions $\mathbf{R}^{(1)}$. The incomplete permutation (7.25) has four states of incompletion: 1° Neither $h - 1$ nor h has been used in the second row; 2° $h - 1$ but not h has been used; 3° h but not $h - 1$ has been used; 4° both have been used.

If we are in state 1°, the only possible choice is $\pi(h) = h - 1$ and we get a new incomplete permutation of state 1°. Continuing this reasoning we get the adjacency matrix A of G:

$$A = \begin{pmatrix} 1 & 0 & 0 & 0 \\ 0 & 1 & 1 & 0 \\ 0 & 1 & 0 & 0 \\ 0 & 0 & 0 & 1 \end{pmatrix}.$$

The characteristic polynomial of G is $(\lambda^2 - \lambda - 1)(\lambda - 1)^2$ and we get

$$\mu(\mathbf{R}^{(1)}) = \frac{1 + \sqrt{5}}{2}.$$

7.6. Determination of the number of spanning trees

The spectral method for determining the number of spanning trees $t(G)$ of a graph G is based on Propositions 1.3 and 1.4 (Section 1.5). By means of these spectral techniques almost all known results from this domain of graph theory (see, for example, [Moo 2]) can be obtained.

Primarily, we shall describe the possibilities of the application of Proposition 1.4. In Section 2.6 spectra and characteristic polynomials of several regular graphs were determined. For these graphs Proposition 1.4 immediately gives the number of spanning trees.

1° For the complete graph K_n with n vertices we have

$$t(K_n) = n^{n-2}.$$

This is the famous Cayley formula.

2° For the regular graph G with $n = 2k$ vertices of degree $n - 2$ (coctail party graph) we have $t(G) = 2^{2k-2}(k - 1)^k k^{k-2}$.

3⁰ Since $P_{K_{n,n}}(\lambda) = (\lambda^2 - n^2)\,\lambda^{2n-2}$,

$$t(K_{n,n}) = n^{2n-2}.$$

4⁰ In Section 2.6 the graph G of the k-dimensional lattice was defined and its spectrum was determined. By means of Proposition 1.4 we obtain (see [Cve 10])

$$t(G) = n^{n^k-k-1} \prod_{j=1}^{k} j^{\binom{k}{j}(n-1)^j}.$$

5⁰ The spectrum of the Möbius ladder M_n was determined in Section 2.6, too. There from we have (see [Moo 2])

$$t(M_n) = \frac{1}{2n} \prod_{j=1}^{2n-1} \left(3 - 2\cos\frac{\pi j}{n} - (-1)^j\right).$$

6⁰ Consider the sum $G^* = G + K_2$ of an arbitrary graph G and of the complete graph K_2 with two vertices. The spectrum of K_2 contains numbers 1, −1. If $\lambda_1, \ldots, \lambda_n$ are the eigenvalues of the graph G, the eigenvalues of the graph G^* are $\lambda_1 + 1, \ldots, \lambda_n + 1, \lambda_1 - 1, \ldots, \lambda_n - 1$, i.e.,

$$P_{G^*}(\lambda) = P_G(\lambda - 1)\, P_G(\lambda + 1).$$

If G is a regular graph of degree r with n vertices, G^* is a regular graph of degree $r + 1$ with $2n$ vertices. Proposition 1.4 now yields

$$t(G^*) = \frac{1}{2n} P'_{G^*}(r + 1) = \frac{1}{2n} P'_G(r)\, P_G(r + 2) = \frac{1}{2}\, t(G)\, P_G(r + 2).$$

For the circuit C_n of length n, obviously $t(C_n) = n$. Since $P_{C_n}(\lambda) = 2\left(T_n\left(\frac{\lambda}{2}\right) - 1\right)$, where $T_n(x)$ is the Chebyshev polynomial of the first kind, we obtain (see [Cve 10])

$$t(C_n^*) = nT_n(2) - n.$$

The graph C_n^* consists of all vertices and edges of an n-sided prism.

Using the spectral method we can also prove some theorems about the number $t(G)$. See also Section 7.7.

Theorem 7.24: *If G is a regular graph of degree r with n vertices and m edges, then*

$$t\big(L(G)\big) = 2^{m-n+1} r^{m-n-1} t(G).$$

Proof. $L(G)$ is a regular graph of degree $2r - 2$. According to Proposition 1.4 and (2.30) we have

$$t\big(L(G)\big) = \frac{1}{m} P'_{L(G)}(2r - 2) = \frac{1}{m} \left\{\frac{\mathrm{d}}{\mathrm{d}\lambda} \left[(\lambda + 2)^{m-n} P_G(\lambda - r + 2)\right]\right\}_{\lambda = 2r-2}$$

$$= \frac{n}{m} (2r)^{m-n} \frac{P'_G(r)}{n} = 2^{m-n+1} r^{m-n-1} t(G),$$

which proves the theorem.

This result was obtained in [Кел 2] by the use of the function $B_\lambda^n(G)$. In [Кел 3] the analogous relations for graphs $S(G)$ and $R(G)$ (for definitions see Section 2.4) have been determined, too.

Using the function $B_\lambda^n(G)$ the following results can be obtained. For example:

1° It is readily verified that $B_\lambda^1(K_1) = 1$. Since $\overline{K}_n$ is a direct sum of n graphs K_1, from (2.20) we get $B_\lambda^n(\overline{K}_n) = \lambda^{n-1}$. Further, we have $K_{n_1,n_2} = \overline{K}_{n_1} \nabla \overline{K}_{n_2}$ and by the use of (2.21) we obtain

$$B_\lambda^{n_1+n_2}(K_{n_1,n_2}) = (\lambda + n_1 + n_2)(\lambda + n_1)^{n_2-1}(\lambda + n_2)^{n_1-1},$$

which together with Proposition 1.3 implies

$$t(K_{n_1,n_2}) = n_1^{n_2-1} n_2^{n_1-1}.$$

Note that if $n_1 = n_2$, we have the same result as above. (See also Section 1.9, no. 11.)

2° Let G (see [Moo 2]) be a graph obtained from K_n by removing m $(2m \leqq n)$ non-adjacent edges. Obviously, $\overline{G} = mK_2 \dot{+} (n - 2m) K_1$, where, for instance, kH denotes the k-fold direct sum of the graph H with itself. Since $B_\lambda^2(K_2) = -B^2_{-(\lambda+2)}(\overline{K}_2) = \lambda + 2$, we get $B_\lambda^n(\overline{G}) = \lambda^{n-m-1}(\lambda + 2)^m$.

Further we have

$$B_\lambda^n(G) = (\lambda + n)^{n-m-1}(\lambda + n - 2)^m,$$

$$t(G) = n^{n-2}\left(1 - \frac{2}{n}\right)^m.$$

3° Let G (see [Moo 2]) be the graph obtained from K_n by removing p edges adjacent to the same vertex. Then $\overline{G} = K_{p,1} \dot{+} (n - p - 1) K_1$. From 1° we have the expression for $B_\lambda^{p+1}(K_{p,1})$ and using (2.20), (2.19) and Proposition 1.3 we obtain

$$t(G) = n^{n-2}\left(1 - \frac{p+1}{n}\right)\left(1 - \frac{1}{n}\right)^{p-1}.$$

4° According to Section 2.6, (1.17) and (1.33) we have

$$B_\lambda^n(C_n) = \frac{2}{\lambda}\left(T_n\left(\frac{\lambda+2}{2}\right) - 1\right),$$

where $T_n(x)$ is the Chebyshev polynomial of the first kind.

Let G (see [Moo 2]) be the graph obtained from K_n by removing all edges of a circuit C_m of length m $(m \leqq n)$. Then $\overline{G} = C_m \dot{+} (n - m) K_1$ and $B_\lambda^n(\overline{G}) = \lambda^{n-m} B_\lambda^m(C_m)$. It follows further

$$B_\lambda^n(G) = 2(\lambda + n)^{n-m-1}\left(T_m\left(\frac{\lambda + n - 2}{2}\right) - (-1)^m\right),$$

$$t(G) = 2n^{n-m-2}\left(T_m\left(\frac{n-2}{2}\right) - (-1)^m\right).$$

5⁰ Similarly we obtain the number of spanning trees in the graph G of an n-sided pyramid (see [Moo2], [Rohl]).

Obviously, $G = C_n \bigtriangledown K_1$ and we get $B_\lambda^{n+1}(G) = \frac{2(\lambda + n + 1)}{\lambda + 1}\left(T_n\left(\frac{\lambda + 3}{2}\right) - 1\right)$, $t(G) = 2\left(T_n\left(\frac{3}{2}\right) - 1\right)$.

For generalizations see [BoFS], [JoJo], [Wal3].

6⁰ We shall determine the function $B_\lambda^n(P_n)$ of a path P_n on n vertices starting from the following result from [Mui1]. For the determinant $\alpha(x)$ of order n the following development holds:

$$\alpha(x) := \begin{vmatrix} 1+x & 1 & & & 0 \\ 1 & x & \ddots & & \\ & \ddots & \ddots & \ddots & \\ & & \ddots & x & 1 \\ 0 & & & 1 & 1+x \end{vmatrix} = \prod_{k=1}^{n}\left(x - 2\cos\frac{k\pi}{n}\right).$$

It is easy to see that

$$\alpha(x) = (x + 2)\, U_{n-1}\left(\frac{x}{2}\right), \tag{7.26}$$

where $U_n(x) = \frac{\sin[(n+1)\arccos x]}{\sqrt{1 - x^2}}$ is the Chebyshev polynomial of the second kind. Further we have

$$B_\lambda^n(P_n) = \frac{1}{\lambda}\begin{vmatrix} 1+\lambda & -1 & & & 0 \\ -1 & 2+\lambda & \ddots & & \\ & \ddots & \ddots & \ddots & \\ & & \ddots & 2+\lambda & \lambda-1 \\ 0 & & & -1 & 1+\lambda \end{vmatrix} = \frac{(-1)^n}{\lambda}\,\alpha(-\lambda - 2).$$

On the basis of (7.26) and by the relation $U_n(-x) = (-1)^n U_n(x)$ we have

$$B_\lambda^n(P_n) = U_{n-1}\left(\frac{\lambda + 2}{2}\right).$$

Let G (see [Moo2]) be the graph obtained from K_n by removing the edges constituting a path on s vertices. Then $\overline{G} = P_s \dotplus (n - s) K_1$ and

$$B_\lambda^n(G) = (\lambda + n)^{n-s}\, U_{s-1}\left(\frac{\lambda + n - 2}{2}\right), \quad t(G) = n^{n-s-1} U_{s-1}\left(\frac{n-2}{2}\right)$$

A general algorithm for obtaining the C-spectrum (and, with it, also the complexity) of a graph G, which is composable by means of the operations $\dotplus$ and $\bigtriangledown$ from one-vertex graphs, is given in [Кел 1], [Кел 2] (see also [KeCh]).

7.7. Extremal problems

In the theory of graphs extremal problems are frequently encountered requiring the determination of bounds for a numerical characteristic of a graph as functions of some other numerical characteristics, using some additional conditions which the graph must satisfy. Utilization of graph spectra in solving such problems is possible. We shall show a typical example according to [Wil2]. (In Chapter 3 bounds for some numerical characteristics as functions of eigenvalues are given.)

We shall determine an upper bound for the chromatic number of a graph having n vertices and m edges.

If a graph G contains n vertices and m edges, then its spectrum contains n eigenvalues $\lambda_1, \ldots, \lambda_n$, for which the following relations hold:

$$\lambda_1 + \cdots + \lambda_n = 0, \tag{7.27}$$

$$\lambda_1^2 + \cdots + \lambda_n^2 = 2m. \tag{7.28}$$

Determining the maximum value of the function $f = \lambda_1$ under conditions (7.27) and (7.28), we get

$$\lambda_1 \leqq \sqrt{2m\left(1 - \frac{1}{n}\right)}. \tag{7.29}$$

Then Theorem 3.14 yields the following bound for the chromatic number $\gamma(G)$ of G:

$$\gamma(G) \leqq 1 + \sqrt{2m\left(1 - \frac{1}{n}\right)}. \tag{7.30}$$

Equality holds in (7.30) if G is a complete graph or a graph without edges. This bound is not the best possible; such bounds for $\gamma(G)$ under certain conditions are given in [EpKo].

Nevertheless, spectral methods can be of some interest in various areas of graph theory as the following theorems indicate.

Theorem 7.25 (P. Turán [Turá]): *A graph G with n vertices and more than $\left[\frac{n^2}{4}\right]$ edges contains at least one triangle. The only graphs without triangles having n vertices and $\left[\frac{n^2}{4}\right]$ edges are $K_{l,l}$ for $n = 2l$ and $\dot{K}_{l+1,l}$ for $n = 2l + 1$.*

Proof. According to Theorem 3.8 we have for the greatest eigenvalue λ_1 of G the inequality $\lambda_1 \geqq \frac{2m}{n}$, where m is the number of edges of G. If $m > \frac{n^2}{4} \left(\geqq \left[\frac{n^2}{4}\right]\right)$, we get $\sqrt{m} > \frac{n}{2}, \frac{2m}{n} = \left(\frac{2}{n}\sqrt{m}\right)\sqrt{m} > \sqrt{m}$ and $\lambda_1 > \sqrt{m}$. By Corollary of Theorem 3.9 we conclude that G contains at least one triangle.

If $n = 2l$, $m = l^2$ and if G has no triangle, then according to Theorem 3.8, $\lambda_1 \geqq l$ and according to Corollary of Theorem 3.9, $\lambda_1 \leqq l$. Hence, $\lambda_1 = l$ and $\sum_{i=2}^{n} \lambda_i^2 = l^2$. Since G has no triangles, $\sum_{i=1}^{n} \lambda_i^3 = 0$ and $\left|\sum_{i=2}^{n} \lambda_i^3\right| = l^3$. Using Lemma 3.1 we get $\lambda_2 = \lambda_3 = \cdots = \lambda_{n-1} = 0$ and $\lambda_n = -l$. According to Theorem 6.6, G is $K_{l,l}$.

Suppose now that $n = 2l + 1$, $m = l^2 + l$ and that G has no triangles. Then the mean value $\bar{d}$ of the vertex degrees in G is equal to $\frac{2l(l+1)}{2l+1}$. G contains no vertex of degree $< l$, since by removing that vertex a graph with $2l$ vertices, more than l^2 edges and no triangles would appear. Since $\bar{d} < l + 1$, there is a vertex in G of degree l. Removing this vertex we obtain a graph without triangles having $2l$ vertices and l^2 edges. This graph must be $K_{l,l}$. Therefore it is clear that G is equal to $K_{l+1,l}$. This completes the proof.

Theorem 7.26: *If a graph G has $2n$ vertices and $n^2 + p$ edges, then it does not contain less than $\frac{pn}{2} + \frac{p^2}{2n} + \frac{p^3}{6n^3}$ triangles.*

Proof. According to Theorem 3.8 we have $\lambda_1 \geqq \frac{2(n^2+p)}{2n} = n + \frac{p}{n}$. Since $\sum_{i=1}^{2n} \lambda_i^2 = 2(n^2 + p)$, we get $\sum_{i=2}^{2n} \lambda_i^2 \leqq n^2$. Using Lemma 3.1 we have $\left|\sum_{i=1}^{2n} \lambda_i^3\right| \leqq n^3$ and for the number of triangles in G we get

$$t = \frac{1}{6}\lambda_1^3 + \frac{1}{6}\sum_{i=2}^{2n} \lambda_i^3 \geqq \frac{1}{6}\left(n + \frac{p}{n}\right)^3 - \frac{1}{6}n^3 = \frac{np}{2} + \frac{p^2}{2n} + \frac{p^3}{6n^3}.$$

This completes the proof.

The proofs of these two theorems are given according to [Nos 1].

Now we shall give a lower and an upper bound for the number of spanning trees of a regular graph. The problem of finding these bounds was presented in [Sed 2].

Theorem 7.27 (E. Nosal [Nos 1]): *Let G be a regular graph of degree r with n vertices $\left(r \geqq \frac{1}{2}(n-2),\ n \geqq 2\right)$. Then for the number $t(G)$ of spanning trees of G the following inequality holds:*

$$\left(\frac{2r+2-n}{n}\right)^{n-1} n^{n-2} \leqq t(G) \leqq \left(\frac{r}{n-1}\right)^{n-1} n^{n-2}.\dagger$$

† Note that the right-hand inequality follows also from a more general inequality proved in [Кел 3]. Furthermore, for arbitrary graphs G also

$$t(G) \leqq n^{n-2}\left(1 - \frac{2}{n}\right)^m,$$

where m is the number of edges of $\overline{G}$ (A. K. Kel'mans, private communication, Sept. 1979).

Proof. Let $\lambda_1 = r, \ldots, \lambda_n$ be the eigenvalues of G. In Section 2.7 we have seen that $\lambda_i \leqq n - 2 - r$ $(i = 2, 3, \ldots, n)$. According to (2.55) we get

$$t(G) = \frac{1}{n} \prod_{i=2}^{n} (r - \lambda_i) \geqq \frac{1}{n} \prod_{i=2}^{n} (2r + 2 - n) = \left(\frac{2r + 2 - n}{n}\right)^{n-1} n^{n-2}.$$

The upper bound follows from the relation between the geometric and the arithmetic means of the numbers $r - \lambda_2, r - \lambda_3, \ldots, r - \lambda_n$. This completes the proof of the theorem.

See also [Grim].

7.8. Miscellaneous results and problems

1. All regular graphs of degree r with n vertices for which $n > 2r$ are ∇-prime (see [FiGr]).
Proof. If such graph is not ∇-prime, the number $r - n$ must belong to its spectrum (Theorem 3.29). But this implies $r - n \geqq -r$, and we get $n \leqq 2r$, which contradicts the hypothesis.

2. We shall determine all regular graphs G being not ∇-prime and belonging to the set of line graphs (see [Cve 9]).

First, G is connected. According to Theorem 6.11 we have $q \geqq -2$ for the least eigenvalue q of G. Thus, G is not ∇-prime if and only if either $r - n = -1$ or $r - n = -2$. In the first case we have $r = n - 1$, i.e., G is the complete graph with $n > 1$. In the second case, n must be even and clearly G does not belong to the set of line graphs for $n > 6$. Hence, G can be, except for complete graphs, a circuit of length 4 and the regular graph of degree 4 with six vertices, $H(3) \cong L(K_4)$. (For the definition of $H(n)$, see Section 6.3, p. 170.)

3. Prove that a complete multipartite graph G cannot be represented as a product of graphs [CvL 2].

4. In [BeCh 1] the following result is mentioned (for the proof see [BeCN]): The total graph G of K_n is isomorphic to the line graph of K_{n+1}.
Prove this statement by the use of graph spectra [Cve 13].
Moreover, using the graph spectra one can solve the "graph equation"

$$T(G) = L(H), \tag{7.31}$$

G being a regular graph, i.e., find all pairs of (connected) graphs (G, H) satisfying (7.31) [Cve 13].

The graph equations (7.31) has been completely solved in [CvS2] by non-spectral means.

5. The graph equation $L(G) = G_1 + G_2$ has only the following non-trivial connected solutions:

$$G = K_{m,n}, \quad G_1 = K_m, \quad G_2 = K_n \qquad (m, n = 1, 2, \ldots).$$

Proof. Let q, q_1, q_2 be respectively the least eigenvalues of the graphs G, G_1, G_2. Clearly, $q = q_1 + q_2$. According the Theorem 6.11 we have $q \geqq -2$ and according to Theorem 0.13 the inequalities $q_1, q_2 \leqq -1$ for non-trivial graphs G_1, G_2 hold. It follows $q = -2, q_1 = q_2 = -1$. According to Theorem 0.13, G_1 and G_2 are complete graphs, say, K_m and K_n, and G is equal to $K_{m,n}$.

(M. Doob [Doo 11])

6. A. Ádám [Ádám] has proposed the following problem: Let $\mathcal{M} = \{k_1, \ldots, k_m\}$ be a subset of $\{1, \ldots, n\}$ and let $G_n(k_1, \ldots, k_m)$ be the digraph having vertices $x_1, \ldots, x_n$, where an arc leads from x_i to x_j if and only if there is an element k_t of $\mathcal{M}$ such that $k_t \equiv j - i \pmod{n}$.

A sufficient condition for two such graphs $G_n(k_1, \dots, k_n)$ and $G'_n(k'_1, \dots, k'_n)$ to be isomorphic is that there exists a number r $(0 < r < n)$, relatively prime to n, and a permutation α of $\{1, \dots, n\}$ such that $k'_t \equiv rk_{\alpha(t)} \pmod n$ for $1 \leq t \leq m$. Is this sufficient condition necessary, too?

Using Theorem 6.10 B. ELPAS and J. TURNER proved that this conjecture is true if the number of vertices is prime [ElTu]. The same result was proved in a similar way by D. Ž. DJOKOVIĆ [Djok]. In [ElTu] some counterexamples to ÁDÁM's conjecture are given such that a new problem arises: In which cases, except for n being a prime number, is the conjecture true?

In [ElTu] it was also shown that this conjecture holds for graphs having non-repeated eigenvalues. Some sufficient conditions for the validity of the conjecture are given in [Djok].

Some similar problems have been considered in [Bab 1].

7. We have noticed in Section 3.2 that $\det \boldsymbol{A} \equiv k \pmod 2$, where k is the number of factors in a graph G and $\boldsymbol{A}$ is the adjacency matrix of G. If k is even, we have $\det \boldsymbol{A} \equiv 0 \pmod 2$. Hence, the row vectors of $\boldsymbol{A}$ must be linearly dependent under addition modulo 2. Translating this condition on the adjacency matrix into terms of graphs we get the following result [Lit 1]:

A graph G has an even number of factors if and only if there exists a non-empty subset $\mathscr{S}$ of the vertex set of G, such that every vertex of G is adjacent to an even number of vertices of $\mathscr{S}$.

8. J. SEDLÁČEK [Sed 2] asked whether there is a graph G such that G and its complement have the same number of spanning trees but are non-isomorphic.

Now we can say that the answer is positive. In Chapter 6 we have mentioned examples where regular graphs G and $\overline{G}$ are non-isomorphic but have the same spectrum. Since the number of spanning trees is determined by the spectrum for regular graphs, we have the desired result.

9. Let $\mathscr{G}_{n,m}$ be the set of graphs whose complements contain n vertices and m $\left(m \leq \frac{n}{2}\right)$ edges. $G \in \mathscr{G}_{n,m}$ has the maximal number of spanning trees among all graphs from $\mathscr{G}_{n,m}$ if and only if $\overline{G} = mK_2 \dotplus (n - 2m) K_1$. The corresponding "minimal" graph is $K_{1,m} \dotplus (n - m - 1) K_1$. Other related results have been obtained by the use of the function $B^n_\lambda(G)$. Some inequalities for the number of spanning trees are also given.

(A. K. KEL'MANS, V. M. CHELNOKOV [KeCh])

10. Show that the graph of the square lattice on the surface of a torus, described in Section 2.6, has

$$\frac{4^{mn-1}}{mn} \prod_{i=0}^{m-1} \prod_{j=0}^{n-1} \left(\sin^2 \frac{\pi i}{m} + \sin^2 \frac{\pi j}{n}\right), \qquad (i, j) \neq (0, 0)$$

spanning trees.

(D. M. CVETKOVIĆ [Cve 10])

11. Determine the number of spanning trees of the graphs G and H which are defined as follows.

The vertex set of G is the union of m $(m \geq 3)$ pairwise non-intersecting subsets $\mathscr{S}_1, \mathscr{S}_2, \dots, \mathscr{S}_m$ each containing n vertices. Every vertex of $\mathscr{S}_i$ is adjacent to every vertex of $\mathscr{S}_{i-1}$ and to every vertex of $\mathscr{S}_{i+1}$, where the subscript is to be taken modulo m, and there are no other adjacencies.

The $2n$ vertices and $n(n - 3)$ edges of H are realized by the vertices and interior diagonals, respectively, of the n-sided prism.

(D. M. CVETKOVIĆ [Cve 10])

12. Show that the graphs of the Platonic solids have the following complexities:

tetrahedron: $t = 2^4$

cube and octahedron: $t = 2^7 \cdot 3$

icosahedron and dodecahedron: $t = 2^9 \cdot 3^4 \cdot 5^3$.

13. As an illustration of Theorem 3.27 consider its application to the graphs P (the 5-cage) and Q (the 6-cage), Fig. 7.3. P is Petersen's graph; Q is Heawood's graph and can be realized on the surface **S** of the torus without intersection of edges: Q divides **S** into seven hexagonal faces any two of which have exactly one boundary edge in common.

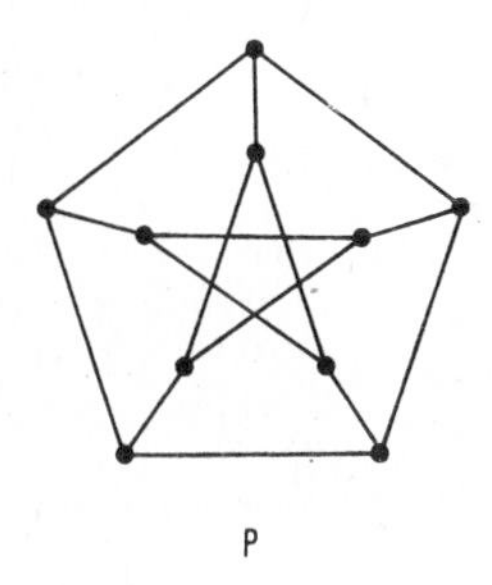

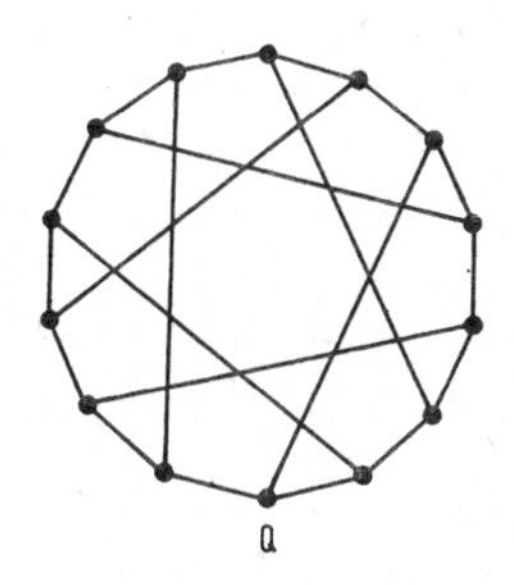

Fig. 7.3

For regular graphs of degree 3 with sufficiently large girth, the numbers $E_{c+2,c}$ and $E_{c+4,c}$, defined in the proof of Theorem 3.27, satisfy the following relation:

$$E_{c+2,c} = m - 2c, \quad E_{c+4,c} = \frac{1}{2}(m^2 - (5 + 4c)m + 4c^2 + 12c),$$

where m is the number of edges. For the same graphs the first numbers b_q are

$$b_1 = m, \quad b_2 = \frac{1}{2} m(m - 5), \quad b_3 = \frac{1}{2} m(m^2 - 15m + 58),$$

$$b_4 = \frac{1}{24} m(m^3 - 30m^2 + 307m - 1086),$$

$$b_5 = \frac{1}{120} m(m^4 - 50m^3 + 995m^2 - 8330m + 28334)$$

(see [Sac5]).

The characteristic polynomial of P has the following coefficients (see Appendix, Table 3):

$$a_1 = 0, \quad a_2 = -15, \quad a_3 = 0, \quad a_4 = 75, \quad a_5 = -24, \quad a_6 = -165, \quad a_7 = 120,$$
$$a_8 = 120, \quad a_9 = -160, \quad a_{10} = 48.$$

Since $m = -a_2 = 15$, we obtain $b_1 = 15$, $b_2 = 75$, $b_3 = 145$, $b_4 = 90$, and $\tilde{a}_1 = \tilde{a}_2 = \tilde{a}_3 = \tilde{a}_4 = 0$, $\tilde{a}_5 = -24$. Let D_i be the number of circuits of length i. According to Theorem 3.26, $g = 5$ and $D_5 = 12$. According to (3.24) and (3.25), $D_6 = -\frac{1}{2}(a_6 + b_3) = 10$, $D_7 = -\frac{1}{2} \times (a_7 - 2E_{7,5}D_5) = 0$, $D_8 = -\frac{1}{2} a_8 + \frac{1}{2} b_4 + E_{8,6}D_6 = 15$, $D_9 = -\frac{1}{2} a_9 - E_{9,5} + E_{9,7}D_7 = 20$.

Put in a more suggestive form:

P: i	1	2	3	4	5	6	7	8	9
D_i	0	0	0	0	12	10	0	15	20

The characteristic polynomial of Q is (see Appendix, Table 4)

$$P_G(\lambda) = \lambda^{14} - 21\lambda^{12} + 168\lambda^{10} - 700\lambda^8 + 1680\lambda^6 - 2352\lambda^4 + 1792\lambda^2 - 576,$$

and, in a similar way as above, the following result is obtained:

Q: i	1	2	3	4	5	6	7	8	9	10	11	12	13
D_i	0	0	0	0	0	28	0	21	0	84	0	?	0

(H. SACHS [Sac 3])

14. The ranking of the participants of a round-robin tournament can be carried out by means of the eigenvector belonging to the largest positive eigenvalue of a suitably chosen multi-digraph.

If the tournament has players $1, 2, \ldots, n$, we construct a multi-digraph G with vertices $1, 2, \ldots, n$, where for each pair i, j of players there are exactly two arcs connecting vertices i and j: both of them directed from i to j (from j to i) if i beats j (j beats i), and one directed from i to j and the other one from j to i if the play ends in a draw. Besides, exactly one loop is added to each vertex of G. Then the number N_{ki} of walks of length k starting at vertex i can be taken as the *strength of order* k of player i. The *relative strength of order* k can be defined as $S_{ki} = \dfrac{N_{ki}}{\sum_{j=1}^{n} N_{kj}}$.

There are reasons for believing that the real relative strength of player i is given by $\lim_{k \to +\infty} S_{ki}$. As noted in Section 3.5, the vector $(S_{k1}, \ldots, S_{kn})$ tends to the eigenvector belonging to the largest positive eigenvalue of G.

(T. H. WEI [Wei], [Ber 1])

15. Check whether the following graphs are strongly regular:

1° The NEPS of three copies of K_4 with the basis $\mathscr{B} = \{(1, 1, 0), (1, 0, 1), (0, 1, 1)\}$;

2° The NEPS of four copies of K_3 with basis containing all quadruples with three 1's;

3° The NEPS of five copies of K_2 with basis containing all 5-tuples with four 1's.

16. Given ϱ_2, there are only finitely many possible ϱ_1 such that a regular two-graph with the spectrum $\{\varrho_1, \varrho_2\}$ exists. For $\varrho_2 = -3$ the only possibilities for regular two-graphs are:

$$\varrho_1 = 1, 3, 5, 9, 21$$

$$n = 4, 10, 16, 28, 64.$$

For $\varrho_2 = -5$ we have

$$\varrho_1 = 1, 3, 5, 7, 15, 19, 25, 35, 55, 115$$

$$n = 6, 16, 26, 36, 76, 96, 126, 176, 276, 576.$$

(J. J. SEIDEL [Sei 4])

17. Let $(\mathscr{X}, \Delta)$ be a regular two-graph on n vertices with the parameter a and the spectrum $\{\varrho_1, \varrho_2\}$. Its complement $(\mathscr{X}, \mathscr{X}^{(3)} \setminus \Delta)$ is a regular two-graph with the parameter $n - 2 - a$ and the spectrum $\{-\varrho_2, -\varrho_1\}$.

(J. J. SEIDEL [Sei 4])

18. The eigenvalues ϱ_1 and ϱ_2 of any non-void and non-complete regular two-graph are odd integers, unless $\varrho_1 + \varrho_2 = 0$.

(J. J. SEIDEL [Sei 4])

19. Let $(\mathscr{X}, \Delta)$ be a regular two-graph with the spectrum $\{\varrho_1, \varrho_2\}$. Let $(\mathscr{X}, \mathscr{E})$ be any graph in its switching class which has an isolated vertex $x \in \mathscr{X}$. Then $(\mathscr{X} \setminus \{x\}, \mathscr{E})$ is a strongly regular graph with the eigenvalues $\varrho_0 = \varrho_1 + \varrho_2, \varrho_1, \varrho_2$.

(J. J. SEIDEL [Sei4])

20. A Steiner triple system of order n consists of $\frac{1}{6}n(n-1)$ unordered triples out of n symbols such that every unordered pair of symbols occurs in exactly one triple. It is known that Steiner triple systems exist for all $n \equiv 1, 3 \pmod 6$. Let us define a *Steiner graph* as a graph whose vertices are the triples of a Steiner triple system, two vertices being adjacent if and only if the corresponding triples have one symbol in common. Show that for $n > 9$ all Steiner graphs are strongly regular.

(J. J. SEIDEL [Sei3])

21. A *Latin square of order n* can be defined as the set of n^2 ordered triples out of n symbols such that for each pair of coordinates every ordered pair of symbols occurs exactly once. A *Latin square graph* has the mentioned triples as vertices, any two triples being adjacent if and only if they have one symbol in common. Show that for $n \geqq 3$ all Latin square graphs are strongly regular.

(J. J. SEIDEL [Sei3])

22. Let G be a strongly regular graph with $e = f$. Using spectra, show that $K_2 \times G$ is the graph of a symmetric block design.

23. Some bounds for the eigenvalues of $(-1, 1)$ matrices are given in [Hof20] and [Hof21].

24. The C-eigenvalues of $G = G_1 + G_2$ are the sums of the C-eigenvalues of the graphs G_1 and G_2, which enable the calculation of the complexity $t(G)$ by Proposition 1.3 (p. 39). For example, for the square lattice Graph G mentioned in Section 2.6 (p. 74), the complexity can be determined in this way.

(G. KREWERAS [Krew])

8. Applications in Chemistry and Physics

In Chapter 7 the spectra of graphs were used as a means for proving several theorems not containing any statement about spectra. In this chapter the applications of graph spectra in which spectra themselves have a significance are described. For example, the eigenvalues of a graph can be interpreted in some situations as the energy levels of an electron in a molecule or as the possible frequencies of the tone of a vibrating membrane. Nevertheless, this chapter contains several theorems of the type from the preceding chapters.

In describing some chemical or physical models in which graph spectra appear, we shall not be rigorous in all details since this book is of mathematical character. The reader who wants to know more facts from Chemistry or Physics is directed to the literature. For HÜCKEL's theory we refer especially to the book [McWS].

8.1. Hückel's theory

According to quantum theory, the properties of microobjects (electrons, atoms, molecules) in a stationary state are described by wave functions Ψ, representing solutions of Schrödinger's equation $\hat{H}\Psi = E\Psi$, in which $\hat{H}$ is the energy operator and E is the energy of the object under consideration. The energy operator $\hat{H}$ for a particle, also called *Hamilton's operator* or *Hamiltonian*, is of the form

$$\hat{H} = -\frac{\hbar^2}{2m}\left(\frac{\partial^2}{\partial x^2} + \frac{\partial^2}{\partial y^2} + \frac{\partial^2}{\partial z^2}\right) + V(x, y, z),$$

where $\hbar$ is Planck's constant, m is the mass of the particle, and $V(x, y, z)$ is the potential energy (depending on space coordinates) of the particle. The complex function Ψ depends also on coordinates x, y, z, and the quantity $\Psi\overline{\Psi} = |\Psi|^2$ represents a probability density, $|\Psi|^2 \, d\tau$ being the probability of finding the given particle within the volume element $d\tau$.

For a system of n particles Hamilton's operator becomes

$$-\sum_{i=1}^{n} \frac{\hbar^2}{2m_i}\left(\frac{\partial^2}{\partial x_i^2} + \frac{\partial^2}{\partial y_i^2} + \frac{\partial^2}{\partial z_i^2}\right) + V(x_1, y_1, z_1, \ldots, x_n, y_n, z_n),$$

x_i, y_i, z_i being the coordinates and m_i the mass of the i-th particle. A more precise description requires consideration of relativistic effects and the introduction of spin coordinates into the equation, but in this book we shall not consider such details.

For complex multi-electron systems (such as molecules) Schrödinger's equation cannot be solved in closed form and therefore approximate solutions are sought. In a certain class of compounds of carbon and hydrogen, the conjugated hydrocarbons, the situation is as follows.

The valency of carbon is equal to 4, i.e., a carbon atom forms bonds with neighbouring atoms in the molecule by direct use of four of its electrons. In the particular class of compounds referred to, three of these electrons (called *σ-electrons* because they are symmetric with respect to reflection in the molecular plane) form localized bonds with other carbon atoms and hydrogen atoms (the valency of hydrogen being equal to one). The fourth electron (called a *π-electron* which is antisymmetric with respect to reflection in the molecular plane) may form bonds with π-electrons of other carbon atoms so that, in the usual graphical representation of the molecule (i.e., in the structural formula of the molecule), two-fold bonds may occur between some carbon atoms (see Fig. 8.1). According to the quantum theory, these π-electrons are not localized in certain carbon atoms (or in bonds between two carbon atoms) as are the σ-electrons, but are delocalized over the entire skeleton of the molecule which is formed by the carbon atoms via the carbon-carbon σ-bonds. The skeleton of the molecule can be represented in a natural manner by an undirected graph without loops or multiple edges, as shown in Fig. 8.1.

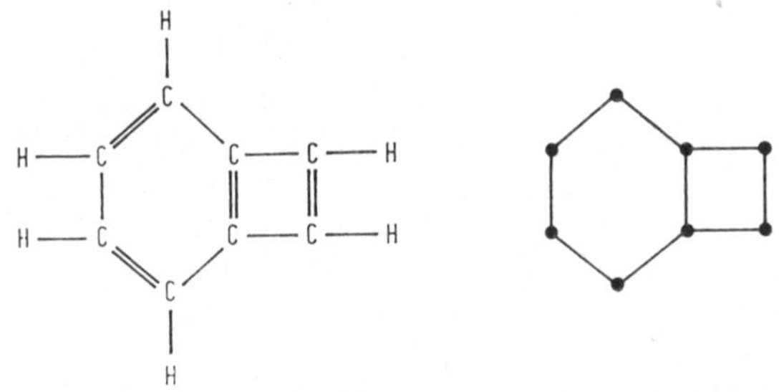

Fig. 8.1

The vertices of the graph associated with a given molecule are in one-to-one correspondence with the carbon atoms of the hydrocarbon system. Two vertices in the graph are adjacent if and only if there is a σ-electron bond between the corresponding carbon atoms.

Let a molecule have n carbon atoms and let χ_j $(j = 1, \ldots, n)$ be the wave function describing a p-electron[†] of the j-th carbon atom, considered, initially, to be isolated from other atoms in the molecule. The functions χ_j are called $(2p_z)$ *atomic orbitals.* Let Ψ_i $(i = 1, \ldots, n)$ be wave functions which describe the total π-electron system in the molecule and which extend over the whole of the carbon atom skeleton of the molecule (these are the so called *molecular orbitals*).

† The electron in an isolated carbon atom, which becomes the electron of π-symmetry when the carbon atom is in a conjugated molecule, is in an orbital denoted by $2p$. For details see [McWS].

The customary approximation used in quantum chemistry is the LCAO-MO (*linear combination of atomic orbitals — molecular orbital*) method. In this approach molecular orbitals are taken to be of the form

$$\Psi_i = \sum_{j=1}^{n} C_{ij}\chi_j \qquad (i = 1, \ldots, n),$$

where the coefficients C_{ij} must be chosen so that the π-electron energy

$$E_i = \frac{\int \Psi_i \hat{H} \Psi_i \, d\tau}{\int \Psi_i^2 \, d\tau}$$

attains its minimal value (integration in the above expression is taken over all coordinates and over all space). Starting from the conditions $\frac{\partial E_i}{\partial C_{ij}} = 0$ $(j = 1, \ldots, n)$, we arrive at the characteristic equation or what, in a chemical context, is often called the *secular equation*

$$\det(\boldsymbol{H} - E\boldsymbol{S}) = 0,$$

where $\boldsymbol{H} = (h_{ij})_1^n$ and $\boldsymbol{S} = (s_{ij})_1^n$ with $h_{ij} = \int \chi_i \hat{H} \chi_j \, d\tau$ and $s_{ij} = \int \chi_i \chi_j \, d\tau$.

The eigenvalues of the above eigenvalue problem represent possible values for the energy of π-electrons, and the coordinates of the corresponding eigenvectors determine the molecular orbitals which are characterized by such energies.

For further details see, for example, [McWS].

For conjugated hydrocarbons it has become customary to adopt the following approximations:

$$\boldsymbol{H} = \alpha\boldsymbol{I} + \beta\boldsymbol{A}, \qquad \boldsymbol{S} = \boldsymbol{I} + \sigma\boldsymbol{A},$$

where $\boldsymbol{A}$ is the adjacency matrix of the graph representing the carbon-atom skeleton of the atoms comprising the given conjugated system, $\boldsymbol{I}$ is the unit matrix, and α, β, σ are constants which are assumed known. These approximations form the basis of the so-called *Hückel theory* ([Hück] is E. Hückel's pioneering paper in this field).

The spectrum of the corresponding graph clearly bears a simple algebraic relation to the set of possible π-electron energy levels. If λ is an eigenvalue from the spectrum of the molecular graph in question and E is the energy of the corresponding molecular orbital, we have the relation

$$-\lambda = \frac{\alpha - E}{\beta - E\sigma}.$$

In this way the task of determining molecular-orbital energies is reduced to the determination of the spectrum of the corresponding molecular graph. Similarly, since $\boldsymbol{H}$ and $\boldsymbol{A}$ commute, they have the same eigenvectors and so the weighting coefficients of the atomic orbitals in a given molecular orbital are simply the components of the corresponding eigenvectors of $\boldsymbol{A}$.

The approximations which have been introduced are very crude and cannot be given any strict theoretical justification. On the other hand, Hückel's method agrees with experimental data, the results often being even better than those given by other methods of calculation which make fewer approximations (see, for example, [HeSc]).

Though the relation between Hückel's theory and the theory of graph spectra was observed in [GüPr] (see also [CvG 1]), for a long time it was scarcely capitalized upon. Recently, a group of chemists and physicists (together with a mathematician) have exploited this relation in a series of papers (see the papers of I. GUTMAN).

We give below a short list of corresponding terms from chemistry and from graph theory:

graph	*conjugated hydrocarbon, conjugated polyene, polyene, aromatic hydrocarbon* (i.e. skeleton of their σ-electron carbon-carbon bonds)
vertex	*carbon atom*
edge	*bond*
vertex degree, valency	*valency*
adjacency matrix	*topological matrix, Hückel matrix*
characteristic equation	*secular equation*
bipartite graph	*alternant hydrocarbon*
cycle with N vertices	$[N]$-*annulene*

The theory of graph spectra thus has a nice application in chemistry. The graphs which are of interest in chemistry (i.e., "molecular graphs" considered in the theory of conjugated hydrocarbons) belong, however, to a rather restricted class of graphs. We shall quote some of these restrictions. All graphs considered are connected, planar, with maximal vertex degree of 3. Furthermore, in the majority of cases molecular graphs must be such that they can be represented in the plane so that all edges are approximately of the same length and two edges with a common vertex form an angle of about $\frac{2\pi}{3}$. This is why triangles seldom appear in chemical graphs while cycles of length $\geqq 4$ can do so. Regular graphs, however, arise very rarely and, therefore, many theorems which hold for regular graphs have no particular applications in simple molecular orbital theory. Nevertheless, the set of graphs which are of interest in this context is sufficiently large, so that many nontrivial problems can be posed.

The basic problem in Hückel theory is to determine the eigenvalues and eigenvectors of the graph representing the carbon atom connectivity of a given conjugated system and, on the basis of these quantities, to calculate other quantities of interest in chemistry. Therefore, any methods for determining the spectra of graphs (numerical methods as well as those yielding formulas for eigenvalues as functions of certain graph parameters) are of interest. In practice, chemists use: 1^0 computers, 2^0 extensive tables of graph spectra (see Appendix) and 3^0 hand-calculation, where factorization of the characteristic polynomials by use of the symmetry (or automorphism) group of the graph in question plays an important role (see Chapter 5).

In [GrGTŽ], Theorem 1.3 was used to deduce all the important facts known in Hückel theory via a graph theoretical formalism. Theorem 3.11 plays an important role in chemistry and is called by chemists the *Coulson-Rushbrooke pairing theorem*.

In this section we shall discuss some mathematically interesting problems which are related to Hückel theory.

1⁰ In [CoSi 1] L. Collatz and U. Sinogowitz posed the problem of characterizing all graphs which have the number zero in their spectrum. This question is a very interesting one in chemistry, because, as has been shown in [Long], the occurrence of a zero eigenvalue in the spectrum of a bipartite graph (corresponding to an alternant hydrocarbon) indicates chemical instability of the molecule which such a graph represents. The question is of interest also for non-alternant hydrocarbons (non-bipartite graphs), but a direct connection with chemical stability is in these cases not so straightforward.

The problem has not yet been solved completely, but some particular results are known [CvGT 1], [CvG 1]. We shall consider only bipartite graphs, although some of the theorems detailed below can be extended to the case of non-bipartite graphs [CvGT 6].

Let $\eta(G)$ be the algebraic multiplicity of the eigenvalue 0 in the spectrum of the (bipartite) graph G. The problem is to find a connection between the graph structure and the number $\eta(G)$. This connection can perhaps be expressed by a set of rules by which we can, after a finite number of steps, determine $\eta(G)$. However, it seems that the problem is not a simple one. For example, $\eta(G)$ is not determined by the set of vertex degrees of G (Fig. 8.2). For some classes of bipartite graphs the spectrum is known and hence so is the number $\eta(G)$ (see Section 2.6).

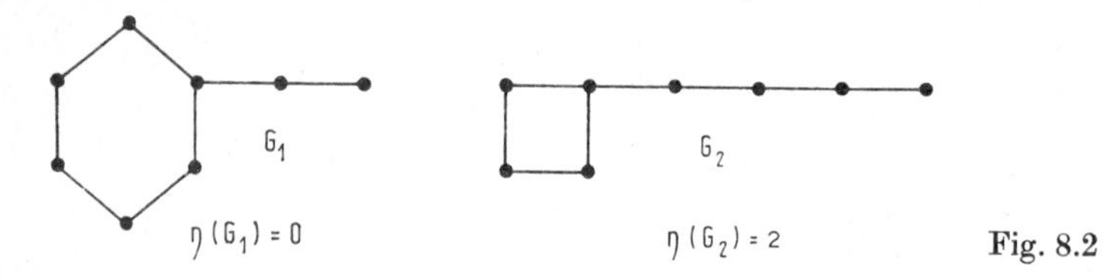

Fig. 8.2

For even circuits C_{2l} we have $\eta(C_{2l}) = 1 + (-1)^l$, and for the path P_n we have $\eta(P_n) = \frac{1}{2}\left(1 - (-1)^n\right)$.

The graph G of Fig. 8.3 with $4N + 2$ vertices has the following eigenvalues:

$$1, \quad -1, \quad \text{and} \quad \frac{1}{2}\left(\pm 1 \pm \sqrt{9 + 8\cos\frac{\pi j}{N+1}}\right), \qquad j = 1, \ldots, N,$$

where all four combinations of signs have to be considered ([Cou 1], [Hei 2], [Ruth]). Thus, $\eta(G) = 0$.

For some special classes of bipartite graphs it is possible to find relatively easily the relation between the structure of G and $\eta(G)$. The problem is solved for trees by the following theorem [CvG 1].

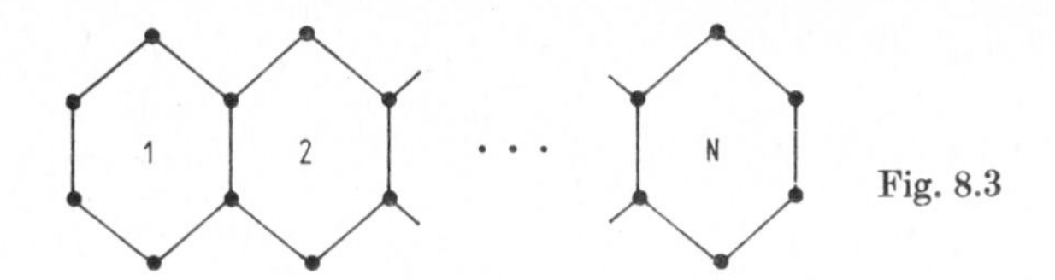

Fig. 8.3

Theorem 8.1: *If q is the maximal number of mutually non-adjacent edges of the tree G having n vertices, then $\eta(G) = n - 2q$.*

This theorem is an immediate consequence of a statement concerning the coefficients of the characteristic polynomial of the adjacency matrix of a tree (see Proposition 1.1). See also Theorem 8.14.

The problem considered can be reduced to another problem which can be solved in certain special cases.

It is clear that the vertices of a bipartite graph may be numbered so that the adjacency matrix has the following form:

$$\boldsymbol{A} = \begin{pmatrix} \boldsymbol{O} & \boldsymbol{B} \\ \boldsymbol{B}^{\mathsf{T}} & \boldsymbol{O} \end{pmatrix}. \tag{8.1}$$

On the main diagonal there are square zero-matrices. The matrix $\boldsymbol{B}$ is the "incidence matrix" between the two sets $\mathscr{X}$ and $\mathscr{Y}$ of vertices of the bipartite graph $G = (\mathscr{X}, \mathscr{Y}; \mathscr{U})$ ($\mathscr{U}$ is the set of edges).

Theorem 8.2: *For the bipartite graph G with n vertices and incidence matrix* $\boldsymbol{B}$, $\eta(\boldsymbol{G}) = n - 2 \cdot \operatorname{rk} \boldsymbol{B}$.

This simple theorem was proved in [Long].

Since for $G = (\mathscr{X}, \mathscr{Y}; \mathscr{U})$ we have $\operatorname{rk} \boldsymbol{B} \leqq \min(|\mathscr{X}|, |\mathscr{Y}|)$, Theorem 8.2 yields the following corollary.

Corollary: $\eta(G) \geqq \max(|\mathscr{X}|, |\mathscr{Y}|) - \min(|\mathscr{X}|, |\mathscr{Y}|)$.

If the number of vertices n is odd, $|\mathscr{X}| \neq |\mathscr{Y}|$ and $\eta(G) > 0$. Thus a necessary condition for a molecule to be stable (i.e., to have no zeros in the spectrum) is that it must have an even number of atoms.

In the subsequent discussion we shall use the following notion. The *square*† $G^{\{2\}}$ of the graph G is the graph defined on the same set of vertices as G having the following property: two vertices are adjacent in $G^{\{2\}}$ if and only if there exists in G at least one path of length 2 joining these two vertices.

Note that the square of the connected graph $G = (\mathscr{X}, \mathscr{Y}; \mathscr{U})$ has two components corresponding to the sets $\mathscr{X}$ and $\mathscr{Y}$. (See Section 7.4, p. 208, in particular Theorem 7.17.)

Theorem 8.3: *Let $G = (\mathscr{X}, \mathscr{Y}; \mathscr{U})$, with $|\mathscr{X}| = n_1$ and $|\mathscr{Y}| = n_2$, be a connected bi-*

† See Section 7.4, p. 208; in Section 2.1 another definition of the square of a graph is given.

partite graph. Let H be the component of $G^{\{2\}}$ corresponding to $\mathcal{X}$. If all vertices of $\mathcal{Y}$ have degree 2, we have

$$\eta(G) = \begin{cases} n_2 - n_1 + 2 & \textit{if } H \textit{ is bipartite}, \\ n_2 - n_1 & \textit{otherwise}. \end{cases}$$

According to Theorem 8.2, the proof of this theorem (which was mentioned in [CvG 1]) is reduced to the determination of the rank of $\boldsymbol{B}$. In a way analogous to the proof of Theorem 4 of [Sac 8] (see Section 3.6, no. 16), it can be easily proved that $\operatorname{rk} \boldsymbol{B} = n_1 - 1$ if H is bipartite, and $\operatorname{rk} \boldsymbol{B} = n_1$ in the opposite case.

Corollary: *Let G be a graph having n vertices and m edges and let $S(G)$ be its subdivision graph* (see Section 0.2). *Then $\eta(S(G)) = m - n + 2$ if G is bipartite, and $\eta(S(G)) = m - n$ otherwise.*

The following three theorems ([CvG 1], [CvGT 1]) enable, in special cases, the reduction of the problem of determining $\eta(G)$ for some graphs to the same problem for simpler graphs.

Theorem 8.4 (D. M. Cvetković, I. Gutman [CvG 1]): *Let $G_1 = (\mathcal{X}_1, \mathcal{Y}_1; \mathcal{U}_1)$ and $G_2 = (\mathcal{X}_2, \mathcal{Y}_2; \mathcal{U}_2)$, where $|\mathcal{X}_1| = n_1$, $|\mathcal{Y}_1| = n_2$, $n_1 \leqq n_2$, and $\eta(G_1) = n_2 - n_1$. If the graph G is obtained from G_1 and G_2 by arbitrarily joining vertices from $\mathcal{X}_1$ to vertices in $\mathcal{Y}_2$ (or $\mathcal{X}_2$), then the relation $\eta(G) = \eta(G_1) + \eta(G_2)$ holds.*

Proof. Let $\boldsymbol{B}_1$, $\boldsymbol{B}_2$, $\boldsymbol{B}$ be the incidence matrices of graphs G_1, G_2, G. We may assume

$$\boldsymbol{B} = \begin{pmatrix} \boldsymbol{B}_1 & \boldsymbol{M} \\ \boldsymbol{O} & \boldsymbol{B}_2 \end{pmatrix},$$

where $\boldsymbol{B}_1$ is an $n_1 \times n_2$ matrix, $\boldsymbol{O}$ is a zero matrix, and $\boldsymbol{M}$ is an arbitrary matrix with entries from the set $\{0, 1\}$.

From $\eta(G_1) = n_2 - n_1$ we have $\operatorname{rk} \boldsymbol{B}_1 = n_1$. Thus $\boldsymbol{B}_1$ contains n_1 linearly independent columns. Consequently, each column of the matrix $\boldsymbol{M}$ can be expressed as a linear combination of the aforementioned columns of $\boldsymbol{B}_1$. Hence, the matrix $\boldsymbol{B}$ can be reduced by operations not changing the rank to the form

$$\boldsymbol{B}' = \begin{pmatrix} \boldsymbol{B}_1 & \boldsymbol{O} \\ \boldsymbol{O} & \boldsymbol{B}_2 \end{pmatrix},$$

whence $\operatorname{rk} \boldsymbol{B} = \operatorname{rk} \boldsymbol{B}_1 + \operatorname{rk} \boldsymbol{B}_2$. Theorem 8.2 then gives $\eta(G) = \eta(G_1) + \eta(G_2)$.

Corollary 1: *If the bipartite graph G contains a vertex of degree 1, and if the induced subgraph H (of G) is obtained by deleting this vertex together with the vertex adjacent to it, then the relation $\eta(G) = \eta(H)$ holds.*

This corollary of Theorem 8.4 is proved in the following way: we take the complete graph with two vertices as G_1 and the graph H as G_2. This way of deleting the vertices and the edges from a graph is called, according to [Bax 2], *dismantling* of graphs. Some features of this procedure are described in [Bax 2].

Corollary 2: *Let G_1 and G_2 be bipartite graphs. If $\eta(G_1) = 0$, and if the graph G is obtained by joining an arbitrary vertex of G_1 by an edge with an arbitrary vertex of G_2, then the relation $\eta(G) = \eta(G_2)$ holds.*

Example 1. See Fig. 8.4.

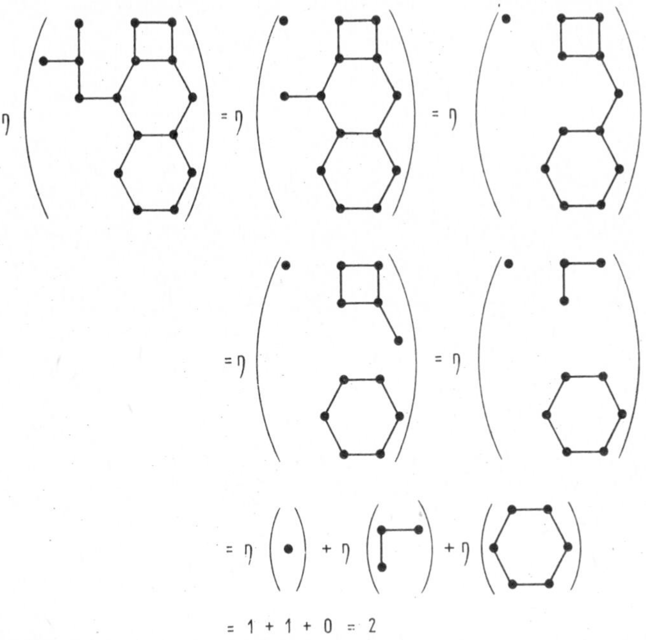

$= 1 + 1 + 0 = 2$

Fig. 8.4

Theorem 8.5 (D. M. Cvetković, I. Gutman, N. Trinajstić [CvGT1]): *A path with four vertices of valency* 2 *in a bipartite graph G can be replaced by an edge* (see Fig. 8.5) *without changing the value of $\eta(G)$.*

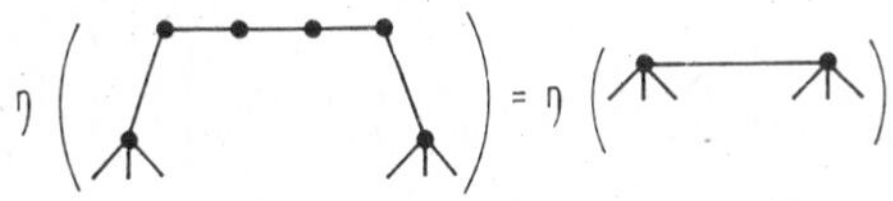

Fig. 8.5

Proof. In this case the matrix $\boldsymbol{B}$ from (8.1) has the form

$$\boldsymbol{B} = \left(\begin{array}{cc|cccc} 1 & 1 & 0 & 0 & \cdots & 0 \\ 1 & 0 & 1 & 0 & \cdots & 0 \\ \hline 0 & 1 & 0 & & \boldsymbol{b} & \\ 0 & 0 & & & & \\ \cdot & \cdot & & & & \\ \cdot & \cdot & & & & \\ \cdot & \cdot & \boldsymbol{c} & & \boldsymbol{B}'' & \\ 0 & 0 & & & & \end{array}\right)$$

and we get

$$\operatorname{rk} \boldsymbol{B} = 2 + \operatorname{rk} \begin{pmatrix} 1 & \boldsymbol{b} \\ \boldsymbol{c} & \boldsymbol{B}'' \end{pmatrix},$$

from which, by the use of Theorem 8.2, we can deduce Theorem 8.5.

Theorem 8.6 (D. M. Cvetković, I. Gutman, N. Trinajstić [CvGT 1]): *Two vertices and the four edges of a cycle of length 4, which are positioned in a bipartite graph G as shown in* Fig. 8.6, *can be removed without changing the value of $\eta(G)$.*

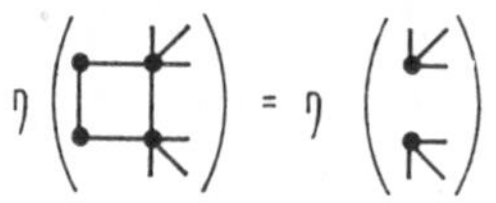

Fig. 8.6

The proof is similar to the previous one and is based on the fact that

$$\operatorname{rk} \boldsymbol{B} = \operatorname{rk} \begin{pmatrix} 1 & 1 & 0 \cdots 0 \\ 1 & 1 & \boldsymbol{b} \\ 0 & & \\ \vdots & \boldsymbol{c} & \boldsymbol{B}'' \\ 0 & & \end{pmatrix} = 1 + \operatorname{rk} \begin{pmatrix} 0 & \boldsymbol{b} \\ \boldsymbol{c} & \boldsymbol{B}'' \end{pmatrix},$$

where $\boldsymbol{B}$ and the last matrix are the incidence matrices of the graph before and after application of the reduction procedure, respectively.

Example 2. See Fig. 8.7.

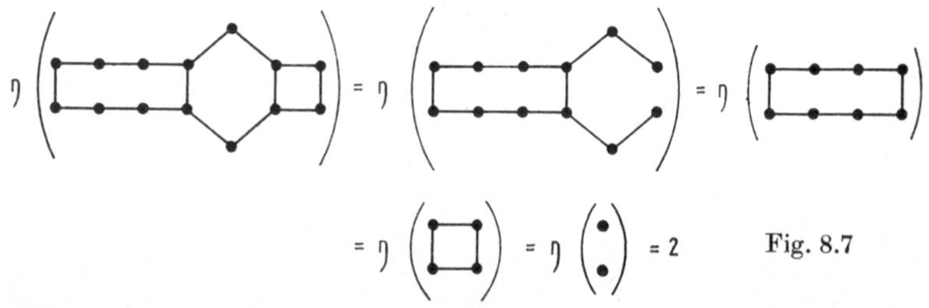

= 2

Fig. 8.7

Some related problems will be considered in Section 8.2. See also [Živk], [CvGT5], [CvGT6]. Note that Corollary 1 of Theorem 8.4 and Theorems 8.5 and 8.6 hold also in the case when G is not a bipartite graph [Gut 10], [CvGT6].

2⁰ An interesting quantity in Hückel theory is the sum of the energies of all the electrons in a molecule, i.e. the so-called *total π-electron energy* E_π. The electrons occupy those molecular orbitals which correspond to greater eigenvalues (minimum energy principle). But according to *Pauli's principle* one orbital can be occupied by at most two electrons. Therefore, for a molecule with n atoms (n being even) we can define the total π-electron energy as†

$$E_\pi = 2 \sum_{i=1}^{\frac{n}{2}} \lambda_i .$$

For a bipartite graph, because of the symmetry of the spectrum, we can write†

$$E_\pi = \sum_{i=1}^{n} |\lambda_i| .$$

Apart from some inequalities and some empirical rules and formulas, nothing is known about the dependence of E_π on the graph structure. We shall now deduce a lower and an upper bound for E_π.

Theorem 8.7 (B. J. McClelland [McC1]): *Let G be a bipartite graph with n vertices, m edges, adjacency matrix $\boldsymbol{A}$ and total π-electron energy E_π. Then the following relation holds:*

$$\sqrt{2m + n(n-1)\,|\det \boldsymbol{A}|^{\frac{2}{n}}} \leqq E_\pi \leqq \sqrt{2mn}. \tag{8.2}$$

Proof. By application of the relation $\sum_{i=1}^{n} \lambda_i^2 = 2m$, and from the well-known inequality

$$\frac{1}{n} \sum_{i=1}^{n} |\lambda_i| \leqq \left(\frac{1}{n} \sum_{i=1}^{n} |\lambda_i|^2 \right)^{\frac{1}{2}}$$

the upper bound follows. Applying the inequality between the algebraic and the geometric means, we have

$$\frac{1}{n(n-1)} \sum_{\substack{i,j=1 \\ i \neq j}}^{n} |\lambda_i| \cdot |\lambda_j| \leqq \left(\prod_{i=1}^{n} |\lambda_i|^{2(n-1)} \right)^{\frac{1}{n}} .$$

Using the relations

$$\prod_{i=1}^{n} |\lambda_i| = |\det \boldsymbol{A}| \quad \text{and} \quad E_\pi^2 = \left(\sum_{i=1}^{n} |\lambda_i| \right)^2 = \sum_{i=1}^{n} \lambda_i^2 + \sum_{\substack{i,j=1 \\ i \neq j}}^{n} |\lambda_i| \cdot |\lambda_j|$$

leads to the lower bound in (8.2). This completes the proof of the theorem.

The meaning of det $\boldsymbol{A}$ for bipartite graphs will be described in the next section.

---

† It should be noticed that the validity of this argument depends on whether the *Aufbau-Prinzip* of Pauli determines that all molecular orbitals are doubly occupied. But this is not always true: the Aufbau-Prinzip sometimes leads to singly occupied orbitals.

The last theorem shows that the total π-electron energy depends to some degree on the number of vertices and edges. However, there are many graphs which have the same number of vertices and the same number of edges but different energy. Observation has shown that circuits of length $4s + 2$, which are the boundaries of faces in the (planar) graph considered, enhance the energy, whereas those of length $4s$ diminish it. These empirical rules are justified in a paper of I. GUTMAN and N. TRINAJSTIĆ [GuT 10]. Some approximate formulas for E_π were given in [CrT 1], [CrT 2], [GuT 1], [GuTW], [WiGT], [HoHG], [GrGT 1], [GrGT 2], [Gut 11]. An improvement of McClelland's bounds for E_π is given in [Gut 2]. The experience of chemists shows that among all trees with a fixed number of vertices, the path has the maximum value of the total π-electron energy. But this statement was proved only very recently [Gut 9], [Gut 12].†

3⁰ Let C_{ij} ($j = 1, \ldots, n$) be the components of an eigenvector belonging to the eigenvalue λ_i and let g_i be the number of electrons in the i-th molecular orbital. The matrix $\boldsymbol{P} = (p_{ij})_1^n$, where

$$p_{ij} = \sum_{k=1}^{n} g_k C_{ki} C_{kj},$$

is called *Coulson's bond-order matrix* or *density matrix*. If indices i and j correspond to two adjacent vertices, the bond order p_{ij} is closely related to the distance between the two corresponding atoms. The quantity p_{ii} represents the π-electron charge on the i-th atom. If vertices i and j are not adjacent, the quantity p_{ij} is of less importance physically; it can be related to the stability of some molecular systems.

For bipartite graphs with n vertices (n even) we can write††

$$p_{ij} = 2 \sum_{k=1}^{\frac{n}{2}} C_{ki} C_{kj}.$$

The problem is again: "How does $\boldsymbol{P}$ depend on the structure of the graph?"

For bipartite graphs $(\mathscr{X}, \mathscr{Y}; \mathscr{U})$ with $|\mathscr{X}| = |\mathscr{Y}|$, the matrix $\boldsymbol{P}$ is of the form

$$\boldsymbol{P} = \begin{pmatrix} \boldsymbol{I} & \boldsymbol{R} \\ \boldsymbol{R}^\mathrm{T} & \boldsymbol{I} \end{pmatrix}$$

(see [HeSt]), where $\boldsymbol{I}$ is a unit matrix and where $\boldsymbol{R}$ is a matrix which can be computed from the adjacency matrix [Ha,G], [CoRu].

Also, it can be proved that

$$E_\pi = 2 \sum_{(r,s)} p_{rs},$$

where the summation is taken over all edges (r, s) of the graph considered.

---

† For further references see I. GUTMAN, Total pi-electron energy and molecular topology, Bibliography. match **4** (1978), 195–200.

†† See footnote on p. 237.

8.2. Graphs related to benzenoid hydrocarbons

Let H be a bipartite digraph. H does not contain cycles of odd length, and any spanning linear directed subgraph of H, if it exists, also contains no cycles of odd length.

Let the vertices of H be coloured by two colours in the usual manner. A necessary condition for the existence of a spanning linear directed subgraph of H is that H contains the same number of vertices of each colour. In what follows we shall usually consider bipartite digraphs (or graphs) with $n = 2\nu$ vertices, so that ν of them are coloured, say, red while the other ν vertices are coloured blue. We shall say that such a digraph (or graph) has $\nu + \nu$ vertices.

A spanning directed subgraph of H such that exactly one arc goes out from every red (blue) vertex and exactly one arc terminates in every blue (red) vertex is called a *red* (*blue*) *separation*.

Every spanning linear directed subgraph of H can be represented in a unique way as the union of one red and one blue separation. Conversely, the union of one red and one blue separation yields a spanning linear directed subgraph of H [GuT6].

If s_1 is the number of red, and if s_2 is the number of blue separations in H, then, according to the foregoing, the number of spanning linear directed subgraphs of H is equal to $s_1 s_2$. From formula (1.40), p. 34, we obtain the following relation for the permanent of the adjacency matrix $\boldsymbol{A}$ of H:

$$\operatorname{per} \boldsymbol{A} = s_1 s_2 . \tag{8.3}$$

Let G be an (undirected) bipartite graph with $n = \nu + \nu$ vertices. A regular spanning subgraph of G of degree 1 is called a 1-*factor* (or, briefly, *factor*) of G. The number of factors will be denoted by k.

A digraph H, obtained from G by replacing every edge of G by two directed edges of opposite orientations between the same pair of vertices, can be associated with G. Obviously, to every factor of G there corresponds, in a natural way, exactly one red and exactly one blue separation in H. Therefore, $s_1 = s_2 = k$. Since the adjacency matrices of G and H are equal, formula (8.3) becomes

$$\operatorname{per} \boldsymbol{A} = k^2 . \tag{8.4}$$

This is a well-known formula (see, for example, [Ore]); a generalization of it for non-bipartite graphs was given in [CvGT2].

In a digraph H, let the red vertices be numbered $1, \ldots, \nu$ and the blue vertices $\nu + 1, \ldots, 2\nu$. A red separation of H can be represented as a permutation $j_1, \ldots, j_\nu$ of numbers $\nu + 1, \ldots, 2\nu$. ($j_1, \ldots, j_\nu$ denote the blue vertices on which the arcs, going out from the red vertices $1, \ldots, \nu$, terminate.) In a similar way, a blue separation can be represented by a permutation $i_1, \ldots, i_\nu$ of numbers $1, \ldots, \nu$.

Consider the spanning linear directed subgraph of H obtained by the union of the red separation $j_1, \ldots, j_\nu$ and of the blue separation $i_1, \ldots, i_\nu$. The sign of the term in the development of $\det \boldsymbol{A}$ (see (1.42)', p. 35) corresponding to the linear directed subgraph considered can be determined on the basis of the number of inversions in the permutation

$$j_1, \ldots, j_\nu, i_1, \ldots, i_\nu. \tag{8.5}$$

Clearly, the numbers $j_1, \ldots, j_\nu$ form, with the numbers $i_1, \ldots, i_\nu$, ν^2 inversions. If the number of inversions of the permutation $j_1, \ldots, j_\nu$ is j, and if the number of inversions of the permutation $i_1, \ldots, i_\nu$ is i, then the number of inversions of the permutation (8.5) is equal to $i + j + \nu^2$, and the sign of the term in the development of $\det \boldsymbol{A}$ corresponding to the linear directed subgraph in question is $(-1)^{i+j+\nu}$ since ν^2 and ν are of the same parity. The quantity $(-1)^i$ will be called the *parity of the blue separation*, and $(-1)^j$ will be called the *parity of the red separation*.

Let K be a factor of G and let R and B be the corresponding red and blue separations in H, respectively. Let par X be the parity of the separation X. It can then easily be seen that par R = par B. The parity par K of the factor K is defined by the relation

$$\text{par } K = \text{par } R = \text{par } B.$$

Lemma 8.1: *For a bipartite graph G with $n = \nu + \nu$ vertices and adjacency matrix $\boldsymbol{A}$ we have*

$$\det \boldsymbol{A} = (-1)^\nu \sum_{U \subset G} (-1)^{r(U)} 2^{c(U)}, \tag{8.6}$$

where $r(U) = r$ is the number of circuits of lengths $4s$ ($s = 1, 2, \ldots$) and $c(U)$ is the total number of circuits of all kinds contained in U, the subgraph U running through all spanning basic figures of G.

Proof. In order to get (8.6) from (1.42)'' it is necessary (and sufficient) to prove that $(-1)^{2\nu+p(U)} = (-1)^{\nu+r(U)}$, i.e. $p(U) \equiv \nu + r(U) \pmod 2$, for each basic figure U. For a particular basic figure U, let $q(U) = q$ be the total number of circuits of lengths $4s + 2$ ($s = 1, 2, \ldots$) and of graphs K_2 in U and let $4t_i + 2$ ($i = 0, 1, \ldots, q$) be the numbers of vertices contained in these circuits or graphs K_2. If $4s_i$ ($i = 1, \ldots, r$) are the lengths of the other circuits in U, we get

$$\sum_{i=1}^{q} (4t_i + 2) + \sum_{i=1}^{r} 4s_i = 2\nu,$$

$$2 \sum_{i=1}^{q} t_i + q + 2 \sum_{i=1}^{r} s_i = \nu,$$

$$q \equiv \nu \pmod 2.$$

Since $p(U) = q(U) + r(U)$, we get $p(U) \equiv \nu + r(U) \pmod 2$, which was to be proved.

Note that the corresponding formula for the bipartite digraph H reads

$$\det \boldsymbol{A} = (-1)^\nu \sum_{L \subset H} (-1)^{r(L)}, \tag{8.7}$$

where the summation runs over all spanning linear directed subgraphs L of H with $r(L)$ being the number of cycles of length $4s$ ($s = 1, 2, \ldots$) in L.

Theorem 8.8 (A. GRAOVAC, I. GUTMAN, N. TRINAJSTIĆ, T. ŽIVKOVIĆ [GrGTŽ]): *In a bipartite graph G two factors K_1 and K_2 are of the same parity if and only if the union of the sets of edges of K_1 and K_2 forms an even number of circuits of length $4s$ $(s = 1, 2, \ldots)$.*

Proof. Together with factors K_1 and K_2, consider the corresponding red and blue separations R_1, R_2 and B_1, B_2. Let $\text{par } R_1 = \text{par } K_1 = (-1)^j$ and $\text{par } B_2 = \text{par } K_2 = (-1)^i$. The sign of the term in the development of $\det A$ corresponding to the linear subgraph $R_1 \cup B_2$ is equal to $(-1)^{i+j+\nu}$. Comparing this with (8.7), we see that $i + j \equiv r(R_1 \cup B_2) \pmod 2$. Thus, K_1 and K_2 are of the same parity if and only if the number of cycles of length $4s$ $(s = 1, 2, \ldots)$ in the linear subgraph $R_1 \cup B_2$ is even. The statement of the theorem now follows immediately.

According to this theorem the property of two factors "being of the same parity" depends neither on the labelling of the vertices nor on the choice of colours in the graph colouring. This binary relation is an equivalence relation and, in a natural way, subdivides the set of factors into two equivalence classes. However, the parity of a factor depends on the labelling of the vertices, as can be seen from examples.

Let k_+ be the number of factors of positive parity and k_- the number of factors of negative parity in a bipartite graph G with $\nu + \nu$ vertices having the adjacency matrix A.

Theorem 8.9 (M. S. J. DEWAR, H. C. LONGUET-HIGGINS [DeLo]):

$$\det A = (-1)^\nu (k_+ - k_-)^2.$$

Proof. It has already been established that $(-1)^{r(R_i \cup B_j)} = \text{par } R_i \cdot \text{par } B_j$. Therefore (8.7) becomes

$$\det A = (-1)^\nu \sum_{i=1}^{k} \sum_{j=1}^{k} (-1)^{r(R_i \cup B_j)}$$

$$= (-1)^\nu \sum_{i=1}^{k} \text{par } R_i \cdot \sum_{j=1}^{k} \text{par } B_j = (-1)^\nu (k_+ - k_-)^2.$$

The value of the determinant of the adjacency matrix naturally does not depend on the labelling of vertices — a fact which is again confirmed by this formula.

In the preceding section, the problem of determining the number $\eta(G)$ for a bipartite graph G has been considered. We shall show that the number η is closely connected with the number of factors, k, that are contained in G and in its subgraphs.

If the graph G contains no factor, i.e., if $k = 0$ $(k_+ = k_- = 0)$, then from the last theorem it follows that $\det A = 0$, i.e., $\eta(G) > 0$. From this observation the following theorem is obtained.

Theorem 8.10 (H. C. LONGUET-HIGGINS [Long]): *If G is a bipartite connected graph with $\eta(G) = 0$, then G has a factor.*

However, if $k \neq 0$, $\eta(G) = 0$ is not always valid since, according to Theorem 8.9, $k_+ = k_-$ $(\neq 0)$ and $\det A = 0$ may occur.

We shall now define a class of graphs (we shall call it class $\mathscr{A}$) which is of great interest in chemistry and in which, for every graph G, the implication $k \neq 0 \Rightarrow \eta(G) = 0$ holds. This will be done in such a way that a graph will have all factors of the same parity (say, positive) in the constructed class. Then $k_- = 0$, $k_+ = k$, and $\det A = (-1)^\nu k^2$.

Before defining the class $\mathscr{A}$, we introduce some notion which will be of use in the discussion of it. A graph embedded in a plane in such a way that no pair of its edges intersect will be called a *plane graph*. A graph is *planar* if it is isomorphic to some plane graph. A plane graph divides the plane into one infinite and some finite regions. The finite regions will be called *faces*. Vertices of the graph lying on the boundary of the infinite region are referred to as *peripheral vertices*.

In order to define the class $\mathscr{A}$ consider first planar graphs without bridges or cutpoints, which can be represented in the plane (without crossing of edges) so that every face-boundary is a circuit of length of the form $4s + 2$ $(s \in \mathscr{N})$.† Subsequently we shall always assume that these graphs are embedded in the plane in the manner described.

The graphs described and all trees belong to the class $\mathscr{A}$. All other graphs from $\mathscr{A}$ are obtainable from the aforementioned graphs by a finite number of applications of the following two operations:

1° joining two graphs from $\mathscr{A}$ by an edge between two peripheral vertices of these graphs (so that the new edge represents a bridge);

2° identifying two peripheral vertices of different graphs (or of different copies of the same graph) from $\mathscr{A}$ (so that the vertex obtained by the identification is a cut point).

From the manner of construction we see that graphs belonging to $\mathscr{A}$ are connected, bipartite, plane†† graphs in which every face-boundary is a circuit of length of the form $4s + 2$ $(s \in \mathscr{N})$.

Lemma 8.2 (D. M. Cvetković, I. Gutman, N. Trinajstić [CvGT4]): *Let $G \in \mathscr{A}$. Then in the interior of every circuit of G of length $4s$ $(s \in \mathscr{N})$ there is an odd number of vertices and in the interior of every circuit of length $4s + 2$ $(s \in \mathscr{N})$ there is an even number of vertices.*

Proof. Let C be a circuit of G. Consider the subgraph G' of G induced by the vertices lying in the interior of C or on the circuit C. G' does not contain bridges or cut points and, naturally, all of its faces are bounded by circuits. Let G' have n vertices, m edges, and f $(= m - n + 1)$ faces $\varphi_1, \varphi_2, \ldots, \varphi_f$. Let $d_j = 4s_j + 2$ $(s_j \in \mathscr{N}$; $j = 1, \ldots, f)$ be the length of the boundary circuit of face φ_j and let d be the length of C. Then

$$d + \sum_{j=1}^{f} d_j = 2m.$$

† $\mathscr{N} = \{1, 2, 3, \ldots\}$.

†† Note that although the atoms of some molecules by no means lie in a plane, the structure of such molecules may nevertheless be represented by planar graphs, as defined above.

If $s = \sum_{j=1}^{f} s_j$, we get $d + 4s + 2f = 2m$. According to Euler's theorem for plane graphs we have $f = m - n + 1$, which, together with the last relation, implies $d + 4s = 2(n - 1)$ or

$$n - d = 2s + 1 - \frac{d}{2}.$$

Since $n - d$ is the number of vertices in the interior of the circuit C, the statement of the Lemma is immediately obtained from this relation.

Lemma 8.3 (D. M. Cvetković, I. Gutman, N. Trinajstić [CvGT4]): *Let $G \in \mathscr{A}$. No basic figure of G which contains all vertices of G contains a circuit of length $4s$ ($s \in \mathscr{N}$).*

Proof. Suppose the contrary, i.e. that a spanning basic figure with at least one circuit of length $4s$ exists. The vertices from the interior of such a circuit are covered by elementary figures, i.e., by circuits of even length or by graphs K_2, and therefore their number is even. But this contradicts Lemma 8.2.

Theorem 8.11 (D. M. Cvetković, I. Gutman, N. Trinajstić [CvGT4]): *Let $G \in \mathscr{A}$. All factors of G are of the same parity.*

This theorem is an immediate corollary of Theorem 8.8 and Lemma 8.3 if we keep in mind that the union of sets of edges of two factors of G forms a basic figure of G.

According to the foregoing we obtain the following theorem.

Theorem 8.12: *Let G be a bipartite graph with $\nu + \nu$ vertices, k factors and adjacency matrix $\boldsymbol{A}$, belonging to class $\mathscr{A}$. Then*

1^0 $\det \boldsymbol{A} = (-1)^\nu \cdot k^2$,

2^0 $\det \boldsymbol{A} = (-1)^\nu \cdot \operatorname{per} \boldsymbol{A}$,

3^0 $\eta(G) = 0$ *if and only if* $k > 0$.

We emphasize the relation between the number of factors and the spectrum of a graph from $\mathscr{A}$ in the form of a special theorem.

Theorem 8.13: *Let $G \in \mathscr{A}$. The number of factors of G is equal to the product of all non-negative eigenvalues of G.*

Proof. If G is a bipartite graph having $n_1 + n_2$ ($n_1 \neq n_2$) vertices, then, naturally, $k = 0$; but then, according to Corollary of Theorem 8.2, the number zero belongs to the spectrum of G. In the case when $n_1 = n_2$, Theorem 8.13 immediately follows from Theorems 8.12 and 3.11.

Other relations between the number of factors and spectra of graphs will be described in the next section.

In [CvGT4] a subclass $\mathscr{B}$ of class $\mathscr{A}$ was considered. A graph G is in $\mathscr{B}$ if it satisfies the following conditions:

1⁰ $G \in \mathscr{A}$,

2⁰ the degrees of all vertices of G are not greater than 3,

3⁰ the boundaries of any two faces in G have at most one edge in common.

Class $\mathscr{B}$ is of great significance in chemistry, since graphs from $\mathscr{B}$ represent molecules of benzenoid hydrocarbons.

According to Theorem 8.12, the question of stability of benzenoid hydrocarbons (i.e., the investigation whether or not $\eta(G) = 0$) is reduced to establishing the existence of at least one factor in the corresponding graph. For a large number of molecules the existence of a factor can easily be established by direct inspection. (Note that chemists are very skilful in such inspections!)

If we consider only those graphs from $\mathscr{B}$ which contain no circuit of length $4s$ ($s = 1, 2, \ldots$) (irrespective of whether or not it is the boundary of a face), then we get a more narrow class of graphs which shall be denoted by $\mathscr{C}$. This class contains all graphs corresponding to cata-fuzed benzenoid hydrocarbons (e.g. naphthalene, phenanthrene), unfuzed (e.g. benzene, biphenyl, styrene) and acyclic hydrocarbons (e.g. ethylene, butadiene). It can be shown ([CvGT4]) that those graphs from $\mathscr{B}$ which belong to class $\mathscr{C}$ are those for which every vertex is peripheral. Both classes $\mathscr{B}$ and $\mathscr{C}$ possess interesting graph theoretical properties and it may be expected that at least some of them are responsible for the rather specific chemical behaviour of the aromatic hydrocarbons.

According to the definition of the class $\mathscr{C}$, an arbitrary graph from this class does not contain any basic figure (with an arbitrary number of vertices) with circuits of lengths $4s$ ($s = 1, 2, \ldots$). Therefore, according to (1.36) and (8.7), all terms (summands) in the expression for the coefficient a_k of the characteristic polynomial have the same sign. Because of this $a_k \neq 0$ if and only if there is at least one basic figure with k vertices. However, the existence of a basic figure with k vertices implies the existence of a linear subgraph with the same number of vertices (since each factor of a basic figure is a linear subgraph) and, conversely, the existence of a linear subgraph implies the existence of a basic figure, since a linear subgraph is itself a basic figure. Note that these considerations hold not only for the class $\mathscr{C}$ but also for all bipartite graphs not containing circuits of lengths $4s$ ($s = 1, 2, \ldots$).

According to the foregoing we can formulate the following theorem, a special case of which is Theorem 8.1 from the preceding section.

Theorem 8.14 (D. M. Cvetković, I. Gutman, N. Trinajstić [CvGT1]): *If a bipartite graph G with n vertices does not contain any circuits of lengths $4s$ ($s = 1, 2, \ldots$), then $\eta(G) = n - 2q$ where q is the maximum number of mutually non-adjacent edges in G.*

8.3. The dimer problem

The spectra of graphs, or the spectra of certain matrices which are closely related to adjacency matrices appear in a number of problems in statistical physics (see, for example, [Kas2], [Mont], [Perc]). We shall describe in some detail the so-called *dimer problem.*

The dimer problem is related to the investigation of the thermodynamic properties of a system of diatomic molecules ("dimers") adsorbed on the surface of a crystal. The most favourable points for the adsorption of atoms on such a surface form a two-dimensional lattice, and a dimer can occupy two neighbouring points. It is necessary to count all ways in which dimers can be arranged on the lattice without overlapping each other, so that every lattice point is occupied.

The dimer problem on a square lattice is equivalent to the problem of enumerating all ways in which a chess-board of dimension $n \times n$ (n being even) can be covered by $\frac{1}{2} n^2$ dominoes, so that each domino covers two adjacent squares of the chess-board and that all squares are so covered.

A graph can be associated with a given adsorption surface. The vertices of the graph represent the points which are the most favourable for adsorption. Two vertices are adjacent if and only if the corresponding points can be occupied by a dimer. In this manner an arrangement of dimers on the surface determines a factor in the corresponding graph, and vice versa. Thus, the dimer problem is reduced to the task of determining the number of factors in a graph.

Since all graphs which are to be considered are bipartite, we can apply formula (8.4). However, we do not know how to calculate the permanent of a matrix in a convenient manner and it is from this fact that the difficulties arise.

Several techniques have been developed in an attempt to avoid these difficulties. All these are based on the following idea. Let us multiply the elements of the adjacency matrix $\boldsymbol{A} = (a_{ij})_1^n$ by suitable numbers α_{ij} and denote the new matrix by $\boldsymbol{A}^* = (\alpha_{ij}a_{ij})$. It has been proved that the α_{ij}'s can be chosen such that per $\boldsymbol{A} = \det \boldsymbol{A}^*$. The influence of multiplying elements of $\boldsymbol{A}$ by α_{ij} can be such that only some 1's from $\boldsymbol{A}$ have their signs altered. In particular, P. W. Kasteleyn [Kas1] has shown that in this way the adjacency matrix $\boldsymbol{A}$ of a planar graph can be transformed into a skew-symmetric matrix $\boldsymbol{A}^*$ such that per $\boldsymbol{A} = \det \boldsymbol{A}^*$. Recently, H. C. H. Little [Lit2] has extended Kasteleyn's results to non-planar graphs. Naturally, for graphs belonging to class $\mathscr{A}$ (see the preceding section), no such procedures are necessary and we can immediately replace the permanent by the determinant of the adjacency matrix and then calculate the latter by means of graph spectra (see Theorem 8.12).

We shall now enumerate dimer arrangements, i.e. factors, for the $m \times n$ square lattice, i.e. for the graph $G_{m,n} = P_m + P_n$ (see Fig. 2.2) using one of many possible variants (see, e.g., [Perc]) for transforming the permanent into the determinant. Consider the graph $G_{m,n}$ of a square lattice as shown in Fig. 2.2. According to the situation in that figure, $G_{m,n}$ contains horizontal and vertical edges. Let $H_{m,n}$ denote a digraph obtained from $G_{m,n}$ by replacing every edge by a corresponding pair of oriented edges of opposite orientation. The oriented edges of $H_{m,n}$ can also be horizontal or vertical. Let $\boldsymbol{A}_{m,n}$ be the adjacency matrix of $G_{m,n}$, i.e. of $H_{m,n}$. We have

$$k^2 = \operatorname{per} \boldsymbol{A}_{m,n} = \sum_{L \subset H_{m,n}} 1, \qquad \det \boldsymbol{A}_{m,n} = \sum_{L \subset H_{m,n}} (-1)^{c(L)},$$

where k is the number of factors and the summation runs over all linear spanning subgraphs L of $H_{m,n}$.

The next lemma is given without proof (see, for example, [Perc]).

Lemma 8.4: *For every linear spanning subgraph L of $H_{m,n}$ we have*

$$(-1)^{c(L)} = i^{h(L)},$$

where $i = \sqrt{-1}$ and $h(L)$ is the number of horizontal edges in the linear spanning subgraph L.

According to Lemma 8.4, we can easily prove the following theorem (see, for example, [Perc]).

Theorem 8.15: *The number k of factors in $G_{m,n}$ is given by*

$$k^2 = \det (\boldsymbol{A}_m \otimes \boldsymbol{I}_n + i\boldsymbol{I}_m \otimes \boldsymbol{A}_n),$$

where $\boldsymbol{A}_s$ is the adjacency matrix of a path with s vertices.

Proof. According to Theorem 2.21, $\boldsymbol{A}_{m,n} = \boldsymbol{A}_m \otimes \boldsymbol{I}_n + \boldsymbol{I}_m \otimes \boldsymbol{A}_n$. 1's from $\boldsymbol{A}_m \otimes \boldsymbol{I}_n$ correspond to vertical and 1's from $\boldsymbol{I}_m \otimes \boldsymbol{A}_n$ correspond to horizontal edges of $G_{m,n}$ or $H_{m,n}$. The matrix

$$\boldsymbol{A}^*_{m,n} = \boldsymbol{A}_m \otimes \boldsymbol{I}_n + i\boldsymbol{I}_m \otimes \boldsymbol{A}_n$$

differs from $\boldsymbol{A}_{m,n}$ in that 1's corresponding to horizontal edges are multiplied by i. According to Lemma 8.4,

$$\operatorname{per} \boldsymbol{A}_{m,n} = \det \boldsymbol{A}^*_{m,n}$$

and this completes the proof of the theorem.

The eigenvalues of $\boldsymbol{A}_{m,n}$ are

$$2 \cos \frac{\pi}{m+1} j + 2 \cos \frac{\pi}{n+1} l \qquad (j = 1, \ldots, m;\ l = 1, \ldots, n)$$

(see Section 2.6). Bearing in mind the proof of Theorem 2.23 (p. 69) it is easy to see that the eigenvalues of $\boldsymbol{A}^*_{m,n}$ are

$$2 \cos \frac{\pi}{m+1} j + 2i \cos \frac{\pi}{n+1} l \qquad (j = 1, \ldots, m;\ l = 1, \ldots, n).$$

So we have

$$k^2 = \prod_{j=1}^{m} \prod_{l=1}^{n} \left(2 \cos \frac{\pi}{m+1} j + 2i \cos \frac{\pi}{n+1} l\right).$$

If both m and n are odd, $k = 0$. Without loss of generality let m be even. Then

$$k^2 = 2^{mn} \prod_{j=1}^{\frac{m}{2}} \prod_{l=1}^{n} \left(\cos^2 \frac{\pi}{m+1} j + \cos^2 \frac{\pi}{n+1} l\right).$$

For square $n \times n$ lattices with $n = 2, 4, 6, 8$ we have $k = 2, 36, 6728, 12988816$, respectively. The last number is $2^4 \cdot 901^2$ and this is the number of ways in which an 8×8 chess-board can be covered by 32 dominoes.

We have also

$$k \sim e^{\frac{mn}{\pi} G} \qquad (m \to +\infty, n \to +\infty),$$

where G is Catalan's constant.

There is another quite different approach to the dimer problem; this is the so-called *transfer matrix method.*

A dimer arrangement on a square lattice can be specified by listing the type of dimer at each vertex, i.e. "up", "down", "to the left" or "to the right". We shall use abbreviations U, D, L, R for these four cases, respectively. Then the situation regarding a column of vertices of $G_{m,n}$ can be represented by an m-tuple of symbols U, D, L, R. For example, the column of eight vertices on Fig. 8.8 is represented by the 8-tuple (D, U, R, L, D, U, R, R).

Note that the number of m-tuples representing all possible columns of vertices in dimer arrangements is equal to the number of walks of length $m - 1$ in the graph on Fig. 8.9, starting from vertices D, L, R and terminating at vertices U, L, R. Such m-tuples will be called *feasible.*

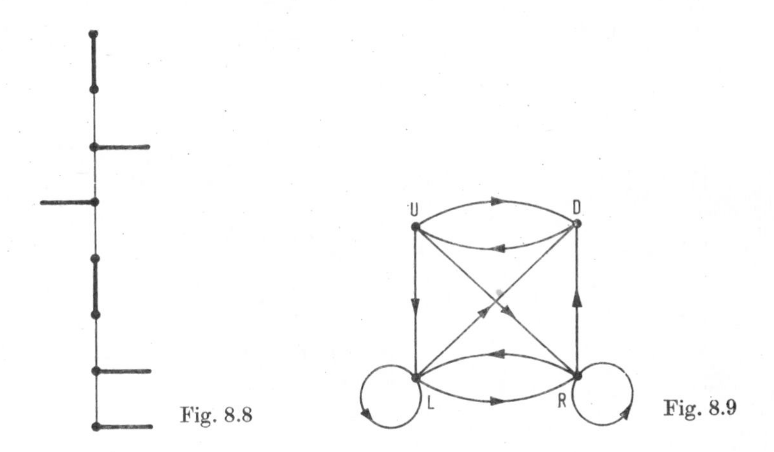

Fig. 8.8 Fig. 8.9

Consider now the graph F whose vertices are all feasible m-tuples of symbols U, D, L, R and whose edges are defined in the following way. For every two feasible m-tuples x and y an oriented edge goes from x to y if and only if m-tuple x has symbols R exactly in those positions in which m-tuple y has symbols L. In this way, if an edge goes from x to y, the column represented by y can be situated immediately to the right of the column represented by x, in a dimer covering of $G_{m,n}$.

Let $\mathcal{S}$ ($\mathcal{Q}$) be the set of m-tuples containing no symbol L (R). Each m-tuple from $\mathcal{S}$ can be the first (from the left) in a dimer covering of $G_{m,n}$. Similarly, $\mathcal{Q}$ is the set of the possible last m-tuples in dimer coverings.

It is easy to see that the number of dimer arrangements in $G_{m,n}$ is equal to the number of walks of length $n - 1$ in F, starting from a vertex in $\mathcal{S}$ and terminating at a vertex in $\mathcal{Q}$.

If we consider, instead of $G_{m,n}$ $(= P_m + P_n)$, the graph $P_m + C_n$, we see that the number of dimer coverings is equal to the number of closed walks of length n in the graph F.

Similar conclusions hold for the square lattice on the torus, which is represented by the graph $C_m + C_n$. Instead of F we must consider a graph E whose vertices are m-tuples generated by closed walks of length m in the graph of Fig. 8.9. The edges of E are defined in the same way as in F.

In the case of $C_m + C_n$ (similarly also for $P_m + C_n$) we can simplify the graph E in the following way: consider the set $\mathcal{W}$ of all m-tuples (vertices) from E having a given arrangement of symbols R. The set $\mathcal{W}$ can be substituted by a unique vertex w in such a way that all edges starting from (terminating at) $\mathcal{W}$ now start from (terminate at) w. After such simplifications, for each arrangement of R's, we get a vertex of the graph E^*. It is not very difficult to see that the numbers of closed walks of a given length are equal in E and E^*.

The adjacency matrix $\boldsymbol{T}$ of E^* is called the *transfer matrix* and it is given by (see, for example, [Perc])

$$\boldsymbol{T} = \prod_{i=1}^{m} (\boldsymbol{I}_{2^m} + \boldsymbol{H}_{i,i+1}) \prod_{j=1}^{m} \boldsymbol{V}_j,$$

where

$$\boldsymbol{H}_{i,i+1} = \underbrace{\boldsymbol{I}_2 \otimes \boldsymbol{I}_2 \otimes \cdots \otimes \boldsymbol{I}_2}_{i-1 \text{ times}} \otimes \begin{pmatrix} 0 & 0 \\ 1 & 0 \end{pmatrix} \otimes \begin{pmatrix} 0 & 0 \\ 1 & 0 \end{pmatrix} \otimes \underbrace{\boldsymbol{I}_2 \otimes \cdots \otimes \boldsymbol{I}_2}_{m-i-1 \text{ times}},$$

$i = 1, \ldots, m;$

$$\boldsymbol{H}_{m,m+1} = \begin{pmatrix} 0 & 0 \\ 1 & 0 \end{pmatrix} \otimes \underbrace{\boldsymbol{I}_2 \otimes \cdots \otimes \boldsymbol{I}_2}_{m-2 \text{ times}} \otimes \begin{pmatrix} 0 & 0 \\ 1 & 0 \end{pmatrix};$$

$$\boldsymbol{V}_j = \underbrace{\boldsymbol{I}_2 \otimes \boldsymbol{I}_2 \otimes \cdots \otimes \boldsymbol{I}_2}_{j-1 \text{ times}} \otimes \begin{pmatrix} 0 & 1 \\ 1 & 0 \end{pmatrix} \otimes \underbrace{\boldsymbol{I}_2 \otimes \cdots \otimes \boldsymbol{I}_2}_{m-j \text{ times}}, \quad j = 1, \ldots, m,$$

$\boldsymbol{I}_s$ being the unit matrix of order s.

Let r be the maximum eigenvalue of $\boldsymbol{T}$. For large n we then have the following asymptotic relation for k, the number of factors (i.e. dimer coverings) of $C_m + C_n$:

$$k \sim cr^n \qquad (n \to +\infty)$$

where c is a positive number.

It can be proved [Perc] that

$$k \sim \mathrm{e}^{\frac{mn}{\pi}} \qquad (m \to +\infty, n \to +\infty).$$

We shall now explain the transfer matrix method with a simple example. We have seen that also in chemistry the number of factors of a graph plays an important role. In [CvG2] the problem of determining the number of factors of some graphs is considered. We shall now interpret the results in a different way.

We consider those graphs which can be realized as a sequence of circuits $(C^{(1)}, C^{(2)}, \ldots, C^{(n)})$ of even length, where two adjacent circuits $C^{(i)}, C^{(i+1)}$ of the sequence have exactly one edge, e_i, in common and where no vertex belongs to more than two circuits. Such a graph is represented in Fig. 8.10. The endpoints of e_i are connected by exactly two (disjoint) paths with the endpoints of e_{i+1}. The numbers of edges contained in these paths are of the same parity, since $C^{(i+1)}$ has an even length. If the numbers of edges referred to are even we shall say that the circuit $C^{(i+1)}$ is of type α; otherwise it is of type β.

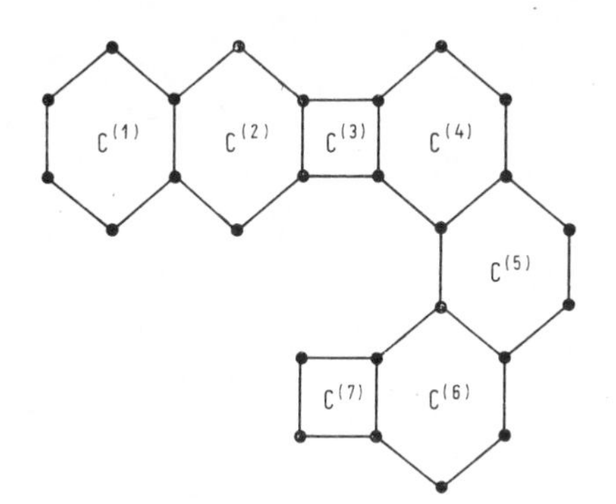

Fig. 8.10

Consider the circuit $C^{(1)}$ as a graph itself. It has two factors: the edge e_1, common to $C^{(1)}$ and $C^{(2)}$, can be in two positions: it can either belong to a specified factor (position $\boldsymbol{a}$) or not (position $\boldsymbol{b}$). Consider now the graph $(C^{(1)}, C^{(2)})$, $C^{(2)}$ being of type α. Behind position $\boldsymbol{a}$ of e_1, the edge e_2 can be in position $\boldsymbol{a}$ as well as in position $\boldsymbol{b}$, and behind position $\boldsymbol{b}$ of e_1, only position $\boldsymbol{b}$ of e_2 can appear (see Figs. 8.11, 8.12).

It is clear that, if circuits $C^{(2)}, \ldots, C^{(n)}$ are of type α, the number of factors k_n in the graph $(C^{(1)}, C^{(2)}, \ldots, C^{(n)})$ is equal to the number of walks of length $n - 1$ in the graph G_1 of Fig. 8.13. Obviously, $k_n = n + 1$. In the case when circuits $C^{(2)}, \ldots, C^{(n)}$ are of type β, the graph G_2 from Fig. 8.13 plays the role of G_1.

The adjacency matrices of G_1 and G_2 are

$$\boldsymbol{T}_1 = \begin{pmatrix} 1 & 1 \\ 0 & 1 \end{pmatrix}, \qquad \boldsymbol{T}_2 = \begin{pmatrix} 1 & 1 \\ 1 & 0 \end{pmatrix}.$$

They can be interpreted as transfer matrices in the cases of factor enumeration described above.

The characteristic polynomial of $\boldsymbol{T}_2$ is $\lambda^2 - \lambda - 1$, and for the numbers of walks N_n of G_2 we obtain the relations

$$N_{n+2} = N_{n+1} + N_n \qquad (n = 0, 1, \ldots).$$

Since $N_0 = 2$ and $N_1 = 3$, we see that $N_n = f_{n+3}$ where f_n is the n-th Fibonacci

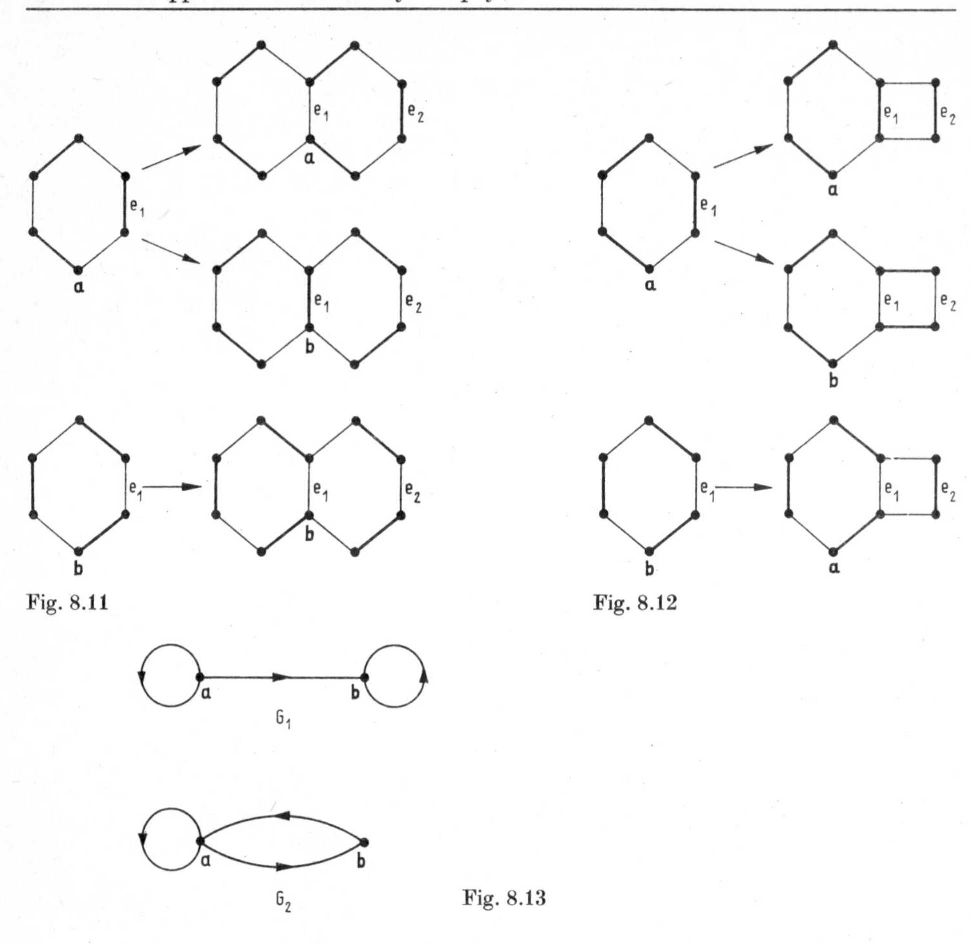

Fig. 8.11

Fig. 8.12

Fig. 8.13

number (Fibonacci's sequence is 1, 1, 2, 3, 5, 8, ...). Thus, the number of factors k_n in the graph consisting of n circuits of type β (the type of the first and of the n-th circuit is unimportant) is given by $k_n = f_{n+2}$.

If we consider a graph in which circuits $C^{(1)}, C^{(2)}, \ldots, C^{(n)}$ are of different types then the number k_n of factors is equal to the sum of all elements of the matrix

$$\boldsymbol{T} = \boldsymbol{T}_{i_2}\boldsymbol{T}_{i_3} \cdots \boldsymbol{T}_{i_n},$$

where for $j = 2, 3, \ldots, n$ we have $i_j = 1$ or 2 depending on whether $C^{(j)}$ is of type α or of type β, respectively.

Consider, as an example, the graph having $n = 2\nu + 1$ circuits, whose types form the sequence $\alpha, \beta, \alpha, \beta, \alpha, \ldots, \alpha, \beta, \alpha$. Now we have $\boldsymbol{T} = (\boldsymbol{T}_2\boldsymbol{T}_1)^\nu$ and the number of factors is equal to the number of walks of length ν of the multigraph G^* having adjacency matrix

$$\boldsymbol{T}_2\boldsymbol{T}_1 = \begin{pmatrix} 1 & 2 \\ 1 & 1 \end{pmatrix}.$$

Obviously, $G^* = G_2 \cdot G_1$ (see Section 2.1). The generating function for the number of walks of G^* is equal to $\frac{2+t}{1-2t-t^2}$ (see Section 1.8).

Graphs E, E^*, G_1, G_2, and G^* can be called *transfer graphs*. From the examples given we see that, with each sequence of graphs G_n $(n = 1, 2, \ldots)$ with a "periodic" structure, a corresponding transfer graph G_T can be associated so that the number of factors in G_n is equal to the number of walks of length n (or of another length, possibly with additional requirements which the walks must satisfy). Examples from chemistry show that a "periodic" structure does not necessarily mean the isomorphism of parts which are "repeated". It is only important that the transfer matrix between the adjacent parts should be the same.

The transfer matrix method is applicable not only to the enumeration of factors but also to other covering problems. The problem of arranging kings on a chess-board has been outlined in [Cve9], and the transfer matrix method was suggested. The transfer matrix method can also be applied in enumerating coverings of a chess-board by "polyminoes".

Spectra of transfer graphs can play an important role in some problems as the following example will show.

Let $G_1, G_2, \ldots, G_s, \ldots$ be a sequence of graphs with a "periodic" structure belonging to class $\mathscr{A}$ (see the preceding section) and let G_T be the corresponding transfer graph. Further, let $\boldsymbol{A}_s$, k_s, n_s be the adjacency matrix, the number of factors and the number of vertices, respectively, of G_s. In the expression for the lower bound of the total π-electron energy of G_s (see Theorem 8.7), the quantity $(\det \boldsymbol{A}_s)^{\frac{2}{n_s}}$ appears. If $\lim_{s\to+\infty} \frac{n_s}{s} = p$, then, according to Theorem 8.12,

$$\lim_{s\to+\infty} |\det \boldsymbol{A}_s|^{\frac{2}{n_s}} = \lim_{s\to+\infty} k_s^{\frac{1}{n_s}} = \lim_{s\to+\infty} (c\lambda_1^s)^{\frac{1}{n_s}} = \lambda_1^{\frac{1}{p}}, \tag{8.8}$$

since for $s \to +\infty$ we have $k_s \sim c\lambda_1^s$, where λ_1 is the largest eigenvalue of G_T and c is a positive number.

We see that for sufficiently large s we can substitute $\sqrt{2m + n(n-1)\,\lambda_1^{\frac{1}{p}}}$ for the lower bound from Theorem 8.7.

Some other problems can be reduced to the enumeration of factors (i.e. dimer arrangements). The most well-known is the famous *Ising problem* arising in the theory of ferromagnetism (see, for example, [Mont], [Kas2]).

The graph-walk problem is of interest in physics not only because of the factor enumeration problem. The numbers of walks of various kinds in a lattice graph appear in several other problems; the random-walk and self-avoiding-walk problems (see [Mont], [Kas2], [Perc]) are but two examples, and some problems outlined in Section 8.4 are related to those discussed here.

8.4. Vibration of a membrane

In the approximate numerical solution of certain partial differential equations, graphs and their spectra arise quite naturally.

Consider, for example, the partial differential equation

$$\frac{\partial^2 z}{\partial x^2} + \frac{\partial^2 z}{\partial y^2} + \lambda z = 0 \tag{8.9}$$

(or $\Delta z + \lambda z = 0$; $\Delta = \frac{\partial^2}{\partial x^2} + \frac{\partial^2}{\partial y^2}$ being the Laplacian operator), where the unknown function $z = z(x, y)$ is subject to the boundary condition $z(x, y) = 0$ on a simple closed curve Γ lying in the xy-plane. It is known that equation (8.9) has a solution only for an infinite sequence $\lambda_1 \leqq \lambda_2 \leqq \cdots \leqq \lambda_n \leqq \cdots$ of (discrete) values of λ which are called the eigenvalues of the equation. The sequence of eigenvalues is called the spectrum of the equation, and to each eigenvalue there corresponds a (set of) solution(s) of (8.9) called its eigenfunction(s).

In an approximate determination of z we consider the values only for a set of points (x_i, y_i) which form a regular lattice (square, triangular, hexagonal) in the xy-plane. A corresponding (infinite) graph can be associated, in a straightforward and natural way, with this lattice. Points (x_i, y_i) are the vertices of the graph and the edges connect pairs of points of minimal distance. The points (or vertices) lying in the interior of Γ are called *internal points* (or *vertices*) and the other points (vertices) of the lattice (of the corresponding lattice graph) are called *external*. Let $z_i = z(x_i, y_i)$. Because of the boundary condition, we can take $z_i = 0$ for all external points.

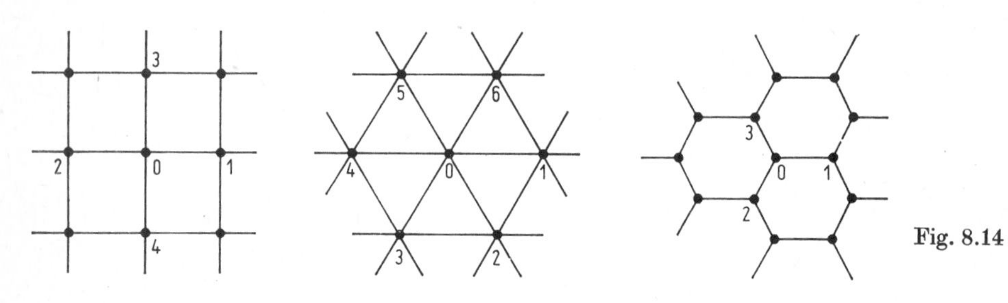

Fig. 8.14

In the case of a square lattice (Fig. 8.14) let $z_0 = z(x_0, y_0)$, (x_0, y_0) being a fixed point of the lattice, and let $z_1 = z(x_0 + h, y_0)$, $z_2 = z(x_0 - h, y_0)$, $z_3 = z(x_0, y_0 + h)$, $z_4 = z(x_0, y_0 - h)$ be the values of z for the neighbouring points (we assume that the points of the lattice lie on lines which are parallel with the coordinate axes and that the distance between any two neighbouring points is h). The value of $\frac{\partial^2 z}{\partial x^2} + \frac{\partial^2 z}{\partial y^2}$ at the point (x_0, y_0) can, as usual, be approximated by

$$\frac{1}{h^2}(z_1 + z_2 + z_3 + z_4 - 4z_0);$$

(8.9) then becomes

$$\frac{1}{h^2}(z_1 + z_2 + z_3 + z_4 - 4z_0) + \lambda z_0 = 0,$$

$$(4 - \lambda h^2)\, z_0 = z_1 + z_2 + z_3 + z_4.$$

Now let the internal points be labelled by 1, 2, ..., n. Taking $\nu = 4 - \lambda h^2$ and writing the equations which correspond to the last equation for all internal points (x_i, y_i), $i = 1, \ldots, n$, of the lattice, we obtain

$$\nu z_i = \sum_{j \cdot i} z_j \qquad (i = 1, \ldots, n), \tag{8.10}$$

where the summation is taken over all indices j corresponding to internal points (x_j, y_j) neighbouring (x_i, y_i).

For some external point neighbouring (x_i, y_i) it is not necessary to include it in the sum of (8.10) since the value of z for this point is zero. Let G be the subgraph of the lattice graph induced by the internal vertices. If we interpret ν as an eigenvalue of G and $(z_1, \ldots, z_n)^\mathsf{T}$ as the corresponding eigenvector, we see that (8.10) just defines the eigenvalue problem for G. The graph G will be called the "membrane graph".

If ν_i $(i = 1, \ldots, n)$ are the eigenvalues of G, the approximate eigenvalues of equation (8.9) are given by $\lambda_i^* = \frac{4 - \nu_i}{h^2}$. The corresponding eigenvector of G represents an approximate solution of (8.9). Note that the λ_i^* $(i = 1, \ldots, n)$ do not necessarily represent approximate values for the first n eigenvalues $\lambda_1, \ldots, \lambda_n$ of (8.9), but for some eigenvalues $\lambda_{i_1}, \ldots, \lambda_{i_n}$.

For the triangular and hexagonal lattices (see Fig. 8.14) we have, respectively, the following approximate expressions for Δz in the point (x_0, y_0):

$$\frac{2}{3h^2}(z_1 + z_2 + z_3 + z_4 + z_5 + z_6 - 6z_0),$$

$$\frac{4}{3h^2}(z_1 + z_2 + z_3 - 3z_0).$$

We again obtain (8.10), but now the connection between the eigenvalues of G an of (8.9) is given by $\lambda_i^* = \frac{2}{3}\,\frac{6 - \nu_i}{h^2}$ and $\lambda_i^* = \frac{4}{3}\,\frac{3 - \nu_i}{h^2}$, respectively.

The procedure described for approximately solving a partial differential equation is often used in technical problems (see, for example, [Col1]). In this way the theory of graph spectra can be very useful in practical calculations. For example, theorems giving relations between graph spectra and the automorphism group of a graph (see Chapter 5 and especially the procedure for the factorization of characteristic polynomials) play an important role which can also be seen from [CoSi1], [Col1].

The most interesting problem which can be treated by such a procedure is that of *membrane vibration*. There are some other problems of this kind, for example, air oscillations in space, etc. (see [CoSi1], [Col1], [Kac], [Ruth]). These problems motivated the authors of [CoSi1] to consider graph spectra.

If a vibrating membrane Ω is held fixed along its boundary Γ, its displacement $F(x, y, t)$ in the direction orthogonal to its plane is a function of the coordinates x, y and of the time t and satisfies the wave equation

$$\frac{\partial^2 F}{\partial t^2} = c^2 \left(\frac{\partial^2 F}{\partial x^2} + \frac{\partial^2 F}{\partial y^2} \right), \tag{8.11}$$

where c is a constant depending on the physical properties of the membrane and of the tension under which the membrane is held.

The solutions of the form $F(x, y, t) = z(x, y)\, \mathrm{e}^{i\omega t}$ are of particular interest. If we substitute this expression in (8.11), we obtain

$$-\omega^2 z(x, y) = c^2 \left(\frac{\partial^2 z(x, y)}{\partial x^2} + \frac{\partial^2 z(x, y)}{\partial y^2} \right). \tag{8.12}$$

Setting $\lambda = \frac{\omega^2}{c^2}$ reduces (8.12) to (8.9).

M. Kac [Kac] has posed the problem of determining the shape of a membrane (i.e. the curve Γ) if the spectrum of the corresponding partial differential equation (8.9) is known. Kac entitled his paper "*Can one hear the shape of a drum?*", since the human ear, by its operation, "hears" the spectrum (the frequences of the tones, produced by the membrane, are simply related to the eigenvalues of the equation). Instead of the spectrum, the generating function for the sums of powers of eigenvalues (or the moments of eigenvalues)

$$G(t) = \sum_{n=1}^{+\infty} \mathrm{e}^{-\lambda_n t}$$

was considered. It has been shown that

$$G(t) \sim \frac{|\Omega|}{2\pi t} - \frac{L}{4} \frac{1}{\sqrt{2\pi t}} \qquad (t \to 0), \tag{8.13}$$

where $|\Omega|$ is the area of the membrane and L the length of its boundary. Further, for a membrane with a smooth boundary the next term in the asymptotic relation (8.13) is $+\frac{1}{6}(C - H)$, where C is the number of separate components of the membrane (the membrane may be disconnected) and H is the number of holes in the membrane (the boundary curve Γ need not be a simple curve). It is not known in which way further information can be obtained from the spectrum, i.e., from $G(t)$ and, in particular, it is not known, whether or not the curve Γ really can be determined from $G(t)$.

M. E. Fisher [Fish] has considered the discrete analogue to Kac's problem. In his model the membrane consists of a set of atoms which in the equilibrium state lie on the vertices of a regular lattice graph embedded in a plane. Each atom acts on its neighbouring atoms by elastic forces. We assume that all atoms have the same mass and that the elastic forces are of the same intensity for all neighbouring pairs of atoms. If $z_i(t)$ and $z_j(t)$ are displacements of neighbouring atoms i and j at time t, the elastic force tending to reduce the relative displacement between these atoms is

$$F_{ij} = -K\big(z_i(t) - z_j(t)\big),$$

where K is a constant characteristic of the elastic properties of the membrane.

The equation of motion of the k-th atom is

$$m\,\frac{\mathrm{d}^2 z_k(t)}{\mathrm{d}t^2} = -K \sum_{j\cdot k} \big(z_k(t) - z_j(t)\big), \tag{8.14}$$

where m is the mass of an atom and where the summation is taken over the nearest neighbours, j, of the k-th atom. For a vertex j of the lattice graph in which there is no atom of the membrane, we have $z_j(t) = 0$ (just as before, such vertices are called *external*). The area $|\Omega|$ of the membrane Ω in this discrete model is proportional to the number n of atoms (*internal* vertices), and the length of the boundary L is proportional to the number B of edges of the lattice graph which connect external and internal vertices. So we have $|\Omega| = an$, $L = bB$, where a and b are certain constants.

We can again consider pure harmonic oscillations and take $z_k(t) = z_k \mathrm{e}^{i\omega t}$ (where $i = \sqrt{-1}$). If we insert this expression into (8.14) and do so for each atom k, then we again obtain the graph eigenvalue problem (8.10). Thus, a solution of Fisher's discrete model is equivalent to an approximate solution of the continuum model.

It was shown in [Fish] that the same amount of information which can be obtained from the spectrum in a continuum model can also be obtained from the spectrum of the corresponding graph in the discrete model. As in the continuous case, we may consider the moments $M_k = \sum_{\nu=1}^{n} \lambda_\nu^k$ $(k = 0, 1, 2, \ldots)$ of the eigenvalues $\lambda_1, \ldots, \lambda_n$ of the membrane graph and the corresponding generating function $G(t) = \sum_{\nu=1}^{n} \mathrm{e}^{\lambda_\nu t}$.

Naturally, knowledge of the moments $M_0, M_1, \ldots, M_n$ is equivalent, according to Newton's formulas, to knowing the characteristic polynomial of the membrane graph, i.e., to a knowledge of the corresponding spectrum.

Trivially, we have $M_0 = n$, $M_1 = 0$, and $M_2 = rn - B$, where r is the degree of the corresponding lattice graph ($r = 6, 4, 3$ for the triangular, square, and hexagonal lattice, respectively). The third moment is equal to zero for square and hexagonal lattices (since the corresponding graphs are bipartite) and for the triangular lattice we have, according to [Fish],

$$M_3^{\text{triangular}} = 12n - 3B + 6(C - H),$$

where, as earlier, C is the number of components of the membrane graph, and H is the number of holes in the membrane (i.e., the number of components of the subgraph of the lattice graph induced by the external vertices, minus one).

According to the above we have the following formulas:

$$|\Omega| = aM_0,$$
$$L = b(rM_0 - M_2),$$
$$C - H = \frac{1}{6} M_3 - \frac{1}{2} M_2 + M_0 \text{ (triangular lattice)}.$$

Furthermore, the expressions for M_4^{square} and $M_4^{\text{triangular}}$ are given in [Fish]. Similar relations including higher moments can be found in [BeJa]. Some formulas for the moments, which are also applicable to the graphs under consideration, were developed in [GuT1], [GuT5].

These facts show that for some classes of planar graphs, graph spectra can provide some information about certain details related to the embedding of graphs in the plane, although in general the problem of whether a graph is planar (or not) cannot be resolved simply by examination of the spectrum (see Section 6.1).

M. E. Fisher [Fish] also considered the problem of "hearing" the shape of a membrane. Naturally, this question reduces to the question of whether or not the graph spectrum determines the graph uniquely. As we know, the answer is negative, and therefore Fisher imposed a further condition on a membrane graph: such a

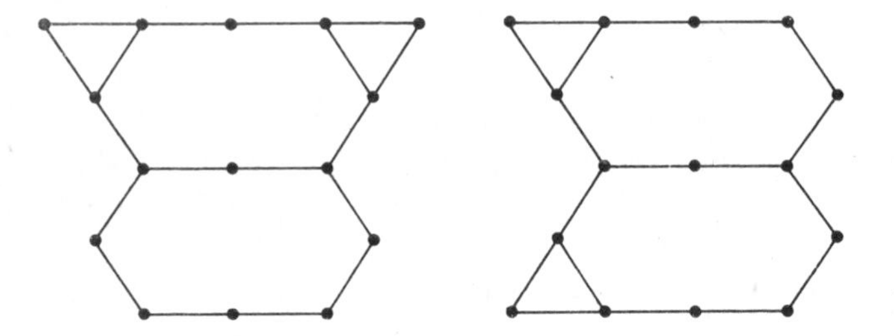

Fig. 8.15

graph (being an induced subgraph of a regular lattice graph and, therefore, planar) is also required to be 2-connected. But, G. Baker [Bak2] found a PING consisting of two 2-connected, induced subgraphs of the triangular lattice graph (see Fig. 8.15) which, together with the discussion of Section 6.1, shows that "one cannot always hear the shape of a drum" (in the discrete model).

Remark (H. S.). The problem of approximately solving the partial differential equation (8.9) — in particular, the problem of the vibrating membrane — is in fact the starting point of the fundamental paper of L. Collatz and U. Sinogowitz [CoSi1] (1957), and the consideration of a vibrating finite membrane may indeed serve as a practically motivated starting point of the whole theory of graph spectra. The assumption that G is a part of a plane lattice may be dropped: A purely *combinatorial membrane* or drum is any graph G with the same elastic tension in each of its edges (all vertices of G are supposed to have "mass" 1, and all edges to have "length" 1):

G may either be supposed to be a subgraph of a graph H, where every vertex of H that belongs to G has the same valency r and where every vertex of H not belonging to G is kept at zero level; then, replacing in (8.9) the Δ-operator applied to vertex i of G by $\sum_{j \cdot i} (z_j - z_i) = \sum_{j \cdot i} z_j - r \cdot z_i$, we again arrive at equation (8.10), namely $\nu z_i = \sum_{j \cdot i} z_j$ $(i = 1, 2, \ldots, n)$ with $\nu = r - \lambda$ or, shorter, $\nu \boldsymbol{z} = \boldsymbol{A}\boldsymbol{z}$, leading to the compatibility condition $P_G(\nu) = |\nu \boldsymbol{I} - \boldsymbol{A}| = 0$;

or G per se may be considered as a finite space without any boundary; then from (8.9) the equation

$$\lambda z_i + \sum_{j \cdot i} (z_j - z_i) = \lambda z_i - d_i z_i + \sum_{j \cdot i} z_j = 0 \qquad (i = 1, 2, \ldots, n), \tag{8.15}$$

where d_i is the valency of vertex i, is obtained. Using valency matrix $\boldsymbol{D}$ and adjacency matrix $\boldsymbol{A}$, (8.15) may be given the shorter form

$$\lambda \boldsymbol{z} - \boldsymbol{D}\boldsymbol{z} + \boldsymbol{A}\boldsymbol{z} = \lambda \boldsymbol{z} - \boldsymbol{C}\boldsymbol{z} = \boldsymbol{o}, \tag{8.15'}$$

where $\boldsymbol{C} = \boldsymbol{D} - \boldsymbol{A}$, the matrix of admittance (see Section 1.2, p. 27), is the finite matrix-analogue of the Δ-operator. Thus the polynomial $C_G(\lambda) = |\lambda \boldsymbol{I} - \boldsymbol{C}|$ turns out to be a natural concept of graph theory and its applications, perhaps of greater significance than has been realized so far.

Eventually, a third version of a vibrating graph (viz., with mass m_i of vertex i being proportional to the valency d_i of i) leads to the equation

$$\lambda z_i = \frac{1}{d_i} \sum_{j \cdot i} z_j \qquad (i = 1, 2, \ldots, n), \tag{8.16}$$

equivalent with

$$\lambda \boldsymbol{D}\boldsymbol{z} = \boldsymbol{A}\boldsymbol{z} \tag{8.16'}$$

with corresponding compatibility condition

$$Q_G(\lambda) = \frac{1}{|\boldsymbol{D}|} |\lambda \boldsymbol{D} - \boldsymbol{A}| = 0.$$

We see that we are led to some concepts of a graph spectrum in a very natural way whenever the wave equation, or some version of it, is involved.

At the same time, the problem of the vibrating membrane motivates the divisor concept: consider a graph G and a divisor D of G as vibrating membranes. Since the elastic forces connected with a vertex will directly affect only the positions of the neighbours of this vertex (see equations (8.10), (8.15), (8.16); see also Section 1.2, equations (1.1), (1.3), (1.7), (1.15)), it is quite clear that if D can vibrate with a certain frequency then so can G; in other words: every eigenvalue of D (with multiplicity m) is at the same time an eigenvalue of G (with multiplicity $\geqq m$). This applies to each of the spectra considered except for the Seidel spectrum (in this case, each vertex is supposed to be in direct elastic contact with all of the other vertices, see Section 1.2, equation (1.12), p. 27), thus

$$D \mid G \quad \text{implies} \quad F_D(\lambda) \mid F_G(\lambda),$$

where F stands for any one of the letters C, P, Q, or R (see Section 1.2).

8.5. Miscellaneous results and problems

1. Let G be a tree with r vertices of degree 1 and s vertices adjacent to vertices of degree 1. Then

$$r - s \leq \eta(G) \leq r - 1.$$

(J. H. SMITH [Sm,J]; E. NOSAL [Nos1])

2. Let $\mathcal{Y}$ be a subset of the set of vertices of a graph G and let the vertices x_1 and x_2, not belonging to $\mathcal{Y}$, be adjacent to each vertex of $\mathcal{Y}$, and only to these (x_1, x_2 may be adjacent). Then the spectrum of G contains the number -1 if x_1 and x_2 are adjacent, and the number 0 if x_1 and x_2 are non-adjacent. The case when $\mathcal{Y}$ contains two non-adjacent vertices and x_1, x_2 are non-adjacent is of particular interest in chemistry.

(E. V. VAHOVSKIJ [Bax1]; I. GUTMAN [Gut1])

3. If $\alpha(G) > \frac{n}{2}$, where $\alpha(G)$ is the internal stability number and n is the number of vertices of the graph G, then $\eta(G) > 0$ [Bax1]. This fact follows immediately from Theorem 3.14. A particular case of this fact was noticed in [Rou2]: If $\chi(G) = 3$ ($\chi(G)$ = chromatic number of G) and if more vertices are coloured by one colour than by the two other colours taken together, then $\eta(G) > 0$.

4. Let T be a tree and $T - v$ the forest obtained by the removal of a vertex v from T. Then $\eta(T) = 0$ if and only if every forest $T - v$ has exactly one component with an odd number of vertices.

(I. GUTMAN, private communication)

5. Define a bipartite graph $G = G(m; n_1, \ldots, n_k)$ as follows: G contains a circuit C_{2m} whose vertices are labelled by $1'$, 1, $2'$, 2, ..., m', m; vertex $1'$ is adjacent to vertices $n_1, \ldots, n_k$ of C_{2m}. Let $\nu = 1 - (-1)^m + (-1)^m \sum_{i=1}^{k} (-1)^{n_i}$. Then

$$\eta(G(m; n_1, \ldots, n_k)) = \begin{cases} 0 & \text{if} \quad \nu \neq 0, \\ 2 & \text{if} \quad \nu = 0. \end{cases}$$

(D. M. CVETKOVIĆ, I. GUTMAN [CvG1])

6. Let G be a graph with n vertices whose adjacency matrix A has rank r ($r = n - \eta(G)$). Suppose that there is a sequence $\Delta = (D_n, D_{n-1}, \ldots, D_0)$ with $D_n = |A|$ and $D_0 = 1$, where D_{j-1} is a principal subdeterminant of D_j ($j = 2, 3, \ldots, n$), satisfying the conditions $D_r \neq 0$ and $D_{j-1}^2 + D_j^2 > 0$ for $j = 2, 3, \ldots, r - 1$. Then the number of negative eigenvalues of G is equal to the number of changes of sign in Δ.

(I. GUTMAN [Gut4], [Gut6])

7. Let A be the adjacency matrix of a bipartite graph G and let A^{-1} exist. Then

$$(A^{-1})_{ij} = \pm \frac{k_{ij}^{+} - k_{ij}^{-}}{k_{+} - k_{-}},$$

where k_+ and k_- are defined in Section 8.2 and k_{ij}^{+}, k_{ij}^{-} are the analogous quantities for the graph obtained from G by removing vertices i and j. The sign of $(A^{-1})_{ij}$ depends in a somewhat complicated way on the graph structure and on i, j (but is, of course, independent of the vertex numbering).

(D. M. CVETKOVIĆ, I. GUTMAN, N. TRINAJSTIĆ [CvGT4])

8. Let A be the adjacency matrix of a bipartite graph G having no circuit of length $4m$, and let A^{-1} exist. Then

$$(A^{-1})_{ij} = (-1)^{\frac{d(i,j) - 1}{2}} \cdot \frac{k_{ij}}{k},$$

where k and k_{ij} are the numbers of factors in G and in the subgraph of G obtained by removing the vertices i and j, respectively; $d(i, j)$ is the length of an arbitrary elementary path P connecting the vertives i, j and having the property that the graph $G - P$ (obtained by removing all vertices of P from G) contains at least one basic figure covering all vertices of $G - P$.

(E. HEILBRONNER [Hei4]; D. M. CVETKOVIĆ, I. GUTMAN, N. TRINAJSTIĆ [CvGT4])

9. The inequalitites (8.2) can be improved. Let $E_\pi = \sum_{i=1}^{n} |\lambda_i|$. For an arbitrary graph,

$$D \leqq 2nm - E_\pi^2 \leqq (n-1)\,D$$

and for a bipartite graph,

$$2D \leqq 2nm - E_\pi^2 \leqq (n-2)\,D,$$

where $D = 2m - n\sqrt[n]{|\boldsymbol{A}|^2}$, the other quantities being defined as earlier.

(I. GUTMAN [Gut2])

10. Let G be a subgraph of a hexagonal lattice graph which is 2-connected (every vertex having valency either 2 or 3) and such that the boundaries of all of its faces are circuits of length 6. Then the following relations hold:

$$2n - n_p = 4R + 2, \qquad m + n - n_p = 5R + 1, \qquad T = 2(R-1),$$

where n is the number of vertices, m is the number of edges, R is the number of faces, n_p is the number of peripheral vertices, and T is the number of vertices of degree 3. All these quantities can be calculated from the graph spectrum. Moreover,

$$\sum_{i=1}^{n} \lambda_i^4 = 6(7R - n + n_p - 1).$$

(I. GUTMAN [Gut5])

11. Let G be a bipartite graph without circuits of lengths $4s$ and with $\eta(G) = 1$. Let G have vertices $1, 2, \ldots, n$ and let $(C_1, C_2, \ldots, C_n)^{\mathrm{T}}$ be an eigenvector belonging to the eigenvalue zero. Let the graph $G - p$ have k_p factors ($p = 1, 2, \ldots, n$). Then $|C_p|\,k_q = |C_q|\,k_p$.

(M. S. J. DEWAR, H. C. LONGUET-HIGGINS [DeLo], see also D. M. CVETKOVIĆ, I. GUTMAN, N. TRINAJSTIĆ [CvGT4])

12. Let $\boldsymbol{A}$ be the adjacency matrix of a graph and let $\lambda_1, \ldots, \lambda_n$ be the corresponding eigenvalues. Let $\boldsymbol{U}$ be such that $\boldsymbol{UAU}^{-1} = (\lambda_i\delta_{ij})$, and put $\boldsymbol{A}' = \boldsymbol{U}^{-1}(|\lambda_i|\,\delta_{ij})\,\boldsymbol{U}$. Then

$$\boldsymbol{AP} = \boldsymbol{A} + \boldsymbol{A}', \qquad \boldsymbol{P}^2 = 2\boldsymbol{P}$$

if $\lambda_{\frac{n}{2}} \geqq 0 \geqq \lambda_{\frac{n}{2}+1}$ when n is even, or if $\lambda_{\frac{n+1}{2}} = 0$ when n is odd, the density matrix $\boldsymbol{P}$ being defined in Section 8.1, p. 238.

(K. RUEDENBERG [Rued], [McWS])

13. The polynomial $Q_G(\lambda) = \frac{1}{|\boldsymbol{D}|}\,|\lambda\boldsymbol{D} - \boldsymbol{A}|$ was used and investigated by O. BOTTEMA already in 1935 in a paper [Bott] concerning random walks in a connected graph G. Under the hypothesis that for every vertex j all edges incident with j have the same chance of being chosen by a perambulator having arrived at j, he proved the following proposition.

If G is not bipartite, then the probability of reaching a fixed vertex k in a walk of length l starting at a fixed vertex i tends, for $l \to \infty$, towards a limit that does not depend on i and is proportional to the valency of k.

If $G = (\mathscr{X}, \mathscr{Y}; \mathscr{U})$ is bipartite, and if only walks are taken into consideration that start at a vertex of $\mathscr{X}$ and have even (odd) length, then an analogous statement holds for all $k \in \mathscr{X}$ (or $k \in \mathscr{Y}$, respectively).

The results can be extended to multigraphs and to graphs with weighted adjacencies.

(O. BOTTEMA [Bott])

9. Some Additional Results

This chapter contains some additional material which did not fit into the classification of the other chapters. The last section gives a list of some unsolved problems.

9.1. Eigenvalues and imbeddings

The change in the least eigenvalue of a graph under imbedding is considered in some papers of A. J. HOFFMAN and others ([Hof6], [Hof7], [Hof10], [Hof14], [Hof17], [HoOs], [How1], [How2], [Doo10]). Although some important results have been obtained, the whole question is not yet clarified. We shall briefly present problems and results in connection with this and also some related topics.

If G is an induced subgraph of H, we shall write $G \subset H$. Let H_{R} denote a regular graph and H_{RC} a regular connected graph; $d(F)$, $\operatorname{diam}(F)$, and $\lambda(F)$ denote the smallest vertex degree, the diameter, and the least eigenvalue of a graph F, respectively.

A. J. HOFFMAN [Hof6], [Hof7] defined the following invariants for a graph G:

$$\lambda_{\mathrm{R}}(G) = \sup \{\lambda(H_{\mathrm{R}}) \mid G \subset H_{\mathrm{R}}\},$$

$$\mu(G) = \limsup_{d \to +\infty} \{\lambda(H) \mid G \subset H,\ d(H) > d\},$$

$$\mu_{\mathrm{R}}(G) = \limsup_{d \to +\infty} \{\lambda(H_{\mathrm{R}}) \mid G \subset H_{\mathrm{R}},\ d(H_{\mathrm{R}}) > d\}.$$

M. DOOB [Doo10] defined the quantity

$$\varkappa_{\mathrm{R}}(G) = \limsup_{d \to +\infty} \{\lambda(H_{\mathrm{RC}}) \mid G \subset H_{\mathrm{RC}},\ \operatorname{diam}(H_{\mathrm{RC}}) > d,\ d(H_{\mathrm{RC}}) > d\}.$$

(The finiteness of the limits is guaranteed by facts which will be described below.)

These quantities express several connections between spectral and structural properties of graphs. For example, if H is a regular graph for which $\lambda(H) > \lambda_{\mathrm{R}}(G)$, then H does not contain G as an induced subgraph. If $G \subset H$, it is clear that $\lambda(G) \geqq \lambda(H)$ (see Theorem 0.10). The reason for defining the mentioned graph invariants is to see how close can $\lambda(H)$ come to $\lambda(G)$ under several restrictions on H.

Unfortunately, not much is known about the quantities λ_{R}, μ, μ_{R}, $\varkappa_{\mathrm{R}}$.

Primarily, we have the following inequality due to R. M. KARP (see [Hof7]):

$$\mu_{\mathrm{R}}(G) \geqq \lambda(G) - 1.$$

M. DOOB [Doo10] has shown that

$$\varkappa_{\mathrm{R}}(G) \geqq \lambda(G) - 2.$$

Together with this inequality the following theorem was proved:

Let G be a (not necessarily connected) graph and let m and n be arbitrary positive integers. Then there exist graphs $H_1, \ldots, H_n$ such that

(i) *$H_1, \ldots, H_n$ are mutually non-isomorphic,*

(ii) *$H_1, \ldots, H_n$ are all regular and connected,*

(iii) *$H_1, \ldots, H_n$ all have the same spectrum,*

(iv) *G is an induced subgraph of H_i, $i = 1, \ldots, n$, and*

(v) *$d(H_1) = d(H_2) = \cdots = d(H_n) > m$ and*
$\operatorname{diam}(H_1) = \operatorname{diam}(H_2) = \cdots = \operatorname{diam}(H_n) > m$.

A. J. HOFFMAN [Hof7] has characterized all graphs G for which $\mu_{\mathrm{R}}(G) = -2$.

The problem of determining $\mu_{\mathrm{R}}(G)$ in general seems to be very difficult.

The problem of determining $\lambda_{\mathrm{R}}(G)$ is even harder. Trivially, $\lambda_{\mathrm{R}}(G) = \lambda(G)$ if G is regular. But† in the case of non-regular graphs $\lambda_{\mathrm{R}}(G)$ is known only for $G = K_{1,2}$ and $G = K_{1,3}$. We have $\lambda_{\mathrm{R}}(K_{1,2}) = 2\cos\frac{4\pi}{5}$ and $\lambda_{\mathrm{R}}(K_{1,3}) = -2$ (see [Hof7]). A sufficient condition for $\lambda_{\mathrm{R}}(G) < \lambda(G)$, which is based on Theorem 0.9, is also given in [Hof7]. A bit more is known about $\mu(G)$. A. J. HOFFMAN and A. M. OSTROWSKI [HoOs] have proved the following result:

*Let G be a graph with n vertices and with adjacency matrix **A**. Further, let Γ be the set of n-row (0, 1) matrices **C** each of which has the property that every row sum is positive, but loses that property if any column is deleted. If $\lambda(\boldsymbol{X})$ denotes the least eigenvalue of a square matrix **X**, we have* $\mu(G) = \max_{\boldsymbol{C}\in\Gamma} \lambda(\boldsymbol{A} - \boldsymbol{C}\boldsymbol{C}^{\mathrm{T}})$.

A sketch of the proof of this theorem can be found in [Hof6], [Hof7], [Hof9].

Several corollaries of this theorem were discussed in [Hof7]. In particular, $\mu(K_{1,n})$ is determined.

Further progress in connection with $\mu(G)$ has been made by A. J. HOFFMAN [Hof 10] who proved that $\mu(G) > -1 - \sqrt{2}$ if and only if $G = L(G_n; a_1, \ldots, a_n)$ for some G_n and for some $a_1, \ldots, a_n$. (For the definition of the generalized line graph $L(G_n; a_1, \ldots, a_n)$ see Section 6.3.) This result is a corollary of the following theorem [Hof18].

---

† See [29] of the supplementary bibliography where $\lambda_{\mathrm{R}}(G)$ has been determined for a large class of non-regular graphs and, in particular, for all line graphs.

Let $\lambda \in [-1 - \sqrt{2}, -1]$. There exists a function $d(\lambda)$ such that

1° *If $-2 < \lambda \leqq -1$, $\lambda(G) \geqq \lambda$, $d(G) \geqq d(\lambda)$, then $\lambda(G) = -1$, and G is the direct sum of some complete graphs. Conversely, if G is the direct sum of complete graphs and has at least one edge, then $\lambda(G) = -1$.*

2° *If $-1 - \sqrt{2} < \lambda \leqq -2$, $\lambda(H) \geqq \lambda$, $d(H) \geqq d(\lambda)$, then $\lambda(H) \geqq -2$, and $H = L(G_n; a_1, \ldots, a_n)$ for some graph G_n and for some non-negative integers $a_1, \ldots, a_n$. Conversely, if $H = L(G_n; a_1, \ldots, a_n)$, then $\lambda(H) \geqq -2$.*

The whole investigation of the graph invariants considered has been inspired by the fact that the knowledge of a lower bound for $\lambda(G)$ is very useful in proving several structural properties of G (see Section 6.3.). Therefore, A. J. Hoffman [Hof10] suggested the following two problems:

Problem 1. Let $\mathcal{G}$ be an infinite set of graphs. It may or may not be true that there exists a λ such that

$$\lambda(G) \geqq \lambda \quad \text{for all} \quad G \in \mathcal{G}. \tag{9.1}$$

Characterize those sets $\mathcal{G}$ for which (9.1) holds.

Problem 2'. Let λ be given, define $\mathcal{G}(\lambda)$ to be the set of all graphs G such that $\lambda(G) \geqq \lambda$, and characterize $\mathcal{G}(\lambda)$.

Problem 2. Let $\tilde{\mathcal{G}}(d)$ be the set of all graphs G such that $d(G) \geqq d$. Find a set of graphs $\mathcal{G}^*$ such that $\mathcal{G}^* \subset \mathcal{G}(\lambda)$, and for some $d(\lambda)$, $\mathcal{G}(\lambda) \cap \tilde{\mathcal{G}}(d(\lambda)) \subset \mathcal{G}^*$.

Problem 2' is too difficult and Problem 2 is a more modest version of it. The theorem mentioned above is related to Problem 2. A solution of Problem 1 is given by A. J. Hoffman [Hof10] in the form of the following theorem.

Let $\mathcal{G}$ be an infinite set of graphs. Then the following statements about $\mathcal{G}$ are either all true or all false:

(i) *there exists a number λ such that, for all $G \in \mathcal{G}$, $\lambda(G) \geqq \lambda$;*

(ii) *there exists a positive integer l such that, for all $G \in \mathcal{G}$, $K_{1,l} \not\subset G$ and $H_l \not\subset G$, where H_t denotes a graph on $2t + 1$ vertices, $2t$ of which form a K_{2t}, while the remaining vertex is adjacent to exactly t of these $2t$ vertices;*

(iii) *there exists a positive integer q such that, for each $G \in \mathcal{G}$, there exist graphs $\hat{G}$ and H with $G \cup \hat{G} = H$, every vertex degree of $\hat{G}$ is at most q and H contains a family of complete subgraphs $K^1, K^2, \ldots$ with the properties:*
   (a) *each edge of H is contained in at least one K^i,*
   (b) *each vertex of H is contained in at most q of the complete subgraphs $K^1, K^2, \ldots$,*
   (c) *$|\mathcal{V}(K^i) \cap \mathcal{V}(K^j)| \leqq q$ for $i \neq j$.*

The proof of this theorem is given in [Hof14].

This theorem has been used by L. Howes [How1] to characterize the sets $\mathcal{G}$ of graphs for which there exists a uniform upper bound on $\lambda_2(G)$ for all $G \in \mathcal{G}$, where $\lambda_1(G) \geqq \lambda_2(G) \geqq \cdots$ are the eigenvalues of G. A summary of these results can be found in [How2].

Let us define

$$\mu_2(G) = \liminf_{d \to +\infty} \{\lambda_2(H) \mid G \subset H, d(H) > d\}.^\dagger$$

A. J. Hoffman [Hof17] proved the following result.

Let G be a graph with n vertices and with adjacency matrix $\boldsymbol{A}$. Let Γ be the set of all (0, 1) *matrices $\boldsymbol{C}$ with n rows and at least two columns such that every row sum of $\boldsymbol{C}$ is positive, and if $\boldsymbol{C}$ has more than two columns, no column can be deleted without destroying the property that $\boldsymbol{C}$ has positive row sums. Then*

$$\mu_2(G) = \min_{\boldsymbol{C} \in \Gamma} \lambda_1\left(\boldsymbol{A} - \boldsymbol{C}(\boldsymbol{J} - \boldsymbol{I})^{-1} \boldsymbol{C}^{\mathrm{T}}\right).$$

9.2. The distance polynomial

So-called *distance polynomials* and corresponding spectra have been considered in [GrP1], [GrP2], [HoMC].

Let G be a connected graph whose vertices are numbered by $1, \ldots, n$. The distance d_{ij} between the vertices i and j in G is, as usual, defined to be the minimum number of edges in any path between i and j. The matrix $\boldsymbol{D}(G) = \boldsymbol{D} = (d_{ij})_1^n$ is called the *distance matrix of* G and the corresponding characteristic polynomial is the *distance polynomial of* G.

The general formula for the distance polynomial of a path as well as tables for distance polynomials of several graphs are given in [HoMC]. It is established in [GrP2] and [HoMC] that the determinant of the distance matrix of a tree depends only on the number of vertices, i.e., it is independent of the tree structure. For a tree T with n vertices we have

$$|\boldsymbol{D}(T)| = (-1)^{n-1} (n - 1) 2^{n-2}. \tag{9.2}$$

In [GrP1] and [GrP2] the distance polynomials are related to the following embedding problem.

Let $\mathscr{S} = \{0, 1, *\}$ and let d be a function from $\mathscr{S} \times \mathscr{S}$ to the set of non-negative integers defined by

$$d(s, s') = \begin{cases} 1 & \text{if } \{s, s'\} = \{0, 1\}, \\ 0 & \text{otherwise}. \end{cases}$$

This function can be extended to a mapping of $\mathscr{S}^n \times \mathscr{S}^n$ to the non-negative integers in the following way:

$$d\left((s_1, \ldots, s_n), (s'_1, \ldots, s'_n)\right) = \sum_{k=1}^{n} d(s_k, s'_k).$$

† $\lambda_i(\boldsymbol{X})$ and $\lambda_i(H)$ denote the i-th greatest eigenvalue of a matrix $\boldsymbol{X}$ and of a graph H, respectively.

Given a connected graph G, find the least integer $N(G)$ for which there is a function A from the vertex set of G to $\mathcal{S}^{N(G)}$ such that $d_{ij} = d(A(i), A(j))$ for all pairs of vertices i and j in G.

The following result has been proved.

Let n_+, n_- be the number of positive and negative eigenvalues of the distance matrix $\boldsymbol{D}$ for a graph G, respectively. Then $N(G) \geqq \max(n_+, n_-)$.

In addition, the number $N(G)$ has been determined for complete graphs, circuits and trees.

Recently R. L. Graham, A. J. Hoffman, and H. Hosoya [GrHH] proved that the determinant $|\boldsymbol{D}(G)|$ of the distance matrix of any strongly connected digraph G depends only on the blocks† of G and not on how the blocks are interconnected. They gave the formulas

$$\operatorname{cof} \boldsymbol{D}(G) = \prod_{i=1}^{r} \operatorname{cof} \boldsymbol{D}(G_i),$$

$$\boldsymbol{D}(G) = \sum_{i=1}^{r} \boldsymbol{D}(G_i) \prod_{j \neq i} \operatorname{cof} \boldsymbol{D}(G_j), \tag{9.3}$$

where $G_1, \ldots, G_r$ are the blocks of G and (for any square matrix $\boldsymbol{X}$) $\operatorname{cof} \boldsymbol{X}$ denotes the sum of cofactors of $\boldsymbol{X}$. (9.3) is a generalization of (9.2). This can easily be seen if we replace each edge in a tree by two directed edges of opposite orientations and then apply (9.3) to the obtained digraph.

Other generalizations of (9.2) have been given in [EdGG], [GrLo], where the coefficients α_i of the distance polynomial of a tree have been studied (remember that $\alpha_n = (-1)^n |\boldsymbol{D}(G)|$). It is proved in [EdGG] that $\alpha_{n-1} = 2^{n-3}(2nN_{K_2} - 2N_{K_{1,2}} - 4)$, where N_H is the number of induced subgraphs isomorphic to H contained in the considered tree. Corresponding expressions for the coefficients α_{n-2} and α_{n-3} have also been given, and a conjecture concerning the general shape of the coefficients of the distance polynomial has been given. This conjecture, which states that α_i can be expressed as a linear combination of numbers N_H for various H, has been confirmed in [GrLo].

The question whether the distance polynomial determines the graph uniquely has also been discussed in [EdGG] and [GrLo].

If G is metrically regular (as defined in 6.4) with $\boldsymbol{A}_0 = \boldsymbol{I}, \boldsymbol{A}_1, \ldots, \boldsymbol{A}_m$ as adjacency matrices of the induced association scheme, then we obviously have $\boldsymbol{D}(G) = \sum_{k=0}^{m} k\boldsymbol{A}_k$. Hence if two such graphs give rise to association schemes with the same parameters, then by Theorem 6.23 they must have cospectral distance matrices. On the other hand, no pair of trees with the same distance polynomial is known, and there is some hope that for trees, at least, the spectrum of the distance matrix determines the graph.††

---

† Two arcs of a digraph are in the same *block* if and only if there is a cycle in the digraph containing both of them.

†† Recently this conjecture has been disproved by B. McKay [McKa].

9.3. The algebraic connectivity of a graph

In Section 1.2 the C-spectrum $\boldsymbol{Sp}_C(G)$ of a graph G was defined as the spectrum of the matrix $\boldsymbol{C}(G) = \boldsymbol{C} = \boldsymbol{D} - \boldsymbol{A}$. The coefficients of the corresponding characteristic polynomial $C_G(\lambda)$ were interpreted in Section 1.5 in terms of the tree structure of G. Now we shall pay some attention to $\boldsymbol{Sp}_C(G) = [\lambda_1, \ldots, \lambda_n]_C$. The matrix $\boldsymbol{C}$ can be represented in the form $\boldsymbol{C} = \boldsymbol{V}\boldsymbol{V}^\top$, where $\boldsymbol{V}$ is the vertex-edge (0, 1, −1)-incidence matrix of a digraph obtained from G by giving to each edge of G an (arbitrary) orientation. Hence, $\boldsymbol{C}$ is symmetric, singular (since all row sums are 0) and positive semidefinite. In other words, the eigenvalues are non-negative, the least one always being 0. The second smallest C-eigenvalue λ_{n-1} will be denoted by $a(G)$. This quantity shares many properties with the vertex- or edge-connectivity† and, according to M. Fiedler [Fie 1], is called the *algebraic connectivity* of a graph. Let us describe some properties of $a(G)$.

The matrix $(n - 1)\,\boldsymbol{I} - \boldsymbol{C}$ is non-negative with constant row sums. Hence, the multiplicity of the largest eigenvalue is equal to the number of components of G (see Theorem 3.23). That means that $a(G) = 0$ if and only if G is disconnected.

If G_1, G_2 are edge-disjoint graphs on the same set of vertices, then $a(G_1 \cup G_2) \geqq a(G_1) + a(G_2)$. In other words, introducing new edges in a graph does not diminish the algebraic connectivity. To prove this property consider the set $\mathscr{W}$ of n-dimensional vectors in which the sum of the coordinates is 0. Then we have

$$\begin{aligned} a(G_1 \cup G_2) &= \min_{\boldsymbol{x}\in\mathscr{W}} \left(\boldsymbol{x}^\top \boldsymbol{C}(G_1)\,\boldsymbol{x} + \boldsymbol{x}^\top \boldsymbol{C}(G_2)\,\boldsymbol{x}\right) \\ &\geqq \min_{\boldsymbol{x}\in\mathscr{W}} \boldsymbol{x}^\top \boldsymbol{C}(G_1)\,\boldsymbol{x} + \min_{\boldsymbol{x}\in\mathscr{W}} \boldsymbol{x}^\top \boldsymbol{C}(G_2)\,\boldsymbol{x} = a(G_1) + a(G_2). \end{aligned}$$

It can easily be seen that $a(G') + k$ is an eigenvalue of $\boldsymbol{C}(G' \nabla G_k)$, where G_k is any graph on k vertices (see (2.18)). If G' is an induced subgraph of a graph G on $n + k$ vertices, then G can be realized as a spanning subgraph of $G' \nabla G_k$ for some G_k. Therefore we have $a(G) \leqq a(G' \nabla G_k) \leqq a(G') + k$. Suppose now that G has vertex connectivity k and G_k is a cut set†† for G. Then G' is disconnected, $a(G') = 0$, and $a(G) \leqq k$.

M. Fiedler [Fie 1] proved the inequality

$$a(G) \geqq 2e\left(1 - \cos\frac{\pi}{n}\right),$$

where e is the edge connectivity of G, as well as several other related inequalities. M. Fiedler [Fie 3] describes how one can construct a cut set in a graph G by means of the eigenvector of $\boldsymbol{C}$ corresponding to $a(G)$. In [Fie 2] the concept of algebraic connectivity is extended to graphs with positively weighted edges.

---

† A graph has the *vertex-(edge-)connectivity* k if k is the least number of vertices (edges) which may be removed so that the remaining graph is disconnected or consists of an isolated vertex.

†† The set of vertices (edges) whose removal makes a connected graph disconnected is called a *cut set*.

Some results of M. Fiedler have been obtained also by W. N. Anderson Jr. and T. D. Morley [AnMo].

The assumption that $a(G)$ is a very useful parameter for describing the "shape" of a graph has been confirmed by a computer search of cubic graphs [BuČCS] (see the description of Table 3 in the Appendix).

9.4. Integral graphs

A graph is called *integral* if its spectrum consists entirely of integers. F. Harary and A. J. Schwenk [HaS2], [Har3] posed the problem of characterizing integral graphs. The problem seems to be very hard and only some partial results have been obtained up to now. Some graph operations, which when applied on integral graphs produce again integral graphs, are described in [HaS2]. It is proved in [Cve16] that the set of regular, connected, integral graphs of a fixed degree is finite. F. C. Bussemaker and D. M. Cvetković [Cve16], [BuCv] found all cubic, connected, integral graphs. There are exactly 13 such graphs: K_4, $K_{3,3}$, the Petersen graph, the cube graph, C_6, Tutte's 8-cage, the graph on 10 vertices obtained from $K_{3,3}$ by specifying a pair of non-adjacent vertices and replacing each of them by a triangle, the line graph of the subdivision graph of K_4, Desargues' graph and a graph cospectral to it, a bipartite graph on 12 vertices, the 6-sided prism and a bipartite graph on 24 vertices of girth 6. The same result has independently been rediscovered by A. J. Schwenk [Schw6] whose proofs are in a more condensed form. In addition, there are exactly 7 connected integral graphs whose vertex degrees do not exceed 3 and are not all equal to 3 [CvGT3].

9.5. Some problems

At the end of this chapter we shall mention some open problems concerning graph spectra.

1. *Which graphs have distinct eigenvalues* [HaS2]?

2. *Can we say anything about spectral properties of* a) *trees,* b) *planar graphs,* c) *tournaments* [Wi,RJ1]?

3. Let r be the index and let $\bar{d}$ be the average of the vertex degrees of a graph. The quantity $\delta = r - \bar{d}$ can be considered as a measure for the deviation from regularity. *Find the best possible upper bound for δ as a function of the number of vertices n. For $n \leq 5$ the best bound is* $\sqrt{n-1} - 2\left(1 - \frac{1}{n}\right)$ *and is attained by* $K_{n-1,1}$ [CoSi1].

4. *Find the best possible lower bound for eigenvalues of graphs with n vertices.*

5. *Find the relation between the theory of Markov chains and the theory of graph spectra.*

6. *Which (multi-)digraphs have real eigenvalues?*

7. *Find a graph G for which the equation $P_G(\lambda) = 0$ cannot be solved by radicals* (I. GUTMAN)

8. *Which parts of the theory of graph spectra can be extended to hypergraphs?* (Concerning the number of spanning trees see [RuSa]; for a general approach see also [Rung].)

9. Let $\mathscr{X} = \{x_1, \ldots, x_n\}$ be the vertex set of a graph G. G_i denotes the subgraph of G induced by the set $\mathscr{X} \setminus \{x_i\}$. *Is it true that for $n > 2$ the characteristic polynomial $P_G(\lambda)$ of G is uniquely determined by the collection of the characteristic polynomials $P_{G_i}(\lambda)$ $(i = 1, \ldots, n)$* [GuCv]? (The well-known conjecture of ULAM states that, for $n > 2$, G is reconstructable from the collection of graphs G_i $(i = 1, \ldots, n)$. The present problem is solved in [GuCv] in the affirmative for regular graphs and some other classes of graphs. It is proved in [Tut 1] that the polynomial $P_G(\lambda)$ is reconstructable from the collection of graphs G_i. That means, in fact, that possible counter examples to ULAM's conjecture should be looked for in cospectral graphs. For some related problems see [Pouz], [Tut 2].) Among other things, it is proved that $P_{\bar{G}}(\lambda)$ is also reconstructable from the collection of G_i's. Having in mind no. 15 from Section 1.9 (p. 50), it follows that Seidel's spectrum is also reconstructable from the same collection. A natural conjecture is then the following one: *If two graphs have the same collection of vertex-deleted subgraphs, then they are switching equivalent.*)

Appendix

Tables of Graph Spectra

In Section 2.6. analytic expressions for characteristic polynomials and spectra of some classes of graphs are given. The present tables contain numerical data related to characteristic polynomials, spectra, and similar topics for some specified graphs. There are eight tables we shall now list and comment upon.

Table 1. *Characteristic polynomials and spectra of connected graphs with n vertices, $2 \leqq n \leqq 5$.*

This table is taken from [CoSi 1], its accuracy was checked by computer. Characteristic polynomials $P_G(\lambda)$ are factored into irreducible factors over the field of rational numbers. For a fixed n the graphs are ordered by decreasing indices.

Table 2. *Characteristic polynomials and spectra of all trees with n vertices, $2 \leqq n \leqq 10$.*

In [MiKH] and [Mow 5] the characteristic polynomials $P_G(\lambda)$ of trees with up to 10 vertices were published. In order to check the accuracy we have compared both these tables and no difference was observed. We reproduce here the table from [Mow 5] wherein the trees are ordered so that their polynomials appear in lexicographic order; for $n = 2, 3, \ldots, 9$ the eigenvalues are taken from [Nos 1] (unpublished), for $n = 10$ they were calculated by D. Maksimović who also checked and corrected in a few positions the list given in [Nos 1]. Characteristic polynomials for trees with $2, 3, \ldots, 8$ vertices are also given in [CoSi 1].

There is no PING among trees on less than 8 vertices, there is exactly one PING on 8 vertices, and there are exactly five PINGs on 9 and four on 10 vertices.

Table 3. *Characteristic polynomials and spectra of connected cubic graphs with n vertices for $n = 4, 6, 8, 10, 12$.*

This table contains the coefficients of the characteristic polynomials $P_G(\lambda)$ and spectra for all connected cubic graphs G with up to 12 vertices. Integer (i.e., rational) eigenvalues are given exactly, non-rational eigenvalues are given to an accuracy of two decimal places.

19 connected cubic graphs with 10 vertices are displayed, for example, in [BuS 1] and [Imri]. The 85 connected cubic graphs with 12 vertices given in this table were found by S. Čobeljić who also calculated the corresponding characteristic polynomials and spectra by means of a computer. Recently [ПеПе] was published providing 86 cubic connected graphs with 12 vertices! Comparing his results with those of [ПеПе], S. Čobeljić established that among the 86 graphs given in [ПеПе] there are two isomorphic graphs, namely, graphs no. 35 and no. 41. (There are also some technical mistakes in graphs no. 24 and no. 26.) Finally, J. J. Seidel and F. C. Bussemaker ([BuČCS] and private communication) generated by means of a computer all cubic graphs with up to 14 vertices; they confirmed number 85. There are exactly 540 cubic graphs with 14 vertices, 509 of which being connected.

Several other authors considered the problem of enumerating cubic graphs, but we shall not go into further detail in this book.

Table 3 contains the Petersen graph under no. 3.26.

In a recent paper by F. C. BUSSEMAKER, S. ČOBELJIĆ, D. M. CVETKOVIĆ and J. J. SEIDEL [BuČCS] all of the 621 connected cubic graphs G with not more than 14 vertices — together with their characteristic polynomial, spectrum, numbers of circuits of length 3, 4, ..., 14, diameter, connectivity, order of the automorphism group, and a statement corcerning planarity — are displayed. The sequence of eigenvalues is given in non-increasing order, and for a fixed number of vertices the graphs are ordered lexicographically with respect to their sequences of eigenvalues. There are no PINGs among all cubic graphs with less than 14 vertices, and there are exactly three PINGs among all connected cubic graphs with 14 vertices. The authors observed that there is a noteworthy relation between the second largest eigenvalue and the connectivity of the graph and, moreover, that the lexicographical order reveals a strong correspondence between the spectrum and the "shape" (described in terms of diameter, girth, connectivity, etc.) of a (regular) graph which, however, still lacks precise formulation. This paper thus provides a lot of material for further investigations in the spectral theory of (regular) graphs.

Table 4. *Characteristic polynomials and spectra of miscellaneous (multi-)graphs.*

This table was compiled partly on the basis of data taken from [CoSi 1]; part of the material was computed by L. L. KRAUS and S. K. SIMIĆ.

All 49 (multi-)graphs of this table have certain symmetry properties, for 36 of them the automorphism group has only one or two orbits. Graphs 4.42 and 4.46 are self-complementary.

The *graphs of the five Platonic solids* (regular polyhedra) are contained in Table 3 (3.1: tetrahedron; 3.7: cube) and Table 4 (4.20: dodecahedron; 4.37: octahedron; 4.49: icosahedron).

The *nine forbidden subgraphs* in the well-known Beineke [Bein] characterization of line graphs are contained in Table 1 (1.8, 1.11, 1.17) and Table 4 (4.23, 4.24, 4.25, 4.33, 4.35, 4.36).

Table 5. *Characteristic polynomials and non-zero eigenvalues of graphs $K_{n_1,\ldots,n_k}$ $(n_1 + \cdots + n_k = n)$ for $2 \leqq k \leqq n \leqq 10$.*

The table was computed by S. K. SIMIĆ and is published in [Cve 11]. The characteristic polynomial of $K_{n_1,\ldots,n_k}$ can be written in the form (see Section 2.6.)

$$P_{K_{n_1,\ldots,n_k}}(\lambda) = \lambda^{n-k}\left(\lambda^k - \sum_{i=2}^{k} b_i\lambda^{k-i}\right).$$

In the table before every group of graphs, for which k and n are constant, the values of k and n are indicated. A graph $K_{n_1,\ldots,n_k}$ is determined by the partition of n into $n_1, \ldots, n_k$ (for example, $3^2 1^3$ denotes 3, 3, 1, 1, 1). After this, the coefficients b_i $(i = 2, \ldots, k)$ and finally the approximate values of the roots of $\lambda^k - \sum_{i=2}^{k} b_i\lambda^{k-i} = 0$ in non-increasing order are given. The table includes, naturally, complete graphs K_n with $n = 2, \ldots, 10$ vertices.

The four following short tables include some graphs of interest in chemistry and they are taken from [Hei 1].

Table 6.1. *Characteristic polynomials of graphs consisting of a circuit C_5 and a path P_s, $1 \leqq s \leqq 7$.*

Table 6.2. *Characteristic polynomials of graphs consisting of a circuit C_6 and a path P_s, $1 \leqq s \leqq 7$.*

Table 6.3. *Characteristic polynomials of graphs consisting of a circuit C_7 and a path P_s, $1 \leqq s \leqq 7$.*

Table 6.4. *Characteristic polynomials of graphs consisting of two circuits C_s and C_r with an edge in common, $5 \leqq s \leqq r \leqq 7$.*

Table 7. *Simple eigenvalues of graphs whose automorphism group has two orbits of degrees* r_1, r_2, $1 \leqq r_2 \leqq r_1 \leqq 5$.

As was proved in Section 5.1, all simple eigenvalues of a graph whose automorphism group has t orbits, with $r_1, r_2, \ldots, r_t$ being the vertex degrees of the different orbits, are elements of a finite set $\mathscr{S}_{r_1,r_2,\ldots,r_t}$. In this table the sets $\mathscr{S}_{r_1,r_2}$ ($r_1, r_2 = 1, 2, \ldots, 5$) are given.

Naturally, $\mathscr{S}_{r_1,r_2} = \mathscr{S}_{r_2,r_1}$ and, according to Section 5.1, $\lambda \in \mathscr{S}_{r_1,r_2}$ implies $-\lambda \in \mathscr{S}_{r_1,r_2}$. Therefore, only the non-negative numbers contained in $\mathscr{S}_{r_1,r_2}$ (first column), in increasing order, are given (for every pair of r_1, r_2). For irrational numbers $\lambda \in \mathscr{S}_{r_1,r_2}$ the second and the third columns provide coefficients p and q (respectively) of a second order equation $\lambda^2 + p\lambda + q = 0$ which is satisfied by λ. The table was published in [KrC2].

The following two tables are in connection with the operation of Seidel switching described in Section 6.5 and are taken from [LiSe]:

Table 8.1. *Equivalence classes under Seidel switching of graphs with n vertices*, $2 \leqq n \leqq 6$.

Table 8.2. *Equivalence classes under Seidel switching of graphs with seven vertices.*

Each equivalence class is given by a representative graph. In the first table classes are arranged according to the partial order of inclusion.

A class represented by a graph G is said to include each of the classes that contain a subgraph of G. The number of inclusions is indicated for each class. Further, for each class approximate values of the eigenvalues are given. Two classes are complementary if they contain complementary representative graphs. The self-complementary classes are indicated. For graphs with 7 vertices there are 54 classes, complementary in pairs. Half of the classes are tabulated in the second table.

There are several other tables in the literature including numerical data about graph spectra and similar topics.

Primarily, we mention a complete catalogue of the characteristic polynomials of graphs with seven vertices [King] including disconnected graphs, too.

Further, we refer to two extensive books of tables [CoSt], [StBr]. These books contain numerical values for spectra, eigenvectors, and some other quantities (which are of interest in chemistry in connection with total π-electron energies and bond orders, etc.) of a great number of graphs related to the most important chemical compounds. Hetero-molecules (i.e., graphs whose adjacency matrix has some non-zero elements on the main diagonal) are also dealt with.

Papers [HoDKP], [HoDKT], [HoDT], [HoKZ], [TiH1], [TiH2], [TiH3], [ZaM1], [ZaM2], [ZaM3], [ZaMK1], [ZaMK2], [ZaPá] contain tables of a similar character. Some graphs, treated in them, are included in [CoSt] or [StBr].

Apart from Tables 6.1.—6.4. of this Appendix, some other tables can be found in [Hei1]. They also contain characteristic polynomials for some classes of graphs which are of interest in chemistry.

In [MiKH] and [KaMH] tables of characteristic polynomials and so-called *topological indices*† of graphs were published. [MiKH] was mentioned in connection with Table 2. [KaMH] is related to the connected mono- and bicyclic graphs up to eight vertices in which the vertex degrees are not greater than 4. Among the considered graphs there are 13 PINGs.

---

† The *topological index of a graph G* ([Hos1]) is defined as the number of regular subgraphs of degree 1 contained in G. In trees the topological index is equal to the sum of the absolute values of the coefficients of the characteristic polynomial. The topological index of a graph which corresponds in the natural way (i.e., not as in the Hückel theory) to a molecule of a saturated hydrocarbon is related to the boiling point and to other thermodynamic properties of the considered compound. Generally speaking, the boiling point increases with the increase of the topological index. This fact was empirically established [Hos1].

[CoSi 1] contains a table of characteristic polynomials of miscellaneous graphs. The majority of them is included in our Table 4.

[Col 1] contains characteristic polynomials and spectra of all connected subgraphs, having up to six vertices, of the graph $P_m + P_n$ (see Section 2.6.). All such graphs are included in our tables.

In [Wal 1] the eigenvalues of the matrix $\boldsymbol{A} - \boldsymbol{D} + (n - 1)\,\boldsymbol{I}$ ($\boldsymbol{A}$ adjacency matrix, $\boldsymbol{D}$ valency matrix) have been calculated for all graphs with up to five vertices.

A list of all connected graphs with up to five vertices together with their characteristic and minimal polynomials, with respect to the Galois field $GF(2)$, is given in [Mow 6].

Finally, distance polynomials (see Section 9.2) for some graphs are tabulated in [HoMC], [EdGG].

Numerical data on spectra or characteristic polynomials of some particular graphs can be found in several mathematical, chemical, or physical papers, too. For data regarding graphs contained in some PINGs see Section 6.1.

Table 1. Characteristic polynomials and spectra of connected graphs with n vertices, $2 \leqq n \leqq 5$

n	graph	characteristic polynomial $P_G(\lambda)$	spectrum λ_1	λ_2	λ_3	λ_4	λ_5
2	1.1	$\lambda^2 - 1 = (\lambda - 1)(\lambda + 1)$	1	-1			
3	1.2	$\lambda^3 - 3\lambda - 2 = (\lambda - 2)(\lambda + 1)^2$	2	-1	-1		
	1.3	$\lambda^3 - 2\lambda = (\lambda^2 - 2)\lambda$	1.4142	0	-1.4142		
4	1.4	$\lambda^4 - 6\lambda^2 - 8\lambda - 3 = (\lambda - 3)(\lambda + 1)^3$	3	-1	-1	-1	
	1.5	$\lambda^4 - 5\lambda^2 - 4\lambda = (\lambda^2 - \lambda - 4)(\lambda + 1)\lambda$	2.5616	0	-1	-1.5616	
	1.6	$\lambda^4 - 4\lambda^2 - 2\lambda + 1 = (\lambda^3 - \lambda^2 - 3\lambda + 1)(\lambda + 1)$	2.17011	0.3111	-1	-1.4812	
	1.7	$\lambda^4 - 4\lambda^2 = (\lambda - 2)\lambda^2(\lambda + 2)$	2	0	0	-2	
	1.8	$\lambda^4 - 3\lambda^2 = (\lambda^2 - 3)\lambda^2$	1.7321	0	0	-1.7321	
	1.9	$\lambda^4 - 3\lambda^2 + 1 = (\lambda^2 - \lambda - 1)(\lambda^2 + \lambda - 1)$	1.6180	0.6180	-0.6180	-1.6180	
5	1.10	$\lambda^5 - 10\lambda^3 - 20\lambda^2 - 15\lambda - 4 = (\lambda - 4)(\lambda + 1)^4$	4	-1	-1	-1	-1
	1.11	$\lambda^5 - 9\lambda^3 - 14\lambda^2 - 6\lambda = (\lambda^2 - 2\lambda - 6)\lambda(\lambda + 1)^2$	3.6458	0	-1	-1	-1.6458
	1.12	$\lambda^5 - 8\lambda^3 - 10\lambda^2 - \lambda + 2 = (\lambda^3 - 2\lambda^2 - 5\lambda + 2)(\lambda + 1)^2$	3.3234	0.3579	-1	-1	-1.6813
	1.13	$\lambda^5 - 8\lambda^3 - 8\lambda^2 = (\lambda^2 - 2\lambda - 4)\lambda^2(\lambda + 2)$	3.2361	0	0	-1.2361	-2
	1.14	$\lambda^5 - 7\lambda^3 - 8\lambda^2 + 2 = (\lambda^3 - 2\lambda^2 - 4\lambda + 2)(\lambda + 1)^2$	3.0861	0.4280	-1	-1	-1.5141
	1.15	$\lambda^5 - 7\lambda^3 - 6\lambda^2 = (\lambda - 3)\lambda^2(\lambda + 1)(\lambda + 2)$	3	0	0	-1	-2
	1.16	$\lambda^5 - 7\lambda^3 - 6\lambda^2 + 3\lambda + 2 = (\lambda^3 - \lambda^2 - 5\lambda - 2)(\lambda^2 + \lambda - 1)$	2.9354	0.6180	-0.4626	-1.4728	-1.6180
	1.17	$\lambda^5 - 7\lambda^3 - 4\lambda^2 + 2\lambda = (\lambda^3 - \lambda^2 - 6\lambda + 2)\lambda(\lambda + 1)$	2.8558	0.3216	0	-1	-2.1774
	1.18	$\lambda^5 - 6\lambda^3 - 4\lambda^2 + 2\lambda = (\lambda^4 - 6\lambda^2 - 4\lambda + 2)\lambda$	2.6855	0.3349	0	-1.2713	-1.7491
	1.19	$\lambda^5 - 6\lambda^3 - 4\lambda^2 + 3\lambda + 2 = (\lambda^4 - \lambda^3 - 5\lambda^2 + \lambda + 2)(\lambda + 1)$	2.6412	0.7237	-0.5892	-1	-1.7757

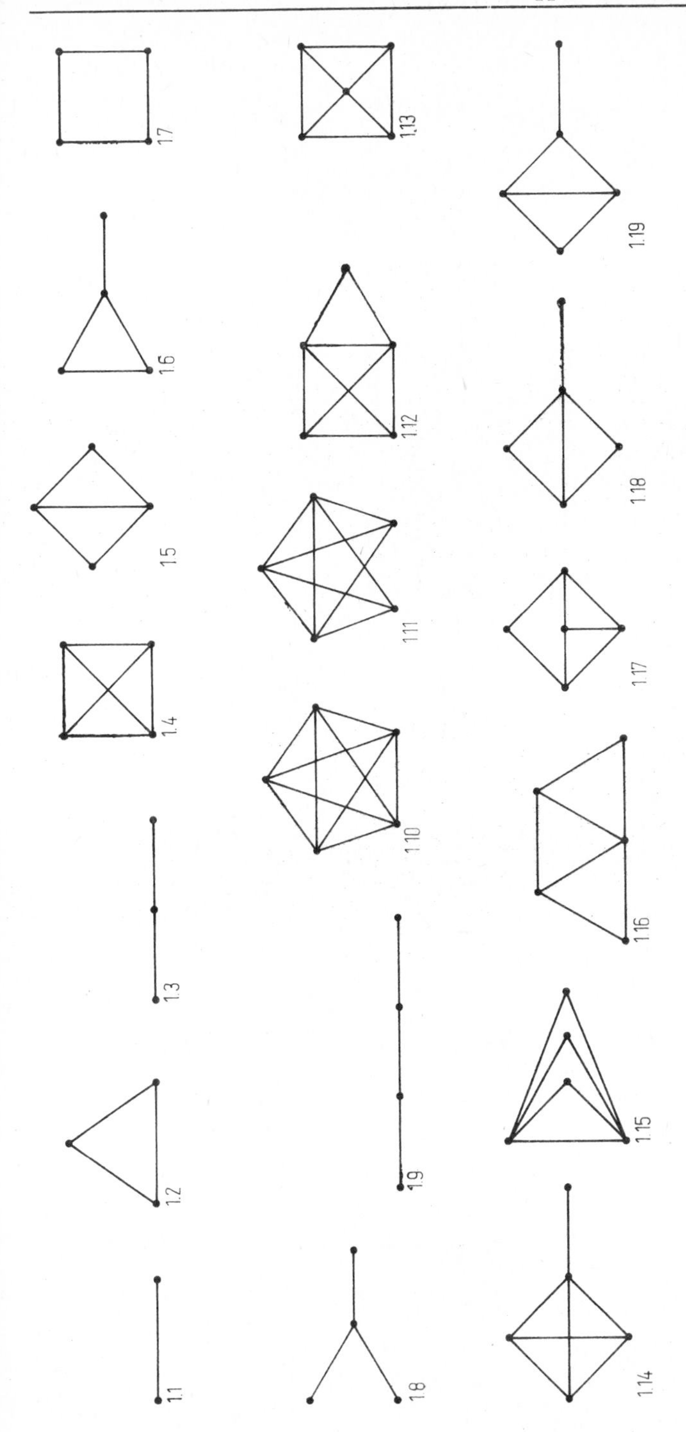
1.1
1.2
1.3
1.4
1.5
1.6
1.7
1.8
1.9
1.10
1.11
1.12
1.13
1.14
1.15
1.16
1.17
1.18
1.19

Table 1 (continued)

n	graph	characteristic polynomial $P_G(\lambda)$		spectrum λ_1	λ_2	λ_3	λ_4	λ_5
5	1.20	$\lambda^5 - 6\lambda^3 - 4\lambda^2 + 5\lambda + 4$	$= (\lambda^2 - \lambda - 4)(\lambda - 1)(\lambda + 1)^2$	2.5616	1	−1	−1	−1.5616
	1.21	$\lambda^5 - 6\lambda^3 - 2\lambda^2 + 4\lambda$	$= (\lambda^3 - 2\lambda^2 - 2\lambda + 2)\lambda(\lambda + 2)$	2.4812	0.6889	0	−1.1701	−2
	1.22	$\lambda^5 - 6\lambda^3$	$= (\lambda^2 - 6)\lambda^3$	2.4495	0	0	0	−2.4495
	1.23	$\lambda^5 - 5\lambda^3 - 2\lambda^2 + 2\lambda$	$= (\lambda^3 - \lambda^2 - 4\lambda + 2)\lambda(\lambda + 1)$	2.3429	0.4707	0	−1	−1.8136
	1.24	$\lambda^5 - 5\lambda^3 - 2\lambda^2 + 3\lambda$	$= (\lambda^2 - \lambda - 3)(\lambda^2 + \lambda - 1)\lambda$	2.3028	0.6180	0	−1.3028	−1.6180
	1.25	$\lambda^5 - 5\lambda^3 - 2\lambda^2 + 4\lambda + 2$	$= (\lambda^3 - 4\lambda - 2)(\lambda - 1)(\lambda + 1)$	2.2143	1	−0.5392	−1	−1.6751
	1.26	$\lambda^5 - 5\lambda^3 + 2\lambda$	$= (\lambda^4 - 5\lambda^2 + 2)\lambda$	2.1358	0.6622	0	−0.6622	−2.1358
	1.27	$\lambda^5 - 5\lambda^3 + 5\lambda - 2$	$= (\lambda - 2)(\lambda^2 + \lambda - 1)^2$	2	0.6180	0.6180	−1.6180	−1.6180
	1.28	$\lambda^5 - 4\lambda^3$	$= (\lambda - 2)\lambda^3(\lambda + 2)$	2	0	0	0	−2
	1.29	$\lambda^5 - 4\lambda^3 + 2\lambda$	$= (\lambda^4 - 4\lambda^2 + 2)\lambda$	1.8478	0.7654	0	−0.7654	−1.8478
	1.30	$\lambda^5 - 4\lambda^3 + 3\lambda$	$= (\lambda^2 - 3)(\lambda - 1)\lambda(\lambda + 1)$	1.7321	1	0	−1	−1.7321

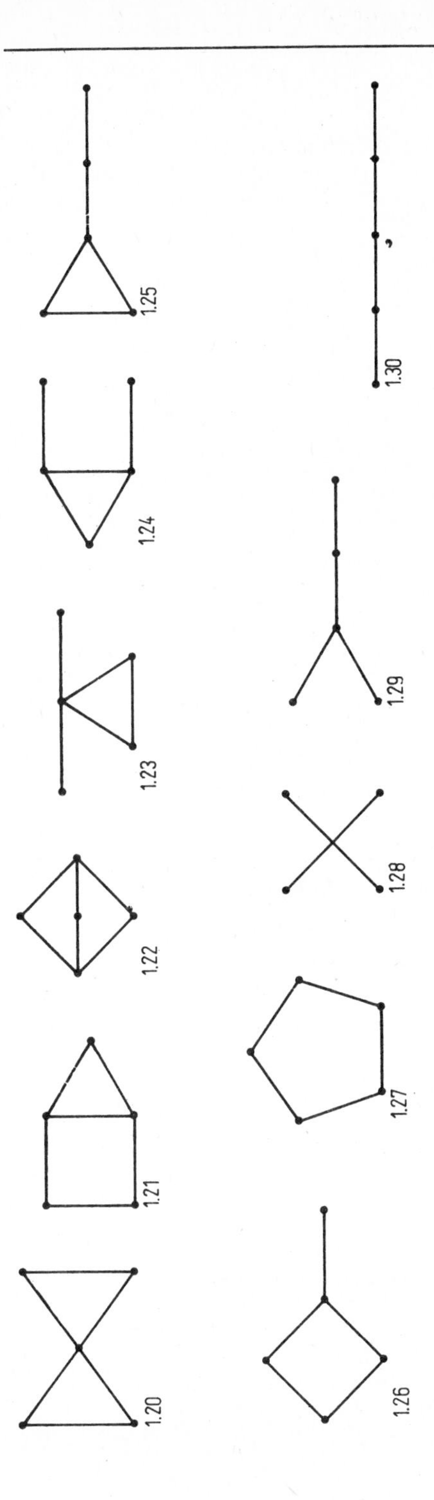
1.20
1.21
1.22
1.23
1.24
1.25
1.26
1.27
1.28
1.29
1.30

Table 2. Characteristic polynomials and spectra of all trees with n vertices, $2 \leqq n \leqq 10$

n	tree	coefficients				spectrum*			
		a_2	a_4	a_6	a_8	λ_1	λ_2	λ_3	λ_4
2	2.1	−1				1			
3	2.2	−2				1.414			
4	2.3	−3	0			1.732	0		
	2.4	−3	1			1.618	0.618		
5	2.5	−4	0			2	0		
	2.6	−4	2			1.848	0.765		
	2.7	−4	3			1.732	1		
6	2.8	−5	0	0		2.236	0	0	
	2.9	−5	3	0		2.074	0.835	0	
	2.10	−5	4	0		2	1	0	
	2.11	−5	5	0		1.902	1.176	0	
	2.12	−5	5	−1		1.932	1	0.518	
	2.13	−5	6	−1		1.802	1.247	0.445	
7	2.14	−6	0	0		2.449	0	0	
	2.15	−6	4	0		2.288	0.874	0	
	2.16	−6	6	0		2.175	1.126	0	
	2.17	−6	7	0		2.101	1.259	0	
	2.18	−6	7	−2		2.136	1	0.662	
	2.19	−6	8	0		2	1.414	0	
	2.20	−6	8	−2		2.053	1.209	0.570	
	2.21	−6	9	−2		1.932	1.414	0.518	
	2.22	−6	9	−3		1.970	1.286	0.684	
	2.23	−6	9	−4		2	1	1	
	2.24	−6	10	−4		1.848	1.414	0.765	
8	2.25	−7	0	0	0	2.646	0	0	0
	2.26	−7	5	0	0	2.488	0.899	0	0
	2.27	−7	8	0	0	2.358	1.199	0	0

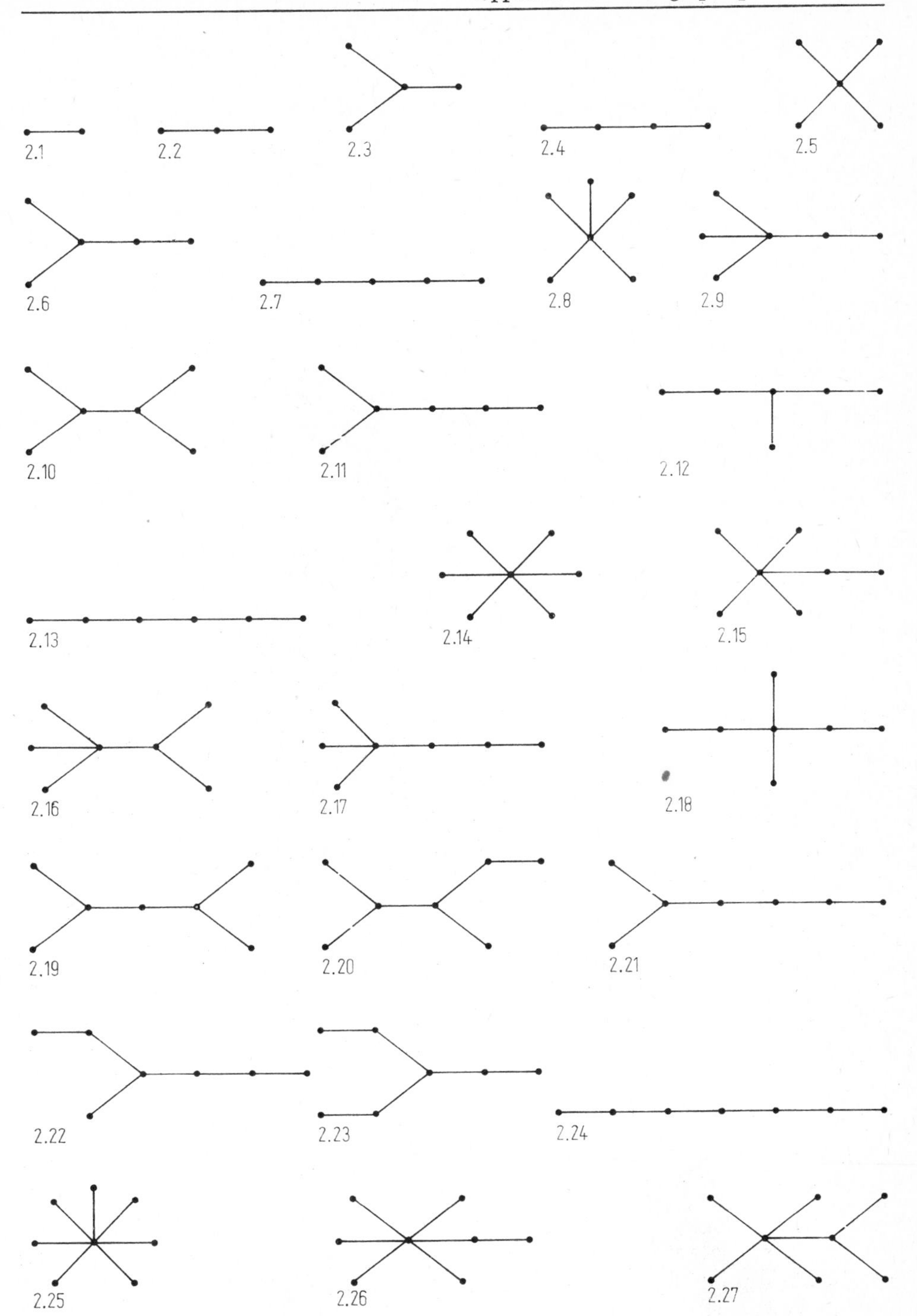
2.1
2.2
2.3
2.4
2.5
2.6
2.7
2.8
2.9
2.10
2.11
2.12
2.13
2.14
2.15
2.16
2.17
2.18
2.19
2.20
2.21
2.22
2.23
2.24
2.25
2.26
2.27

Table 2 (continued)

n	tree	coefficients				spectrum*			
		a_2	a_4	a_6	a_8	λ_1	λ_2	λ_3	λ_4
8	2.28	−7	9	0	0	2.303	1.303	0	0
	2.29	−7	9	0	0	2.303	1.303	0	0
	2.30	−7	9	−3	0	2.334	1	0.742	0
	2.31	−7	11	0	0	2.149	1.543	0	0
	2.32	−7	11	−3	0	2.206	1.338	0.587	0
	2.33	−7	11	−4	0	2.222	1.240	0.726	0
	2.34	−7	12	−3	0	2.112	1.496	0.548	0
	2.35	−7	12	−4	0	2.136	1.414	0.662	0
	2.36	−7	12	−5	0	2.157	1.314	0.789	0
	2.37	−7	12	−7	1	2.189	1	1	0.457
	2.38	−7	13	−4	0	2	1.618	0.618	0
	2.39	−7	13	−5	0	2.042	1.520	0.720	0
	2.40	−7	13	−6	0	2.074	1.414	0.835	0
	2.41	−7	13	−7	0	2.101	1.259	1	0
	2.42	−7	13	−7	1	2.095	1.356	0.738	0.477
	2.43	−7	14	−7	0	1.950	1.564	0.868	0
	2.44	−7	14	−8	0	2	1.414	1	0
	2.45	−7	14	−8	1	1.989	1.486	0.813	0,416
	2.46	−7	14	−9	1	2.029	1.321	1	0.373
	2.47	−7	15	−10	1	1.879	1.532	1	0,347
9	2.48	−8	0	0	0	2.828	0	0	0
	2.49	−8	6	0	0	2.676	0.915	0	0
	2.50	−8	10	0	0	2.540	1.245	0	0
	2.51	−8	11	0	0	2.497	1.328	0	0
	2.52	−8	11	−4	0	2.524	1	0.792	0
	2.53	−8	12	0	0	2.449	1.414	0	0
	2.54	−8	14	0	0	2.327	1.608	0	0
	2.55	−8	14	−4	0	2.376	1.414	0.595	0

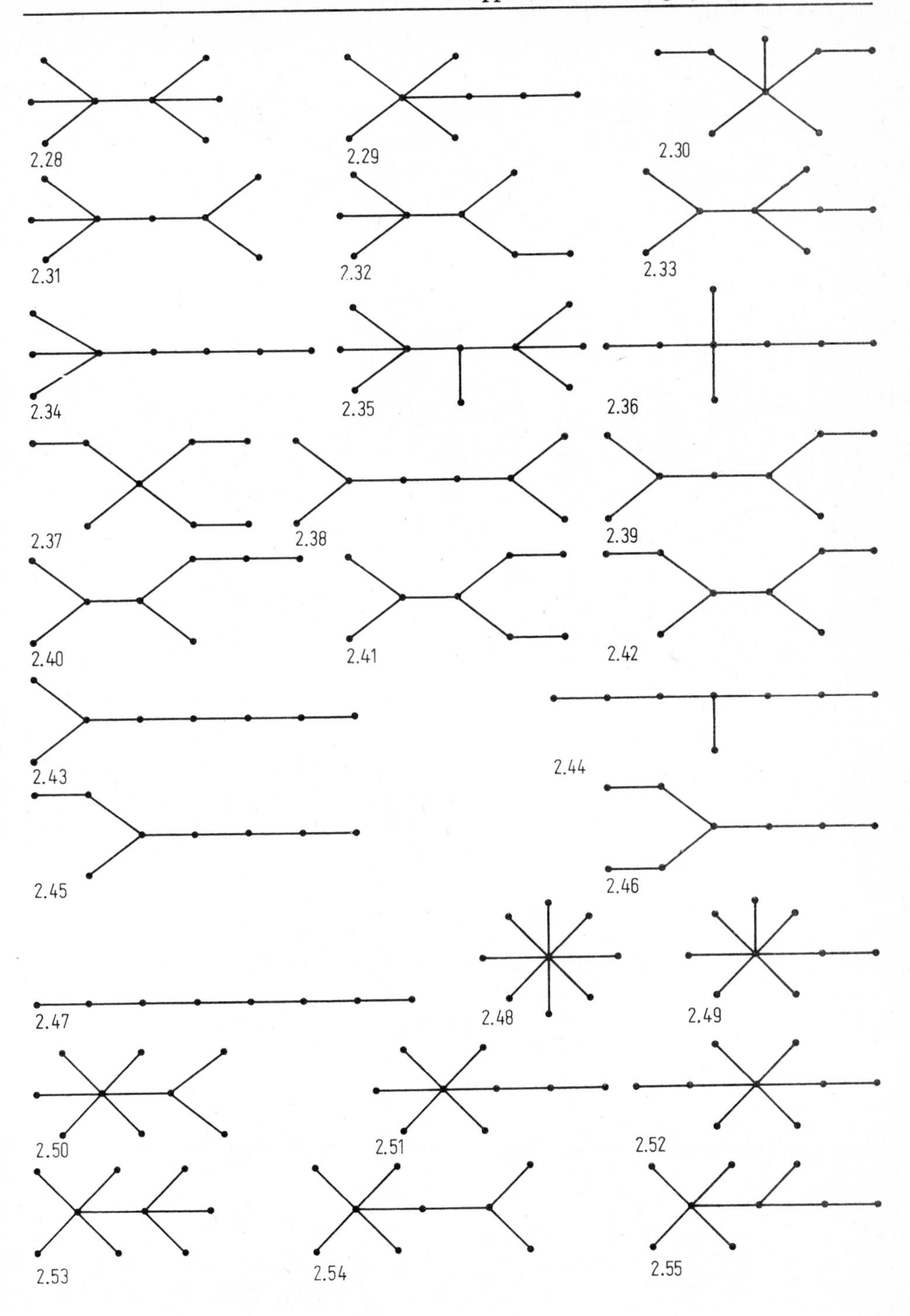
2.28
2.29
2.30
2.31
2.32
2.33
2.34
2.35
2.36
2.37
2.38
2.39
2.40
2.41
2.42
2.43
2.44
2.45
2.46
2.47
2.48
2.49
2.50
2.51
2.52
2.53
2.54
2.55

Table 2 (continued)

n	tree	coefficients				spectrum*			
		a_2	a_4	a_6	a_8	λ_1	λ_2	λ_3	λ_4
9	2.56	−8	14	−6	0	2.397	1.267	0.807	0
	2.57	−8	15	0	0	2.236	1.732	0	0
	2.58	−8	15	−4	0	2.307	1.536	0.565	0
	2.59	−8	15	−6	0	2.334	1.414	0.742	0
	2.60	−8	15	−7	0	2.347	1.333	0.845	0
	2.61	−8	15	−10	2	2.376	1	1	0.595
	2.62	−8	16	−6	0	2.255	1.558	0.697	0
	2.63	−8	16	−8	0	2.288	1.414	0.874	0
	2.64	−8	17	−6	0	2.136	1.732	0.662	0
	2.65	−8	17	−7	0	2.168	1.662	0.735	0
	2.66	−8	17	−8	0	2.194	1.590	0.811	0
	2.67	−8	17	−9	0	2.216	1.512	0.895	0
	2.68	−8	17	−10	0	2.236	1.414	1	0
	2.69	−8	17	−10	0	2.236	1.414	1	0
	2.70	−8	17	−11	2	2.247	1.414	0.802	0.555
	2.71	−8	17	−12	2	2.264	1.279	1	0.488
	2.72	−8	18	−10	0	2.117	1.640	0.911	0
	2.73	−8	18	−10	0	2.117	1.640	0.911	0
	2.74	−8	18	−12	0	2.175	1.414	1.126	0
	2.75	−8	18	−12	0	2.175	1.414	1.126	0
	2.76	−8	18	−12	2	2.165	1.528	0.854	0.501
	2.77	−8	18	−12	2	2.165	1.528	0.854	0.501
	2.78	−8	18	−14	3	2.206	1.338	1	0.587
	2.79	−8	18	−16	5	2.236	1	1	1
	2.80	−8	19	−12	0	2	1.732	1	0
	2.81	−8	19	−13	0	2.061	1.598	1.095	0
	2.82	−8	19	−13	2	2.036	1.691	0.884	0.465
	2.83	−8	19	−14	2	2.084	1.572	1	0.432

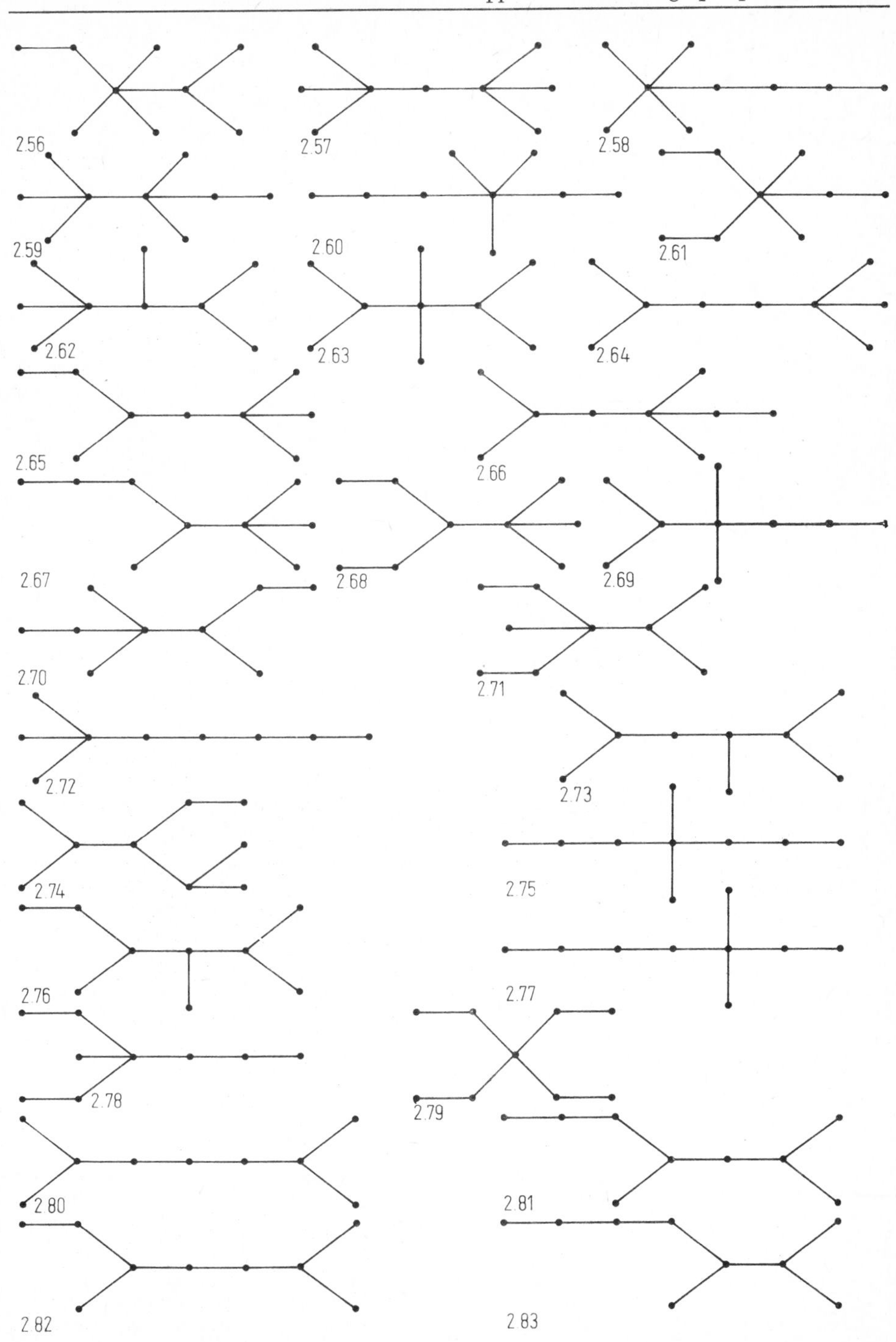
2.56
2.57
2.58
2.59
2.60
2.61
2.62
2.63
2.64
2.65
2.66
2.67
2.68
2.69
2.70
2.71
2.72
2.73
2.74
2.75
2.76
2.77
2.78
2.79
2.80
2.81
2.82
2.83

Table 2 (continued)

n	tree	coefficients					spectrum*				
		a_2	a_4	a_6	a_8	a_{10}	λ_1	λ_2	λ_3	λ_4	λ_5
9	2.84	−8	19	−14	2		2.084	1.572	1	0.432	
	2.85	−8	19	−14	3		2.074	1.618	0.835	0.618	
	2.86	−8	19	−15	2		2.119	1.414	1.159	0.407	
	2.87	−8	19	−15	3		2.112	1.496	1	0.548	
	2.88	−8	19	−16	4		2.136	1.414	1	0.662	
	2.89	−8	20	−16	2		1.962	1.663	1.111	0.390	
	2.90	−8	20	−17	3		2.015	1.548	1.143	0.486	
	2.91	−8	20	−17	4		2	1.618	1	0.618	
	2.92	−8	20	−18	4		2.053	1.414	1.209	0.570	
	2.93	−8	20	−18	5		2.042	1.520	1	0.720	
	2.94	−8	21	−20	5		1.902	1.618	1.176	0.618	
10	2.95	−9	0	0	0	0	3	0	0	0	0
	2.96	−9	7	0	0	0	2.853	0.927	0	0	0
	2.97	−9	12	0	0	0	2.715	1.276	0	0	0
	2.98	−9	13	0	0	0	2.682	1.344	0	0	0
	2.99	−9	13	−5	0	0	2.705	1	0.827	0	0
	2.100	−9	15	0	0	0	2.606	1.486	0	0	0
	2.101	−9	16	0	0	0	2.562	1.562	0	0	0
	2.102	−9	17	0	0	0	2.511	1.642	0	0	0
	2.103	−9	17	−5	0	0	2.550	1.461	0.600	0	0
	2.104	−9	17	−8	0	0	2.571	1.288	0.854	0	0
	2.105	−9	18	−5	0	0	2.499	1.557	0.575	0	0
	2.106	−9	18	−9	0	0	2.532	1.347	0.879	0	0
	2.107	−9	18	−13	3	0	2.558	1	1	0.677	0
	2.108	−9	19	0	0	0	2.370	1.839	0	0	0
	2.109	−9	19	−8	0	0	2.470	1.529	0.749	0	0
	2.110	−9	19	−9	0	0	2.479	1.477	0.819	0	0
	2.111	−9	20	−8	0	0	2.404	1.646	0.715	0	0

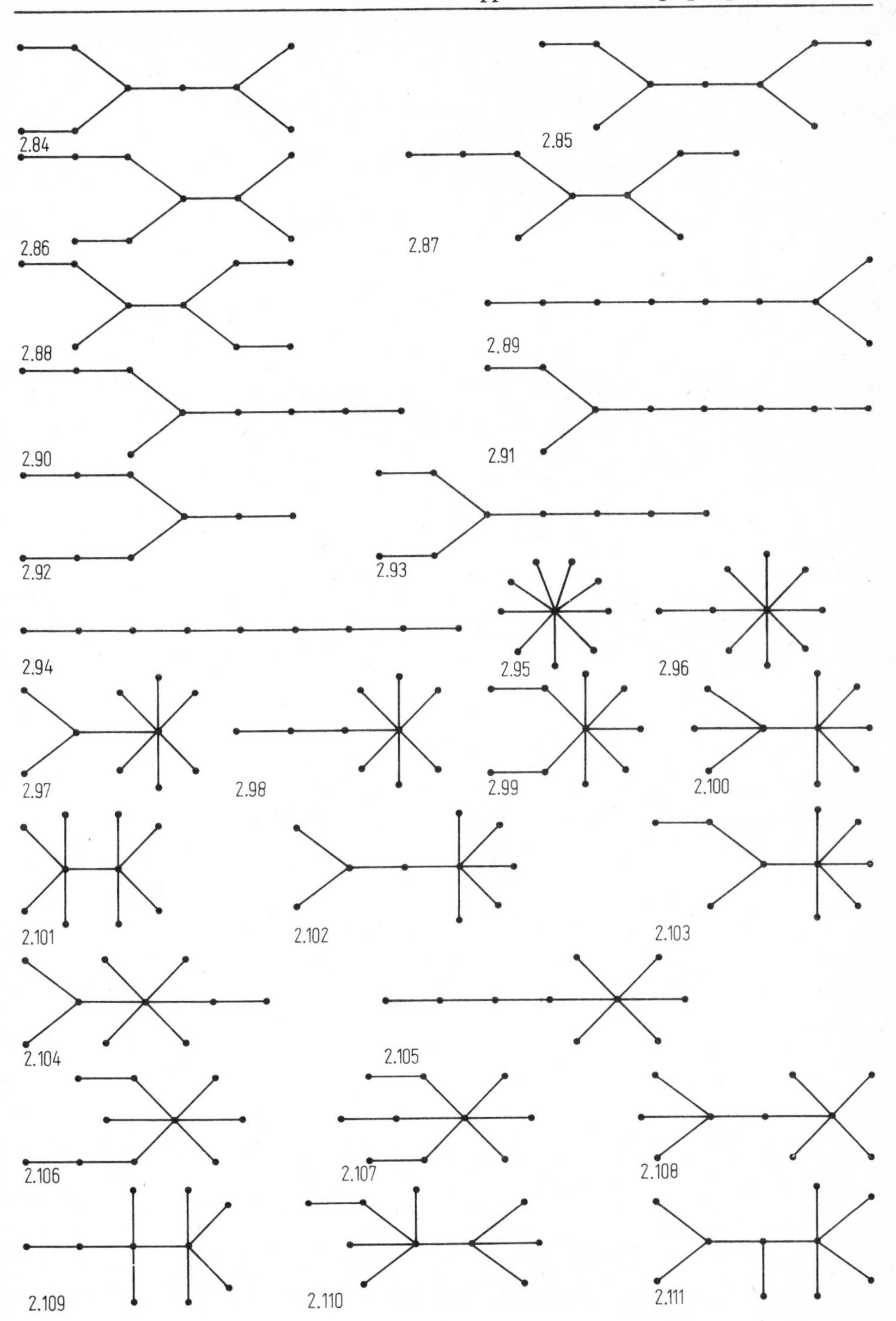
2.84
2.85
2.86
2.87
2.88
2.89
2.90
2.91
2.92
2.93
2.94
2.95
2.96
2.97
2.98
2.99
2.100
2.101
2.102
2.103
2.104
2.105
2.106
2.107
2.108
2.109
2.110
2.111

Table 2 (continued)

n	tree	coefficients					spectrum*				
		a_2	a_4	a_6	a_8	a_{10}	λ_1	λ_2	λ_3	λ_4	λ_5
10	2.112	−9	20	−12	0	0	2.450	1.414	1	0	0
	2.113	−9	21	−8	0	0	2.315	1.780	0.686	0	0
	2.114	−9	21	−9	0	0	2.334	1.732	0.742	0	0
	2.115	−9	21	−9	0	0	2.334	1.732	0.742	0	0
	2.116	−9	21	−11	0	0	2.367	1.631	0.859	0	0
	2.117	−9	21	−12	0	0	2.381	1.574	0.925	0	0
	2.118	−9	21	−12	0	0	2.381	1.574	0.925	0	0
	2.119	−9	21	−13	0	0	2.394	1.506	1	0	0
	2.120	−9	21	−14	0	0	2.407	1.414	1.099	0	0
	2.121	−9	21	−15	3	0	2.412	1.457	0.849	0.580	0
	2.122	−9	21	−17	4	0	2.433	1.296	1	0.634	0
	2.123	−9	22	−9	0	0	2.199	1.912	0.714	0	0
	2.124	−9	22	−11	0	0	2.265	1.789	0.818	0	0
	2.125	−9	22	−13	0	0	2.309	1.673	0.934	0	0
	2.126	−9	22	−15	0	0	2.343	1.531	1.080	0	0
	2.127	−9	22	−16	0	0	2.358	1.414	1.199	0	0
	2.128	−9	22	−16	3	0	2.350	1.552	0.883	0.538	0
	2.129	−9	22	−17	3	0	2.365	1.469	1	0.498	0
	2.130	−9	22	−17	4	0	2.362	1.508	0.826	0.680	0
	2.131	−9	22	−19	5	0	2.387	1.350	1	0.694	0
	2.132	−9	22	−22	9	−1	2.414	1	1	1	0.414
	2.133	−9	23	−14	0	0	2.205	1.804	0.941	0	0
	2.134	−9	23	−15	0	0	2.236	1.732	1	0	0
	2.135	−9	23	−16	0	0	2.262	1.657	1.067	0	0
	2.136	−9	23	−17	0	0	2.283	1.569	1.151	0	0
	2.137	−9	23	−17	3	0	2.271	1.664	0.904	0.507	0
	2.138	−9	23	−18	4	0	2.288	1.618	0.874	0.618	0
	2.139	−9	23	−19	4	0	2.307	1.536	1	0.565	0

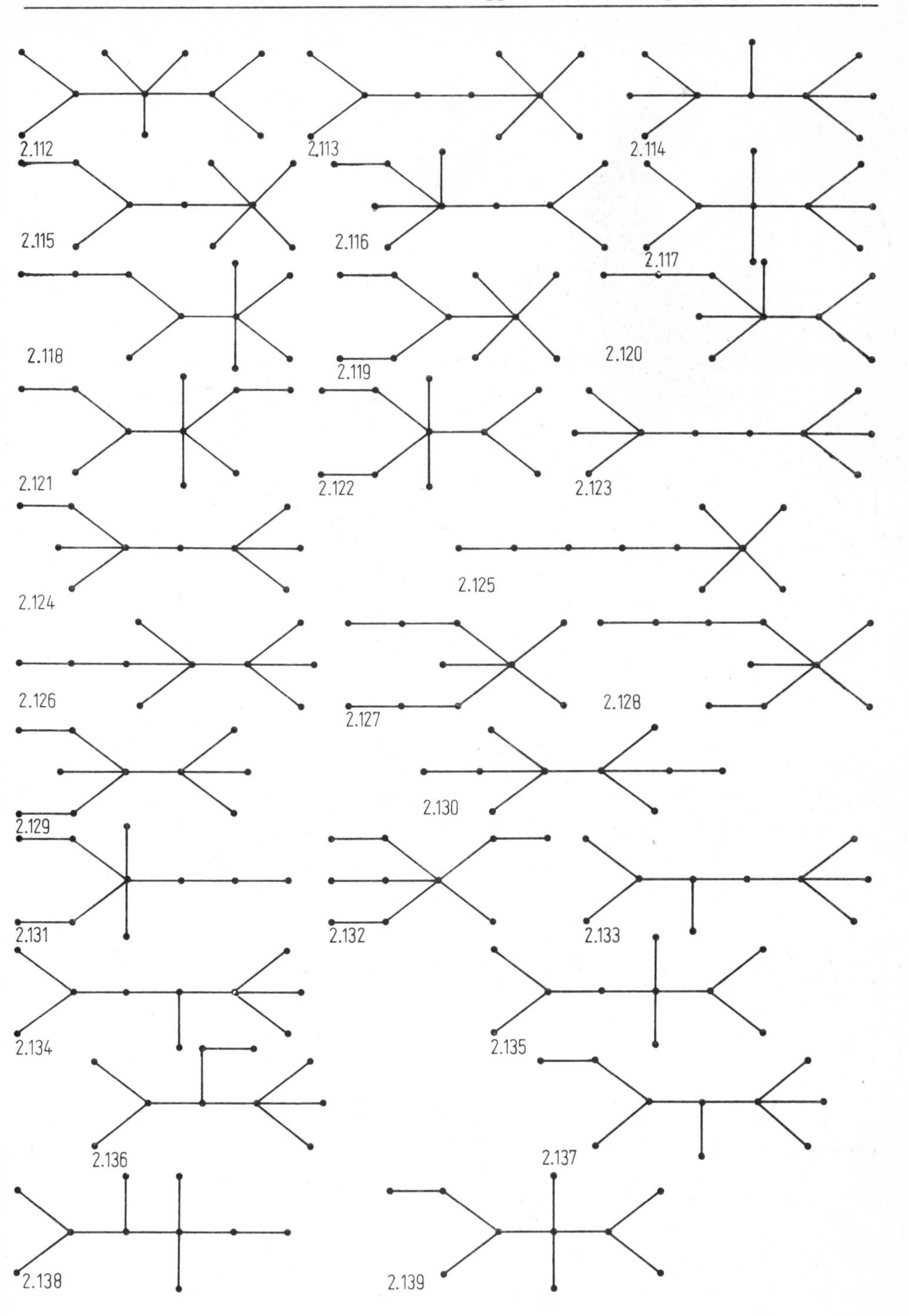
2.112
2.113
2.114
2.115
2.116
2.117
2.118
2.119
2.120
2.121
2.122
2.123
2.124
2.125
2.126
2.127
2.128
2.129
2.130
2.131
2.132
2.133
2.134
2.135
2.136
2.137
2.138
2.139

Table 2 (continued)

n	tree	coefficients					spectrum*				
		a_2	a_4	a_6	a_8	a_{10}	λ_1	λ_2	λ_3	λ_4	λ_5
10	2.140	−9	23	−20	4	0	2.324	1.414	1.147	0.530	0
	2.141	−9	24	−17	0	0	2.129	1.830	1.059	0	0
	2.142	−9	24	−18	0	0	2.175	1.732	1.126	0	0
	2.143	−9	24	−18	3	0	2.147	1.820	0.919	0.482	0
	2.144	−9	24	−19	0	0	2.209	1.629	1.212	0	0
	2.145	−9	24	−19	3	0	2.189	1.732	1	0.457	0
	2.146	−9	24	−19	4	0	2.181	1.760	0.899	0.579	0
	2.147	−9	24	−20	0	0	2.236	1.414	1.414	0	0
	2.148	−9	24	−20	3	0	2.220	1.642	1.089	0.436	0
	2.149	−9	24	−20	4	0	2.214	1.675	1	0.539	0
	2.150	−9	24	−20	5	0	2.208	1.705	0.861	0.690	0
	2.151	−9	24	−21	3	0	2.246	1.526	1.205	0.419	0
	2.152	−9	24	−21	4	0	2.241	1.578	1.107	0.511	0
	2.153	−9	24	−21	5	0	2.236	1.618	1	0.618	0
	2.154	−9	24	−22	5	0	2.260	1.507	1.135	0.579	0
	2.155	−9	24	−22	6	0	2.255	1.558	1	0.697	0
	2.156	−9	24	−23	6	0	2.276	1.414	1.186	0.642	0
	2.157	−9	24	−23	7	0	2.272	1.492	1	0.780	0
	2.158	−9	24	−24	9	−1	2.285	1.453	1	0.688	0.438
	2.159	−9	24	−25	9	0	2.303	1.303	1	1	0
	2.160	−9	25	−21	0	0	2.101	1.732	1.259	0	0
	2.161	−9	25	−22	3	0	2.119	1.732	1.159	0.407	0

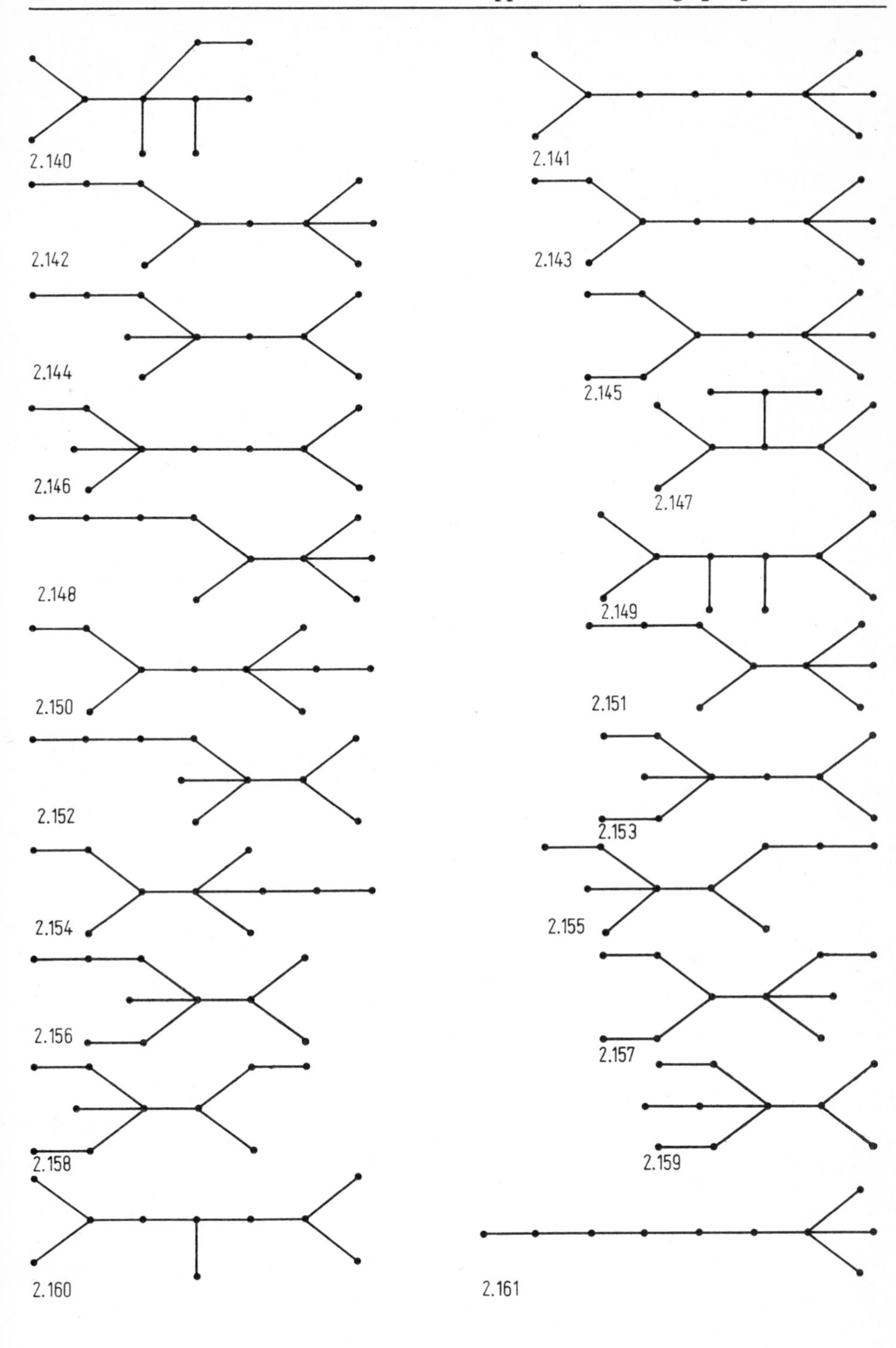

2.140
2.141
2.142
2.143
2.144
2.145
2.146
2.147
2.148
2.149
2.150
2.151
2.152
2.153
2.154
2.155
2.156
2.157
2.158
2.159
2.160
2.161

Table 2 (continued)

n	tree	coefficients					spectrum*				
		a_2	a_4	a_6	a_8	a_{10}	λ_1	λ_2	λ_3	λ_4	λ_5
10	2.162	−9	25	−22	4	0	2.107	1.772	1.086	0.493	0
	2.163	−9	25	−23	4	0	2.154	1.650	1.188	0.474	0
	2.164	−9	25	−23	5	0	2.145	1.694	1.103	0.558	0
	2.165	−9	25	−23	6	0	2.136	1.732	1	0.662	0
	2.166	−9	25	−24	4	0	2.189	1.414	1.414	0.457	0
	2.167	−9	25	−24	5	0	2.182	1.561	1.232	0.533	0
	2.168	−9	25	−24	6	0	2.175	1.618	1.126	0.618	0
	2.169	−9	25	−24	7	0	2.168	1.662	1	0.735	0
	2.170	−9	25	−25	7	0	2.200	1.534	1.163	0.674	0
	2.171	−9	25	−25	8	0	2.194	1.590	1	0.818	0
	2.172	−9	25	−25	9	−1	2.189	1.618	1	0.618	0.457
	2.173	−9	25	−26	8	0	2.222	1.414	1.240	0.726	0
	2.174	−9	25	−26	10	−1	2.212	1.546	1	0.751	0.389
	2.175	−9	25	−28	12	−1	2.250	1.352	1	1	0.329
	2.176	−9	26	−25	4	0	2	1.802	1.247	0.445	0
	2.177	−9	26	−26	5	0	2.070	1.669	1.297	0.499	0
	2.178	−9	26	−26	6	0	2.053	1.732	1.209	0.570	0
	2.179	−9	26	−26	7	0	2.031	1.788	1.119	0.651	0
	2.180	−9	26	−27	7	0	2.101	1.618	1.259	0.618	0
	2.181	−9	26	−27	8	0	2.089	1.681	1.149	0.701	0
	2.182	−9	26	−27	8	0	2.089	1.681	1.149	0.701	0
	2.183	−9	26	−27	9	0	2.074	1.732	1	0.835	0

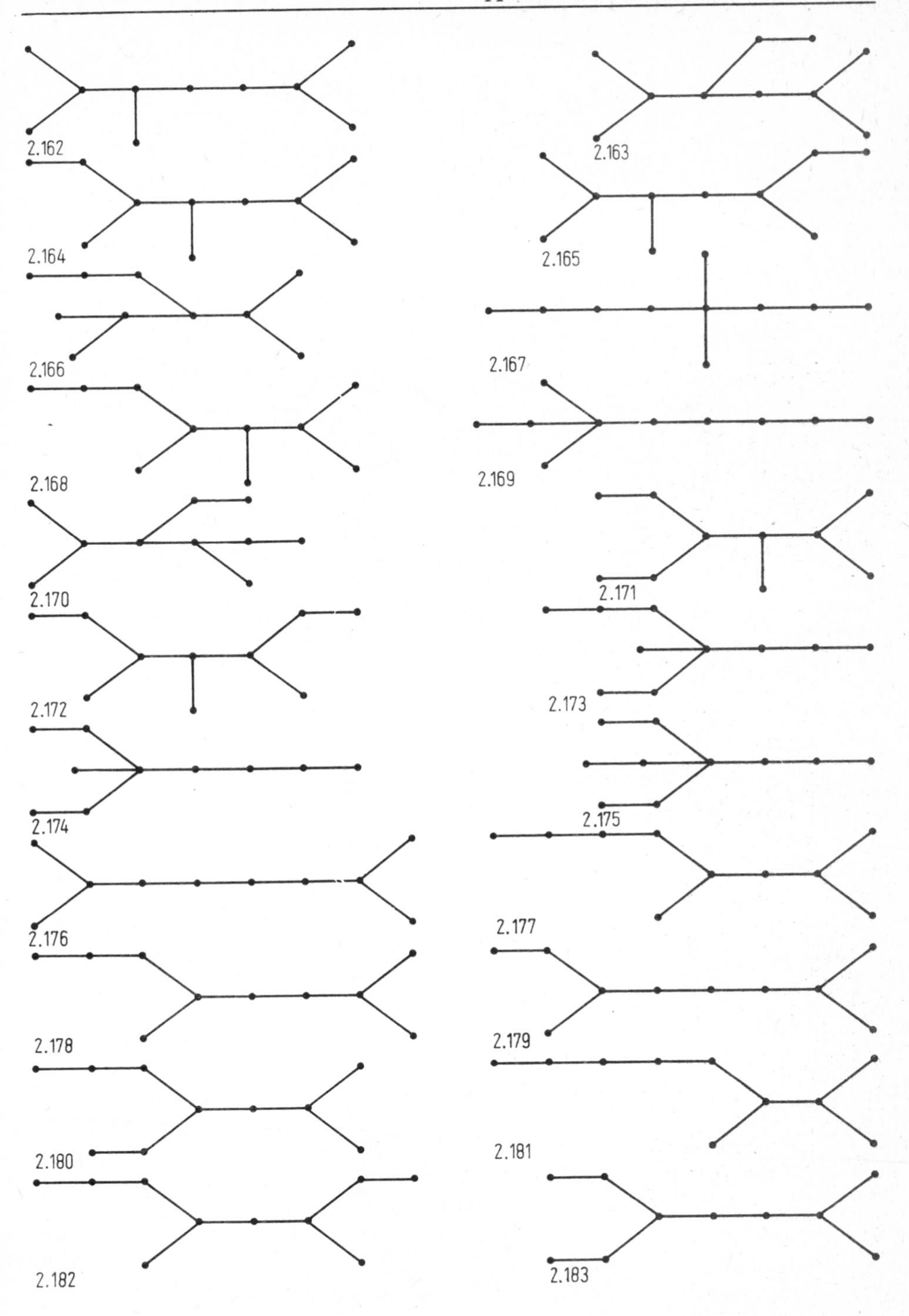
2.162
2.163
2.164
2.165
2.166
2.167
2.168
2.169
2.170
2.171
2.172
2.173
2.174
2.175
2.176
2.177
2.178
2.179
2.180
2.181
2.182
2.183

Table 2 (continued)

n	tree	coefficients					spectrum*				
		a_2	a_4	a_6	a_8	a_{10}	λ_1	λ_2	λ_3	λ_4	λ_5
10	2.184	−9	26	−27	10	−1	2.061	1.764	1	0.694	0.396
	2.185	−9	26	−28	8	0	2.136	1.414	1.414	0.662	0
	2.186	−9	26	−28	9	0	2.127	1.576	1.197	0.747	0
	2.187	−9	26	−28	9	0	2.127	1.576	1.197	0.747	0
	2.188	−9	26	−28	10	−1	2.119	1.618	1.159	0.618	0.407
	2.189	−9	26	−28	11	−1	2.109	1.672	1	0.794	0.357
	2.190	−9	26	−29	11	0	2.149	1.543	1	1	0
	2.191	−9	26	−29	11	−1	2.151	1.505	1.221	0.699	0.362
	2.192	−9	26	−30	13	−1	2.170	1.481	1	1	0.311
	2.193	−9	27	−30	9	0	1.970	1.732	1.286	0.684	0
	2.194	−9	27	−31	11	0	2.024	1.655	1.234	0.803	0
	2.195	−9	27	−31	11	−1	2.029	1.618	1.321	0.618	0.373
	2.196	−9	27	−31	12	−1	2.007	1.707	1.190	0.729	0.337
	2.197	−9	27	−32	12	0	2.074	1.414	1.414	0.835	0
	2.198	−9	27	−32	13	−1	2.064	1.557	1.268	0.779	0.315
	2.199	−9	27	−32	14	−1	2.049	1.647	1	1	0.296
	2.200	−9	28	−35	15	−1	1.919	1.683	1.310	0.831	0.285

* Only eigenvalues $\lambda_1, \lambda_2, \ldots, \lambda_{\left[\frac{n}{2}\right]}$ are given since any bipartite graph with n vertices satisfies $\lambda_{n-i+1} = -\lambda_i$.

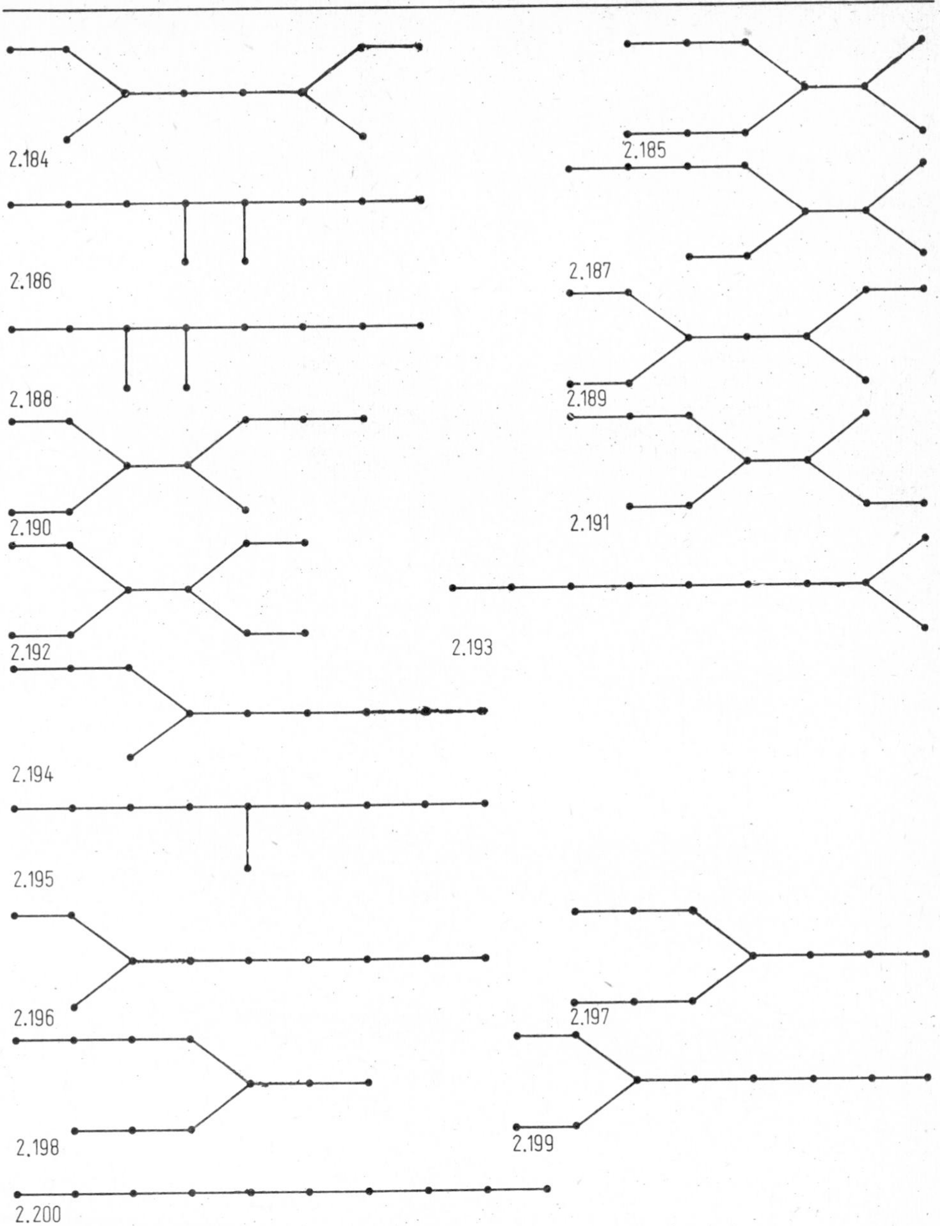

2.184
2.185
2.186
2.187
2.188
2.189
2.190
2.191
2.192
2.193
2.194
2.195
2.196
2.197
2.198
2.199
2.200

Table 3. Characteristic polynomials and spectra of connected cubic graphs with n vertices for $n = 4, 6, 8, 10, 12$

n	graph	coefficients $1\ \ a_1\ \ a_2\ \ \cdots\ \ a_n$	spectrum $\lambda_1\ \ \lambda_2\ \ \cdots\ \ \lambda_n$
4	3.1	1 0 −6 −8 −3	3 −1 −1 −1
6	3.2	1 0 −9 −4 12 0 0	3 1 0 0 −2 −2
	3.3	1 0 −9 0 0 0 0	3 0 0 0 0 −3
8	3.4	1 0 −12 0 34 −16 −20 16 −3	3 1 1 0.41 0.41 −1 −2.41 −2.41
	3.5	1 0 −12 −4 38 16 −36 −12 9	3 1.73 1 0.41 −1 −1 −1.73 −2.41
	3.6	1 0 −12 −2 36 0 −31 12 0	3 1.56 0.61 0.61 0 −1.61 −1.61 −2.56
	3.7	1 0 −12 0 30 0 −28 0 9	3 1 1 1 −1 −1 −1 −3
	3.8	1 0 −12 −8 38 48 −12 −40 −15	3 2.23 1 −1 −1 −1 −1 −2.23
10	3.9	1 0 −15 0 65 0 −105 0 55 0 −9	3 1.61 1.61 0.61 0.61 −0.61 −0.61 −1.61 −1.61 −3
	3.10	1 0 −15 0 69 −12 −117 36 59 −12 −9	3 1.61 1.30 1 0.61 −0.38 −0.61 −1.61 −2.30 −2.61
	3.11	1 0 −15 −4 73 28 −141 −52 99 16 −21	3 1.93 1.61 0.61 0.61 −0.61 −1.46 −1.61 −1.61 −2.47
	3.12	1 0 −15 0 71 −16 −133 64 76 48 0	3 1.56 1 1 1 0 −1 −2 −2 −2.56
	3.13	1 0 −15 −2 71 8 −132 −2 91 −8 −12	3 1.87 1.26 1 0.51 −0.34 −1.18 −1.53 −2 −2.59
	3.14	1 0 −15 0 65 −4 −85 −20 35 20 3	3 1.61 1.61 1 −0.38 −0.38 −0.61 −0.61 −2.61 −2.61
	3.15	1 0 −15 −4 69 32 −105 −64 23 20 3	3 2.11 1.61 0.61 −0.25 −0.38 −0.61 −1.61 −1.86 −2.61
	3.16	1 0 −15 −4 75 24 −157 −36 144 16 −48	3 2 1 1 1 −1 −1 −2 −2 −2
	3.17	1 0 −15 −2 67 12 −96 −22 35 12 0	3 2.07 1.30 0.80 0 −0.42 −0.55 −1.29 −2.24 −2.66
	3.18	1 0 −15 −6 75 48 −144 −114 75 68 12	3 1.87 1.87 1 −0.34 −0.34 −1.53 −1.53 −2 −2
	3.19	1 0 −15 −2 69 12 −116 −24 54 26 3	3 1.90 1.24 1.24 −0.19 −0.44 −0.44 −1.80 −1.80 −2.70
	3.20	1 0 −15 0 63 0 −85 0 36 0 0	3 2 1 1 0 0 −1 −1 −2 −3

3.1

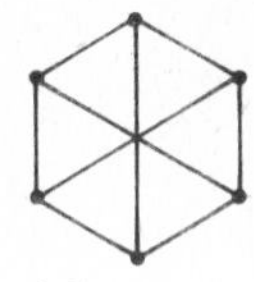
3.2

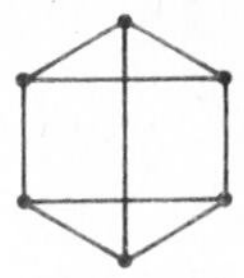
3.3

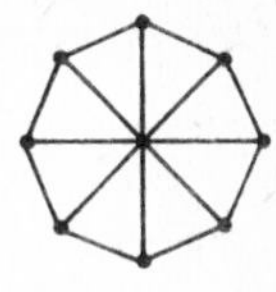
3.4

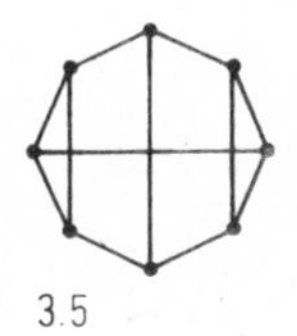
3.5

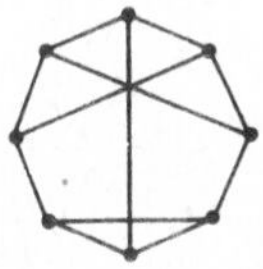

3.6

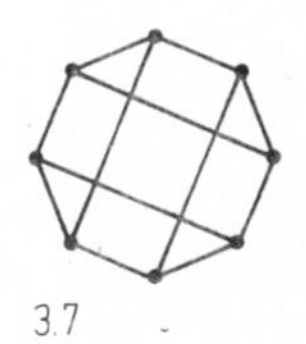
3.7

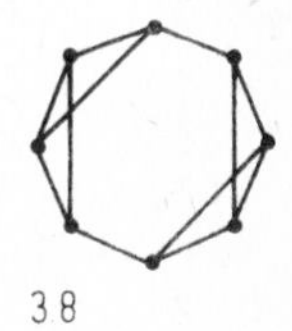
3.8

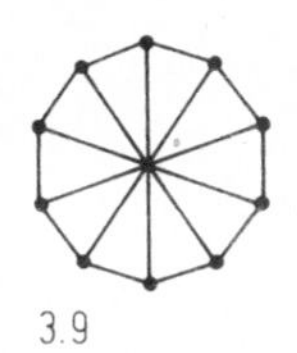
3.9

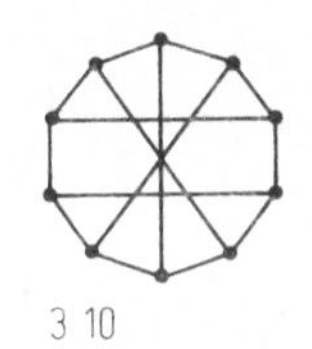
3.10

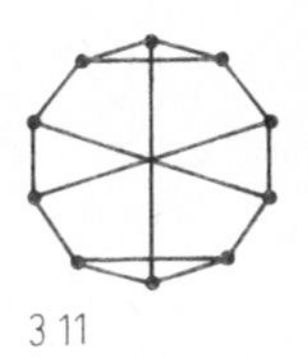
3.11

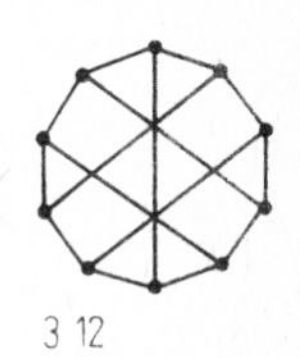
3.12

3.13

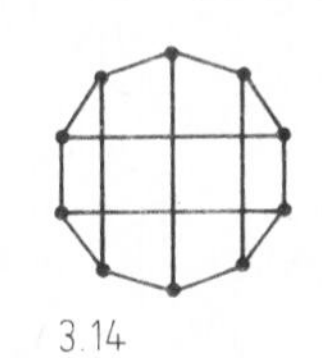
3.14

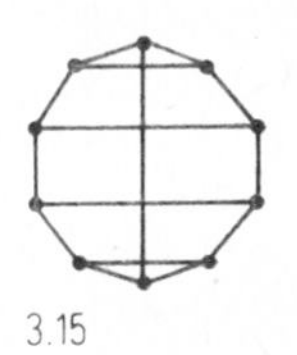
3.15

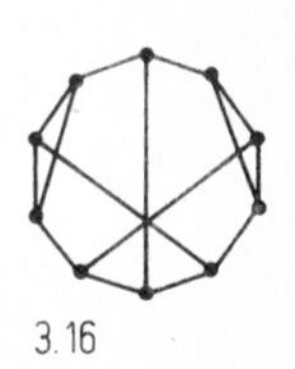
3.16

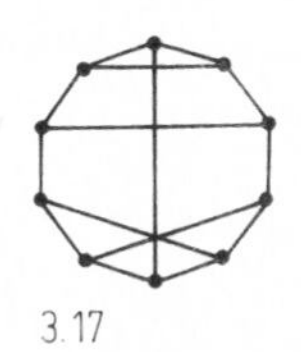
3.17

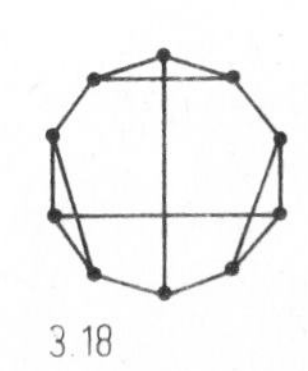
3.18

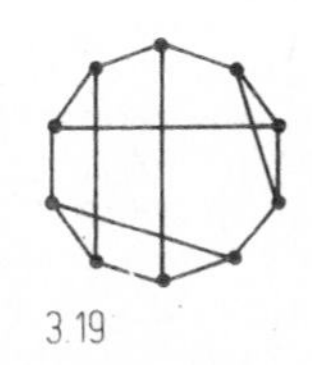
3.19

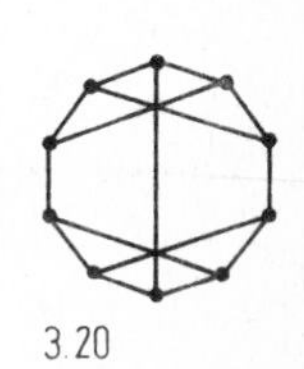
3.20

Table 3 (continued)

n	graph	coefficients 1 a_1 a_2 $\cdots$ a_n	spectrum λ_1 λ_2 $\cdots$ λ_n
10	3.21	1 0 −15 −8 71 64 −101 −104 44 48 0	3 2.56 1 1 0 −1 −1 −1.56 −2 −2
	3.22	1 0 −15 −4 71 28 −121 −48 64 24 0	3 2.14 1.28 1 0 −0.36 −1 −1.60 −2 −2.45
	3.23	1 0 −15 −6 69 48 −96 −76 30 26 3	3 2.43 1.24 0.72 −0.14 −0.44 −1 −1.53 −1.80 −2.48
	3.24	1 0 −15 −4 63 36 −61 −56 −12 0 0	3 2.41 1.34 0 0 −0.41 −0.52 −1 −2 −2.81
	3.25	1 0 −15 −8 71 68 −93 −132 −36 0 0	3 2.41 1.73 0 0 −0.41 −1 −1.73 −2 −2
	3.26†	1 0 −15 0 75 −24 −165 120 120 −160 48	3 1 1 1 1 1 −2 −2 −2 −2
	3.27	1 0 −15 −8 63 64 −37 −56 −12 0 0	3 2.77 1 0 0 −0.28 −1 −1 −2 −2.48
12	3.28	1 0 −18 0 105 0 −228 −24 180 16 −48 0 0	3 2 2 0.73 0.73 0 0 −1 −1 −1 −2.73 −2.73
	3.29	1 0 −18 0 109 −8 −264 40 220 −32 −48 0 0	3 2 1.81 1 0.73 0 0 −0.47 −1 −2 −2.34 −2.73
	3.30	1 0 −18 0 109 −4 −272 4 284 8 −96 0 0	3 2 1.56 1.41 0.73 0 0 −1 −1 −1.41 −2.56 −2.73
	3.31	1 0 −18 −4 113 48 −308 −188 348 264 −112 −96 0	3 2 2 1.41 0.73 0 −1 −1 −1 −1.41 −2 −2.73
	3.32	1 0 −18 0 111 −10 −286 54 277 −54 −63 0 0	3 1.96 1.57 1.36 0.74 0 0 −0.43 −1.19 −2.12 −2.21 −2.67
	3.33	1 0 −18 −2 113 16 −307 −42 354 36 −135 0 0	3 2.05 1.73 1.30 0.76 0 0 −1 −1.23 −1.73 −2.30 −2.58
	3.34	1 0 −18 −2 113 20 −315 −78 410 120 −227 −60 36	3 2 1.69 1.30 1 0.32 −1 −1 −1 −1.32 −2.30 −2.69
	3.35	1 0 −18 −6 117 72 −339 −306 414 532 −99 −324 −108	3 2 2 1.30 1.30 −1 −1 −1 −1 −1 −2.30 −2.30
	3.36	1 0 −18 0 111 −8 −292 40 327 −56 −138 24 9	3 1.90 1.73 1 1 0.41 −0.19 −1 −1 −1.73 −2.41 −2.70

† The Petersen graph.

3.21

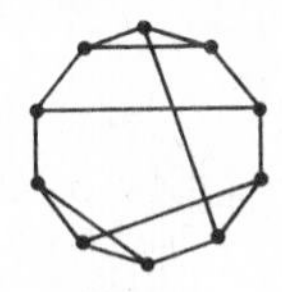
3.22

3.23

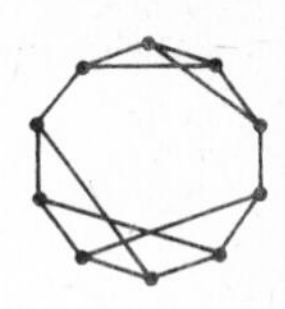
3.24

3.25

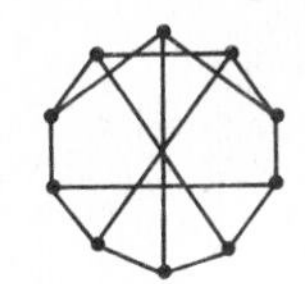
3.26

3.27

3.28

3.29

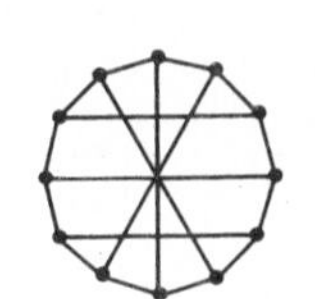
3.30

3.31

3.32

3.33

3.34

3.35

3.36

Table 3 (continued)

n	graph	coefficients 1 a_1 a_2 $\cdots$ a_n	spectrum λ_1 λ_2 $\cdots$ λ_n
12	3.37	1 0 −18 −4 115 44 −328 −164 419 244 −198 −120 9	3 2.13 1.73 1.35 1 0.06 −1 −1 −1 −1.73 −1.94 −2.61
	3.38	1 0 −18 0 115 −16 −328 104 387 −176 −102 24 9	3 1.73 1.48 1.48 1 0.41 −0.31 −0.31 −1.73 −2.17 −2.17 −2.41
	3.39	1 0 −18 −8 111 96 −268 −336 207 416 30 −168 −63	3 2.64 1.73 1 1 −1 −1 −1 −1 −1 −1.73 −2.64
	3.40	1 0 −18 0 111 −8 −292 40 323 −48 −118 0 9	3 1.96 1.57 1.18 1 0.29 −0.39 −0.47 −1.36 −1.66 −2.38 −2.72
	3.41	1 0 −18 −4 115 40 −320 −128 375 136 −154 −36 9	3 2.08 1.96 1.18 0.75 0.16 −0.39 −1 −1.36 −1.79 −2.20 −2.38
	3.42	1 0 −18 −4 113 40 −304 −116 360 128 −152 −48 0	3 2.39 1.41 1.22 1 0 −0.30 −1 −1.41 −1.71 −2 −2.59
	3.43	1 0 −18 0 113 −12 −312 76 368 −128 −136 48 0	3 1.93 1.41 1.32 1 0.35 0 −0.77 −1.41 −2 −2.15 −2.67
	3.44	1 0 −18 −2 115 12 −327 −12 413 −16 −193 18 9	3 2.06 1.60 1.19 1 0.29 −0.18 −1 −1.29 −2.09 −2.19 −2.39
	3.45	1 0 −18 0 113 −10 −314 54 386 −76 −179 30 9	3 1.81 1.53 1.30 1.13 0.34 −0.16 −1 −1.11 −1.87 −2.30 −2.67
	3.46	1 0 −18 −2 113 18 −313 −56 390 74 −184 −36 9	3 2.09 1.58 1.23 1.07 0.14 −0.37 −1 −1.26 −1.65 −2.14 −2.70
	3.47	1 0 −18 −4 109 44 −256 −128 188 64 −48 0 0	3 2.34 2 0.73 0.47 0 0 −1 −1 −1.81 −2 −2.73
	3.48	1 0 −18 −2 115 10 −325 10 397 −76 −148 48 0	3 2.12 1.50 1.34 0.67 0.38 0 −0.82 −1.66 −2 −2.18 −2.36
	3.49	1 0 −18 −2 111 18 −285 −50 277 40 −48 0 0	3 2.12 1.76 1.34 0.38 0 0 −0.56 −1.48 −1.66 −2.18 −2.71
	3.50	1 0 −18 −4 113 44 −300 −152 300 160 −48 0 0	3 2.06 2 1.41 0.22 0 0 −1 −1.41 −1.65 −2 −2.63
	3.51	1 0 −18 −2 107 18 −237 −42 153 0 0 0 0	3 2.33 1.81 0.87 0 0 0 0 −1.34 −1.52 −2.53 −2.63
	3.52	1 0 −18 −8 117 96 −316 −384 240 512 192 0 0	3 2 2 2 0 0 −1 −1 −1 −2 −2 −2
	3.53	1 0 −18 0 105 −8 −216 40 96 0 0 0 0	3 2.32 1.56 1 0 0 0 0 −0.64 −2 −2.56 −2.68

3.37

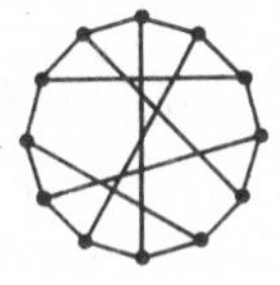
3.38

3.39

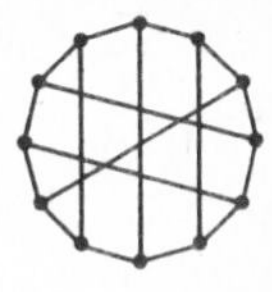
3.40

3.41

3.42

3.43

3.44

3.45

3.46

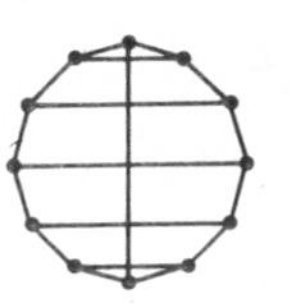
3.47

3.48

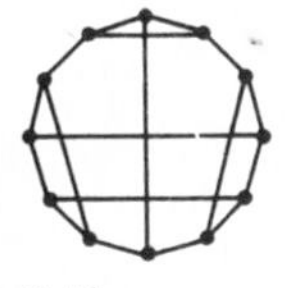
3.49

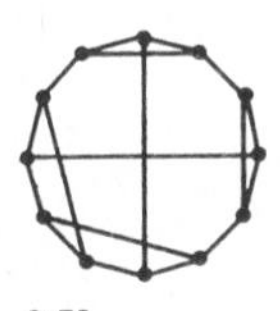
3.50

3.51

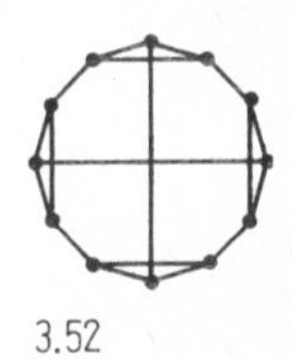
3.52

3.53

Table 3 (continued)

n	graph	coefficients 1 a_1 a_2 $\cdots$ a_n	spectrum λ_1 λ_2 $\cdots$ λ_n
12	3.54	1 0 –18 0 113 –12 –308 68 340 –88 –96 0 0	3 1.81 1.56 1.41 1 0 0 –0.47 –1.41 –2 –2.34 –2.56
	3.55	1 0 –18 0 109 –8 –260 32 192 0 0 0 0	3 2 1.56 1.56 0 0 0 0 –1 –2 –2.56 –2.56
	3.56	1 0 –18 0 113 –16 –304 112 304 –192 0 0 0	3 2 1.56 1 1 0 0 0 –2 –2 –2 –2.56
	3.57	1 0 –18 –2 111 20 –291 –64 317 72 –121 –18 9	3 2.12 1.76 1.22 0.69 0.23 –0.39 –1 –1 –1.78 –2.06 –2.78
	3.58	1 0 –18 –4 115 40 –320 –128 371 136 –126 –12 9	3 2.17 1.73 1.48 0.41 0.31 –0.31 –1 –1.48 –1.73 –2.17 –2.41
	3.59	1 0 –18 –10 113 120 –263 –434 90 468 209 –48 –36	3 2.61 2 1.30 0.38 –1 –1 –1 –1 –1 –2 –2.30
	3.60	1 0 –18 –6 117 68 –335 –262 398 392 –127 –192 –36	3 2.11 2 1.30 1 –0.25 –1 –1 –1 –1.86 –2 –2.30
	3.61	1 0 –18 –6 115 68 –311 –248 317 308 –57 –66 9	3 2.27 1.89 1.43 0.42 0.13 –1 –1 –1 –1.66 –2.14 –2.36
	3.62	1 0 –18 –2 109 20 –267 –58 250 40 –75 0 0	3 2.23 1.79 1 0.61 0 0 –1 –1 –1.61 –2.23 –2.79
	3.63	1 0 –18 –4 113 38 –294 –98 290 44 –95 –6 9	3 2.26 1.94 0.80 0.61 0.37 –0.43 –0.55 –1.61 –1.78 –2.24 –2.36
	3.64	1 0 –18 –2 111 16 –287 –32 309 20 –117 6 9	3 2.27 1.43 1.32 0.54 0.42 –0.27 –1 –1 –1.90 –2.14 –2.69
	3.65	1 0 –18 –2 111 14 –281 –18 269 –4 –60 0 0	3 2.26 1.60 1.16 0.59 0 0 –0.53 –1.30 –2 –2.20 –2.59
	3.66	1 0 –18 0 107 –6 –246 26 201 –14 –39 0 0	3 2.22 1.44 1.24 0.56 0 0 –0.48 –1 –1.70 –2.52 –2.75
	3.67	1 0 –18 –4 111 46 –282 –154 257 142 –39 0 0	3 2.22 1.95 1.24 0.20 0 0 –1 –1.33 –1.70 –1.82 –2.75
	3.68	1 0 –18 –2 109 20 –267 –62 254 60 –63 0 0	2 2.27 1.59 1.30 0.44 0 0 –1 –1 –1.55 –2.30 –2.76
	3.69	1 0 –18 –4 115 38 –322 –110 401 122 –179 –48 0	3 2.30 1.50 1.16 1.09 0 –0.26 –1 –1.47 –1.78 –2.19 –2.35
	3.70	1 0 –18 –4 113 42 –302 –134 334 140 –123 –30 9	3 2.28 1.74 1.24 0.61 0.19 –0.42 –1 –1.37 –1.61 –2.07 –2.60

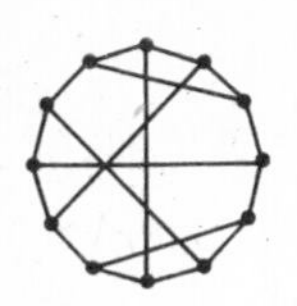
3.54

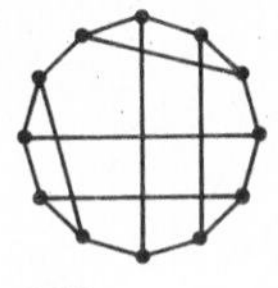
3.55

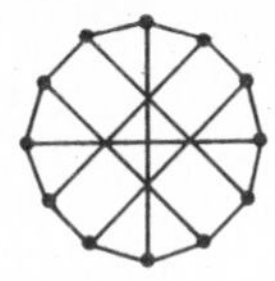
3.56

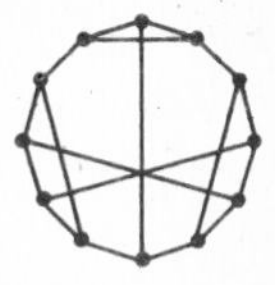
3.57

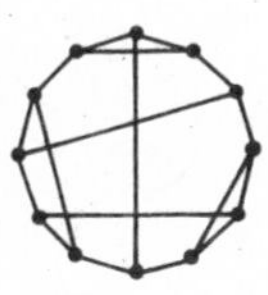
3.58

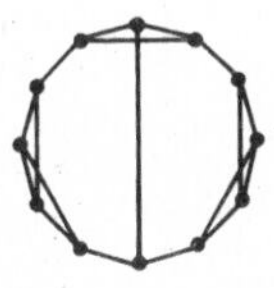
3.59

3.60

3.61

3.62

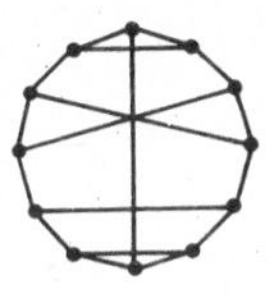
3.63

3.64

3.65

3.66

3.67

3.68

3.69

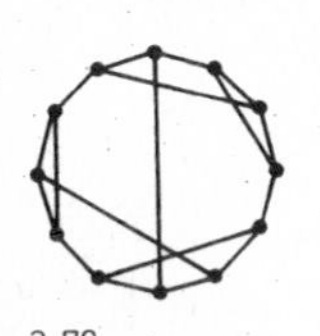
3.70

Table 3 (continued)

n	graph	coefficients 1 a_1 a_2 ⋯ a_n	spectrum λ_1 λ_2 ⋯ λ_n
12	3.71	1 0 −18 −6 111 68 −275 −220 257 236 −61 −54 9	3 2.55 1.63 1.25 0.47 0.15 −1 −1 −1 −1.47 −1.96 −2.63
	3.72	1 0 −18 −4 117 38 −346 −118 482 148 −283 −66 45	3 2.19 1.53 1.30 1.06 0.34 −0.69 −1 −1.45 −1.87 −2.10 −2.30
	3.73	1 0 −18 0 117 −18 −354 126 486 −272 −207 162 −27	3 1.53 1.53 1.30 1.30 0.34 0.34 −1 −1.87 −1.87 −2.30 −2.30
	3.74	1 0 −18 −2 115 14 −333 −26 453 12 −256 0 36	3 2.08 1.41 1.24 1.15 0.45 −0.44 −1 −1.41 −1.80 −2.10 −2.58
	3.75	1 0 −18 −6 113 64 −295 −202 334 252 −135 −108 0	3 2.56 1.30 1.30 1 0 −1 −1 −1 −1.56 −2.30 −2.30
	3.76	1 0 −18 −4 109 40 −260 −100 248 72 −72 0 0	3 2.54 1.41 1.18 0.49 0 0 −1 −1.33 −1.41 −2.25 −2.64
	3.77	1 0 −18 −4 111 42 −278 −126 261 102 −63 0 0	3 2.37 1.76 1.15 0.37 0 0 −1 −1.31 −1.53 −2.20 −2.61
	3.78	1 0 −18 −4 113 38 −298 −102 326 88 −119 −6 9	3 2.38 1.53 1.30 0.47 0.34 −0.30 −1 −1.21 −1.87 −2.30 −2.34
	3.79	1 0 −18 −2 109 16 −263 −26 234 4 −39 0 0	3 2.36 1.50 1.19 0.46 0 0 −0.45 −1.33 −1.76 −2.24 −2.71
	3.80	1 0 −18 0 109 0 −288 0 340 0 −144 0 0	3 2 1.41 1.41 1 0 0 −1 −1.41 −1.41 −2 −3
	3.81	1 0 −18 −6 113 64 −291 −198 294 204 −83 −48 0	3 2.51 1.64 1.22 0.61 0 −0.43 −1 −1.44 −1.61 −2.10 −2.39
	3.82	1 0 −18 0 111 0 −316 0 447 0 −306 0 81	3 1.73 1.73 1 1 1 −1 −1 −1 −1.73 −1.73 −3
	3.83	1 0 −18 0 115 −12 −340 76 479 −148 −282 84 45	3 1.73 1.48 1.21 1 1 −0.31 −1 −1.53 −1.73 −2.17 −2.67
	3.84	1 0 −18 0 117 −16 −360 112 532 −256 −304 192 0	3 1.56 1.41 1.41 1 1 0 −1.41 −1.41 −2 −2 −2.56
	3.85	1 0 −18 −4 107 48 −248 −152 219 144 −70 −36 9	3 2.51 1.65 1 0.57 0.21 −1 −1 −1 −1 −2.08 −2.86
	3.86	1 0 −18 −8 115 92 −300 −332 263 420 30 −108 −27	3 2.51 1.73 1.48 0.57 −0.31 −1 −1 −1 −1.73 −2.08 −2.17
	3.87	1 0 −18 −6 115 66 −309 −226 309 244 −68 −48 0	3 2.27 2 1.24 0.51 0 −0.44 −1 −1.45 −1.80 −2 −2.33

3.71

3.72

3.73

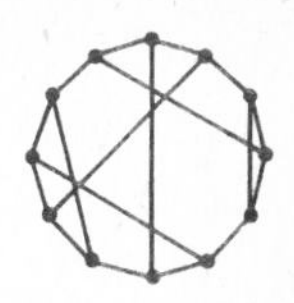
3.74

3.75

3.76

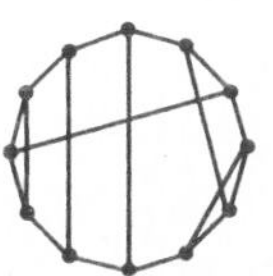
3.77

3.78

3.79

3.80

3.81

3.82

3.83

3.84

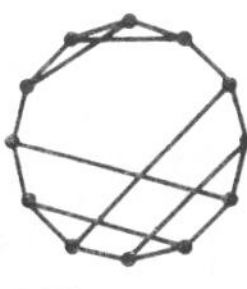
3.85

3.86

3.87

Table 3 (continued)

n	graph	coefficents 1 a_1 a_2 $\cdots$ a_n	spectrum λ_1 λ_2 $\cdots$ λ_n
12	3.88	1 0 −18 0 105 0 −236 0 180 0 0 0 0	3 2.23 1.41 1.41 0 0 0 0 −1.41 −1.41 −2.23 −3
	3.89	1 0 −18 −2 111 18 −293 −42 333 44 −120 −36 0	3 2.27 1.24 1.24 1.15 0 −0.44 −0.44 −1.62 −1.80 −1.80 −2.80
	3.90	1 0 −18 −4 117 36 −344 −96 468 80 −240 0 0	3 2.23 1.41 1.41 1 0 0 −1.41 −1.41 −2 −2 −2.23
	3.91	1 0 −18 −8 111 88 −260 −264 199 232 −42 −48 9	3 2.70 1.73 1 0.41 0.19 −1 −1 −1 −1.73 −1.90 −2.41
	3.92	1 0 −18 −8 113 92 −276 −312 188 300 16 −48 0	3 2.52 2 1.11 0.36 0 −1 −1 −1 −1.65 −2 −2.34
	3.93	1 0 −18 −6 111 60 −271 −152 273 124 −97 −18 9	3 2.66 1.36 1.19 0.49 0.29 −0.40 −1 −1.29 −1.76 −2.19 −2.34
	3.94	1 0 −18 −4 111 36 −276 −76 279 44 −106 0 9	3 2.51 1.48 1 0.57 0.41 −0.31 −1 −1 −2.08 −2.17 −2.41
	3.95	1 0 −18 −4 109 40 −256 −100 216 56 −60 0 0	3 2.50 1.67 0.86 0.53 0 0 −1 −1 −1.75 −2.21 −2.62
	3.96	1 0 −18 −8 113 88 −280 −280 244 296 −36 −72 0	3 2.65 1.67 1.21 0.53 0 −1 −1 −1 −1.86 −2 −2.21
	3.97	1 0 −18 −4 105 44 −228 −104 184 72 −36 0 0	3 2.65 1.27 1.21 0.31 0 0 −1 −1 −1.70 −1.86 −2.89
	3.98	1 0 −18 −4 109 36 −256 −64 228 16 −48 0 0	3 2.58 1.41 1 0.54 0 0 −0.54 −1.41 −2 −2 −2.58
	3.99	1 0 −18 −6 109 68 −247 −198 146 88 −39 0 0	3 2.50 2.01 0.61 0.37 0 0 −1 −1.39 −1.61 −1.83 −2.68
	3.100	1 0 −18 −2 103 18 −201 −26 105 0 0 0 0	3 2.57 1.49 0.81 0 0 0 0 −1 −1.65 −2.38 −2.83
	3.101	1 0 −18 −6 111 62 −265 −166 213 92 −60 0 0	3 2.57 1.80 0.81 0.44 0 0 −1 −1.24 −2 −2 −2.38
	3.102	1 0 −18 0 97 0 −144 0 0 0 0 0 0	3 2.56 1.56 0 0 0 0 0 0 −1.56 −2.56 −3
	3.103	1 0 −18 −8 111 92 −252 −292 119 180 −34 −36 9	3 2.51 2.17 0.57 0.41 0.31 −1 −1 −1 −1.48 −2.08 −2.41
	3.104	1 0 −18 −8 113 88 −272 −272 176 192 0 0 0	3 2.56 2 1 0 0 0 −1 −1.56 −2 −2 −2

3.88

3.89

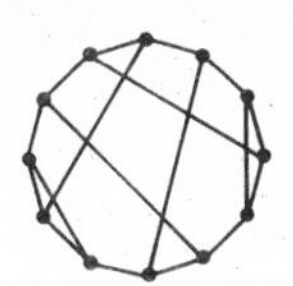
3.90

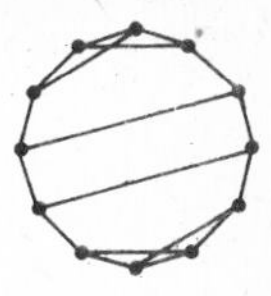
3.91

3.92

3.93

3.94

3.95

3.96

3.97

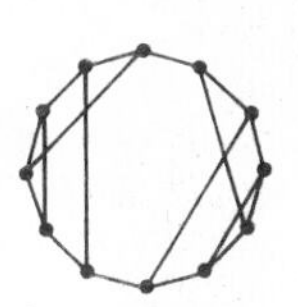
3.98

3.99

3.100

3.101

3.102

3.103

3.104

Table 3 (continued)

n	graph	coefficients 1 a_1 a_2 $\cdots$ a_n	spectrum λ_1 λ_2 $\cdots$ λ_n
12	3.105	1 0 −18 0 105 0 −232 0 144 0 0 0 0	3 2 2 1 0 0 0 0 −1 −2 −2 −3
	3.106	1 0 −18 −6 105 60 −211 −122 146 52 −39 0 0	3 2.82 1.43 0.61 0.56 0 0 −1 −1 −1.61 −2.18 −2.62
	3.107	1 0 −18 −4 101 36 −176 −40 84 0 0 0 0	3 2.81 1.24 0.73 0 0 0 0 −1 −1.67 −2.39 −2.73
	3.108	1 0 −18 −8 109 84 −240 −220 172 168 0 0 0	3 2.81 1.41 1.24 0 0 0 −1 −1.41 −1.67 −2 −2.39
	3.109	1 0 −18 −10 109 112 −223 −326 58 196 9 −36 0	3 2.83 1.90 0.61 0.50 0 −1 −1 −1 −1.61 −1.88 −2.35
	3.110	1 0 −18 −2 117 12 −355 −18 534 8 −387 0 108	3 2 1.30 1.30 1 1 −1 −1 −1 −2 −2.30 −2.30
	3.111	1 0 −18 −4 105 44 −216 −104 96 0 0 0 0	3 2.56 1.84 0.50 0 0 0 0 −1.50 −1.56 −2 −2.84
	3.112	1 0 −18 −12 111 144 −216 −480 −117 256 138 −36 −27	3 2.51 2.51 0.57 0.57 −1 −1 −1 −1 −1 −2.08 −2.08

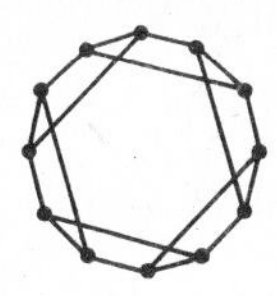
3.105

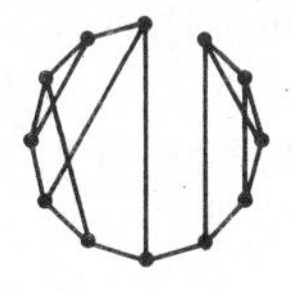
3.106

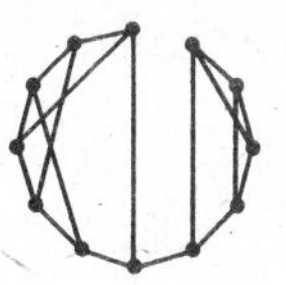
3.107

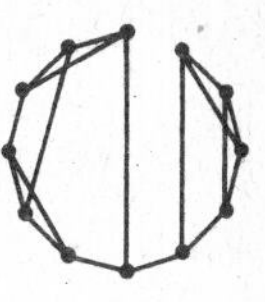
3.108

3.109

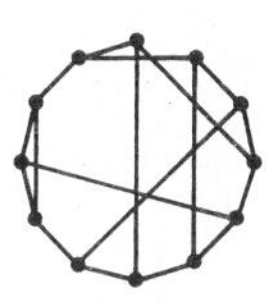
3.110

3.111

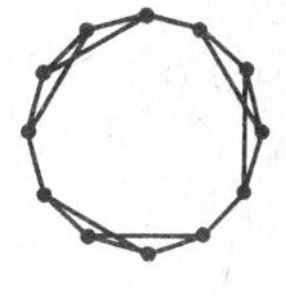
3.112

Table 4. Characteristic polynomials and spectra of miscellaneous (multi-)graphs

n	graph	coefficients $1 \; a_1 \; a_2 \; \cdots \; a_n$	spectrum $\lambda_1 \; \lambda_2 \; \cdots \; \lambda_n$
Circuits			
6	4.1	1 0 −6 0 9 0 −4	2 1 1 −1 −1 −2
7	4.2	1 0 −7 0 14 0 −7 −2	2 1.247 1.247 −0.445 −0.445 −1.802 −1.802
8	4.3	1 0 −8 0 20 0 −16 0 0	2 $\sqrt{2}$ $\sqrt{2}$ 0 0 $-\sqrt{2}$ $-\sqrt{2}$ −2
n	4.4	$a_i = \begin{cases} 0 & \text{for odd } i < n \\ -2 & \text{for odd } i = n \\ (-1)^m \frac{n}{m}\binom{n-m-1}{m-1} & \text{for } i = 2m < n \\ 2[(-1)^m - 1] & \text{for } i = 2m = n \end{cases}$	$2\cos\frac{2\pi j}{n}$, $j = 1, \ldots, n$ (not in non-increasing order)

Some more cubic multigraphs

a) Prisms

n	graph	coefficients	spectrum
10	4.5	1 0 −15 0 65 −4 −85 −20 35 20 3	3 1 $\frac{1}{2}(1+\sqrt{5})$ $\frac{1}{2}(1+\sqrt{5})$ $\frac{1}{2}(-3+\sqrt{5})$ $\frac{1}{2}(-3+\sqrt{5})$ $\frac{1}{2}(1-\sqrt{5})$ $\frac{1}{2}(1-\sqrt{5})$ $\frac{1}{2}(-3-\sqrt{5})$ $\frac{1}{2}(-3-\sqrt{5})$
12	4.6	1 0 −18 0 105 0 −232 0 144 0 0 0 0	3 2 2 1 0 0 0 0 −1 −2 −2 −3
$2k, k \geq 3$	4.7		$\left.\begin{array}{l} \lambda_i = 1 + 2\cos\frac{2\pi i}{k} \\ \lambda_{k+i} = -1 + 2\cos\frac{2\pi i}{k} \end{array}\right\} i = 1, 2, \ldots, k$ (not in non-increasing order)

b) Some cubic multigraphs

n	graph	coefficients	spectrum
2	4.8	1 0 9	3 −3
	4.9	1 −4 3	3 1
4	4.10	1 0 −10 0 9	3 1 −1 −3
	4.11	1 −2 −7 8 12	3 2 −1 −2
	4.12	1 −6 9 4 −12	3 2 2 −1

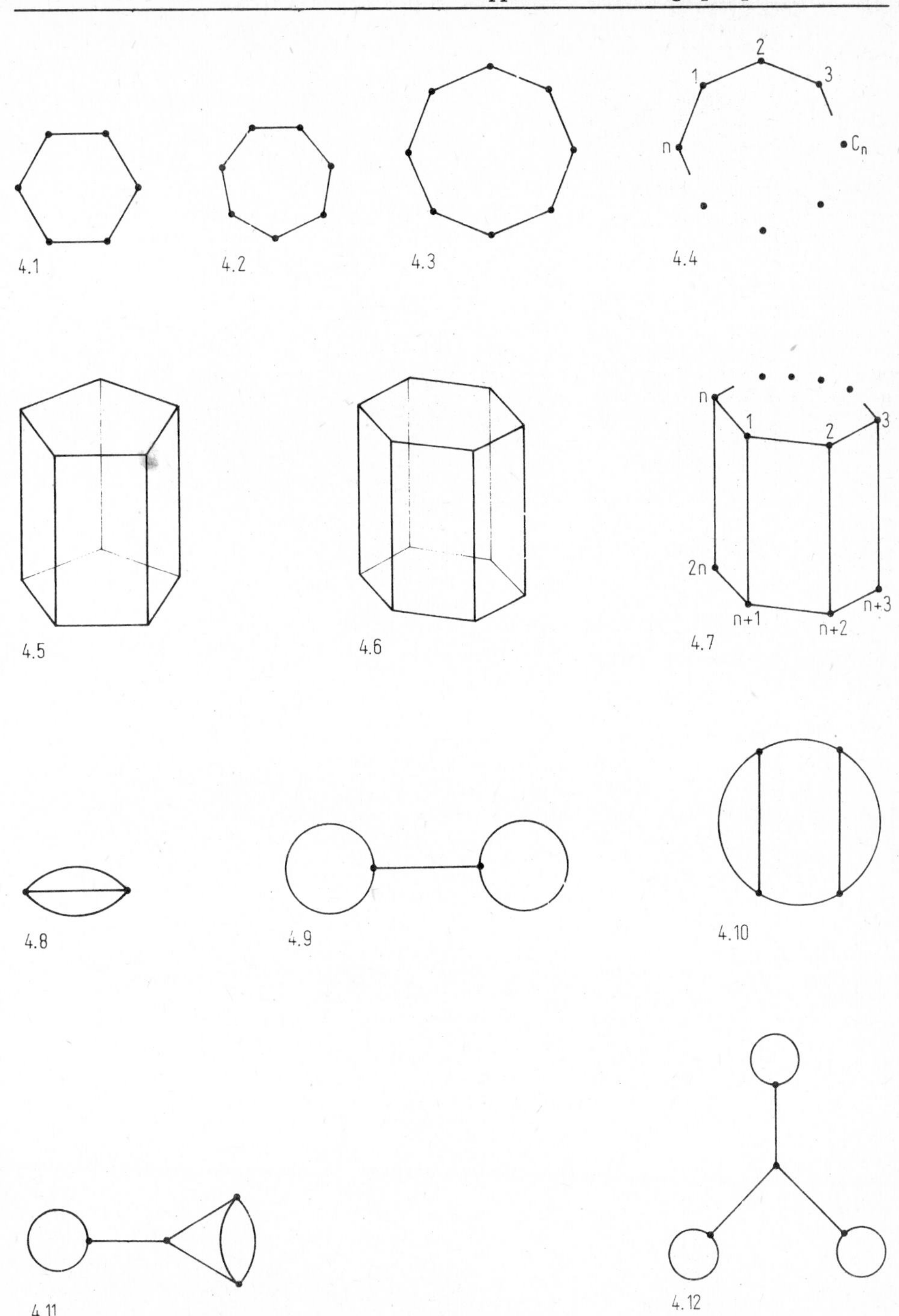

4.1 4.2 4.3 4.4

4.5 4.6 4.7

4.8 4.9 4.10

4.11 4.12

Table 4 (continued)

n	graph	coefficients $1\ a_1\ a_2\ \cdots\ a_n$	spectrum $\lambda_1\ \lambda_2\ \cdots\ \lambda_n$
6	4.13	1 0 −13 0 36 0 0	3 2 0 0 −2 −3
	4.14	1 0 −11 −4 27 12 −9	3 $\sqrt{3}$ $-1+\sqrt{2}$ −1 $-\sqrt{3}$ $-1-\sqrt{2}$
	4.15	1 0 −13 −8 44 48 0	3 $\frac{1}{2}(1+\sqrt{17})$ 0 $\frac{1}{2}(1-\sqrt{17})$ −2 −2
	4.16	1 −2 −8 10 15 0 0	3 $\sqrt{5}$ 0 0 −1 $-\sqrt{5}$
8	4.17	1 0 −18 0 105 0 −232 0 144	3 2 2 1 −1 −2 −2 −3
10	4.18	1 0 −21 −12 156 168 −420 −672 96 512 192	3 $1+\sqrt{3}$ $1+\sqrt{3}$ 1 $1-\sqrt{3}$ $1-\sqrt{3}$ −2 −2 −2 −2

c) The smallest cubic graph of girth 6

14	4.19	1 0 −21 0 168 0 −700 0 1680 0 −2352 0 1792 0 −576	3 $\sqrt{2}$ $\sqrt{2}$ $\sqrt{2}$ $\sqrt{2}$ $\sqrt{2}$ $\sqrt{2}$ $-\sqrt{2}$ $-\sqrt{2}$ $-\sqrt{2}$ $-\sqrt{2}$ $-\sqrt{2}$ $-\sqrt{2}$ −3

d) The dodecahedron graph†

20	4.20a, 4.20b	1 0 −30 0 375 −24 −2540 480 10095 −3760 −23502 14400 28905 −27000 −11400 20000 −6000 0 0 0 0	3 $\sqrt{5}$ $\sqrt{5}$ $\sqrt{5}$ 1 1 1 1 1 0 0 0 0 −2 −2 −2 −2 $-\sqrt{5}$ $-\sqrt{5}$ $-\sqrt{5}$

Some miscellaneous graphs

6	4.21	1 0 −9 −8 9 6 −4	3.236 0.618 0.618 −1.236 −1.618 −1.618
	4.22	1 0 −10 −12 5 12 4	3.562 1 −0.562 −1 −1 −2
	4.23	1 0 −8 −4 12 4 −5	2.791 1 0.618 −1 −1.618 −1.791
	4.24	1 0 −11 −16 3 16 7	3.828 1 −1 −1 −1 −1.828
	4.25	1 0 −9 −8 10 12 3	3.182 1.247 −0.445 −0.594 −1.588 −1.802
	4.26	1 0 −7 0 7 0 −1	2.414 1 0.414 −0.414 −1 −2.414
	4.27	1 0 −7 −2 8 4 0	2.504 1.264 0 −0.577 −1 −2.191
	4.28	1 0 −8 0 4 0 0	2.732 0.732 0 0 −0.732 −2.732

† dodecahedron: $P_G(\lambda) = (\lambda - 3)(\lambda^2 - 5)^3 (\lambda - 1)^5 \lambda^4 (\lambda + 2)^4$

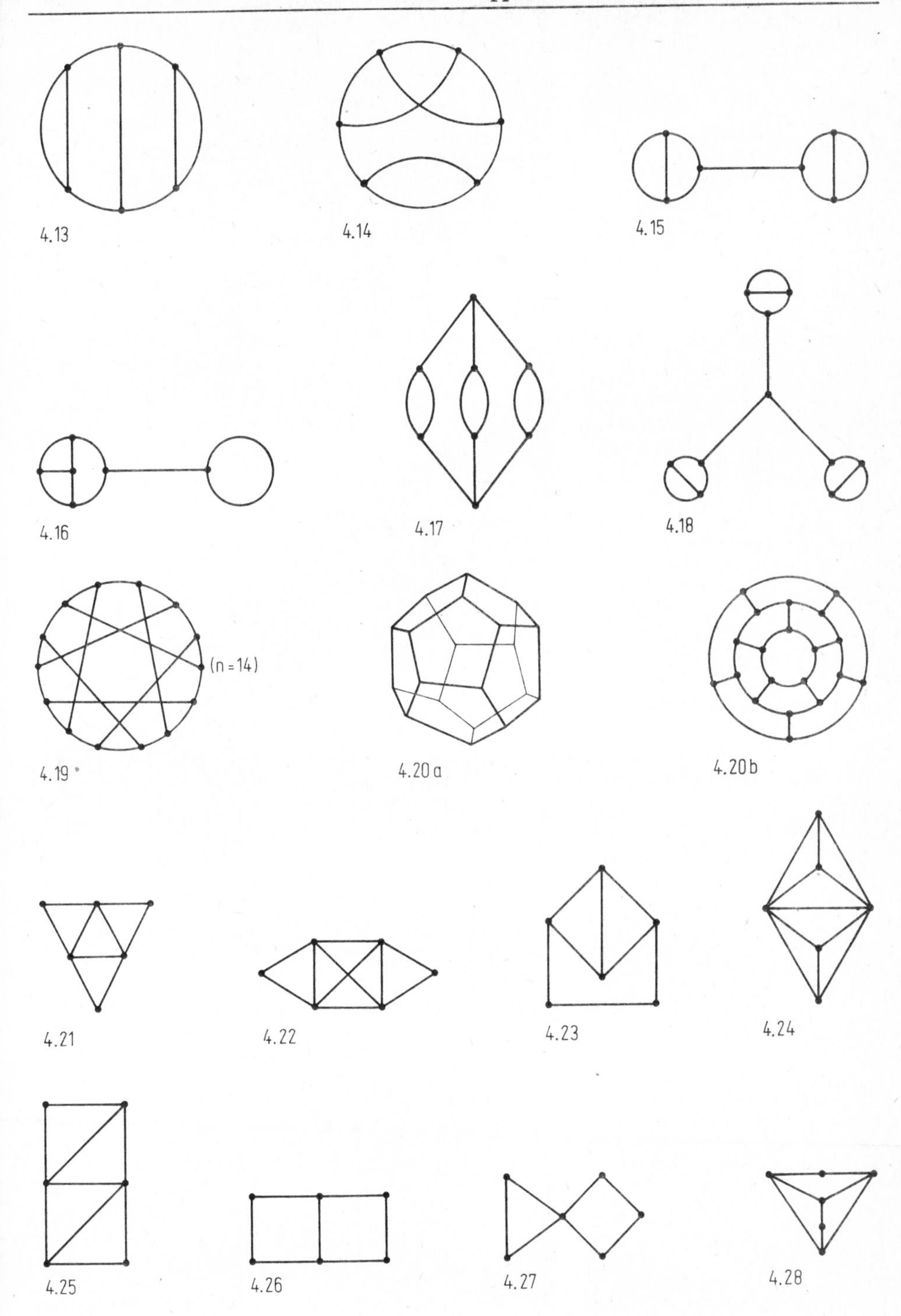
4.13
4.14
4.15
4.16
4.17
4.18
(n = 14)
4.19
4.20 a
4.20 b
4.21
4.22
4.23
4.24
4.25
4.26
4.27
4.28

Table 4 (continued)

n	graph	coefficients 1 a_1 a_2 ⋯ a_n	spectrum λ_1 λ_2 ⋯ λ_n
6	4.29	1 0 −8 −4 12 8 0	2.732 1.414 0 −0.732 −1.414 −2
	4.30	1 0 −10 −8 9 8 0	3.372 1 0 −1 −1 −2.372
	4.31	1 0 −7 −4 11 12 3	2.414 1.732 −0.414 −1 −1 −1.732
	4.32	1 0 −6 −2 8 4 −1	2.228 1.360 0.186 −1 −1 −1.775
	4.33	1 0 −9 −10 5 10 3	3.354 1 −0.476 −1 −1 −1.877
	4.34	1 0 −6 −2 6 0 −1	2.414 0.618 0.618 −0.414 −1.618 −1.618
	4.35	1 0 −7 −4 7 4 −1	2.709 1 0.194 −1 −1 −1.903
	4.36	1 0 −9 −8 9 8 −1	3.223 1 0.112 −1 −1.527 −1.809
	4.37†	1 0 −12 −16 0 0 0	4 0 0 0 −2 −2
7	4.38	1 0 −8 −2 15 8 −2 0	2.378 1.779 0.187 0 −1 −1.150 −2.195
8	4.39	1 0 −10 0 25 0 −16 0 0	2.562 1.562 1 0 0 −1 −1.562 −2.562
	4.40	1 0 −12 −8 36 32 −32 −32 0	3.236 1.414 1.414 0 −1.236 −1.414 −1.414 −2
	4.41	1 0 −8 0 14 0 −8 0 1	2.414 1 1 0.414 −0.414 −1 −1 −2.414
	4.42	1 0 −14 −16 21 40 16 0 0	4 1.562 0 0 −1 −1 −1 −2.562
	4.43	1 0 −10 0 27 −8 −22 16 −3	2.618 1.303 0.618 0.618 0.382 −1.618 −1.618 −2.303
	4.44	1 0 −9 −2 24 8 −19 −4 5	2.359 1.816 0.676 0.618 −0.871 −1 −1.618 −1.980
	4.45	1 0 −10 0 23 0 −10 0 1	2.618 1.618 0.618 0.382 −0.382 −0.618 −1.618 −2.618
	4.46	1 0 −14 −12 31 20 −26 −4 5	3.846 1.068 0.618 0.618 −0.505 −1.618 −1.618 −2.410
9	4.47	1 0 −12 −2 36 0 −27 0 0 0	3 1.303 1.303 0 0 0 −1 −2.303 −2.303
12	4.48	1 0 −12 0 40 0 −48 0 16 0 0 0 0	2.732 1.414 1.414 0.732 0 0 0 0 −0.732 −1.414 −1.414 −2.732
	4.49a, 4.49b††	1 0 −30 −40 255 576 −580 −2640 −1425 3200 5250 3000 625	5 2.236 2.236 2.236 −1 −1 −1 −1 −1 −2.236 −2.236 −2.236

† octahedron: $P_G(\lambda) = (\lambda - 4)\,\lambda^3(\lambda + 2)^2$ †† icosahedron: $P_G(\lambda) = (\lambda - 5)\,(\lambda^2 - 5)^3\,(\lambda + 1)^5$

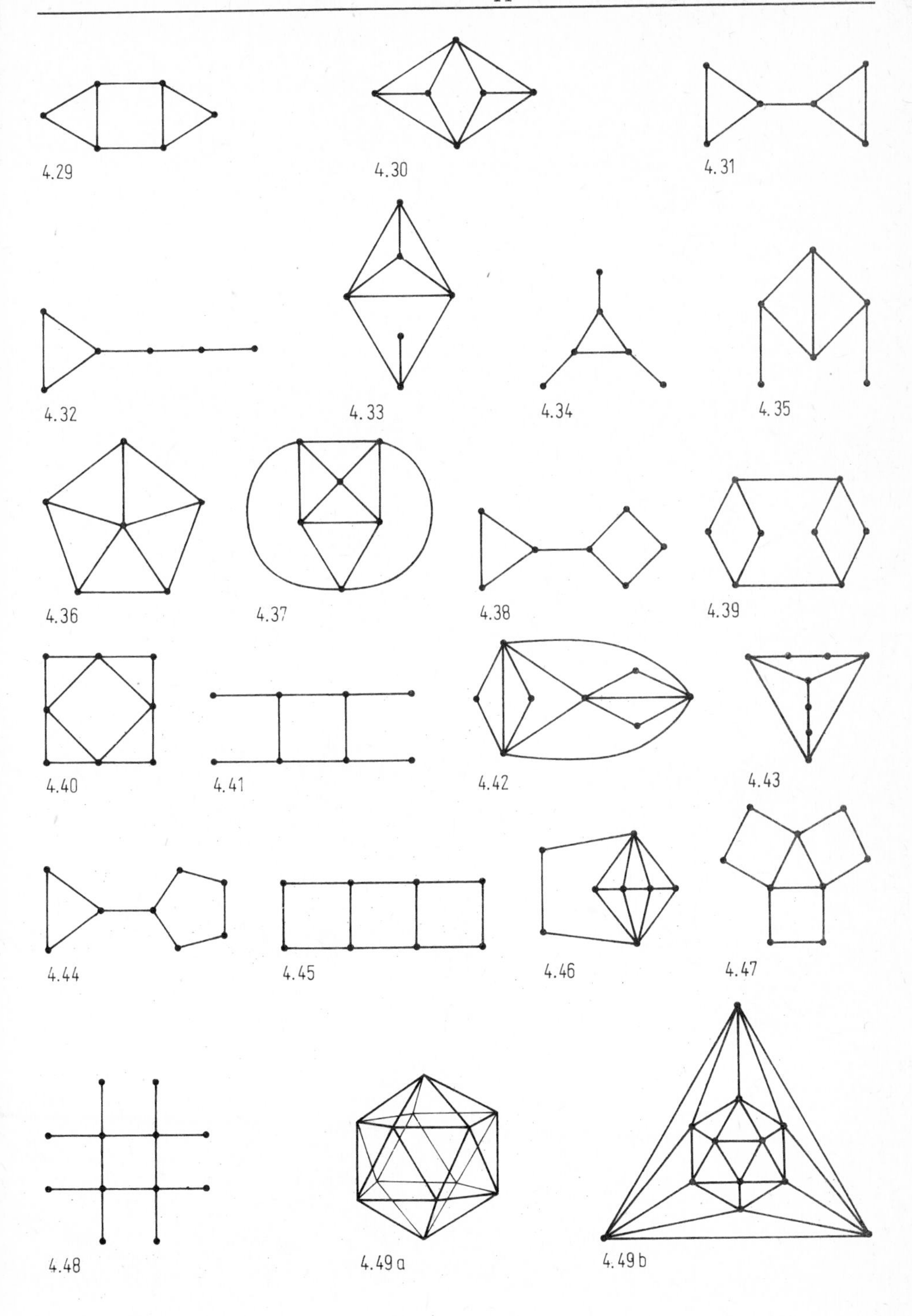
4.29
4.30
4.31
4.32
4.33
4.34
4.35
4.36
4.37
4.38
4.39
4.40
4.41
4.42
4.43
4.44
4.45
4.46
4.47
4.48
4.49 a
4.49 b

Table 5. Characteristic polynomials and non-zero eigenvalues of graphs $K_{n_1,\ldots,n_k}$ $(n_1 + \cdots + n_k = n)$ for $2 \leqq k \leqq n \leqq 10$

$k = 2,\ n = 2$					
1^2			1		
			1	−1	

$k = 2,\ n = 3$					
2	1		2		
			1.41	−1.41	

$k = 2,\ n = 4$					
3	1		3		
			1.73	−1.73	
2^2			4		
			2	−2	

$k = 2,\ n = 5$					
4	1		4		
			2	−2	
3	2		6		
			2.45	−2.45	

$k = 2,\ n = 6$					
5	1		5		
			2.24	−2.24	
4	2		8		
			2.83	−2.83	
3^2			9		
			3	−3	

$k = 2,\ n = 7$					
6	1		6		
			2.45	−2.45	
5	2		10		
			3.16	−3.16	
4	3		12		
			3.46	−3.46	

$k = 2,\ n = 8$					
7	1		7		
			2.65	−2.65	
6	2		12		
			3.46	−3.46	
5	3		15		
			3.87	−3.87	
4^2			16		
			4	−4	

$k = 2,\ n = 9$					
8	1		8		
			2.83	−2.83	
7	2		14		
			3.74	−3.74	
6	3		18		
			4.24	−4.24	
5	4		20		
			4.47	−4.47	

$k = 2,\ n = 10$					
9	1		9		
			3	−3	
8	2		16		
			4	−4	
7	3		21		
			4.58	−4.58	
6	4		24		
			4.90	−4.90	
5^2			25		
			5	−5	

$k = 3,\ n = 3$					
1^3			3	2	
			2	−1	−1

$k = 3,\ n = 4$					
2	1^2		5	4	
			2.56	−1	−1.56

$k = 3,\ n = 5$					
3	1^2		7	6	
			3	−1	−2
2^2	1		8	8	
			3.24	−1.24	−2

$k = 3,\ n = 6$					
4	1^2		9	8	
			3.37	−1	−2.37
3	2	1	11	12	
			3.77	−1.28	−2.48
2^3			12	16	
			4	−2	−2

Table 5 (continued)

$k=3,\ n=7$

5	1^2		11	10	
			3.70	−1	−2.70
4	2	1	14	16	
			4.22	−1.30	−2.92
3^2	1		15	18	
			4.37	−1.37	−3
3	2^2		16	24	
			4.61	−2	−2.61

$k=3,\ n=8$

6	1^2		13	12	
			4	−1	−3
5	2	1	17	20	
			4.62	−1.31	−3.31
4	3	1	19	24	
			4.89	−1.41	−3.48
4	2^2		20	32	
			5.12	−2	−3.12
3^2	2		21	36	
			5.27	−2.27	−3

$k=3,\ n=9$

7	1^2		15	14	
			4.27	−1	−3.27
6	2	1	20	24	
			4.98	−1.31	−3.67
5	3	1	23	30	
			5.35	−1.43	−3.92
5	2^2		24	40	
			5.58	−2	−3.58
4^2	1		24	32	
			5.46	−1.46	−4
4	3	2	26	48	
			5.85	−2.34	−3.51
3^3			27	54	
			6	−3	−3

$k=3,\ n=10$

8	1^2		17	16	
			4.53	−1	−3.53
7	2	1	23	28	
			5.32	−1.32	−4
6	3	1	27	36	
			5.77	−1.45	−4.32
6	2^2		28	48	
			6	−2	−4
5	4	1	29	40	
			5.97	−1.49	−4.48
5	3	2	31	60	
			6.36	−2.36	−4
4^2	2		32	64	
			6.47	−2.47	−4
4	3^2		33	72	
			6.62	−3	−3.62

$k=4,\ n=4$

1^4			6	8	3	
			3	−1	−1	−1

$k=4,\ n=5$

2	1^3		9	14	6	
			3.65	−1	−1	−1.65

$k=4,\ n=6$

3	1^3		12	20	9	
			4.16	−1	−1	−21.6
2^2	1^2		13	24	12	
			4.37	−1	−1.37	−2

$k=4,\ n=7$

4	1^3		15	26	12	
			4.61	−1	−1	−2.61
3	2	1^2	17	34	18	
			4.96	−1	−1.44	−2.52
2^3	1		18	40	24	
			5.16	−1.16	−2	−2

Table 5 (continued)

***k* = 4, *n* = 8**								
5	1^3			18	32	15		
				5	−1	−1	−3	
4	2	1^2		21	44	24		
				5.46	−1	−1.46	−3	
3^2	1^2			22	48	27		
				5.61	−1	−1.61	−3	
3	2^2	1		23	56	36		
				5.81	−1.18	−2	−2.63	
2^4				24	64	48		
				6	−2	−2	−2	
***k* = 4, *n* = 9**								
6	1^3			21	38	18		
				5.36	−1	−1	−3.36	
5	2	1^2		25	54	30		
				5.91	−1	−1.47	−3.44	
4	3	1^2		27	62	36		
				6.16	−1	−1.67	−3.49	
4	2^2	1		28	72	48		
				6.36	−1.19	−2	−3.18	
3^2	2	1		29	78	54		
				6.50	−1.21	−2.30	−3	
3	2^3			30	88	72		
				6.69	−2	−2	−2.69	
***k* = 4, *n* = 10**								
7	1^3			24	44	21		
				5.69	−1	−1	−3.69	
6	2	1^2		29	64	36		
				6.33	−1	−1.48	−3.85	
5	3	1^2		32	76	45		
				6.66	−1	−1.71	−3.96	
5	2^2	1		33	88	60		
				6.86	−1.19	−2	−3.67	
4^2	1^2			33	80	48		
				6.77	−1	−1.77	−4	
4	3	2	1	35	100	72		
				7.11	−1.21	−2.37	−3.53	
4	2^3			36	112	96		
				7.29	−2	−2	−3.29	
3^3	1			36	108	81		
				7.24	−1.24	−3	−3	
3^2	2^2			37	120	108		
				7.42	−2	−2.42	−3	
***k* = 5, *n* = 5**								
1^5				10	20	15	4	
				4	−1	−1	−1	−1

Table 5 (continued)

$k = 5, \quad n = 6$								
2	1^4		14	32	27	8		
			4.70	−1	−1	−1	−1.70	
$k = 5, \quad n = 7$								
3	1^4		18	44	39	12		
			5.27	−1	−1	−1	−2.27	
2^2	1^3		19	50	48	16		
			5.46	−1	−1	−1.46	−2	
$k = 5, \quad n = 8$								
4	1^4		22	56	51	16		
			5.77	−1	−1	−1	−2.77	
3	2	1^3	24	68	69	24		
			6.09	−1	−1	−1.55	−2.55	
2^3	1^2		25	76	84	32		
			6.27	−1	−1.27	−2	−2	
$k = 5, \quad n = 9$								
5	1^4		26	68	63	20		
			6.22	−1	−1	−1	−3.22	
4	2	1^3	29	86	90	32		
			6.64	−1	−1	−1.57	−3.07	
3^2	1^3		30	92	99	36		
			6.77	−1	−1	−1.77	−3	
3	2^2	1^2	31	102	120	48		
			6.95	−1	−1.31	−2	−2.64	
2^4	1		32	112	144	64		
			7.12	−1.12	−2	−2	−2	
$k = 5, \quad n = 10$								
6	1^4		30	80	75	24		
			6.62	−1	−1	−1	−3.62	
5	2	1^3	34	104	111	40		
			7.13	−1	−1	−1.58	−3.56	
4	3	1^3	36	116	129	48		
			7.37	−1	−1	−1.86	−3.51	
4	2^2	1^2	37	128	156	64		
			7.54	−1	−1.32	−2	−3.23	
3^2	2	1^2	38	136	171	72		
			7.67	−1	−1.35	−2.32	−3	
3	2^3	1	39	148	204	96		
			7.84	−1.13	−2	−2	−2.70	
2^5			40	160	240	128		
			8	−2	−2	−2	−2	
$k = 6, \quad n = 6$								
1^6			15	40	45	24	5	
			5	−1	−1	−1	−1	−1

Table 5 (continued)

$k = 6$, $n = 7$									
2	1^5		20	60	75	44	10		
			5.74	−1	−1	−1	−1	−1.74	
$k = 6$, $n = 8$									
3	1^5		25	80	105	64	15		
			6.36	−1	−1	−1	−1	−2.36	
2^2	1^4		26	88	123	80	20		
			6.53	−1	−1	−1	−1.53	−2	
$k = 6$, $n = 9$									
4	1^5		30	100	135	84	20		
			6.90	−1	−1	−1	−1	−2.90	
3	2	1^4	32	116	171	116	30		
			7.19	−1	−1	−1	−1.62	−2.57	
2^3	1^3		33	126	198	144	40		
			7.36	−1	−1	−1.36	−2	−2	
$k = 6$, $n = 10$									
5	1^5		35	120	165	104	25		
			7.39	−1	−1	−1	−1	−3.39	
4	2	1^4	38	144	219	152	40		
			7.78	−1	−1	−1	−1.64	−3.14	
3^2	1^4		39	152	237	168	45		
			7.90	−1	−1	−1	−1.90	−3	
3	2^2	1^3	40	164	273	208	60		
			8.06	−1	−1	−1.40	−2	−2.66	
2^4	1^2		41	176	312	256	80		
			8.22	−1	−1.22	−2	−2	−2	
$k = 7$, $n = 7$									
1^7			21	70	105	84	35	6	
			6	−1	−1	−1	−1	−1	−1
$k = 7$, $n = 8$									
2	1^6		27	100	165	144	65	12	
			6.77	−1	−1	−1	−1	−1	−1.77
$k = 7$, $n = 9$									
3	1^6		33	130	225	204	95	18	
			7.42	−1	−1	−1	−1	−1	−2.42
2^2	1^5		34	140	225	244	120	24	
			7.58	−1	−1	−1	−1	−1.58	−2
$k = 7$, $n = 10$									
4	1^6		39	160	285	264	125	24	
			8	−1	−1	−1	−1	−1	−3

Table 5 (continued)

$k = 7, \quad n = 10$

$3 \quad 2 \quad 1^5$	41	180	345	344	175	36	
	8.27	−1	−1	−1	−1	−1.67	−2.60
$2^3 \quad 1^4$	42	192	387	408	220	48	
	8.42	−1	−1	−1	−1.42	−2	−2

$k = 8, \quad n = 8$

1^8	28	112	210	224	140	48	7	
	7	−1	−1	−1	−1	−1	−1	−1

$k = 8, \quad n = 9$

$2 \quad 1^7$	35	154	315	364	245	90	14	
	7.80	−1	−1	−1	−1	−1	−1	−1.80

$k = 8, \quad n = 10$

$3 \quad 1^7$	42	196	420	504	350	132	21	
	8.48	−1	−1	−1	−1	−1	−1	−2.48
$2^2 \quad 1^6$	43	208	465	585	425	168	28	
	8.62	−1	−1	−1	−1	−1	−1.62	−2

$k = 9, \quad n = 9$

1^9	36	168	378	504	420	216	63	8	
	8	−1	−1	−1	−1	−1	−1	−1	−1

$k = 9, \quad n = 10$

$2 \quad 1^8$	44	224	546	784	700	384	119	16	
	8.82	−1	−1	−1	−1	−1	−1	−1	−1.82

$k = 10, \quad n = 10$

1^{10}	45	240	630	1008	1050	720	315	80	9	
	9	−1	−1	−1	−1	−1	−1	−1	−1	−1

Table 6.1. Characteristic polynomials of graphs consisting of a circuit C_5 and a path P_s, $1 \leqq s \leqq 7$

graph	12	11	10	9	8	7	6	5	4	3	2	1	0
⑤–P_1							1	0	−6	0	8	−2	−1
⑤–P_2						1	0	−7	0	13	−2	−6	2
⑤–P_3					1	0	−8	0	19	−2	−14	4	1
⑤–P_4				1	0	−9	0	26	−2	−27	6	7	−2
⑤–P_5			1	0	−10	0	34	−2	−46	8	21	−6	−1
⑤–P_6		1	0	−11	0	43	−2	−72	10	48	−12	−8	2
⑤–P_7	1	0	−12	0	53	−2	−106	12	94	−20	−29	8	1

Table 6.2. Characteristic polynomials of graphs consisting of a circuit C_6 and a path P_s, $1 \leqq s \leqq 7$

graph	13	12	11	10	9	8	7	6	5	4	3	2	1	0
⑥–P_1							1	0	−7	0	13	0	−7	0
⑥–P_2						1	0	−8	0	19	0	−16	0	4
⑥–P_3					1	0	−9	0	26	0	−29	0	11	0
⑥–P_4				1	0	−10	0	34	0	−48	0	27	0	−4
⑥–P_5			1	0	−11	0	43	0	−74	0	56	0	−15	0
⑥–P_6		1	0	−12	0	53	0	−108	0	104	0	−42	0	4
⑥–P_7	1	0	−13	0	64	0	−151	0	178	0	−98	0	19	0

Table 6.3. Characteristic polynomials of graphs consisting of a circuit C_7 and a path P_s, $1 \leqq s \leqq 7$

graph	14	13	12	11	10	9	8	7	6	5	4	3	2	1	0
⑦–P_1							1	0	−8	0	19	0	−13	−2	1
⑦–P_2						1	0	−9	0	26	0	−27	−2	8	2
⑦–P_3					1	0	−10	0	34	0	−46	−2	21	4	−1
⑦–P_4				1	0	−11	0	43	0	−72	−2	48	6	−9	−2
⑦–P_5			1	0	−12	0	53	0	−106	−2	94	8	−30	−6	1
⑦–P_6		1	0	−13	0	64	0	−149	−2	166	10	−78	−12	10	2
⑦–P_7	1	0	−14	0	76	0	−202	−2	272	12	−172	−20	40	8	−1

Table 6.4. Characteristic polynomials of graphs consisting of two circuits C_s and C_r with an edge in common, $5 \leqq s \leqq r \leqq 7$

graph	12	11	10	9	8	7	6	5	4	3	2	1	0
⑤⑤					1	0	−9	0	24	−4	−20	8	0
⑤⑥				1	0	−10	0	32	−2	−39	6	15	−4
⑤⑦			1	0	−11	0	41	−2	−61	6	31	−2	−4
⑥⑥			1	0	−11	0	41	0	−65	0	43	0	−9
⑥⑦		1	0	−12	0	51	0	−95	−2	76	6	−20	−4
⑦⑦	1	0	−13	0	62	0	−134	−4	129	16	−45	−12	0

Table 7. Simple eigenvalues of graphs whose automorphism group has two orbits of degrees r_1, r_2 for $1 \leqq r_2 \leqq r_1 \leqq 5$

$r_1 = 1,\ r_2 = 1$		
0.000		
1.000		

$r_1 = 2,\ r_2 = 1$		
0.000		
0.618	1	−1
1.000		
1.414	0	−2
1.618	−1	−1
2.000		

$r_1 = 2,\ r_2 = 2$		
0.000		
1.000		
1.414	0	−2
2.000		

$r_1 = 3,\ r_2 = 1$		
0.000		
0.414	2	−1
1.000		
1.732	0	−3
2.000		
2.414	−2	−1
3.000		

$r_1 = 3,\ r_2 = 2$		
0.000		
0.381	−3	1
0.414	2	−1
0.618	1	−1
0.732	2	−2
1.000		
1.302	1	−3
1.414	0	−2
1.561	1	−4
1.618	−1	−1
1.732	0	−3
2.000		
2.302	−1	−3
2.414	−2	−1
2.449	0	−6
2.561	−1	−4
2.618	−3	1
2.732	−2	−2
3.000		

$r_1 = 3,\ r_2 = 3$		
0.000		
0.414	2	−1
1.000		
1.561	1	−4
1.732	0	−3
2.000		
2.236	0	−5
2.414	−2	−1
2.561	−1	−4
3.000		

$r_1 = 4,\ r_2 = 1$		
0.000		
0.302	3	−1
0.381	−3	1
0.618	1	−1
0.732	2	−2
1.000		
1.302	1	−3
1.414	0	−2
1.618	−1	−1
2.000		
2.302	−1	−3
2.618	−3	1
2.732	−2	−2
3.000		
3.302	−3	−1
4.000		

$r_1 = 4,\ r_2 = 2$		
0.000		
0.561	3	−2
0.585	−4	2
0.732	2	−2
1.000		
1.236	2	−4
1.414	0	−2
1.561	1	−4
2.000		
2.561	−1	−4
2.732	−2	−2
2.828	0	−8
3.000		
3.236	−2	−4
3.414	−4	2
3.561	−3	−2
4.000		

$r_1 = 4,\ r_2 = 3$		
0.000		
0.267	−4	1
0.302	3	−1
0.381	−3	1
0.414	2	−1
0.561	3	−2
0.585	−4	2
0.618	1	−1
0.732	2	−2
0.791	3	−3
1.000		
1.236	2	−4
1.302	1	−3
1.381	−5	5
1.414	0	−2
1.449	2	−5
1.561	1	−4
1.618	−1	−1
1.645	2	−6
1.732	0	−3
1.791	1	−5
2.000		
2.192	1	−7
2.302	−1	−3
2.372	1	−8
2.414	−2	−1
2.449	0	−6
2.541	1	−9
2.561	−1	−4
2.618	−3	1
2.645	0	−7
2.732	−2	−2
2.791	−1	−5
3.000		
3.192	−1	−7
3.236	−2	−4
3.302	−3	−1
3.372	−1	−8
3.414	−4	2
3.449	−2	−5
3.464	0	−12
3.541	−1	−9
3.561	−3	−2
3.618	−5	5
3.645	−2	−6
3.732	−4	1
3.791	−3	−3
4.000		

Table 7 (continued)

$r_1 = 4,\ r_2 = 4$

0.000		
0.561	3	−2
0.585	−4	2
0.732	2	−2
1.000		
1.236	2	−4
1.414	0	−2
1.561	1	−4
1.645	2	−6
2.000		
2.372	1	−8
2.561	−1	−4
2.732	−2	−2
2.828	0	−8
3.000		
3.162	0	−10
3.236	−2	−4
3.372	−1	−8
3.414	−4	2
3.561	−3	−2
3.645	−2	−6
4.000		

$r_1 = 5,\ r_2 = 1$

0.000		
0.236	4	−1
0.267	−4	1
0.414	2	−1
0.561	3	−2
1.000		
1.561	1	−4
1.732	0	−3
2.000		
2.236	0	−5
2.414	−2	−1
2.561	−1	−4
3.000		
3.561	−3	−2
3.732	−4	1
4.000		
4.236	−4	−1
5.000		

$r_1 = 5,\ r_2 = 2$

0.000		
0.267	−4	1
0.302	3	−1
0.381	−3	1
0.414	2	−1
0.449	4	−2
0.585	−4	2
0.618	1	−1
0.697	−5	3
0.732	2	−2
0.791	3	−3
1.000		
1.192	3	−5
1.302	1	−3
1.381	−5	5
1.414	0	−2
1.449	2	−5
1.561	1	−4
1.618	−1	−1
1.645	2	−6
1.732	0	−3
1.791	1	−5
2.000		
2.236	0	−5
2.302	−1	−3
2.372	1	−8
2.414	−2	−1
2.449	0	−6
2.561	−1	−4
2.618	−3	1
2.732	−2	−2
2.791	−1	−5
3.000		
3.162	0	−10
3.302	−3	−1
3.372	−1	−8
3.414	−4	2
3.449	−2	−5
3.618	−5	5
3.645	−2	−6
3.732	−4	1
3.791	−3	−3
4.000		
4.192	−3	−5
4.302	−5	3
4.449	−4	−2
5.000		

$r_1 = 5,\ r_2 = 3$

0.000		
0,236	4	−1
0.267	−4	1
0.414	2	−1
0.438	−5	2
0.561	3	−2
0.645	4	−3

$r_1 = 5,\ r_2 = 3$

1.000		
1.372	3	−6
1.449	2	−5
1.561	1	−4
1.585	−6	7
1.732	0	−3
1.828	2	−7
2.000		
2.162	2	−9
2.236	0	−5
2.372	1	−8
2.414	−2	−1
2.561	−1	−4
2.645	0	−7
2.701	1	−10
3.000		
3.372	−1	−8
3.449	−2	−5
3.561	−3	−2
3.701	−1	−10
3.732	−4	1
3.828	−2	−7
3.872	0	−15
4.000		
4.162	−2	−9
4.236	−4	−1
4.372	−3	−6
4.414	−6	7
4.561	−5	2
4.645	−4	−3
5.000		

$r_1 = 5,\ r_2 = 4$

0.000		
0.208	−5	1
0.236	4	−1
0.267	−4	1
0.302	3	−1
0.381	−3	1
0.414	2	−1
0.438	−5	2
0.449	4	−2
0.561	3	−2
0.585	−4	2
0.618	1	−1
0.645	4	−3
0.697	−5	3
0.732	2	−2
0.791	3	−3
0.828	4	−4

Table 7 (continued)

$r_1 = 5$, $r_2 = 4$		
1.000		
1.192	3	−5
1.236	2	−4
1.267	−6	6
1.302	1	−3
1.372	3	−6
1.381	−5	5
1.414	0	−2
1.449	2	−5
1.541	3	−7
1.561	1	−4
1.585	−6	7
1.618	−1	−1
1.645	2	−6
1.701	3	−8
1.732	0	−3
1.791	1	−5
1.828	2	−7
2.000		
2.162	2	−9
2.192	1	−7
2.236	0	−5
2.302	−1	−3
2.316	2	−10
2.372	1	−8
2.381	−7	11
2.414	−2	−1
2.449	0	−6
2.464	2	−11
2.541	1	−9
2.561	−1	−4
2.605	2	−12
2.618	−3	1
2.645	0	−7
2.701	1	−10
2.732	−2	−2
2.791	−1	−5
2.854	1	−11
3.000		
3.140	1	−13
3.162	0	−10
3.192	−1	−7
3.236	−2	−4
3.302	−3	−1
3.316	0	−11

$r_1 = 5$, $r_2 = 4$		
3.372	−1	−8
3.405	1	−15
3.414	−4	2
3.449	−2	−5
3.464	0	−12
3.531	1	−16
3.541	−1	−9
3.561	−3	−2
3.605	0	−13
3.618	−5	5
3.645	−2	−6
3.701	−1	−10
3.732	−4	1
3.791	−3	−3
3.828	−2	−7
3.854	−1	−11
4.000		
4.140	−1	−13
4.162	−2	−9
4.192	−3	−5
4.236	−4	−1
4.302	−5	3
4.316	−2	−10
4.372	−3	−6
4.405	−1	−15
4.414	−6	7
4.449	−4	−2
4.464	−2	−11
4.472	0	−20
4.531	−1	−16
4.541	−3	−7
4.561	−5	2
4.605	−2	−12
4.618	−7	11
4.645	−4	−3
4.701	−3	−8
4.732	−6	6
4.791	−5	1
4.828	−4	−4
5.000		

$r_1 = 5$, $r_2 = 5$		
0.000		
0.236	4	−1
0.267	−4	1
0.414	2	−1
0.438	−5	2
0.561	3	−2
0.645	4	−3
1.000		
1.372	3	−6
1.449	2	−5
1.561	1	−4
1.585	−6	7
1.701	3	−8
1.732	0	−3
1.828	2	−7
2.000		
2.162	2	−9
2.236	0	−5
2.372	1	−8
2.414	−2	−1
2.464	2	−11
2.561	−1	−4
2.645	0	−7
2.701	1	−10
3.000		
3.274	1	−14
3.372	−1	−8
3.449	−2	−5
3.561	−3	−2
3.605	0	−13
3.701	−1	−10
3.732	−4	1
3.828	−2	−7
3.872	0	−15
4.000		
4.123	0	−17
4.162	−2	−9
4.236	−4	−1
4.274	−1	−14
4.372	−3	−6
4.414	−6	7
4.464	−2	−11
4.561	−5	2
4.645	−4	−3
4.701	−3	−8
5.000		

Table 8.1. Equivalence classes under Seidel switching of graphs with n vertices, $2 \leqq n \leqq 6$

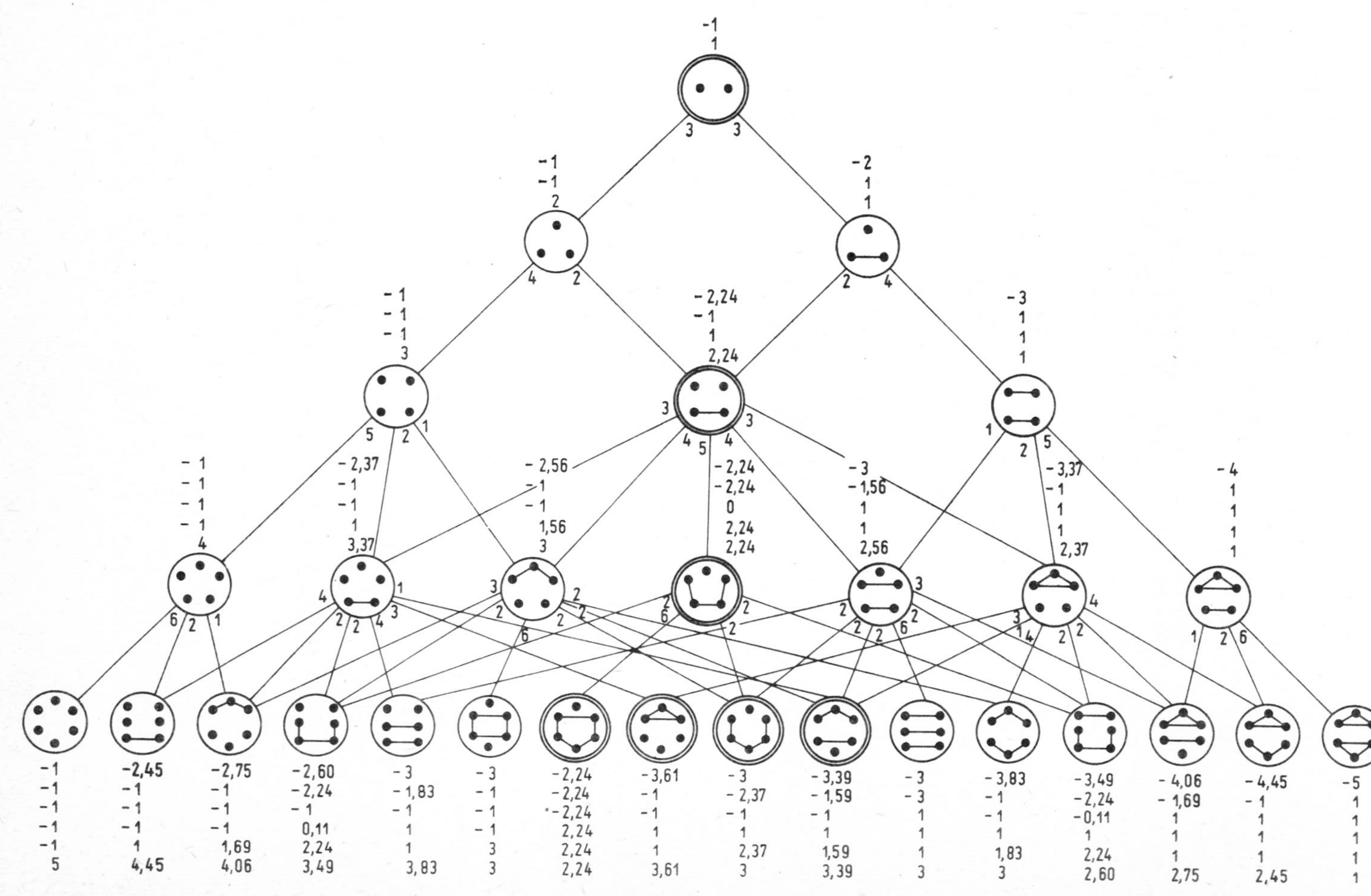

Table 8.2. Equivalence classes under Seidel switching of graphs with seven vertices

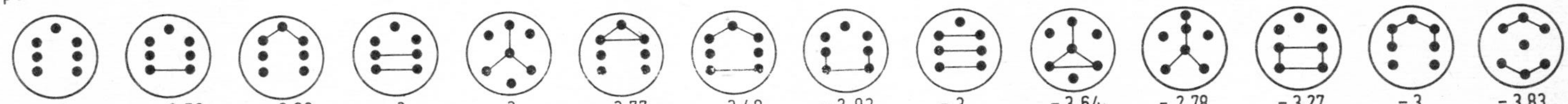

Type	1	2	3	4	5	6	7	8	9	10	11	12	13	14
	−1	−2,53	−2,90	−3	−3	−3,77	−3,40	−2,83	−3	−3,64	−2,78	−3,27	−3	−3,83
	−1	−1	−1	−2	−1	−1	−1,88	−2,24	−3	−1,60	−2,46	−1	−2,59	−1,63
	−1	−1	−1	−1	−1	−1	−1	−1	−1,37	−1	−1	−1	−1	−1
	−1	−1	−1	−1	−1	−1	−1	−1	1	−1	−1	−1	−1	−1
	−1	−1	−1	1	−1	1	1	0,15	1	1	0,29	−1	1	1,48
	−1	1	1,74	1	2	1	1,70	2,24	1	1,76	2,49	3	2,40	1,83
	6	5,53	5,16	5	5	4,77	4,58	4,68	4,37	4,48	4,46	4,27	4,19	4,14

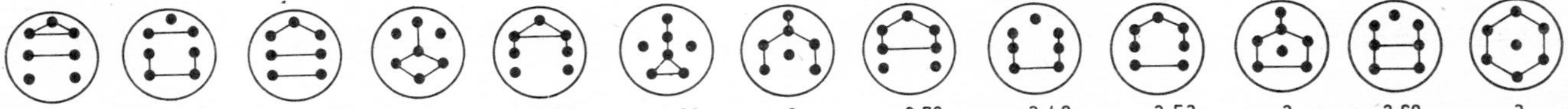

Type	15	16	17	18	19	20	21	22	23	24	25	26	27
	−4,12	−3,52	−3,50	−3	−3,39	−3,63	−3	−2,70	−3,49	−3,53	−3	−2,60	−3
	−2	−2,24	−3	−2,30	−2,24	−2,54	−2,49	−2,24	−2,44	−3	−2,24	−2,60	−3
	−1	−1,58	−1	−1	−1	−1	−1,83	−2,24	−1,51	−1,22	−2,24	−2	−2
	1	0,09	1	−1	−1	−0,02	−0,29	−1	−0,11	1	−0,56	0,11	1
	1	1	1	0,21	1,29	1	1	2,24	1,36	1	2,24	0,11	1
	1	2,24	1,62	3	2,24	2,33	2,78	2,24	2,60	2,32	2,24	3,49	3
	4,12	4,01	3,89	4,08	4,10	3,86	3,83	3,70	3,60	3,43	3,56	3,49	3

Bibliography

The bibliography consists of two parts:

Part I. Papers written in Roman characters;

Part II. Papers written in Cyrillic characters.

In Part I most of the papers are written in English, German, or French; the languages of a few other papers are indicated. In Part II the papers are written in Russian except for a few in which cases the language is again indicated.

The papers are arranged according to the alphabetic order of the authors' names. For papers which have more than one author, each name appears separately in the list.

Books are indicated by an asterisk*. The numbers at the end of the quotations refer to the sections where the papers or books are cited (P means Preface, A Appendix).

The bibliography contains 564 units, 527 of them in part I and 37 in Part II.

Part I

ABERTH, O.

[Aber] On the sum of graphs. Rev. Franç. Rech. Operat. **8** (1964), 353–358. (2.5; 7.4; 7.5)

ABRAMSON, M., *see* [MoAb]

ÁDÁM, Á.

[Ádám] Research problem 2–10. J. Comb. Theory **2** (1967), 393. (7.8)

AIGNER, M.

[Aign] The uniqueness of the cubic lattice graph. J. Comb. Theory **6** (1969), 282–297. (6.4)

AMIN, A. T.

[AmHa] (with S. L. HAKIMI), Upper bounds on the order of a clique of a graph. SIAM J. Appl. Math. **22** (1972), 569–573. (3.2; 3.6)

ANDERSON, S. S.

* [Ande] Graph Theory and Finite Combinatorics. Markham Publ. Co., Chicago 1970.

ANDERSON, Jr., W. N.

[AnMo] (with T. D. MORLEY), Eigenvalues of the Laplacian of a graph. University of Maryland Technical Report TR-71-45, 1971. (2.1; 9.3)

ASCHBACHER, M.

[Asch] The non-existence of rank three permutation groups of degree 3250 and subdegree 57. J. Algebra **19** (1971), 538–540. (6.5)

ASH, R. B.

[Ash] Topology and the solution of linear systems. J. Franklin Inst. **268** (1959), 453–463. (1.4)

BABAI, L.
[Bab1] Spectra of Cayley graphs. J. Comb. Theory (B). (to appear) (5.5; 6.1; 7.8)
[Bab2] Automorphism group and category of cospectral graphs. Acta Math. Acad. Sci. Hung. **31** (1978), 295–306. (5.2; 5.4; 6.1)
[Bab3] Kospektrale Graphen mit vorgegebenen Automorphismengruppen. Wiss. Z. TH Ilmenau **27** (1981) 4, 31–37. (5.4; 6.1)

BAKER, G. A.
[Bak1] Drum shapes and isospectral graphs. BNL 10088, Brookhaven National Laboratory, Long Island, New York 1966.
[Bak2] Drum shapes and isospectral graphs. J. Math. Phys. **7** (1966), 2238–2242. (6.1; 8.4)

BALABAN, A. T.
* [Bala] (ed. A. T. BALABAN) Chemical Applications of Graph Theory. Academic Press, London–New York–San Francisco 1976. (0)
[BaHa] (with F. HARARY), The characteristic polynomial does not uniquely determine the topology of a molecule. J. Chem. Doc. **11** (1971), 258–259. (6.1)

BANNAI, E.
[BaIt] (with T. ITO), On finite Moore graphs. J. Fac. Sci. Univ. of Tokyo **20** (1973), 191–208. (6.2)

BASSALYGO, L. A.
[Bass] Generalization of Lloyd's Theorem to arbitrary alphabet. Problems of Control and Information Theory **2** (1973), 25–28. (4.8)

BEHZAD, M.
[BeCh1] (with G. CHARTRAND), An introduction to total graphs. In: Theory of Graphs (ed. P. ROSENSTIEHL), Dunod, Paris/Gordon and Breach, New York, 1967, pp. 31–33.
* [BeCh2] (with G. CHARTRAND), Introduction to the Theory of Graphs. Allyn and Bacon Inc., Boston 1971. (0)
[BeCN] (with G. CHARTRAND and E. A. NORDHAUS), Triangles in line graphs and total graphs. Indian J. Math. **10** (1968), No. 2, 109–120. (7.8)

BEINEKE, L.
[Bein] Characterization of derived graphs. J. Comb. Theory **9** (1970), 129–135. (6.3; A)

BENSON, C. T.
[Bens] Minimal regular graphs of girth eight and twelve. Canad. J. Math. **18** (1966), 1091–1094.
[BeJa] (with J. B. JACOBS), On hearing the shape of combinatorial drums. J. Comb. Theory (B) **13** (1972), 170–178. (6.1; 8.4)
[BeLo] (with N. E. LOSEY), On a graph of Hoffman and Singleton. J. Comb. Theory (B) **11** (1971), 67–79. (6.2)

BERGE, C.
* [Ber1] Théorie des Graphes et ses Applications. Dunod, Paris 1958. (0; 2.5; 3.5; 7.8)
* [Ber2] Graphes et Hypergraphes. Dunod, Paris 1970. (0)
* [Ber3] Graphs and Hypergraphs. (Translation of [Ber2].) North-Holland Publ. Co., Amsterdam–London–New York 1973. (0)

BERLEKAMP, E. R.
* [BeLS] (with J. H. VAN LINT and J. J. SEIDEL), A strongly regular graph derived from the perfect ternary Golay code. In: A Survey of Combinatorial Theory (ed. J. N. SRIVASTAVA, F. HARARY, C. R. RAO, G.-C. ROTA, S. S. SHRIKHANDE), North-Holland Publ. Co., Amsterdam–London–New York 1973, pp. 25–30.

BHAGAWANDAS
[BhSh] (with S. S. SHRIKHANDE), Seidel-equivalence of strongly regular graphs. Indian J. Statistics, Series A, **30** (1968), 359–368.
See also [ShBh]

BIGGS, N. L.

[Big1] Intersection matrices for linear graphs. In: Combinatorial Mathematics and Its Applications (ed. D. J. A. WELSH), Academic Press, London—New York 1971, pp. 15—23. (7.2)

* [Big2] Finite Groups of Automorphisms. Cambridge University Press, Cambridge 1971. (5.2; 7.2)

[Big3] Perfect codes in graphs. J. Comb. Theory (B) **15** (1973), 289—296. (4.8; 7.2)

[Big4] Three remarkable graphs. Canadian J. Math. **25** (1973), 397—411. (7.2)

* [Big5] Algebraic Graph Theory. Cambridge University Press, Cambridge 1974. (P; 4.2; 5; 5.1; 7.2)

[Big6] Perfect codes and distance-transitive graphs. In: Combinatorics, Proc. British Comb. Conf. 1973, Cambridge 1974, pp. 1—8. (7.2)

[Big7] Designs, factors and codes in graphs. Quart. J. Math. Oxford (2) **26** (1975), 113—119. (7.2)

[Big8] Chromatic and thermodynamic limits. J. Phys. A: Math. Gen. **8** (1975), No. 10, L110—L112.

[Big9] Automorphic graphs and the Krein condition. Geometriae Dedicata **5** (1976), No. 1, 117—127.

[BiMe] (with G. H. J. MEREDITH), A theorem on planar partitions. In: Proc. Fifth British Combinatorial Conference, Aberdeen 1975 (ed. C. ST. J. A. NASH-WILLIAMS and J. SHEEHAN). Utilitas Mathematica Publishing Inc., Winnipeg 1976, pp. 73—78.

[BiSm] (with D. H. SMITH), On trivalent graphs. Bull. London Math. Soc. **3** (1971), 155—158. (7.2)

BOERNER, H.

* [Boer] Darstellungen von Gruppen mit Berücksichtigung der Bedürfnisse der modernen Physik. Springer-Verlag, Berlin—Göttingen—Heidelberg 1955; Representations of Groups. North-Holland Publ. Co., Amsterdam 1963. (5.2; 5.3)

BONDY, J. A.

* [BoMu] (with U. S. R. MURTY), Graph Theory with Applications. American Elsevier Publishing Co., Inc., New York 1976. (0)

BOSE, N. K.

[BoFS] (with R. FEICK and F. K. SUN), General solution to the spanning tree enumeration problem in multigraph wheels. IEEE Trans. Circuit Theory **20** (1973), 69—70. (7.6)

BOSE, R. C.

[Bos1] Strongly regular graphs, partial geometries and partially balanced designs. Pacific J. Math. **13** (1963), 389—419. (3.4; 7.2)

[Bos2] Graphs and designs. In: Finite Geometric Structures and Their Applications, Edizioni Cremonese, Roma 1973, pp. 1—104.

[BoDo] (with T. A. DOWLING), A generalization of Moore graphs of diameter two. J. Comb. Theory (B) **11** (1971), 213—276. (6.2)

[BoLa] (with R. LASKAR), Eigenvalues of the adjacency matrix of tetrahedral graphs. Inst. Statist. mimeo series 571, University of North Carolina, 1968; Aequationes Math. **4** (1970), 37—43. (6.4)

[BoMe] (with D. M. MESSNER), On linear associative algebras corresponding to association schemes of partially balanced designs. Ann. Math. Statist. **30** (1959), 21—36. (6.4; 7.2)

[BoSh] (with T. SHIMAMOTO), Classification and analysis of partially balanced incomplete block designs with two associate classes. J. Amer. Stat. Assn. **47** (1952), 151—184. (6.4)

BOTTEMA, O.

[Bott] Über die Irrfahrt in einem Straßennetz. Math. Z. **39** (1935), 137—145. (8.5)

BRAUER, A.

[BrGe] (with J. C. GENTRY), Some remarks on tournament matrices. Linear Algebra and Appl. **5** (1972), 311—318.

BRAUMAN, J. I., *see* [StBr]

BRIDGES, W. G.

[Bri 1] The polynomial of a non-regular digraph. Pacific J. Math. **38** (1971), 325–341. (3.1)

[Bri 2] A class of normal (0,1)-matrices. Canad. J. Math. **25** (1973), 621–626.

BROOKS, R. L.

[BrSST] (with C. A. B. SMITH, A. H. STONE, and W. T. TUTTE), Dissection of a rectangle into squares. Duke Math. J. **7** (1940), 312–340. (1.5)

BROWN, T. A., *see* [RoBr]

BROWN, W.

[Brow] On the non-existence of a type of regular graphs of girth 5. Canad. J. Math. **19** (1967), 644–648. (7.1)

BRUALDI, R.

[Brua] Kronecker product of fully indecomposible matrices and of ultra strong digraphs. J. Comb. Theory **2** (1967), 135–139. (7.4)

BRUCK, R. H.

[Bruc] Finite nets, II. Uniqueness and imbedding. Pacific J. Math. **13** (1963), 421–457. (6.1)

BUSSEMAKER, F. C.

[BuČCS] (with S. ČOBELJIĆ, D. M. CVETKOVIĆ, and J. J. SEIDEL), Computer investigation of cubic graphs. Technological University Eindhoven, T. H.-Report 76-WSK-01; Cubic graphs on $\leqq 14$ vertices, J. Comb. Theory (B) **23** (1977), 234–235. (6.1; 9.3; A)

[BuCS] (with D. M. CVETKOVIĆ and J. J. SEIDEL), Graphs related to exceptional root systems. In: Combinatorics (Coll. Mat. Soc. J. Bolyai **18**, ed. A. HAJNAL and V. SÓS). North-Holland Publ. Co., Amsterdam–Oxford–New York 1978, Vol. I, pp. 185–191; Technological University Eindhoven, T. H.-Report 76-WSK-05, 1–91. (3.6; 6.1; 6.3)

[BuCv] (with D. M. CVETKOVIĆ), There are exactly 13 connected, cubic, integral graphs. Univ. Beograd Publ. Elektrotehn. Fak., Ser. Mat. Fiz., No. 544–No. 576 (1976), 43–48. (6.1; 7.1; 9.4)

[BuS 1] (with J. J. SEIDEL), Cubical graphs of order $2n \leqq 10$. TH Eindhoven, Onderafdeling der Wiskunde, Notitie nr. 10, September 1968. (A)

[BuS 2] (with J. J. SEIDEL), Symmetric Hadamard matrices of order 36. Technological University Eindhoven, Report 70-WSK-02; Ann. N. Y. Acad. Sci. **175** (1970), 66–79. (6.1; 6.6; 7.2)

CAMERON, P. J.

[Came] Partial quadrangles. Quart. J. Math. Oxford (2) **26** (1975), 61–73.

[CaGSS] (with J. M. GOETHALS, J. J. SEIDEL, and E. E. SHULT), Line graphs, root systems, and elliptic geometry. J. Algebra **43** (1976), 305–327. (6.3)

* [CaLi] (with J. H. VAN LINT), Graph Theory, Coding Theory and Block Designs. London Math. Soc. Lect. Note Ser. **19**, Cambridge University Press, Cambridge–London–New York–Melbourne 1975. (7.2)

[CaSe] (with J. J. SEIDEL), Quadratic forms over GF(2). Indag. Math. **35** (1973), No. 1, 1–8.

CARLITZ, L.

[Carl] Enumeration of certain types of sequences. Math. Nachr. **49** (1971), 125–147. (7.5)

CARTER, R. W.

* [Cart] Simple Groups of Lie Type. John Wiley & Sons, Interscience, Chichester 1972. (6.3)

CHANG, L. C.

[Cha 1] The uniqueness and non-uniqueness of the triangular association scheme. Sci. Record **3** (1959), 604–613.

[Cha 2] Association schemes of partially balanced block designs with parameters $v = 28$, $n_1 = 12$, $n_2 = 15$ and $p_{11}^2 = 4$. Sci. Record **4** (1960), 12–18. (6.1; 6.3)

CHAO, C.-Y.
[Chao] A note on the eigenvalues of a graph. J. Comb. Theory (B) **10** (1971), 301–302. (5.2)

CHARTRAND, G., *see* [BeCH 1], [BeCh 2], [BeCN]

CHELNOKOV, V. M., *see* [KeCh]

CHEN, W.-K.
* [Chen] Applied Graph Theory. Graphs and Electrical Networks North-Holland Publ. Co., Amsterdam–London–New York 1971; Amsterdam–New York–Oxford 1976. (1.4)

CLAPHAM, C. R. J.
[Clap] Triangles in self-complementary graphs. J. Comb. Theory (B) **15**, (1973), 74–76. (3.3)

CLARKE, F. H.
[Clar] A graph polynomial and its applications. Discrete Math. **3** (1972), 305–313. (1.4; 2.3)

CLATWORTHY, W. H., *see* [CoCl]

COATES, C. L.
[Coat] Flow-graph solutions of linear algebraic equations. IRE Trans. Circuit Theory CT-**6** (1959), 170–187. (1.4; 1.9)

ČOBELJIĆ, S., *see* [BuČCS]

COLLATZ, L.
* [Col 1] Eigenwertaufgaben mit technischen Anwendungen. Akad. Verlagsges. Geest & Portig KG, Leipzig 1963. (2.7; 8.4; A)

[Col 2] Einige Beziehungen zwischen Graphen, Geometrie und Kombinatorik. In: Numerische Methoden bei graphentheoretischen und kombinatorischen Problemen (ed. L. COLLATZ, G. MEINARDUS, H. WERNER), Birkhäuser, Basel–Stuttgart **1975**, pp. 27–56.

[CoSi 1] (with U. SINOGOWITZ), Spektren endlicher Grafen. Abh. Math. Sem. Univ. Hamburg **21** (1957), 63–77. (0.1; 0.3; 1.4; 2.7; 3.2; 3.3; 4.6; 6.1; 7.5; 8.1; 8.4; 9.5; A)

[CoSi 2] (with U. SINOGOWITZ), Spektren endlicher Grafen. Fiat Reviews of German Science 1939–1946, Pure Mathematics, part II, 251–252. (1.4)

CONNOR, W. S.
[Conn] The uniqueness of the triangular association scheme. Ann. Math. Statist. **29** (1958), 262–266.

[CoCl] (with W. H. CLATWORTHY), Some theorems for partially balanced designs. Ann. Math. Statist. **25** (1954), 100–112. (7.2)

COOK, C. R.
[Cook] A note on the exceptional graph of the cubic lattice graph characterization. J. Comb. Theory (B) **14** (1973), 132–136. (6.1; 6.4)

COOPER, C. D. H.
[Coop] On the maximum eigenvalue of a reducible non-negative real matrix. Math. Z. **131** (1973), 213–217.

COTOH, M., *see* [HoMC]

COULSON, C. A.
[Cou 1] Exited electronic levels in conjugated molecules. Proc. Phys. Soc. **60** (1948), 257–269. (8.1)

[Cou 2] Notes on the secular determinant in molecular orbital theory. Proc. Cambridge Phil. Soc. **46** (1949), 202–205. (1.4)

[CoJa] (with J. JACOBS), Conjugation across a single bond. J. Chem. Soc. (1949), 2805 to 2812.

[CoLo] (with H. C. LONGUET-HIGGINS), The structure of conjugated systems, II: Unsaturated hydrocarbons and their heteroderivatives. Proc. Roy. Soc. (London), Ser. A, **192** (1947), 16–32. (3.2)

[CoRu] (with G. S. RUSHBROOKE), Note on the method of molecular orbitals. Proc. Cambridge Phil. Soc. **36** (1940), 193–200. (3.2; 8.1)

* [CoSt] (with A. STREITWIESER), Dictionary of π-Electron Calculations. Pergamon Press Inc., San Francisco 1965. (A)

COXETER, H. S. M.
* [Coxe] Regular Polytopes. Methuen, London 1948. (6.3)

ČULIK, K.
[Čuli] Zur Theorie der Graphen. Časopis Pěst. Mat. **83** (1958), 133–155. (2.5)

CURTIS, C. W.
* [CuRe] (with I. REINER), Representation Theory of Finite Groups and Associative Algebras. Interscience Publs., John Wiley & Sons, New York–London 1962. (5.2)

See also [KэPa]

CVETKOVIĆ, D. M.
[Cve 1] Bihromatičnost i spektar grafa (Serbo-Croatian). Mat. biblioteka, No. 41, Beograd 1969; pp. 193–194. (3.2)
[Cve 2] Connectedness of the p-sum of graphs. Univ. Beograd Publ. Elektrotehn. Fak., Ser. Mat. Fiz., No. 274–No. 301 (1969), 96–99. (2.5; 7.4)
[Cve 3] Spectrum of the graph of n-tuples. Univ. Beograd Publ. Elektrotehn. Fak., Ser. Mat. Fiz., No. 274–No. 301 (1969), 91–96. (2.6)
[Cve 4] A note on paths in the p-sum of graphs. Univ. Beograd Publ. Elektrotehn. Fak., Ser. Mat. Fiz., No. 302–No. 319 (1970), 49–51. (7.5)
[Cve 5] The Boolean operations on graphs — spectrum and connectedness. Booleove operacije nad grafovima — spektar i povezanost (summary). V. Kongres mat., fiz. i astr. na Jugoslavija, Ohrid 1970; Zbornik na trudovite, tom I, Skopje 1973; pp. 115–119. (2.5; 2.7)
[Cve 6] Die Zahl der Wege eines Grafen. Glasnik Mat., Ser. III., **5** (25) (1970), 205–210. (7.5)
[Cve 7] New characterizations of the cubic lattice graph. Publ. Inst. Math. (Beograd) **10** (24) (1970), 195–198. (3.3; 6.1; 6.4)
[Cve 8] The generating function for variations with restrictions and paths of the graph and self-complementary graphs. Univ. Beograd Publ. Elektrotehn. Fak., Ser. Mat. Fiz., No. 320–No. 328 (1970), 27–34. (1.8; 3-3; 7.5)
[Cve 9] Graphs and their spectra (Grafovi i njihovi spektri) (Thesis). Univ. Beograd Publ. Elektrotehn. Fak., Ser. Mat. Fiz., No. 354–No. 356 (1971), 1–50. (1.8; 2.2; 2.4; 2.5; 2.6; 3.2; 3.3; 3.6; 6.1; 6.3; 6.6; 7.4; 7.5; 7.8; 8.3)
[Cve 10] The spectral method for determining the number of trees. Publ. Inst. Math. (Beograd) **11** (25) (1971), 135–141. (7.5; 7.8)
[Cve 11] Chromatic number and the spectrum of a graph. Publ. Inst. Math. (Beograd) **14** (28) (1972), 25–38. (3.2; 3.6; A)
[Cve 12] Inequalities obtained on the basis of the spectrum of the graph. Studia Sci. Math. Hung. **8** (1973), 433–436. (3.2; 3.6)
[Cve 13] Spectrum of the total graph of a graph. Publ. Inst. Math. (Beograd) **16** (30) (1973), 49–52. (2.4; 7.8)
[Cve 14] On a graph theory problem of M. Koman. Časopis Pěst. Mat. **98** (1973), 233–236. (7.5)
[Cve 15] The determinant concept defined by means of graph theory. Mat. Vesnik **12** (27) (1975), 333–336. (Dutch translation: Definite en berekening van determinanten met behulp van grafen, Nieuw Tijdschrift voor Wiskunde **63** (1976), 209–215.)
[Cve 16] Cubic integral graphs. Univ. Beograd Publ. Elektrotehn. Fak., Ser. Mat. Fiz., No. 498–No. 541 (1975), 107–113. (6.6; 7.1; 9.4)
[Cve 17] Spectra of graphs formed by some unary operations. Publ. Inst. Math. (Beograd) **19** (33) (1975), 37–41. (2.4)
[Cve 18] The main part of the spectrum, divisors and switching of graphs. Publ. Inst. Math. (Beograd) **23** (37) (1978), 31–38. (4.8)
[CvG 1] (with I. GUTMAN), The algebraic multiplicity of the number zero in the spectrum of a bipartite graph. Mat. Vesnik **9** (24) (1972), 141–150. (8.1; 8.5)

[CvG 2] (with I. GUTMAN), Kekulé structures and topology, II. Cata-condensed systems. Croat. Chem. Acta **46** (1974), 15–23. (8.3)

[CvG 3] (with I. GUTMAN), On spectral structure of graphs having the maximal eigenvalue not greater than two. Publ. Inst. Math. (Beograd) **18** (32) (1975), 39–45. (2.7; 6.1; 7.1)

[CvG 4] (with I. GUTMAN), Note on branching. Croat. Chem. Acta **49** (1977), 115–121. (1.8)

[CvGT 1] (with I. GUTMAN and N. TRINAJSTIĆ), Graph theory and molecular orbitals, II. Croat. Chem. Acta **44** (1972), 365–374. (8.1; 8.2)

[CvGT 2] (with I. GUTMAN and N. TRINAJSTIĆ), Kekulé structures and topology. Chem. Phys. Letters **16** (1972), 614–616. (8.2)

[CvGT 3] (with I. GUTMAN and N. TRINAJSTIĆ), Conjugated molecules having integral graph spectra. Chem. Phys. Letters **29** (1974), 65–68. (2.7; 9.4)

[CvGT 4] (with I. GUTMAN and N. TRINAJSTIĆ), Graph theory and molecular orbitals, VII. The role of resonance structures. J. Chem. Phys. **61** (1974), 2700–2706. (8.2; 8.5)

[CvGT 5] (with I. GUTMAN and N. TRINAJSTIĆ), Graph theory and molecular orbitals, IX. On the stability of cata-condensed hydrocarbons. Theoret. Chim. Acta **34** (1974), 129–136. (8.1)

[CvGT 6] (with I. GUTMAN and N. TRINAJSTIĆ), Graphical studies on the relations between the structure and reactivity of conjugated systems: The role of non-bonding molecular orbitals. J. Mol. Struct. **28** (1975), 289–303. (8.1)

[CvLi] (with J. H. VAN LINT), An elementary proof of Lloyd's theorem. Proc. Kon. Nederl. Acad. Wetensch., Ser. A, **80** (1) (1977), 6–10. (4.8)

[CvL 1] (with R. P. LUČIĆ), A new generalization of the concept of the p-sum of graphs. Univ. Beograd Publ. Elektrotehn. Fak., Ser. Mat. Fiz., No. 302–No. 319 (1970), 67–71. (2.5)

[CvL 2] (with R. P. LUČIĆ), Über die Zerlegung eines Graphen in ein Produkt von Graphen. XVIII. Int. Wiss. Koll. TH Ilmenau 1973, Reihe A2, 57–58. (7.8)

* [CvMi] (with M. MILIĆ), Teorija grafova i njene primene (Serbo-Croatian). Univerzitet u Beogradu, Beograd 1971.

[CvS 1] (with S. K. SIMIĆ), On enumeration of certain type of sequences. Univ. Beograd Publ. Elektrotehn. Fak., Ser. Mat. Fiz., No. 412–No. 460 (1973), 159–164. (1.8; 7.5)

[CvS 2] (with S. K. SIMIĆ), Graph equations for line graphs and total graphs. Discrete Math. **13** (1975), 315–320. (7.8)

See also [BuČCS], [BuCS], [BuCv], [GuCv], [KrC 1], [KrC 2]

DAMERELL, R. M.

[Dam 1] On Moore graphs. Proc. Cambridge Phil. Soc. **74** (1973), 227–236. (6.2)

[Dam 2] Orthogonal transformations of distance-regular graphs. A lecture given at the Conference on Finite Geometries and Designs, Brighton 1975.

DELSARTE, P.

[Del 1] An algebraic approach to the association schemes of coding theory. Philips Res. Repts. Suppl. **10** (1973). (3.6; 4.8)

[Del 2] Association schemes in certain lattices. MBLE Res. Lab., Report 241, Brussels (1974).

[Del 3] Association schemes and t-designs in regular semilattices. J. Comb. Theory (A) **20** (1976), 230–243.

[Del 4] The association schemes of coding theory. In: Combinatorics (ed. M. HALL, Jr., J. H. VAN LINT), Part 1, Math. Centre Tracts **55**, Amsterdam 1974, pp. 139–157. (6.4)

[DeGS 1] (with J. M. GOETHALS and J. J. SEIDEL), Orthogonal matrices with zero diagonal, II. Canad. J. Math. **23** (1971), 816–832.

[DeGS 2] (with J. M. GOETHALS and J. J. SEIDEL), Spherical codes and designs. Geometriae Dedicata **6** (1977), 363–388.

DEMBOWSKI, P.
* [Demb] Finite Geometries. Springer-Verlag, Berlin—Heidelberg—New York 1968. (6.4)

DEO, N.
* [Deo] Graph Theory with Applications to Engineering and Computer Science. Prentice-Hall, Englewood Cliffs, N. J., 1974. (0)

DESOER, C. A.
[Deso] The optimum formula for the gain of a flow-graph or a simple derivation of Coates' formula. Proc. IRE **48** (1960), 883—889. (1.4)

DEWAR, M. S. J.
[DeLo] (with H. C. LONGUET-HIGGINS), The correspondence between the resonance and molecular orbital theories. Proc. Roy. Soc. (London) A **214** (1952), 482—493. (8.2; 8.5)

DJOKOVIĆ, D. Ž.
[Djok] Isomorphism problem for a special class of graphs. Acta Math. Acad. Sci. Hung. **21** (1970), 267—270. (5.1; 6.1; 7.8)

DONATH, W. E.
[DoHo] (with A. J. HOFFMAN), Lower bounds for the partitioning of graphs. IBM J. Res. Develop. **17** (1973), 420—425. (3.6)

DOOB, M.
[Doo 1] On characterizing a line graph by the spectrum of its adjacency matrix. Ph. D. thesis, The City University of New York, 1969. (6.6)
[Doo 2] A geometrical interpretation of the least eigenvalue of a line graph. In: Proc. Second Conference on Comb. Math. and Appl., Chapel Hill, N. C., 1970, pp. 126—135. (3.5; 6.3; 6.6)
[Doo 3] Graphs with a small number of distinct eigenvalues. Ann. New York Acad. Sci. **175** (1970), No. 1, 104—110. (3.5; 6.2; 6.6)
[Doo 4] On characterizing certain graphs with four eigenvalues by their spectra. Linear Algebra and Appl. **3** (1970), 461—482. (6.3)
[Doo 5] On the spectral characterization of the line graph of a BIBD. In: Proc. Second Louisiana Conference on Combinatorics, Graph Theory and Computing, 1970, pp. 225—234.
[Doo 6] On the spectral characterization of the line graph of a BIBD, II. In: Proc. Manitoba Conference on Numerical Mathematics, University of Manitoba 1971; pp. 117—126. (6.1)
[Doo 7] On graph products and association schemes. Utilitas Math. **1** (1972), 291—302. (6.1)
[Doo 8] An interrelation between line graphs, eigenvalues, and matroids. J. Comb. Theory (B) **15** (1973), 40—50. (6.3; 6.6)
[Doo 9] A spectral characterization of the line graph of a BIBD with $\lambda = 1$. Linear Algebra and Appl. **12** (1975), 11—20. (6.3)
[Doo 10] Eigenvalues of a graph and its imbeddings. J. Comb. Theory (B) **17** (1974), 244—248 (6.1; 9.1)
[Doo 11] A note on prime graphs. Utilitas Math. **9** (1976), 297—300. (7.5)
[Doo 12] On imbedding a graph in an isospectral family. In: Proc. 2nd Manitoba Conference Numer. Math., Winnipeg, Man., 1972 (1973), pp. 137—142.
[Doo 13] A note on eigenvalues of a line graph. In: Proc. Conf. on Algebraic Aspects of Combinatorics, Toronto 1975, Utilitas Math., Winnipeg, Man., 1975, pp. 209—211.
[Doo 14] Some asymptotic spectral properties of a graph. In: Proc. 3rd Manitoba Conference Numer. Math., Utilitas Math., Winnipeg, Man., 1974, p. 160.

DOWLING, T. A., *see* [BoDo]

DUBSKÝ, J., *see* [HoDKP], [HoDKT], [HoDT]

DULMAGE, A. L.
[DuMe] (with N. S. MENDELSOHN), Graphs and matrices. In: Graph Theory and Theoretical Physics (ed. F. HARARY), Academic Press, London—New York 1967, pp. 167—227. (0.3; 3.1)

EDELBERG, M.
[EdGG] (with M. R. GAREY and R. L. GRAHAM), On the distance matrix of a tree. Discrete Math. **14** (1976), 23–39. (6.1; 9.2; A)
ELLZEY, JR., M. L., *see* [HeEl]
ELSPAS, B.
[ElTu] (with J. TURNER), Graphs with circulant adjacency matrices. J. Comb. Theory **9** (1970), 297–307. (5.5; 6.2; 7.8)
ERDÖS, P.
[ErSa] (with H. SACHS), Reguläre Graphen gegebener Taillenweite mit minimaler Knotenzahl. Wiss. Z. Univ. Halle **12** (1963), 251–257. (7.1)

FABIAN, J.
[FaHa] (with H. HARTMANN), π-electronic structure of polymethines. J. Mol. Struct. **27** (1975), 67–78.
FARZAN, M.
[Far1] Matrix Methods in Graph Theory. University of Wales thesis, Swansea 1974. (2.5)
[Far2] Automorphisms of double covers of a graph. In: Problèmes Combinatoires et Théorie des Graphes, Coll. Int. C. N. R. S., No. 260, Orsay 1976. C. N. R. S. Publ. 1978 (ed. J.-C. BERMOND, J.-C. FOURNIER, M. LAS VERGNAS, D. SOTTEAU), pp. 137–138. (4.2)
[FaW1] (with D. A. WALLER), Local joins and lexicographic products of graphs. Bull. Iranian Math. Soc. **1** (2) (1974), 1–17. (2.5)
[FaW2] (with D. A. WALLER), Kronecker products and local joins of graphs. Canad. J. Math. **29** (1977), 255–269. (2.5)
FEICK, R., *see* [BoFS]
FEIT, W.
[FeHi] (with G. HIGMAN), The non-existence of certain generalized polygons. J. Algebra **1** (1964), 114–131. (7.1)
FIEDLER, M.
[Fie1] Algebraic connectivity of graphs. Czechoslovak. Math. J. **23** (98) (1973), 298–305. (1.2; 9.3)
[Fie2] Algebraische Zusammenhangszahl der Graphen und ihre numerische Bedeutung. In: Numerische Methoden bei graphentheoretischen und kombinatorischen Problemen (ed. L. COLLATZ, G. MEINARDUS, H. WERNER), Birkhäuser, Basel–Stuttgart 1975, pp. 69–85. (9.3)
[Fie3] An algebraic approach to connectivity of graphs. In: Recent Advances in Graph Theory (ed. M. FIEDLER), Academia Praha 1975, pp. 193–196. (3.6)
[Fie4] A property of eigenvectors of non-negative symmetric matrices and its application to graph theory. Czechoslovak. Math. J. **25** (100) (1975), 619–633.
[Fie5] Eigenvectors of acyclic matrices. Czechoslovak. Math. J. **25** (100) (1975), 607–618.
[FiSe] (with J. SEDLÁČEK), O w-basich orientovaných grafů (Czech). Časopis Pěst. Mat. **83** (1958), 214–225. (1.5; 1.9)
FINCK, H.-J.
[Finc] Vollständiges Produkt, chromatische Zahl und charakteristisches Polynom regulärer Graphen, II. Wiss. Z. TH Ilmenau **11** (1965), 81–87. (3.3; 3.6; 6.2)
[FiGr] (with G. GROHMANN), Vollständiges Produkt, chromatische Zahl und charakteristisches Polynom regulärer Graphen, I. Wiss. Z. TH Ilmenau **11** (1965), 1–3. (2.2; 3.3; 7.8)
[FiSa] (with H. SACHS), Über Beziehungen zwischen Struktur und Spektrum regulärer Graphen. Wiss. Z. TH Ilmenau **19** (1973), 83–99. (4.1; 4.7; 4.8)
FISHER, M.
[Fish] On hearing the shape of a drum. J. Comb. Theory **1** (1966), 105–125. (6.1; 8.4)
FREEDMAN, H. D.
[Free] On the impossibility of certain Moore graphs. J. Comb. Theory (B) **10** (1971), 245–252. (6.2)

FRIED, M.
[FrSm] (with J. H. SMITH), Primitive groups, Moore graphs, and rational curves. Mich. J. Math. **19** (1972), 341–346.
FUJII, Y., *see* [YaFH]

GANTMACHER, F. R.
* [Gant] Theory of Matrices I, II (2 vol.). Chelsea, New York 1960 (Translated from Russian). German edition: Matrizenrechnung, Teil I, II (2 vol.). Third ed., VEB Deutscher Verlag der Wissenschaften, Berlin 1970, 1971.
Second Russian edition: ГАНТМАХЕР, Ф. Р., Теория матриц. Изд. „Наука", Москва 1966. (0; 0,3; 1.4; 2.4; 3.1)
GARDINER, A.
[Gard] Antipodal covering of graphs. J. Comb. Theory (B) **16** (1974), 255–273. (4.2; 4.5)
GAREY, M. R., *see* [EdGG]
GENTRY, J. C., *see* [BrGe]
GEWIRTZ, A.
[Gew1] Graphs with maximal even girth. Thesis, City University of New York, 1967. (7.1)
[Gew2] Graphs with maximal even girth. Canad. J. Math. **21** (1969), 915–934. (7.1)
GIBBS, R. A.
[Gib1] Self-Complementary Graphs: Their Structural Properties and Adjacency Matrices. Ph. D. thesis, Michigan State University, 1970.
[Gib2] Self-complementary graphs. J. Comb. Theory (B) **16** (1974), 106–123. (6.6)
GODSIL, C. D.
[Gods] Graphs, groups and polytopes. In: Combinatorial Mathematics (Lecture Notes in Mathematics 686, ed. D. A. HOLTON, J. SEBERRY), Springer-Verlag, Berlin–Heidelberg–New York 1978, pp. 157–164. (5.2; 5.5)
[GoM1] (with B. MCKAY), Products of graphs and their spectra. In: Combinatorial Mathematics IV (Lecture Notes in Mathematics 560, ed. L. R. A. CASSE, W. D. WALLIS), Springer-Verlag, Berlin–Heidelberg–New York 1976, pp. 61–72. (2.5)
[GoM2] (with B. MCKAY), Some computational results on the spectra of graphs. In: Combinatorial Mathematics IV (Lecture Notes in Mathematics 560, ed. L. R. A. CASSE, W. D. WALLIS), Springer-Verlag, Berlin–Heidelberg–New York 1976, pp. 73–92. (5.2; 6.1)
GOETHALS, J. M.
[GoS1] (with J. J. SEIDEL), Orthogonal matrices with zero diagonal. Canad. J. Math. **19** (1967), 1001–1010.
[GoS2] (with J. J. SEIDEL), Quasisymmetric block designs. In: Combinatorial Structures and Their Applications (ed. R. GUY, H. HANANI, N. SAUER, J. SCHÖNHEIM). Gordon and Breach, Science Publishers, Inc., New York–London–Paris 1970, pp. 111–116.
[GoS3] (with J. J. SEIDEL), Strongly regular graphs derived from combinatorial designs. Canad. J. Math. **22** (1970), 597–614.
[GoS4] (with J. J. SEIDEL), The regular two-graph on 276 vertices. Discrete Math. **12** (1975), 143–158. (3.6)
See also [CaGSS], [DeGS1], [DeGS2]
GRAHAM, R. L.
[GrHH] (with A. J. HOFFMAN and H. HOSOYA), On the distance matrix of a directed graph. J. Graph Theory **1** (1977), 85–88. (9.2)
[GrLo] (with L. LOVÁSZ), Distance matrices of trees. Stanford University, STAN-CS-75-497 (1975). (9.2)
[GrP1] (with H. O. POLLAK), On the adressing problem for loop switching. Bell System Tech. J. **50** (1971), 2495–2519. (9.2)
[GrP2] (with H. O. POLLAK), On embedding graphs in squashed cubes. In: Graph Theory and Applications (Lecture Notes in Mathematics 303, ed. Y. ALAVI, D. R. LICK, A. T. WHITE), Springer-Verlag, Berlin–Heidelberg–New York 1972; pp. 99–110. (9.2)
See also [EdGG]

GRAOVAC, A.
[GrGT1] (with I. GUTMAN and N. TRINAJSTIĆ), On the Coulson integral formula for total π-electron energy. Chem. Phys. Letters **35** (1975), 555—557. (8.1)
[GrGT2] (with I. GUTMAN and N. TRINAJSTIĆ), A linear relationship between the total π-electron energy and the characteristic polynomial. Chem. Phys. Letters **37** (1976), 471—474. (8.1)
[GrGTŽ] (with I. GUTMAN, N. TRINAJSTIĆ, and T. ŽIVKOVIĆ), Graph theory and molecular orbitals. Applications of Sachs theorem. Theoret. Chim. Acta **26** (1972), 67—78. (8.2)

GRIMMETT, G. R.
[Grim] An upper bound for the number of spanning trees of a graph. Discrete Math. **16** (1976), 323—324. (7.7)

GROHMANN, G., see [FiGr]

GÜNTHARD, Hs. H.
[GüPr] (with H. PRIMAS), Zusammenhang von Graphentheorie und MO-Theorie von Molekeln mit Systemen konjugierter Bindungen. Helv. Chim. Acta **39** (1956), 1645—1653. (6.1; 8.1)
See also [WiKG]

GUTMAN, I.
[Gut1] Teorija grafova i molekularne orbitale (Serbo-Croatian). Master thesis, University of Zagreb, 1972. (8.5)
[Gut2] Bounds for total π-electron energy. Chem. Phys. Letters **24** (1974), 283—285. (8.1; 8.5)
[Gut3] Estimating the π-electron energy of very large conjugated systems. Naturwissenschaften **61** (1974), 216—217.
[Gut4] On the number of antibonding MO's in conjugated hydrocarbons. Chem. Phys. Letters **26** (1974), 85—88. (8.5)
[Gut5] Some topological properties of benzenoid systems. Croat. Chem. Acta **46** (1974), 209—215. (8.5)
[Gut6] An algorithm for the enumeration of bonding and antibonding MO's in conjugated hydrocarbons. Chem. Phys. Letters **37** (1976), 475—477. (8.5)
[Gut7] Hückel molecular orbital energies for [O] paracyclophanes — a test for the validity of the perimeter rule. Bull. Soc. Chim. Beograd **41** (1976), 69—74.
[Gut8] Generalizations of a recurrence relation for the characteristic polynomials of trees. Publ. Inst. Math. (Beograd) **21** (35) (1977), 75—80. (2.3; 6.1)
[Gut9] Partial ordering of forests according to their characteristic polynomial. In: Combinatorics (Coll. Math. Soc. J. BOLYAI **18**, ed. A. HAJNAL and V. Sós). North-Holland Publ. Co., Amsterdam—Oxford—New York 1978, Vol. I, pp. 429—436. (8.1)
[Gut10] Investigation of topological properties of conjugated hydrocarbons. (Serbo-Croatian) Thesis, University of Zagreb 1973. (8.1)
[Gut11] A class of approximate topological formulas for total π-electron energy. J. Chem. Phys. **66** (1977), 1652—1655. (8.1)
[Gut12] Acylic systems with extremal Hückel π-electron energy. Theoret. Chim. Acta **45** (1977), 79—87. (8.1)
[GuCv] (with D. M. CVETKOVIĆ), The reconstruction problem for characteristic polynomials of graphs. Univ. Beograd Publ. Elektrotehn. Fak., Ser. Mat. Fiz., No. 498—No. 541 (1975), 45—48. (9.5)
[GuRT] (with M. RANDIĆ and N. TRINAJSTIĆ), Kekulé structures and topology, III. On inseparability of Kekulé structures. Rev. Roumaine Chim. **23** (1978), 383—395.
[GuRTW] (with B. RUŠČIĆ, N. TRINAJSTIĆ, and C. F. WILCOX JR.), Graph theory and molecular orbitals, XII. Acyclic polyenes. J. Chem. Phys. **62** (1975), 3399—3405.
[GuT1] (with N. TRINAJSTIĆ), Graph theory and molecular orbitals. Total π-electron energyof alternant hydrocarbons. Chem. Phys. Letters **17** (1972), 535—538. (8.1; 8.4)
[GuT2] (with N. TRINAJSTIĆ), A graph-theoretical classification of conjugated hydrocarbons. Naturwissenschaften **60** (1973), 475.

[GuT3] (with N. TRINAJSTIĆ), Graph theory and molecular orbitals, IV. Further applications of Sachs formula. Croat. Chem. Acta **45** (1973), 423–429.

[GuT4] (with N. TRINAJSTIĆ), Graph theory and molecular orbitals. In: Topics in current chemistry (Fortschritte der chemischen Forschung) **42**. Springer-Verlag, Berlin–Heidelberg–New York 1973, pp. 49–93.

[GuT5] (with N. TRINAJSTIĆ), Graph theory and molecular orbitals. The loop rule. Chem. Phys. Letters **20** (1973), 257–260. (8.4)

[GuT6] (with N. TRINAJSTIĆ), Graph theory and molecular orbitals, VIII. Kekulé structures and permutations. Croat. Chem. Acta **45** (1973), 539–545. (8.2)

[GuT7] (with N. TRINAJSTIĆ), Violation of the Dewar-Longuet-Higgins conjecture. Z. Naturforsch. **29a** (1974), 1238. (2.7)

[GuT8] (with N. TRINAJSTIĆ), Graph spectral theory of conjugated molecules. Croat. Chem. Acta **47** (1975), 507–533.

[GuT9] (with N. TRINAJSTIĆ), On the parity of Kekulé structures. Croat. Chem. Acta **47** (1975), 35–39.

[GuT10] (with N. TRINAJSTIĆ), Graph theory and molecular orbitals, XV. The Hückel rule. J. Chem. Phys. **64** (1976), 4921–4925. (8.1)

[GuT11] (with N. TRINAJSTIĆ), Graph theory and molecular orbitals, XVI. On π-electron charge distribution. Croat. Chem. Acta **48** (1976), 19–24.

[GuTW] (with N. TRINAJSTIĆ and C. F. WILCOX, Jr.), Graph theory and molecular orbitals, X. The number of Kekulé structures and the thermodynamic stability of conjugated systems. Tetrahedron **31** (1975), 143–146.

[GuTŽ] (with N. TRINAJSTIĆ and T. ŽIVKOVIĆ), Graph theory and molecular orbitals, VI. A discussion of non-alternant hydrocarbons. Tetrahedron **29** (1973), 3449–3454.

See also [CvG1], [CvG2], [CvG3], [CvG4], [CvGT1], [CvGT2], [CvGT3], [CvGT4], [CvGT5], [CvGT6], [GrGT1], [GrGT2], [GrGTŽ], [HoHG], [WiGT]

HAEMERS, W.

[Haem] Partitioning and eigenvalues. Eindhoven University of Technology, Memorandum 1976-11; revised version: A generalization of the Higman-Sims technique. Proc. Kon. Ned. Akad. Wet. A **81** (4) (1978), 445–447. (0.3)

HAKIMI, S. L., *see* [AmHa]

HALL, G. G.

[Ha, G] The bond orders of alternant hydrocarbon molecules. Proc. Roy. Soc. (London) A **229** (1955), 251–259. (8.1)

HALL, K. M.

[Ha, K] r-dimension quadratic placement algorithm. Management Sci. **17** (1970), No. 3, 219–229. (3.6)

HALL, L. H.

* [Ha, L] Group Theory and Symmetry in Chemistry. McGraw-Hill, New York 1969. (5.2)

HALL, M., JR.

* [Ha, M] Combinatorial Theory. Blaisdell Publ. Comp., Waltham–Toronto–London 1967. (6.2; 7.1)

HAMADA, N., *see* [YaFH]

HARARY, F.

[Har1] A graph theoretic method for the complete reduction of a matrix with a view toward finding its eigenvalues. J. Math. Phys. **38** (1959), 104–111. (2.2)

[Har2] The determinant of the adjacency matrix of a graph. SIAM Rev. **4** (1962), 202–210. (1.4; 6.1)

[Har3] Four difficult unsolved problems in graph theory. In: Recent Advances in Graph Theory (ed. M. FIEDLER), Academia Praha 1975, pp. 249–256. (9.4)

* [Har4] Graph Theory. Addison-Wesley Publ. Comp., Reading, Mass.-Menlo Park, Cal.-London-Don Mills, Ontario 1969; German translation: Graphentheorie. R. Oldenbourg Verlag, München–Wien 1974; Japanese translation: Kyoritsu, Tokyo 1971; Russian translation: See [Xapa]. (0; 0.1)

[HaKMR] (with C. KING, A. MOWSHOWITZ, and R. C. READ), Cospectral graphs and digraphs. Bull. London Math. Soc. **3** (1971), 321–328. (2.3; 2.6; 6.1; 6.6)

[HaS1] (with A. J. SCHWENK), The spectral approach to determining the number of walks in a graph. Pacific J. Math. (to appear) (1.8; 3.6)

[HaS2] (with A. J. SCHWENK), Which graphs have integral spectra? In: Graphs and Combinatorics (Lecture Notes in Mathematics 406, ed. R. BARI, F. HARARY), Springer-Verlag, Berlin–Heidelberg–New York 1974, pp. 45–51. (9.4; 9.5)

[HaTr] (with C. A. TRAUTH, JR.), Connectedness of product of two directed graphs. SIAM J. Appl. Math. **14** (1966), 250–254. (7.4)

[HaWi] (with G. W. WILCOX), Boolean operations on graphs. Math. Scand. **20** (1967), 41–51. (2.5; 7.4)

See also [BaHa]

HARTMANN, H., *see* [FaHa]

HAYNSWORTH E. V.

[Hayn] Applications of a theorem on partitioned matrices. J. Res. Nat. Bureau Stand. **62** B (1959), 73–78. (0.3; 4.5)

HEILBRONNER, E.

[Hei1] Das Kompositions-Prinzip: Eine anschauliche Methode zur elektronen-theoretischen Behandlung nicht oder niedrig symmetrischer Molekeln im Rahmen der MO-Theorie. Helv. Chim. Acta **36** (1953), 170–188. (1.4; 2.3; A)

[Hei2] Die Eigenwerte von LCAO-MO's in homologen Reihen. Helv. Chim. Acta **37** (1954), 921–935. (1.4; 2.3; 5.2; 6.3; 8.1)

[Hei3] Ein graphisches Verfahren zur Faktorisierung der Säkulardeterminante aromatischer Ringsysteme im Rahmen der LCAO-MO-Theorie. Helv. Chim. Acta **37** (1954), 913–921. (4.6; 5.2)

[Hei4] Über einen graphentheoretischen Zusammenhang zwischen dem Hückelschel MO-Verfahren und dem Formalismus der Resonanztheorie. Helv. Chim. Acta **45** (1962), 1722–1725. (8.5)

* [HeSt] (with P. A. STRAUB), Hückel molecular orbitals. Springer-Verlag, Berlin–Heidelberg–New York 1966. (8.1)

HERNDON, W. C.

[Her1] Isospectral molecules. Tetrahedron Letters (1974), No. 8, 671–674. (6.1)

[Her2] The characteristic polynomial does not uniquely determine molecular topology. J. Chem. Doc. **14** (1974), 150–151. (6.1)

[HeEl] (with M. L. ELLZEY, JR.), Isospectral graphs and molecules. Tetrahedron **31** (1975), 99–107. (6.1)

HESS, B. A.

[HeSc] (with L. J. SCHAAD), Hückel molecular orbital π-resonance energies. A new approach. J. Amer. Chem. Soc. **93** (1971), 305–310. (8.1)

HESTENES, M. D.

[Hest] On the use of graphs in group theory. In: New Directions in the Theory of Graphs (ed. F. HARARY), Academic Press, New York–London 1973, pp. 97–128. (7.2)

[HeHi] (with D. G. HIGMAN), Rank 3 groups and strongly regular graphs. In: Computers in Algebra and Number Theory; SIAM–AMS Proc. Vol. IV. Providence 1971, pp. 141–159. (0.3; 6.5)

HEYDEMANN, M. C.

[Heyd] Spectral characterization of some graphs. J. Comb. Theory (B) **25** (1978), 307–312. (6.6)

HIGMAN, D. G.

[Hig1] Finite permutation groups of rank 3. Math. Z. **86** (1964), 145–156. (7.2)

[Hig2] Intersection matrices for finite permutation groups. J. Algebra **6** (1967), 22–42. (7.2)

[Hig3] Coherent configurations, I. Rend. Sem. Mat. Univ. Padova **44** (1970), 1–26. (7.2)

[Hig4] A survey of some questions and results about rank 3 permutation groups. In: Actes Congrès Intern. Math. 1970, Tome 1, pp. 361–365. (6.5; 7.2)

[Hig5] Characterization of rank 3 permutation groups by subdegrees, I. Arch. Math. **21** (1970), 151–156. (6.5)

[Hig6] Characterization of rank 3 permutation groups by subdegrees, II. Arch. Math. **21** (1970), 353–361. (6.5)

[Hig7] Partial geometries, generalized quadrangles, and strongly regular graphs. In: Atti Conv. Geom. Comb. e Appl., Perugia 1970 (Ist. Mat., Univ. Perugia 1971), pp. 263–293. (7.2)

[HiSi] (with C. C. SIMS), A simple group of order 44,353,000. Math. Z. **105** (1968), 110–113. (7.2)

See also [FeHi], [HeHi]

HOCHMANN, P.

[HoDKP] (with J. DUBSKÝ, J. KOUTECKÝ, and C. PÁRKÁNYI), Tables of quantum chemical data, VIII. Energy characteristics of some benzenoid hydrocarbons. Coll. Chech. Chem. Comm. **30** (1965), 3560–3565. (A)

[HoDKT] (with J. DUBSKÝ, V. KVASNIČKA, and M. TITZ), Tables of quantum chemical data, X. Energy characteristics of some polyenic hydrocarbons. Coll. Chech. Chem. Comm. **31** (1966), 4172–4175. (A)

[HoDT] (with J. DUBSKÝ and M. TITZ), Tables of quantum chemical data, XI. Energy characteristics of some non-alternant hydrocarbon molecules and ions. Coll. Chech. Chem. Comm. **32** (1967), 1260–1264. (A)

[HoKZ] (with J. KOUTECKÝ and R. ZAHRADNÍK), Tables of quantum chemical data, I. Molecular orbitals of some benzenoid hydrocarbons and benzo derivatives of fluoranthene. Coll. Chech. Chem. Comm. **27** (1962), 3053–3075. (A)

See also [TiH1], [TiH2], [TiH3]

HOFFMAN, A. J.

[Hof1] On the exceptional case in a characterization of the arcs of a complete graph. IBM J. Res. Develop. **4** (1960), 487–496. (0.3; 6.1)

[Hof2] On the uniqueness of the triangular association scheme. Ann. Math. Statist. **31** (1960), 492–497.

[Hof3] On the polynomial of a graph. Amer. Math. Monthly **70** (1963), 30–36. (3.2; 3.3; 6.1; 6.6)

[Hof4] On the line graph of the complete bipartite graph. Ann. Math. Statist. **35** (1964), 883–885. (6.3)

[Hof5] On the line graph of a projective plane. Proc. Amer. Math. Soc. **16** (1965), 297–302. (6.3)

[Hof6] Some recent results on spectral properties of graphs. In: Beiträge zur Graphentheorie (ed. H. SACHS, H.-J. VOSS, H. WALTHER). Int. Koll. Manebach, 1967. Leipzig 1968, pp. 75–80. (6.3; 9.1)

[Hof7] The change in the least eigenvalue of the adjacency matrix of a graph under imbedding. SIAM J. Appl. Math. **17** (1969), 664–671. (6.1; 9.1)

[Hof8] The eigenvalues of the adjacency matrix of a graph. Research note N. C. 689, Thomas J. Watson Research Center, Yorktown Heights, N.Y. 1967.

[Hof9] The eigenvalues of the adjacency matrix of a graph. In: Combinatorial Mathematics and its Applications (ed. R. C. BOSE, T. A. DOWLING), The University of North Carolina Press, Chapel Hill 1969, pp. 578–584. (P; 1.9; 9.1)

[Hof10] $-1 - \sqrt{2}$? In: Combinatorial Structures and Their Applications (ed. R. GUY, H. HANANI, N. SAUER, J. SCHÖNHEIM), Gordon and Breach, Science Publishers, Inc., New York–London–Paris 1970, pp. 173–176. (9.1)

[Hof11] Eigenvalues and partitionings of the edges of a graph. Linear Algebra and Appl. **5** (1972), 137–146. (2.1; 3.2; 3.6; 6.1)

[Hof12] Graphs and eigenvalues. Third Southeastern Conference on Combinatorics, Graph Theory and Computing, 1972, Florida Atlantic University, Boca Raton, Florida.

[Hof13] On limit points of spectral radii of non-negative symmetric integral matrices. In: Graph Theory and Applications (Lecture Notes in Mathematics 303, ed. Y. ALAVI, D. R. LICK, A. T. WHITE), Springer-Verlag, Berlin–Heidelberg–New York 1972, pp. 165–172. (2.7; 3.6)

[Hof14] On spectrally bounded graphs. In: A Survey of Combinatorial Theory (ed. J. N. SRIVASTAVA, F. HARARY, C. R. RAO, G.-C. ROTA, S. S. SHRIKHANDE), North-Holland Publ. Co., Amsterdam—London—New York 1973, pp. 277—283. (9.1)

[Hof15] On vertices near a vertex of a graph. In: Studies in Pure Mathematics, Papers Presented to Richard Rado (ed. L. MIRSKI), London 1971, pp. 131—136. (3.6)

[Hof16] On eigenvalues and colorings of graphs. In: Graph Theory and Its Applications (ed. B. Harris), Academic Press, New York—London 1970, pp. 79—91. (3.1; 3.6)

[Hof17] Applications of Ramsey style theorems to eigenvalues of graphs. In: Combinatorics (ed. M. HALL, Jr., J. H. VAN LINT), Part 2, Math. Centre Tracts, Amsterdam 1974, No. 56, pp. 43—57. (3.6; 9.1)

[Hof18] On graphs whose least eigenvalue exceeds $-1 - \sqrt{2}$. Linear Algebra and Appl. **16** (1977), 153—165. (9.1)

[Hof19] Eigenvalues of graphs. In: Studies in Graph Theory, I, II (ed. D. R. FULKERSON), M. A. A. 1975, pp. 225—245.

[Hof20] On spectrally bounded signed graphs. In: Trans. 21th Conference of Army Mathematics, U. S. Army Research Office, Durham (Abstract), pp. 1—5. (7.8)

[Hof21] On eigenvalues of symmetric $(+1, -1)$ matrices. Israel J. Math. **17** (1974), 69—75. (7.8)

[Hof22] On limit points of the least eigenvalue of a graph. Ars Combinatoria **3** (1977), 3—14.

[Hof23] Spectral functions of graphs. In: Proc. Int. Congr. Math. Vancouver 1974, Vol. 2, S. 1 (1975), pp. 461—463.

[HoHo] (with L. HOWES), On eigenvalues and colorings of graphs, II. Ann. New York Acad. Sci. **175** (1970), No. 1, 238—242. (3.2; 3.6)

[HoJa] (with B. A. JAMIL), On the line graph of the complete tripartite graph. Linear und Multilinear Algebra **5** (1977), 19—25.

[HoJo] (with P. JOFFE), Nearest s-matrices of given rank and the Ramsey problem for eigenvalues of bipartite s-graphs. In: Problèmes Combinatoires et Théorie des Graphes, Coll. Int. C.N.R.S., No. 260, Orsay 1976. C.N.R.S. Publ. 1978 (ed. J.-C. BERMOND, J.-C. FOURNIER, M. LAS VERGNAS, D. SOTTEAU). pp. 237—240.

[HoMc] (with M. H. MCANDREW), The polynomial of a directed graph. Proc. Amer. Math. Soc. **16** (1965), 303—309. (3.1; 6.6)

[HoOs] (with A. M. OSTROWSKI), On the least eigenvalue of a graph of large minimum valence containing a given graph (unpublished). (9.1)

[HoR1] (with D. K. RAY-CHAUDHURI), On the line graph of a finite affine plane. Canad. J. Math. **17** (1965), 687—694. (6.3)

[HoR2] (with D. K. RAY-CHAUDHURI), On the line graph of a symmetric balanced incomplete block design. Trans. Amer. Math. Soc. **116** (1965), No. 4, 238—252. (6.1; 6.3)

[HoR3] (with D. K. RAY-CHAUDHURI), On a spectral characterization of regular line graphs (unpublished). (6.3; 6.6)

[HoSi] (with R. R. SINGLETON), On Moore graphs with diameters 2 and 3. IBM J. Res. Develop. **4** (1960), 497—504. (6.2)

[HoSm] (with J. H. SMITH), On the spectral radii of topologically equivalent graphs. In: Recent Advances in Graph Theory (ed. M. FIEDLER), Academia Praha 1975, pp. 273—281. (2.7)

See also [DoHo], [GrHH]

HONEYBOURNE, C. L.

[Hon1] Topological aspects of odd graphs and their relevance to radical spin densities. J. C. S. Faraday II, **71** (1975), 1343—1351.

[Hon2] Graph theory and free radicals, Validation of a recent assertion and its relation to the pairing theorem. J. C. S. Faraday II, **72** (1976), 34—39.

HOSOI, K., *see* [HoHG]

HOSOYA, H.

[Hos1] Topological index. A newly proposed quantity characterizing the topological nature of structural isomeres of saturated hydrocarbons. Bull. Chem. Soc. Japan **44** (1971), 2332—2339. (A)

[Hos2] Graphical enumeration of the coefficients of the secular polynomial of the Hückel molecular orbital. Theoret. Chim. Acta **25** (1972), 215–222. (1.4)

[Hos3] Topological index and Fibonacci numbers with relation to chemistry. Fibonacci Quart. **11** (1973), No. 3, 255–266.

[HoHG] (with K. Hosoi and I. Gutman), A topological index for the total π-electron energy. Proof of a generalized Hückel rule for an arbitrary network. Theoret. Chim. Acta **38** (1975), 37–47. (8:1)

[HoMC] (with M. Murakami and M. Cotoh), Distance polynomial and characterization of a graph. Natur. Sci. Rept. Ochanumizu Univ. **24** (1973), No. 1, 27–34. (9.2; A)

See also [GrHH], [KaMH], [MiKH]

Howes, L.

[How1] On subdominantly bounded graphs. Thesis, City Univ. of New York 1970. (9.1)

[How2] On subdominantly bounded graphs — summary of results. In: Recent Trends in Graph Theory (Lecture Notes in Mathematics 186, ed. M. Capobianco, J. B. Frechen, M. Krolik), Springer-Verlag, Berlin–Heidelberg–New York 1971, pp. 181–183. (9.1)

See also [HoHo].

Hubaut, X. L.

[Huba] Strongly regular graphs. Discrete Math. **13** (1975), 357–381. (7.2)

Hückel, E.

[Hück] Quantentheoretische Beiträge zum Benzolproblem. Z. Phys. **70** (1931), 204–286. (8.1)

Hutschenreuther, H.

[Huts] Einfacher Beweis des Matrix-Gerüst-Satzes der Netzwerktheorie. Wiss. Z. TH Ilmenau **13** (1967), 403–404. (1.5)

Imrich, W.

[Imri] Zehnpunktige kubische Graphen. Aequationes Math. **6** (1971), No. 1, 6–10. (A)

Ito, T., *see* [BaIt]

Jacobs, J. B., *see* [BeJa]

Jacobs, J., *see* [CoJa]

James, L. O.

[Jame] A combinatorial proof that the Moore (7, 2) graph is unique. Utilitas Math. **5** (1974), 79–84.

Jamil, B. A., *see* [HoJa]

Joffe, P., *see* [HoJo]

John, P. W. M.

* [John] Statistical designs and analysis of experiments. Macmillan Comp., New York 1971.

Johnson, D. E.

[JoJo] (with J. R. Johnson), Comment on "General solution to the spanning tree enumeration problem in multigraph wheels". IEEE Trans. Circuit Theory **20** (1973), 454–455. (7.6)

Johnson, J. R., *see* [JoJo]

Kac, M.

[Kac] Can one hear the shape of a drum? Amer. Math. Monthly **73** (1966), April, Part II, 1–23. (8.4)

Kasteleyn, P. W.

[Kas1] Dimer statistics and phase transitions. J. Math. Phys. **4** (1963), 287–293. (8.3)

[Kas2] Graph theory and crystal physics. In: Graph Theory and Theoretical Physics (ed. F. Harary), Academic Press, London–New York 1967, pp. 43–110. (1.8; 1.9; 8.3)

KAWASAKI, K.
[KaMH] (with K. MIZUTANI and H. HOSOYA), Tables of non-adjacent numbers, characteristic polynomials and topological indices, II. Mono- and bicyclic graphs. Natur. Sci. Rept. Ochanumizu Univ. **22** (1971), No. 2, 181–214. (A)
See also [MiKH]

KELLER, J., *see* [WiKG]

KEL'MANS, A. K.
[Kelm] Comparisons of graphs by their number of spanning trees. Discrete Math. **16** (1976), 241–261.
[KeCh] (with V. M. CHELNOKOV), A certain polynomial of a graph and graphs with an extremal number of trees. J. Comb. Theory (B) **16** (1974), 197–214. Erratum, J. Comb. Theory (B) **24** (1978), 375. (1.5; 2.3; 7.6; 7.8)
See also КЕЛЬМАНС, А. К.

KERSCHBERG, L.
[Kers] The characteristic polynomial of graph products. In: 7th Ann. Asilomar Conf. Circuits, Syst., Comput.; Pacific Grove 1973, Western Periodicals Comp., North Hollywood 1974, pp. 476–481. (2.5)

KING, C.
[King] Characteristic polynomials of 7-node graphs. Sci. Rept., Univ. of West Indies CC6 (AFORS project 1026–66), Kingston 1967. (A)
See also [HaKMR]

KIRCHHOFF, G.
[Kirc] Über die Auflösung der Gleichungen, auf welche man bei der Untersuchung der linearen Verteilung galvanischer Ströme geführt wird. Ann. Phys. Chem. **72** (1847), 497–508. (1.5)

KNOP, J. V.
[KnTŽ] (with N. TRINAJSTIĆ and T. ŽIVKOVIĆ), A graphical study of positional isomers containing bivalent sulphur. Coll. Czech. Chem. Comm. **39** (1974), 2431–2448.

KÖNIG, D.
[Kön1] Über Graphen und ihre Anwendungen auf Determinantentheorie und Mengenlehre. Math. Ann. **77** (1916), 453–465. (1.4)
* [Kön2] Theorie der endlichen und unendlichen Graphen. Akadem. Verlagsges., Leipzig 1936. (1.4)

KOUTECKÝ, J., *see* [HoDKP], [HoKZ], [ZaMK1], [ZaMK2]

KRAUS, L. L.
[KrC1] (with D. M. CVETKOVIĆ), Evaluation of a lower bound for the chromatic number of the complete product of graphs. Univ. Beograd Publ. Elektrotehn. Fak., Ser. Mat. Fiz., No. 357 – No. 380 (1971), 63–68. (3.3)
[KrC2] (with D. M. CVETKOVIĆ), Tables of simple eigenvalues of some graphs whose automorphism group has two orbits. Univ. Beograd Publ. Elektrotehn. Fak., Ser. Mat. Fiz., No. 381 – No. 409 (1972), 89–95. (5.1; A)

KREWERAS, G.
[Krew] Complexité et circuits eulériens dans les sommes tensorielles de graphes. J. Comb. Theory (B) **24** (1978), 202–212. (7.8)

KRISHNAMOORTHY, V.
[KrP1] (with K. R. PARTHASARATHY), A note on non-isomorphic cospectral digraphs. J. Comb. Theory (B) **17** (1974), 39–40. (6.1)
[KrP2] (with K. R. PARTHASARATHY), Cospectral graphs and digraphs with given automorphism group. J. Comb. Theory (B) **19** (1975), 204–213. (5.5; 6.1)

KUHN, W. W.
[Kuhn] Graph isomorphism using vertex adjacency matrix. In: Proc. 25th summer meeting of Can. Math. Congress, Lakehead Univ., Thunder Bay, Ont., 1971, pp. 471–476. (1.8)

Kuich, W.
[KuSa] (with N. Sauer), On the existence of certain minimal regular n-systems with given girth. In: Proof. Techniques in Graph Theory (ed. F. Harary), Academic Press, New York—London 1969, pp. 93—101.
Kvasnička, V., *see* [HoDKT]

Laskar, R.
[Las1] A characterization of cubic lattice graphs. J. Comb. Theory **3** (1967), 386—401. (6.4)
[Las2] Eigenvalues of the adjacency matrix of the cubic lattice graph. Pacific J. Math. **29** (1969), 623—629. (6.4)
See also [BoLa]
Lehmer, D. H.
[Lehm] Permutations with strongly restricted displacements. In: Combinatorial Theory and Its Applications, II. (ed. P. Erdös, A. Rényi, V. T. Sós), Bolyai János Mat. Társulat, Budapest/North-Holland Publ. Co., Amsterdam—London 1970, pp. 755—770. (7.5)
Lemmens, P. W. H.
[LeS1] (with J. J. Seidel), Equi-isoclinic subspaces of euclidean spaces. Indag. Math. **35** (1973), No. 2, 98—107.
[LeS2] (with J. J. Seidel), Equiangular lines. J. Algebra **24** (1973), 494—512. (7.3)
Lenstra, H. W.
[Lens] Two theorems on perfect codes. Discrete Math. **3** (1972), 125—132. (4.8)
Lick, D. R.
[Lick] A class of point partition numbers. In: Recent Trends in Graph Theory (Lecture Notes in Mathematics 186, ed. M. Capobianco, J. B. Frechen, M. Krolik), Springer-Verlag, Berlin—Heidelberg—New York 1971, pp. 184—190. (3.2)
[LiWh] (with A. T. White), k-degenerate graphs. Canad. J. Math. **22** (1970), 1082—1096. (3.2)

Lint, J. H. van
* [Lint] Coding Theory (Lecture Notes in Mathematics 201), Springer-Verlag, Berlin—Heidelberg—New York 1971. (4.8)
[LiSe] (with J. J. Seidel), Equilateral point sets in elliptic geometry. Proc. Nederl. Acad. Wetensch., Ser. A, **69** (1966), 335—348. (1.2; 6.1; 6.5; 7.3; A)
See also [BeLS], [CaLi], [CvLi]
Little, C. H. C.
[Lit1] The parity of the number of 1-factors of a graph. Discrete Math. **2** (1972), 179—181. (7.8)
[Lit2] Kasteleyn's theorem and arbitrary graphs. Canad. J. Math. **25** (1973), 758—764. (8.3)

Lloyd, S. P.
[Lloy] Binary block coding. Bell System Tech. J. **36** (1957), 517—535. (4.8)
Longuet-Higgins, H. C.
[Long] Resonance structures and MO in unsaturated hydrocarbons. J. Chem. Phys. **18** (1950), 265—274. (8.1; 8.2)
See also [CoLo], [DeLo]
Lorens, C. S.
* [Lore] Flowgraphs. McGraw-Hill, New York 1964. (1.4)
Losey, N. E., *see* [BeLo]
Lovász, L.
[Lová] Spectra of graphs with transitive groups. Periodica Math. Hung. **6** (2) (1975), 191—195. (5.1; 5.5)
[LoPe] (with J. Pelikán), On the eigenvalues of trees. Periodica Math. Hung. **3** (1—2), (1973), 175—182. (1.4; 2.7; 3.6)
See also [GrLo]
Lučić, R. P., *see* [CvL1], [CvL2]

MAECHTER, R. T., *see* [StMa]

MALLION, R. B.

[Mal1] Some graph-theoretical aspects of simple ring current calculations on conjugated systems. Proc. Roy. Soc. (London) A **341** (1975), 429–449.

[Mal2] On the number of spanning trees in a molecular graph. Chem. Phys. Letters **36** (1975), 170–174. (1.5)

[MaST1] (with A. J. SCHWENK and N. TRINAJSTIĆ), A graphical study of heteroconjugated molecules. Croat. Chem. Acta **46** (1974), 171–182.

[MaST2] (with A. J. SCHWENK and N. TRINAJSTIĆ), On the characteristic polynomial of a rooted graph. In: Recent Advances in Graph Theory (ed. M. FIEDLER), Academia Praha 1975, pp. 345–350.

[MaTS] (with N. TRINAJSTIĆ and A. J. SCHWENK), Graph theory in chemistry — generalization of Sachs' formula. Z. Naturforsch. **29a** (1974), 1481–1484.

MALLOWS, C. D.

[MaSl] (with N. J. SLOANE), Two-graphs, switching classes and Euler graphs are equal in number. SIAM J. Appl. Math. **28** (1975), 876–880. (7.3)

MARCUS, M.

* [MaMi] (with H. MINC), A Survey of Matrix Theory and Matrix Inequalities. Allyn and Bacon, Inc., Boston 1964. (0; 0.3; 2.1; 2.4; 2.5; 3.2; 3.5)

MARCUS, R. A.

[Marc] Additivity of heats of combustion, LCAO resonance energies and bond orders of conformal sets of conjugated compounds. J. Chem. Phys. **43** (1965), 2643–2654.

MARIMONT, R. B.

[Mari] System connectivity and matrix properties. Bull. Math. Biophys. **31** (1969), 255–274. (3.2)

MARSHALL, C.

* [Mars] Applied Graph Theory. Interscience, John Wiley & Sons, New York–London–Sydney–Toronto 1971.

MASON, S. J.

[Mas1] Feedback theory — some properties of signal flow graphs. Proc. IRE **41** (1953), 1144–1156. (1.4)

[Mas2] Feedback theory — further properties of signal flow graphs. Proc. IRE **44** (1956), 920–926. (1.4)

MASUYAMA, M.

[Masu] A test for graph isomorphism. Repts. Statist. Appl. Res. Union Jap. Sci. Eng. **20** (1973), No. 2, 41–64.

MAYEDA, W.

* [Maye] Graph Theory. Interscience, John Wiley & Sons, Inc., New York–London–Sydney–Toronto 1972. (0)

MCANDREW, M. H.

[McAn] On the product of directed graphs. Proc. Amer. Math. Soc. **14** (1963), 600–606. (7.4)

See also [HoMc]

MCCLELLAND, B. J.

[McC1] Properties of the latent roots of a matrix: The estimation of π-electron energies. J. Chem. Phys. **54** (1971), 640–643. (8.1)

[McC2] Graphical method for factorizing secular determinants of Hückel molecular orbital theory. J.C.S. Faraday II **70** (1974), 1453–1456. (4.6; 5.2)

MCKAY, B.

[McKa] On the spectral characterization of trees. Ars Combinatoria **3** (1977), 219–232. (9.2)

See also [GoM1], [GoM2]

MCWEENY, R.

* [McWS] (with B. T. SUTCLIFFE), Methods of molecular quantum mechanics. Academic Press, London–New York 1969. (8; 8.1; 8.5)

MENDELSOHN, N. S.

[Men1] Structure of Good graphs and related graphs (unpublished).

[Men 2] Directed graphs with the unique path property. In: Combinatorial Theory and Its Applications, II (ed. P. Erdős, A. Rényi, V. T. Sós), Bolyai János Mat. Társulat, Budapest/North-Holland Publ. Co., Amsterdam—London 1970, pp. 783—799.

See also [DuMe]

Meredith, G. H. J., *see* [BiMe]

Messner, D. M., *see* [BoMe]

Meyer, J. F.

[Meye] Algebraic isomorphism invariants for graphs of automata. In: Graph Theory and Computing (ed. R. C. Read), Academic Press, New York—London 1972, pp. 123—152. (6.1)

Michl, J., *see* [ZaM 1], [ZaM 2], [ZaM 3], [ZaMK 1], [ZaMK 2]

Milić, M.

[Mili] Flow-graph evaluation of the characteristic polynomial of a matrix. IEEE Trans. Circuit Theory CT-**11** (1964), 423—424. (1.4)

See also [CvMi]

Miller, D. J.

[Mill] The categorical product of graphs. Canad. J. Math. **20** (1968), 1511—1521. (2.5; 7.4)

Minc, H., *see* [MaMi]

Mitchem, J.

[Mitc] On extremal partitions of graphs. Thesis, Michigan 1970. (3.2)

Mizutani, K.

[MiKH] (with K. Kawasaki and H. Hosoya), Tables of non-adjacent numbers, characteristic polynomials and topological indices, I. Tree graphs. Natur. Sci. Rept. Ochanumizu Univ. **22** (1971), No. 1, 39—58. (A)

See also [KaMH]

Montrol, E. W.

[Mont] Lattice statistics. In: Applied Combinatorial Mathematics (ed. E. F. Beckenbach), Interscience, John Wiley & Sons, New York—London—Sydney 1964, pp. 96—143. Russian translation: Статистика решеток. In: Прикладная комбинаторная математика. Изд. „Мир", Москва 1968, pp. 9—60. (8.3)

Moon, J. W.

[Moo 1] On the line graph of the complete bigraph. Ann. Math. Statist. **34** (1963), 664—667. (6.3)

* [Moo 2] Counting Labelled Trees. Canadian Mathematical Monographs No. 1, Canad. Math. Congress 1970. (1.5; 7.6)

[MoPu] (with N. J. Pullman), On generalized tournament matrices. SIAM Review **12** (1970), 384—399.

Morley, T. D., *see* [AnMo]

Moser, W. O. J.

[MoAb] (with M. Abramson), Enumeration of combinations with restricted differences and cospan. J. Comb. Theory **7** (1969), 162—170. (7.5)

Mowshowitz, A.

[Mow 1] Entropy and complexity of graphs. University of Michigan Techn. Rept., August 1967. (5.5)

[Mow 2] Entropy and complexity of graphs, III. Graphs with prescribed information content. Bull. Math. Biophys. **30** (1968), 387—414. (2.5; 5.5)

[Mow 3] The group of a graph whose adjacency matrix has all distinct eigenvalues. In: Proof Techniques in Graph Theory (ed. F. Harary), Academic Press, New York—London 1969, pp. 109—110. (5.1)

[Mow 4] Graphs, groups and matrices. In: Proc. Canad. Math. Congr., 1971, pp. 509—522. (5.2; 5.5)

[Mow 5] The characteristic polynomial of a graph. J. Comb. Theory (B) **12** (1972), 177—193. (1.4; 2.7; 6.1; 6.6; A)

[Mow6] The adjacency matrix and the group of a graph. In: New Directions in the Theory of Graphs (ed. F. HARARY), Academic Press, New York—London 1973, pp. 129—148. (5.2; A)

See also [HaKMR]

MUIR, T.

* [Mui1] History of the Theory of Determinants, IV. London 1923. (7.6)

* [Mui2] The Theory of Determinants in the Historical Order of Development, I. Dover Publications Inc., New York 1960. (1.4)

MURAKAMI, M., *see* [HoMC]

MURTY, U. S. R., *see* [BoMu]

NASH-WILLIAMS, C. ST. J. A.

[Nash] Unexplored and semiexplored territories in graph theory. In: New Directions in the Theory of Graphs (ed. F. HARARY), Academic Press, New York—London 1973, pp. 149—186. (P)

NATHAN, A.

[Nath] A proof of the topological rules of signal-flow-graph analysis. Proc. IEEE (London) **109C** (1961), 83—85. (1.4)

NOLTEMEIER, H.

* [Nolt] Graphentheorie mit Algorithmen und Anwendungen. Walter de Gruyter, Berlin—New York 1976. (0)

NORDHAUS, E. A.

[Nord] A class of strongly regular graphs. In: Proof Techniques in Graph Theory (ed. F. HARARY), Academic Press, New York—London 1969, pp. 119—123.

See also [BeCN]

NOSAL, E.

[Nos1] Eigenvalues of Graphs. Master thesis, University of Calgary, 1970. (2.7; 3.2; 3.6; 7.7; 8.5; A)

[Nos2] On the number of spanning trees of finite graphs. The University of Calgary, Research paper No. 95 (1970).

NUFFELEN, C. VAN

[Nuf1] On the rank of the incidence matrix of a graph. Cahiers Centre Étud. Rech. Opér. (Bruxelles) **15** (1973), 363—365. (3.6)

[Nuf2] A bound for the chromatic number of a graph. Amer. Math. Monthly **83** (1976), 265—266. (3.2)

OGASAWARA, M.

[Ogas] A necessary condition for the existence of regular and symmetrical PBIB designs of T_m type. Inst. Statist. mimeo series 418, Chapel Hill 1965.

ORE, O.

* [Ore] Theory of Graphs. Amer. Math. Soc. Colloq. Publ. **38**, Providence 1962. (8.2)

OSTROWSKI, A. M., *see* [HoOs]

PÁRKÁNYI, C., *see* [HoDKP], [ZaPá]

PARSONS, T. D.

[Pars] Ramsey graphs and block designs I. Trans. Amer. Math. Soc. **209** (1975), 33—44. (7.1)

PARTHASARATHY, K. R., *see* [KrP1], [KrP2]

PAULUS, A. J. L.

[Paul] Conference matrices and graphs of order 26. Technische Hogeschool Eindhoven, T. H.-Report 73-WSK-06. (5.1; 6.1; 7.2)

PAYNE, S. E.

[Payn] On the non-existence of a class of configurations which are nearly generalized n-gons. J. Comb. Theory (A) **12** (1972), 268—282. (7.1)

PELIKÁN, J., *see* [LoPe]

PERCUS, J. K.
* [Perc] Combinatorial Methods. Springer-Verlag, Berlin—Heidelberg—New York 1971. (8.3)

PETERSDORF, M.
[Pete] Über Zusammenhänge zwischen Automorphismengruppe, Eigenwerten und vorderen Teilern eines Graphen und deren Verwendung zur Interpretation und Lösung gewisser Probleme der Analysis und Algebra. Dissertation TH Ilmenau 1969. (5.1)
[PeS1] (with H. SACHS), Über Spektrum, Automorphismengruppe und Teiler eines Graphen. Wiss. Z. TH Ilmenau **15** (1969), 123—128. (0.3; 1.2; 4.1; 4.4; 4.5)
[PeS2] (with H. SACHS), Spektrum und Automorphismengruppe eines Graphen. In: Combinatorial Theory and Its Applications, III (ed. P. ERDŐS, A. RÉNYI, V. T. SÓS), Bolyai János Mat. Társulat, Budapest/North-Holland Publ. Co., Amsterdam—London 1970, pp. 891—907. (5.2)

POLLAK, H. O., *see* [GrP1], [GrP2]

PONSTEIN, J.
[Pons] Self-avoiding paths and the adjacency matrix of a graph. SIAM J. Appl. Math. **14** (1966), 600—609. (1.4; 1.9; 6.1)

POUZET, M.
[Pouz] Note sur le problème de Ulam. J. Comb. Theory (to appear). (9.5)

PRIMAS, H., *see* [GüPr]

PULLMAN, N. J., *see* [MoPu]

RAGHAVARAO, D.
* [Ragh] Construction and Combinatorial Problems in the Design of Experiments. John Wiley & Sons, New York 1971. (6.4)
See also [ShRa]

RANDIĆ, M.
[RaTŽ] (with N. TRINAJSTIĆ and T. ŽIVKOVIĆ), On molecular graphs having identical spectra. J. C. S. Faraday II, **72** (1976), 244—256. (6.1)
See also [GuRT], [ŽiTR]

RAY-CHAUDHURI, D. K.
[Ray1] Characterization of line graphs. J. Comb. Theory **3** (1967), 201—214. (6.3; 6.6)
[Ray2] On some connections between graph theory and experimental designs and some recent existence results. In: Graph Theory and Its Applications (ed. B. HARRIS), Academic Press, New York—London 1970, pp. 149—166. (7.1)
See also [HoR1], [HoR2], [HoR3]

READ, R. C.
[Rea1] On the number of self-complementary graphs and digraphs. J. London Math. Soc. **38** (1963), 99—104. (3.3)
[Rea2] Teaching graph theory to a computer. In: Recent Progress in Combinatorics (ed. W. T. TUTTE), Academic Press, New York—London 1969, pp. 161—173.
See also [HaKMR]

REID, K. B.
[Reid] Connectivity in products of graphs. SIAM Rev. **18** (1970), 645—651. (7.4)

REINER, I., *see* [CuRe]

REMPEL, J.
[ReSc] (with K.-H. SCHWOLOW), Ein Verfahren zur Faktorisierung des charakteristischen Polynoms eines Graphen. Wiss. Z. TH Ilmenau **23** (1977), Nr. 4, 25—39. (4.1; 4.6)

RINGEL, G.
[Ring] Selbstkomplementäre Graphen. Arch. Math. **14** (1963), 354—358. (3.3)

ROBERTS, F. S.
[RoBr] (with T. A. BROWN), Signed digraphs and the energy crisis. Amer. Math. Monthly **82** (1975), 577—594.

ROHLIČKOVA, I.
[Rohl] Poznámka o počtu koster jednoho typu grafu (Czech). In: Matematika (Geometrie a Teorie Grafů), University Karlova, Praha 1970, pp. 117—120. (7.6)

ROUVRAY, D. H.
[Rou 1] Les valeurs propres des molécules qui possèdent un graph biparti. C. R. Acad. Sci. Paris, Sér. C, **274** (1972), 1561–1563. (3.2)
[Rou 2] Les valeurs propres des molécules qui possèdent un graph triparti. C. R. Acad. Sci. Paris, Sér. C, **275** (1972), 657–659. (8.5)
RUEDENBERG, K.
[Rued] Quantum mechanics of mobile electrons in conjugated bond systems, III. Topological matrix as a generatrix of bond orders. J. Chem. Phys. **34** (1961), 1884–1891. (8.5)
RUNGE, F.
[Rung] Beiträge zur Theorie der Spektren von Graphen und Hypergraphen. Dissertation, TH Ilmenau 1976. (1.3; 1.6; 1.7; 1.9; 3.5; 9.5)
[RuSa] (with H. SACHS), Berechnung der Anzahl der Gerüste von Graphen und Hypergraphen mittels deren Spektren. Mathematica Balkanica **4** (1974), 529–536. (1.9; 9.5)
RUŠČIĆ, B., *see* [GuRTW]
RUSHBROOKE, G. S., *see* [CoRu]
RUTHERFORD, D. E.
[Ruth] Some continuant determinants arising in physics and chemistry. Proc. Roy. Soc. (Edinburgh) A **62** (1947), 229–236. (2.5; 8.1; 8.4)
RYSER, H. J.
[Ryse] A generalization of the matrix equation $A^2 = J$. Linear Algebra and Appl. **3** (1970), 451–460.

SABIDUSSI, G.
[Sabi] Graph multiplication. Math. Z. **72** (1960), 446–457. (2.5; 7.4)
SACHS, H.
[Sac 1] Über selbstkomplementäre Graphen. Publ. Math. Debrecen **9** (1962), 270–288. (2.2; 2.5; 3.3; 4.5; 5.1; 7.1)
[Sac 2] Über die Anzahlen von Bäumen, Wäldern und Kreisen gegebenen Typs in gegebenen Graphen. Habilitationsschrift Univ. Halle, Math.-Nat. Fak., 1963. (1.4)
[Sac 3] Beziehungen zwischen den in einem Graphen enthaltenen Kreisen und seinem charakteristischen Polynom. Publ. Math. Debrecen **11** (1964), 119–134. (1.4; 3.1; 3.2; 3.3; 7.8)
[Sac 4] Abzählung von Wäldern eines gegebenen Typs in regulären und biregulären Graphen. I. Publ. Math. Debrecen **11** (1964), 74–84. (3.3)
[Sac 5] Abzählung von Wäldern eines gegebenen Typs in regulären und biregulären Graphen, II. Publ. Math. Debrecen **12** (1965), 7–24. (7.8)
[Sac 6] Bemerkung zur Konstruktion zyklischer selbstkomplementärer gerichteter Graphen. Wiss. Z. TH Ilmenau **11** (1965), 161–162.
[Sac 7] Über Teiler, Faktoren und charakteristische Polynome von Graphen. Teil I. Wiss. Z. TH Ilmenau **12** (1966), 7–12. (2.5; 3.2; 4.1; 4.2)
[Sac 8] Über Teiler, Faktoren und charakteristische Polynome von Graphen, Teil II. Wiss. Z. TH Ilmenau **13** (1967), 405–412. (2.4; 3.5; 3.6; 8.1)
*[Sac 9] Einführung in die Theorie der endlichen Graphen, Teil I. Teubner Verlagsgesellschaft, Leipzig 1970. (0; 3.2; 3.3)
[Sac 10] Struktur und Spektrum von Graphen. Mitteil. Math. Ges. DDR, Heft 23 (1973), 119–132. (4.1)
[Sac 11] Ein Beitrag zur Theorie der Graphenspektren. XVIII. Int. Wiss. Koll. TH Ilmenau 1973, Reihe A2, 59–60.
[Sac 12] On the number of spanning trees. In: Proc. Fifth British Combinatorial Conference, Aberdeen 1975 (ed. C. ST. J. A. NASH-WILLIAMS and J. SHEEHAN). Winnipeg 1976, pp. 529–535. (1.9)
[Sac 13] Regular graphs with given girth and restricted circuits. J. London Math. Soc. **38** (1963), 423–429. (7.1)

[Sac14] On a theorem connecting the factors of a regular graph with the eigenvectors of its line graph. In: Combinatorics (Coll. Math. Soc. J. Bolyai 18, ed. A. HAJNAL and V. Sós). North-Holland Publ. Co., Amsterdam—Oxford—New York 1978, Vol. II, pp. 947—957. (3.5)

[Sac15] Über einige graphentheoretisch-kombinatorische Problemkreise. In: Beiträge zur Graphentheorie und deren Anwendungen, vorgetragen auf dem Internat. Koll. Oberhof (DDR), April 1977. Math. Ges. DDR / TH Ilmenau 1977, pp. 201—217. (1.2)

[Sac16] Simultane Überlagerungen gegebener Graphen. Publ. Math. Inst. Hung. Acad. Sci. (Budapest) **9** (1964) (Ser. A), 415—427. (4.1; 4.2)

[SaSt] (with M. STIEBITZ), Automorphism group and spectrum of a graph. (To appear in the P. Turán Memorial Volume, Hung. Acad. Sci.) (5.1)

See also [ErSa], [FiSa], [PeS1], [PeS2], [RuSa]

SAMUEL, I.

[Sam1] Résolution d'un déterminant seculaire par la méthode des polygones. C. R. Acad. Sci. Paris **229** (1949), 1236—1237. (1.4)

[Sam2] Méthode des polygones, procédé d'étude graphique des déterminants. Applications aux problèmes de chimie théorique. Thesis, Univ. Paris 1958. (1.4)

SAUER, N.

[Saue] Extremaleigenschaften regulärer Graphen gegebener Taillenweite; I, II. Sitzungsberichte Österr. Akad. Wiss., Math. Nat. Kl., Abt. II, **176** (1967), 9—43.

See also [KuSa]

SCHAAD, L. J., *see* [HeSc]

SCHONLAND, D. S.

* [Scho] Molecular Symmetry. An Introduction to Group Theory and Its Uses in Chemistry. D. van Nostrand Comp., Ltd., London 1965. (5.2)

SCHWENK, A. J.

[Schw1] Almost all trees are cospectral. In: New Directions in the Theory of Graphs (ed. F. HARARY), Academic Press, New York—London 1973, pp. 275—307. (5.4; 6.1)

[Schw2] The spectrum of a graph. Doctoral Dissertation, University of Michigan 1973.

[Schw3] Computing the characteristic polynomial of a graph. In: Graphs and Combinatorics (Lecture Notes in Mathematics 406, ed. R. BARI and F. HARARY), Springer-Verlag, Berlin—Heidelberg—New York 1974, pp. 153—172. (2.7; 4.4)

[Schw4] On moments and coefficients in spectral graph theory. 1975 Winter Meeting A.M.S., Washington, D.C., (to appear).

[Schw5] New derivations of spectral bounds for the chromatic number (abstract), Graph Theory Newsletter 5 (1975), No. 1, 77.

[Schw6] Exactly thirteen connected cubic graphs have integral spectra. In: Theory and Applications of Graphs (Proc. Kalamazoo, 1976, ed. Y. ALAVI, D. LICK), Springer-Verlag, Berlin—Heidelberg—New York 1978, pp. 516—533. (9.4)

See also [HaS1], [HaS2], [MaST1], [MaST2], [MaTS]

SCHWOLOW, K.-H., *see* [ReSc]

SEDLÁČEK, J.

[Sed1] O incidenčnich orientovaných grafů (Czech). Časopis Pěst. Mat. **84** (1959), 303—316 (1.9; 2.2; 3.1)

[Sed2] O kostrách konechých grafů (Czech). Časopis Pěst. Mat. **91** (1966), 221—227. (7.8)

[Sed3] Lucasova čisla v teorii grafů (Czech). In: Matematika (Geometrie a Teorie Grafů), Universita Karlova, Praha 1970, pp. 111—115.

See also [FiSe]

SEIDEL, J. J.

[Sei1] Strongly regular graphs of L_2-type and of triangular type. Indag. Math. **29** (1967), No. 2, 188—196. (6.1)

[Sei2] Strongly regular graphs with (—1, 1, 0) adjacency matrix having eigenvalue 3. Linear Algebra and Appl. **1** (1968), 281—298. (6.1; 6.5; 7.2; 7.8)

[Sei 3] Strongly regular graphs. In: Recent Progress in Combinatorics (ed. W. T. Tutte), Academic Press, New York—London 1969; pp. 185—198. (6.1; 7.2; 7.8)
[Sei 4] A survey of two-graphs. In: Teorie Combinatorie (Coll. Int. Roma, 1973) t. I, Acc. Naz. Lincei, Roma 1976. (Atti dei Convegni Lincei **17** (1976)), pp. 481—511. (6.1; 7.2; 7.3; 7.8)
[Sei 5] On two-graphs and Shult's characterization of symplectic and orthogonal geometries over GF(2). Technische Hogeschool Eindhoven, TH-Report 73-WSK-02, 1—25.
[Sei 6] Graphs and two-graphs. In: Fifth Southeastern Conf. Combinatorics, Graph Theory, Computing. Boca Raton, Flo., 1974. Utilitas Math., Winnipeg 1974, pp. 125—143. (6.1; 6.6; 7.3)
[Sei 7] Metric problems in elliptic geometry. In: The Geometry of Metric and Linear Spaces; Proc. Conf. metric geometry, East Lansing 1974. (Lecture Notes in Mathematics 490, ed. L. M. Kelly), Springer-Verlag, Berlin—Heidelberg—New York 1975, pp. 32—43.
[Sei 8] Quasiregular two-distance sets. Indagationes Math. **31** (1969), No. 1, 64—70.
See also [BeLS], [BuČCS], [BuCS], [BuS 1], [BuS 2], [CaGSS], [CaSe], [DeGS 1], [DeGS 2], [GoS 1], [GoS 2], [GoS 3], [GoS 4], [LeS 1], [LeS 2], [LiSe]

Seiden, E.
[Seid] On a geometrical method of construction of partially balanced designs with two associate classes. Ann. Math. Statist. **32** (1961), 1177—1180. (6.3)

Shannon, C. E.
[Shan] The theory and design of linear differential equation machines. OSRD Rept. 411, Sec. D-2 (Fire Control), U.S. National Defense Research Committee, 1942. (1.4)

Shimamoto, T., *see* [BoSh]

Shrikhande, S. S.
[Shr 1] On a characterization of the triangular association scheme. Ann. Math. Statist. **30** (1959), 39—47.
[Shr 2] The uniqueness of the L_2 association scheme. Ann. Math. Statist. **30** (1959), 781—798. (6.1; 6.3)
[ShBh] (with Bhagawandas), Duals of incomplete block designs. J. Indian Stat. Assoc. **3** (1965), 30—37. (3.4)
[ShRa] (with D. Raghavaraó), Affine α-resolvable incomplete block designs. In: Contributions to Statistics, Calcutta 1964.
See also [BhSh]

Shult, E. E., *see* [CaGSS]

Simić, S. K., *see* [CvS 1], [CvS 2]

Sims, C. C.
[Sim 1] Graphs and finite permutation groups. Math. Z. **95** (1967), 76—86. (6.5)
[Sim 2] Graphs and finite permutation groups, II. Math. Z. **103** (1968), 276—281. (6.5)
[Sim 3] On graphs with rank 3 automorphism group. (unpublished). (6.5)
See also [HiSi]

Singleton, R.
[Sin 1] Regular graphs of even girth. Thesis, Princeton 1963. (7.1)
[Sin 2] On minimal graphs of maximal even girth. J. Comb. Theory **1** (1966), 306—332. (7.1)
See also [HoSi]

Singmaster, D.
[Sing 1] The eigenvalues of the icosahedron and dodecahedron. Notices Amer. Math. Soc. **19** (1972), A-750.
[Sing 2] The eigenvalues of the n-dimensional octahedron. Notices Amer. Math. Soc. **19** (1972), A-684.
[Sing 3] The eigenvalues of the n-dimensional octahedron and of complete n-partite graphs. (unpublished).

Sinogowitz, U., *see* [CoSi 1], [CoSi 2]

SKALA, H. L.
[Skal] A variation of the friendship theorem. SIAM J. Appl. Math. **23** (1972), 214—220. (7.1)

SLOANE, N. J., *see* [MaSl]
SMITH, C. A. B., *see* [BrSST]
SMITH, D. H.
[Sm,D1] On tetravalent graphs. J. London Math. Soc. **6** (1973), 659—662. (7.2)
[Sm,D2] An improved version of Lloyd's theorem. Discrete Math. **15** (1976), 175—184. (4.8; 7.2)

See also [BiSm]

SMITH, J. H.
[Sm,J] Some properties of the spectrum of a graph. In: Combinatorial Structures and Their Applications (ed. R. GUY, H. HANANI, N. SAUER, J. SCHÖNHEIM), Gordon and Breach, Science Publ., Inc., New York—London—Paris 1970, pp. 403—406. (2.7; 5.1; 6.2; 8.5)

See also [FrSm], [HoSm]

SMITH, M. S.
[Sm,M] On rank 3 permutation groups. J. Algebra **33** (1975), 22—42.

ŠOKAROVSKI, R.
[Šoka] A generalized direct product of graphs. Publ. Inst. Math. (Beograd) **22** (36) (1977), 267—269. (2.5)

SPIALTER, L.
[Spia] The atom connectivity matrix characteristic polynomial (ACMCP) and its physico-geometric (topological) significance. J. Chem. Doc. **4** (1964), 269—274. (1.4)

STEWARTSON, K.
[StMa] (with R. T. MAECHTER), On hearing the shape of a drum: further results. Proc. Cambridge Phil. Soc. **69** (1971), 353—363. (6.1)

STIEBITZ, M., *see* [SaSt]
STONE, A. H., *see* [BrSST]
STRAUB, P. A., *see* [HeSt]
STREET, A. P., *see* [WaSW]

STREITWIESER, A.
* [StBr] (with J. I. BRAUMAN), Supplement tables of molecular orbital calculations. I, II. Pergamon Press, Oxford—London—Edinburgh—New York—Paris—Frankfurt 1965. (A)

See also [CoSt]

SUN, F. K., *see* [BoFS]

SUTCLIFFE, B. T., *see* [McWS]

TAYLOR, D. E.
[Tay1] Some topics in the theory of finite groups. Ph. D. thesis, Univ. Oxford 1971. (7.3)
[Tay2] Regular 2-graphs. Proc. London Math. Soc. **35** (1977), 257—274. (7.3)

TEH, H. H.
[TeYa] (with H. D. YAP), Some construction problems of homogeneous graphs. Bull. Math. Soc. Nanyang Univ. (1964), 164—196. (2.5)

TITZ, M.
[TiH1] (with P. HOCHMANN), Tables of quantum chemical data, IX. Energy characteristics of some benzenoid hydrocarbons. Coll. Chech. Chem. Comm. **31** (1966), 4168—4172. (A)

[TiH2] (with P. HOCHMANN), Tables of quantum chemical data, XII. Energy characteristics of some benzoderivatives of acenaphthylene, fluorenthene and azulene. Coll. Chech. Chem. Comm. **32** (1967), 2343—2345. (A)

[TiH3] (with P. HOCHMANN), Tables of quantum chemical data, XIII. Energy characteristics of some benzenoderivatives of fulvene and heptafulvene. Coll. Chech. Chem. Comm. **32** (1967), 3028—3030. (A)

See also [HoDKT], [HoDT]

TRAUTH, Jr., C. A., *see* [HaTr]

TRENT, H. M.

[Tren] A note on the enumeration and listing of all possible trees in a connected linear graph. Proc. Nat. Acad. Sci. USA **40** (1954), 1004—1007. (1.5)

TRINAJSTIĆ, N., *see* [CvGT1], [CvGT2], [CvGT3], [CvGT4], [CvGT5], [CvGT6], [GrGT1], [GrGT2], [GrGTŽ], [GuT1], [GuT2], [GuT3], [GuT4], [GuT5], [GuT6], [GuT7], [GuT8], [GuT9], [GuT10], [GuT11], [GuRT], [GuRTW], [GuTW], [GuTŽ], [KnTŽ], [MaST1], [MaST2], [MaTS], [RaTŽ], [WiGT], [ŽiTR]

TUERO, M.

[Tuer] A contribution to the theory of cyclic graphs. Matrix Tensor Quart. **11** (1961), 74—80. (7.5)

TURÁN, P.

[Turá] Egy gráfelméleti szélsőértékfeladatról (Hungarian). Mat. Fiz. Lapok **48** (1941), 436—452. (7.7)

TURNER, J.

[Turn1] Point-symmetric graphs with a prime number of points. J. Comb. Theory **3** (1967), 136—145. (6.2)

[Turn2] Generalized matrix functions and the graph isomorphism problem. SIAM J. Appl. Math. **16** (1968), 520—526. (1.4; 6.1)

See also [ElTu]

TUTTE, W. T.

[Tut1] The reconstruction problem in graph theory. British Polymer J., September 1977, 180—183. (9.5)

[Tut2] All the king's horses. Univ. Waterloo Res. Report CORR 76-37. In: Graph Theory and Related Topics (Proceedings of the Conference held in honor of W. T. TUTTE on the occasion of his 60th birthday, ed. J. A. BONDY and U. S. R. MURTY), Academic Press, New York (to appear). (9.5)

See also [BrSST]

VAN LINT, J. H., *see* LINT, J. H. VAN

VAN NUFFELEN, C., *see* NUFFELEN, C. VAN

WALLER, D. A.

[Wal1] Eigenvalues of graphs and operations. In: Combinatorics, Proc. British Comb. Conf. 1973, London Math. Soc. Lecture Notes 13, pp. 177—183. (1.5; 2.7; A)

[Wal2] Regular eigenvalues of graphs and enumeration of spanning trees. In: Teorie Combinatorie (Coll. Int. Roma, 1973) t. I, Acc. Naz. Lincei, Roma 1976. (Atti dei Convegni Lincei **17** (1976), pp. 313—320. (1.5)

[Wal3] General solution to the spanning tree enumeration problem in arbitrary multigraph joins. IEEE Circuits and Systems CAS-**23** (1976), 467—469. (1.5; 7.6)

[Wal4] Double covers of graphs. Bull. Australian Math. Soc. **14** (1976), 233—248. (4.2)

[Wal5] Quotient structures in Graph Theory. Graph Theory Newsletter **6** (No. 4) 1977, 12—18. (4.2)

See also [FaW1], [FaW2]

WALLIS, J. S., *see* [WaSW]

WALLIS, W. D.

* [WaSW] (with A. P. STREET and J. S. WALLIS), Combinatorics: Room Squares, Sum-Free Sets, Hadamard Matrices. (Lecture Notes in Mathematics 292), Springer-Verlag, Berlin—Heidelberg—New York 1972. (6.3)

WEI, T. H.

[Wei] The algebraic foundations of ranking theory. Thesis, Cambridge 1952. (3.5; 7.8)

WEICHSEL, P. M.
[Weic] The Kronecker product of graphs. Proc. Amer. Math. Soc. **13** (1962), 47—52. (2.5; 7.4)

WHITE, A. T., *see* [LiWh]

WIGNER, E. P.
* [Wign] Group Theory and Its Application to the Quantum Mechanics of Atomic Spectra. Academic Press, New York—London 1959. (5.2)
See also [Вигн]

WILCOX, JR., C. F.
[WiGT] (with I. GUTMAN and N. TRINAJSTIĆ), Graph theory and molecular orbitals, XI. Aromatic substitution. Tetrahedron **31** (1975), 147—152. (8.1)
See also [HaWi], [GuRTW], [GuTW].

WILD, U.
[WiKG] (with J. KELLER and Hs. H. GÜNTHARD), Symmetry properties of the Hückel matrix. Theoret. Chim. Acta **14** (1969), 383—395. (5.2)

WILF, H. S.
[Wil1] The friendship theorem. In: Combinatorial Mathematics and Its Applications (ed. D. J. A. WELSH), Academic Press, New York—London 1971, pp. 307—309. (7.1)
[Wil2] The eigenvalues of a graph and its chromatic number. J. London Math. Soc. **42** (1967), 330—332. (3.2; 7.7)

WILSON, R. M.
[Wi, RM] Nonisomorphic Steiner triple systems. Math. Z. **135** (1974), 303—313. (6.1)

WILSON, R. J.
[Wi, RJ1] On the adjacency matrix of a graph. In: Combinatorics (Proc. Conf. Comb. Math., Math. Inst., Oxford 1972, ed. D. J. A. WELSH, D. R. WOODALL). The Institute of Mathematics and its Applications, Southend-on-Sea 1972, pp. 295—321. (9.5)
* [Wi, RJ2] Introduction to Graph Theory. Oliver & Boyd, Edinburgh 1972. (0)
See also [Уилс]

WYBOURNE, B. G.
* [Wybo] Classical groups for physicists. John Wiley & Sons, New York 1974. (6.3)

YAMAMOTO, S.
[YaFH] (with Y. FUJII and N. HAMADA), Composition of some series of association algebras. J. Sci. Hiroshima Univ. (A—I) **29** (1965), 181—215.

YAP, H. D., *see* [TeYa]

ZAHRADNÍK, R.
[ZaM1] (with J. MICHL), Tables of quantum chemical data, IV. Molecular orbitals of hydrocarbons of the pentalene, azulene and heptalene series. Coll. Chech. Chem. Comm. **30** (1965), 3173—3188. (A)
[ZaM2] (with J. MICHL), Tables of quantum chemical data, V. Molecular orbitals of peri-condensed tricyclic hydrocarbons. Coll. Chech. Chem. Comm. **30** (1965), 3529—3536. (A)
[ZaM3] (with J. MICHL), Tables of quantum chemical data, VII. Molecular orbitals of indacene-like and some peri-condensed tetracyclic hydrocarbons. Coll. Chech. Chem. Comm. **30** (1965), 3550—3560. (A)
[ZaMK1] (with J. MICHL and J. KOUTECKÝ), Tables of quantum chemical data, II. Energy characteristics of some nonalternant hydrocarbons. Coll. Chech. Chem. Comm. **29** (1964), 1933—1944. (A)
[ZaMK2] (with J. MICHL and J. KOUTECKÝ), Tables of quantum chemical data, III. Molecular orbitals of some fluoranthene-like hydrocarbons, cyclopentadienil, and some of its benzo and naphto derivatives. Coll. Chech. Chem. Comm. **29** (1964), 3184—3210. (A)
[ZaPá] (with C. PÁRKÁNYI), Tables of quantum chemical data, VI. Energy characteristics of some alternant hydrocarbons. Coll. Chech. Chem. Comm. **30** (1965), 3536—3549. (A)
See also [HoKZ]

Živković, T.
[Živk] Calculation of the non-bonding molecular orbitals in the Hückel theory. Croat. Chem. Acta **44** (1972), 351–364. (8.1)
[ŽiTR] (with N. Trinajstić and M. Randić), On conjugated molecules with identical topological spectra. Mol. Phys. **30** (1975), 517–533. (6.1)
See also [GrGTŽ], [GuTŽ], [KnTŽ], [RaTŽ]

Part II

Арлазаров, В. Л.
* [АрЛР] (with А. А. Леман and М. З. Розенфельд), Построение и исследование на ЭВМ графов с 25, 26 и 29 вершинами. Институт проблем управления, Москва 1975. (7.2)
Белоногов, В. А.
* [БеФо] (with А. Н. Фомин), Матричные представления в теории конечных групп. Изд. „Наука", Москва 1976. (5.2)
Беце, А.
[Беце] О коэффициентах характеристического полинома графа. Латв. Матем. Ежегодник **3** (1968), 75–80. (1.4)
Бочвар, Д. А.
[БоС1] (with И. В. Станкевич), Уровни энергии некоторых макромолекул с системой сопряженных двойных и тройных связей. Ж. Структ. Хим. 8 (1967), 943–949.
[БоС2] (with И. В. Станкевич), Качественный анализ топологической матрицы и правило Хюккеля $4n + 2$, сообщение I. Ж. Структ. Хим. **10** (1969), 680–685.
[БоС3] (with И. В. Станкевич), Качественный анализ топологической матрицы и правило Хюккеля $4n + 2$, II. Некоторые углеводородные системы. Ж. Структ. Хим. **12** (1971), 142–146.
[БоС4] (with И. В. Станкевич), Качественный анализ топологической матрицы и правило Хюккеля $4n + 2$, III. Сопряженные системы с большим дефицитом π-электронов. Ж. Структ. Хим. **13** (1972), 1123–1127.

Варвак, Л. П.
[Варв] Узагальнення поняття p-суми графів (Ukrainian). Доповіді АН УССР, 1968, A, No. 11, 965–968. (2.5)
Ваховский, Е. В.
[Вах1] О характеристических числах матриц соседства для неособенных графов. Сибир. Матем. Ж. **6** (1965), 44–49. (1.2; 2.4; 3.2; 8.5)
[Вах2] Об одном способе демонтажа графа. Сибир. Матем. Ж. **9** (1968), 255–263. (8.1)
Вигнер, Е.
* [Вигн] Теория групп и ее приложения к квантовомеханической теории атомных спектров. Изд. иностр. лит., Москва 1961 (Translation of [Wign]). (5.2)
Визинг, В. Г.
[Визи] Декартово произведение графов. Вычисл. Системы **9** (1963), 30–43. (7.4)

Гантмахер, Р. Ф., *see* [Gant]

Дамбит, Я. Я.
[Дамб] Свойства некоторых полиномов матриц циклов и разрезов графа. Латв. Матем. Ежегодник **4** (1968), 59–71.
Дочев, К.
[Доче] Върху решението на едно диференчно уравнение и свързаната с него задача от комбинаториката (Bulgarian). Физ. Мат. Спис. Българ. Акад. Наук **6** (39) (1963), 284–287. (7.5)

Ершов, А. П.
[ЕрКо] (with Г. А. Кожухин), Об оценках хроматического числа связных графов. Доклады АН СССР **142** (1962), 270—273. (7.7)

Зыков, А. А.
[Зык1] О некоторых свойствах линейных комплексов. Матем. Сборник **24** (1949), 163—188. (2.2)
* [Зык2] Теория конечных графов. I. Изд. ,,Наука'', сиб. отд., Новосибирск 1969. (2.2)

Кельманс, А. К.
[Кел1] О числе деревьев графа, I. Автоматика и телемеханика **26** (1965), 2194—2204. (1.2; 2.2; 7.6)
[Кел2] О числе деревьев графа, II. Автоматика и телемеханика 1966, но. 2, 56—65. (7.6)
[Кел3] О свойствах характеристического многочлена графа. In: Кибернетику — на службу коммунизму, Изд. ,,Энергия'', Москва—Ленинград 1967, pp. 27—41. (1.5; 2.4; 7.6; 7.7)
[Кел4] Графы с одинаковым числом путей длины два между смежными и несмежными парами вершин. In: Вопросы Кибернетики, Москва 1973, pp. 70—75.
[Кел5] О ,,разрешающей способности'' характеристического многочлена матрицы проводимости графа. (1964, unpublished.) (6.1)
See also [KeCh]

Кожухин, Г. А., *see* [ЕрКо]

Кэртис, Ч.
* [КэРа] (with И. Райнер), Теория представлений конечных групп и ассоциативных алгебр. Изд. ,,Наука'', Москва 1969 (Translation of [CuRe]). (5.2)

Леман, А. А., *see* [АрЛР]

Лихтенбаум, Л. М.
[Лих1] Характеристические числа неособенного графа. In: Труды 3-го , Всес. Матем. Съезда, Том 1, 1956, pp. 135—136. (0.1; 3.2)
[Лих2] Теорема двойственности для неособенных графов. Усп. Мат. Наук **13** (1958), No. 5, 185—190. (1.2)
[Лих3] Следы степеней матриц соседства вершин и ребер неособенного графа. Изв. Высш. Учебн. Завед. Мат. **5** (1959), 154—163.
[Лих4] Новые теоремы о графах. Сибир. Матем. Ж. **3** (1962), 561—568. (3.6)
[Лих5] О гамильтоновых циклах связного графа. Кибернетика (Киев) 1968, No. 4, 72—75. (3.6)

Мартинова, М. К.
[Март] Върху m-мерния решетъчен граф с характеристика n (Bulgarian). Годишн. Висш. Техн. Учебни Завед. Мат. **8** (1973), 155—163.

Наймарк, М. А.
* [Найм] Теория представлений групп. Изд. ,,Наука'', Москва 1976. (5.2)

Петренюк, А. Я., *see* [ПеПе]

Петренюк, Л. П.
[ПеПе] (with А. Я. Петренюк), О конструктивном перечислении 12-вершинных кубических графов. In: Комбинаторный Анализ, сборник статей, вып. 3, Москва 1974, pp. 72—82. (A)

Райнер, И., *see* [КэРа]

Розенфельд, М. З.
[Розе] О построении и свойствах некоторых классов сильно регулярных графов. Усп. Мат. Наук **28** (1973), No. 3, 197—198. (7.2)
See also [АрЛР]

Станкевич, И. В.
[Стан] Энергия π-электронов в макромолекулах с системой сопряженных двойных связей, I. Обобщенные линейные системы. Ж. Физ. Хим. **43** (1969), 549—555.
See also [БоС1], [БоС2], [БоС3], [БоС4]

Степанов, Н. Ф.
[СтТ1] (with В. М. Татевский), Обоснование разложения энергии π-электронов по связям в простейшем варианте метода молекулярных орбит. Ж. Структ. Хим. **2** (1961), 204—208. (8.1)
[СтТ2] (with В. М. Татевский), Приближенный расчет π-электронной энергии ароматических конденсированных молекул в варианте Хюккеля метод МОЛКАО. Ж. Структ. Хим. **2** (1961), 452—455. (8.1)

Татевский, В. М., *see* [СтТ1], [СтТ2]

Уилсон, Р.
* [Уилс] Введение в теорию графов. Изд. „Мир", Москва 1977. (Translation of [Wi,RJ2].) (0)

Фомин, А. Н., *see* [БеФо]

Харари, Ф.
* [Хара] Теория графов. Изд. „Мир", Москва 1973. (Translation of [Har4].) (0)

Added in proof

The following papers have come to the authors' attention only after the completion of the manuscript.

[1] Aihara, J., General rules for constructing Hückel molecular orbital characteristic polynomials. J. Amer. Chem. Soc. **98** (1976), 6840—6844.
[2] d'Amato, S. S., Eigenvalues of graphs with twofold symmetry. Mol. Phys. **37** (1979), to appear.
[3] d'Amato, S. S., Eigenvalues of graphs with threefold symmetry. To appear.
[4] Tang Au-chin and Kiang Yuan-sun, Graph theory of molecular orbitals. Sci. Sinica **19** (1976), No. 2, 207—226.
[5] Tang Au-chin and Kiang Yuan-sun, Graph theory of molecular orbitals, II: Symmetrical analysis and calculations of MO coefficients. Sci. Sinica **20** (1977), 595—612.
[6] Babai, L., On the isomorphism problem. Preprint appended to Proc. 1977 FCT-Conference, Poznán-Kórlik, Poland, Sept. 19—23, 1977.
[7] Babai, L., Isomorphism testing for graphs with distinct eigenvalues. In preparation.
[8] Beker, H., and W. Haemers, 2-designs having an intersection number $k - n$. J. Combinatorial Theory (A), to appear.
[9] Biggs, N., Coloring square lattice graphs. Bull. London Math. Soc. **9** (1977), No. 1, 54—56.
[10] Biggs, N., Girth, valency and excess. Linear Alg. and its Appl., to appear.
[11] Bílek, O., and P. Kadura, On the tight binding eigenvalues of finite S. C. and F. C. C. crystallites with (100) surfaces. Phys. Stat. Sol. (b) **85** (1978), 225—231.
[12] Bondy, J. A., and R. L. Hemminger, Graph reconstruction — A survey. J. Graph Theory **1** (1977), No. 3, 227—268.
[13] Bose, R. C., and S. S. Shrikhande, Graphs in which each pair of vertices is adjacent to the same number d of other vertices. Studia Sci. Math. Hung. **5** (1970), 181—195.
[14] Bridges, W. G., and M. S. Shrikhande, Special partially balanced incomplete block designs and associated graphs. Discrete Math. **9** (1974), 1—18.

[15] BUMILLER, C., On rank three graphs with a large eigenvalue. Discrete Math. **23** (1978), 183–187.

[16] CAMERON, P. J., J. M. GOETHALS and J. J. SEIDEL, Strongly regular graphs having strongly regular subconstituents. J. of Algebra **55** (1978), 257–280.

[17] COXETER, H. S. M., The product of the generators of a finite group generated by reflections. Duke Math. J. **18** (1951), 765–782.

[18] CVETKOVIĆ, D. M., Some possible directions in further investigations of graph spectra. Proc. Internat. Coll. on Algebraic Methods in Graph Theory, Szeged, August 25–28, 1978, pp. 47–67.

[19] CVETKOVIĆ, D. M., Some topics from the theory of graph spectra. Berichte der Mathematisch-Statistischen Sektion im Forschungszentrum Graz No. 100–No. 105 (1978), Bericht No. 101, 1–5.

[20] CVETKOVIĆ, D. M., M. DOOB and S. K. SIMIĆ, Some results on generalized line graphs. In: Proc. of the 10th South-Eastern Conf. on Combinatorics, Graph Theory, and Computing.

[21] CVETKOVIĆ, D. M., M. DOOB and S. K. SIMIĆ, Generalized line graphs. To appear.

[22] CVETKOVIĆ, D. M., I. GUTMAN and S. K. SIMIĆ, On self pseudo-inverse graphs. Univ. Beograd, Publ. Electrotehn. Fak. Ser. Mat. Fiz., to appear.

[23] CVETKOVIĆ, D. M., and S. K. SIMIĆ, Graph equations. In: Beiträge zur Graphentheorie und deren Anwendungen, vorgetragen auf dem Internat. Koll. Oberhof (DDR), 10. bis 16. April 1977, pp. 40–56.

[24] CVETKOVIĆ, D. M., and S. K. SIMIĆ, A bibliography of graph equations. J. Graph Theory, to appear.

[25] DINIC, E. A., A. K. KELMANS and M. A. ZAITSEV, Nonisomorphic trees with the same T-polynomial. Inform. Process. Lett. **6** (1977), No. 3, 3–8.

[26] DIXON, W. T., Some new theorems and methods for establishing the relationship between the symmetry of a molecular orbital and its energy. J. C. S. Faraday II **74** (1978), 511–520.

[27] DOOB, M., Graphs with a small number of distinct eigenvalues, II. To appear.

[28] DOOB, M., Seidel switching and graphs with four distinct eigenvalues. Ann. New York Acad. Sci., to appear.

[29] DOOB, M., and D. M. CVETKOVIĆ, On spectral characterizations and embeddings of graphs. Linear Algebra and its Appl., to appear.

[30] EICHINGER, B. E., Elasticity theory, I, Distribution functions for perfect phantom networks. Macromolecules **5** (1972), No. 4, 496–503.

[31] EICHINGER, B. E., and J. E. MARTIN, Distribution functions for Gaussian molecules, II, Reduction of the Kirchhoff matrix for large molecules. J. Chem. Phys. **69** (10) (1978), 4595–4599.

[32] EINBU, J. M., The enumeration of bit-sequences that satisfy local criteria. Publ. Inst. Math. (Beograd), to appear.

[33] FARRELL, E. J., On a general class of graph polynomials. J. Combinatorial Theory (B) **26** (1979), 111–122.

[34] FARRELL, E. J., Introduction to matching polynomials. J. Combinatorial Theory (B), to appear.

[35] FARRELL, E. J., On a class of polynomials obtained from the circuits in a graph and its application to characteristic polynomials of graphs. Discrete Mathematics **25** (1979), 121–133.

[36] FORSMAN, W. C., Graph theory and polymer dynamics. J. Chem. Phys. **65** (1976) 4111–4115.

[37] GARDINER, A. D., The classification of symmetric graphs. Berichte der Mathematisch-Statistischen Sektion im Forschungszentrum Graz No. 100–No. 105 (1978), Bericht No. 102, 1–33.

[38] GODSIL, C. D., and I. GUTMAN, On the matching polynomial of a graph. Mathematics Research Report of the University of Melbourne, No. 35 (1978).

[39] GODSIL, C., D. A. HOLTON and B. MCKAY, The spectrum of a graph. In: Combinatorial Mathematics V (Lecture Notes in Mathematics 622, ed. C. H. C. LITTLE), Springer-Verlag, Berlin—Heidelberg—New York 1977, pp. 91—117.

[40] GODSIL, C., and B. MCKAY, A new graph product and its spectrum. Bull. Austral. Math. Soc. **18** (1978), 21—28.

[41] GODSIL, C. D., and B. D. MCKAY, Feasibility conditions for the existence of walk-regular graphs. Linear Algebra and Appl., to appear.

[42] GODSIL, C. D., and B. D. MCKAY, Spectral conditions for the reconstructibility of a graph. J. Combinatorial Theory (B), to appear.

[43] GONDRAN, M., et M. MINOUX, Valeurs propres et vecteurs propres en théorie des graphes. In: Problèmes Combinatoires et Théorie des Graphes, Colloque International C. N. R. S., No. 260, Orsay, 9—13 Juillet 1976, C. N. R. S. Publ. 1978 (ed. J.-C. BERMOND, J.-C. FOURNIER, M. LAS VERGNAS, D. SOTTEAU), pp. 181—183.

[44] GONDRAN, M., and M. MINOUX, Valeurs propres et vecteurs propres dans les semi-modules et leur interprétation en théorie des graphes. Bull. Direction Études Recherches, Ser. C, Math. Informat. 1977, no. 2, i, 25—41.

[45] GRAHAM, R. L., and L. LOVÁSZ, Distance matrix polynomials of trees. In: Problèmes Combinatoires et Théorie des Graphes, Colloque International C. N. R. S., No. 260, Orsay, 9—13 Juillet 1976, C. N. R. S. Publ. 1978 (ed. J.-C. BERMOND, J.-C. FOURNIER, M. LAS VERGNAS, D. SOTTEAU), pp. 189—190.

[46] GRAHAM, R. L., and L. LOVÁSZ, Distance matrix polynomials of trees. Adv. in Math. **29** (1978), No. 1, 60—88.

[47] GRAHAM, R. L., and L. LOVÁSZ, Distance matrix polynomials of trees. In: Theory and Applications of Graphs. Proc. Internat. Conf. Western Michigan Univ., Kalamazoo, Mich., 1976. Springer Lecture Notes Math. 642, Berlin (1978), pp. 186—190.

[48] GRAOVAC, A., and I. GUTMAN, The determinant of the adjacency matrix of the graph of a conjugated molecule. Croat. Chem. Acta **51** (1978), 133—140.

[49] *GRAOVAC, A., I. GUTMAN and N. TRINAJSTIĆ, Topological approach to the chemistry of conjugated molecules. Springer-Verlag, Berlin—Heidelberg—New York 1977.

[50] GRAOVAC, A., O. E. POLANSKY, N. TRINAJSTIĆ and N. TYUTYULKOV, Graph theory in chemistry, II, Graph-theoretical description of heteroconjugated molecules. Z. Naturforsch. **30a** (1975), 1696—1699.

[51] GRAOVAC, A., and N. TRINAJSTIĆ, Möbius molecules and graphs. Croat. Chem. Acta **47** (1975), 95—104.

[52] GRAOVAC, A., and N. TRINAJSTIĆ, Graphical description of Möbius molecules. J. Mol. Struct. **30** (1976), 416—420.

[53] GREENWELL, D. L., R. L. HEMMINGER and J. KLERLEIN, Forbidden subgraphs. In: Proc. 4th South-Eastern Conf. on Combinatorics, Graph Theory, and Computing, Florida Atlantic University, 1972, pp. 389—394.

[54] GUTMAN, I., The acyclic polynomial of a graph. Publ. Inst. Math. (Beograd) **22** (36) (1977), 63—69.

[55] GUTMAN, I., Topological properties of benzenoid systems. Theoret. Chim. Acta (Berlin) **45** (1977), 309—315.

[56] GUTMAN, I., Bounds for total π-electron energy of polymethines. Chem. Phys. Letters **50** (1977), No. 3, 488—490.

[57] GUTMAN, I., The energy of a graph. Berichte der Mathematisch-Statistischen Sektion im Forschungszentrum Graz No. 100—No. 105 (1978), Bericht No. 103, 1—22.

[58] GUTMAN, I., Contribution to the problem of the spectra of compound graphs. Publ. Inst. Math. (Beograd) **24** (38) (1978), 53—60.

[59] GUTMAN, I., On polymethine graphs. match **5** (1979), 161—176.

[60] GUTMAN, I., and A. GRAOVAC, On structural factors causing stability differences between conjugated isomers. Croat. Chem. Acta **49** (1977), 453—459.

[61] HAEMERS, W., Eigenvalues methods. In: Packing and Covering in Combinatorics, Math. Centre Tracts **106** (1979) (ed. A. SCHRIJVER), Amsterdam 1979, pp. 15–38; see also: Stapelen en Overdekken, Syllabus, Stichting Math. Centrum, Amsterdam 1978, pp. 15–33.

[62] HAEMERS, W., On some problem of Lovász concerning the Shannon capacity of a graph. IEEE Trans. Inform. Theory **25** (1979), 231–232.

[63] HAEMERS, W., An upper bound for the Shannon capacity of a graph. In: Proc. Internat. Coll. on Algebraic Methods in Graph Theory, Szeged, August 25–28, 1978, to appear.

[64] HAEMERS, W., Eigenvalue techniques in design and graph theory. Thesis, Tech. Univ. Einhoven, The Netherlands, in preparation.

[65] HAEMERS, W., and C. ROOS, An inequality for generalized hexagons. Geom. Dedic., to appear.

[66] HALL, G. G., On the eigenvalues of molecular graphs. Molec. Phys. **33** (1977), No. 2, 551–557.

[67] HEILBRONNER, E., Some comments on cospectral graphs. match **5** (1979), 105–133.

[68] HEILMANN, O. J., and E. H. LIEB, Theory of monomer-dimer systems. Commun. Math. Phys. **25** (1972), 190–232.

[69] HEINRICH, P., Eine Beziehung zwischen den Weg-Kreis-Systemen in einem gerichteten Graphen und den Unterdeterminanten seiner charakteristischen Matrix. Math. Nachrichten, to appear. Preprint: Bergakademie Freiberg, 1978.

[70] HEINRICH, P., Über die Jordansche Normalform der Adjazenzmatrix für spezielle Klassen gerichteter Graphen. Preprint: Bergakademie Freiberg, 1978.

[71] HEYDEMANN, M. C., Caractérisation spectrale du joint d'un cycle par un stable. In: Problèmes Combinatoires et Théorie des Graphes, Colloque International C. N. R. S., No. 260, Orsay, 9–13 Juillet 1976, C. N. R. S. Publ. 1978 (ed. J.-C. BERMOND, J.-C. FOURNIER, M. LAS VERGNAS, D. SOTTEAU), pp. 225–227.

[72] HIDEO, H., and Y. TAKENAKA, On the spectrum of a graph. Keio Engrg. Rep. **31** (1978) No. 9, 99–105.

[73] HOFFMAN, A. J., On signed graphs and gramians. Geom. Dedic. **6** (1977), No. 4, 445–470.

[74] JOHNSON, C. R., and F. T. LEIGHTON, An efficient linear algebraic algorithm for the determination of isomorphism in pairs of undirected graphs. J. Research NBS, A. Mat. Sci. **80** B (1976), No. 4, 447–483.

[75] JOHNSON, C., and M. NEWMAN, A note on cospectral graphs. J. Combinatorial Theory, to appear.

[76] JONES, O., Contentment in graph theory: covering graphs with cliques. Proc. Kon. Ned. Akad. Wetensch. A **80** (1977), 406–424.

[77] JUHÁSZ, F., and K. MÁLYUSZ, Problems of cluster analysis from the viewpoint of numerical analysis. In: Proc. Conf. Numerical Methods, Keszthely 1977.

[78] KAULGUD, M. K., and V. H. CHITGOPKAR, Polynomial matrix methods for the calculation of π-electron energies for linear conjugated polymers. C. J. S. Faraday II **73** (1977), 1385–1395.

[79] KELMANS, K. A., Comparison of graphs by their number of spanning trees. Discrete Math. **16** (1976), 241–261.

[80] KING, R. B., Symmetry factoring of the characteristic equations of graphs corresponding to polyhedra. Theoret. Chim. Acta (Berlin) **44** (1977), 223–243.

[81] KING, R. B., and D. H. ROUVRAY, Chemical applications of group theory and topology, VII. A graph theoretical interpretation of the bounding topology in polyhedral boranes, carboranes, and metal clusters. J. Amer. Chem. Soc. **99** (1977), 7834–7840.

[82] LAM, C. W. H., and J. H. VAN LINT, Directed graphs with unique paths of fixed length. J. Combinatorial Theory (B) **24** (1978), 331–337.

[83] LIEBLER, R. A., On the uniqueness of the tetrahedral association scheme. J. Combinatorial Theory (B) **22** (1977), 246–262.

[84] LOVÁSZ, L., On the Shannon capacity of a graph. IEEE Trans. Inform. Theory **25** (1979), 1–7.

[85] MALLION, R. B., and D. H. ROUVRAY, Molecular topology and the Aufbau principle. Mol. Phys. **36** (1978), 125–138.

[86] MARTIN, J. E., and B. E. EICHINGER, Distribution functions for Gaussian molecules, I, Stars and random regular nets. J. Chem. Phys. **69** (10) (1978), 4588–4594.

[87] MATHON, R., 3-class association schemes. In: Proc. Conf. on Algebraic Aspects of Comb. 1975, (Ed. D. CORNEIL, E. MENDELSOHN) Utilitas Math., Winnipeg 1975, pp. 123–155.

[88] McKAY, B. D., Transitive graphs with fewer than twenty vertices. Math. Comp., to appear.

[89] McKAY, B. D., The eigenvalue distribution of a random regular labelled graph. To appear.

[90] VAN NUFFELEN, C., On the rank of the adjacency matrix. In: Problèmes Combinatoires et Théorie des Graphes, Colloque International C. N. R. S., No. 260, Orsay, 9–13 Juillet 1976, C. N. R. S. Publ. 1978 (ed. J.-C. BERMOND, J.-C. FOURNIER, M. LAS VERGNAS, D. SOTTEAU), pp. 321–322.

[91] PATTERSON, H. D., R. E. WILLIAMS, Some theoretical results on general block designs. In: Proc. V. British Comb. Conf. 1975, (Ed. C. ST. J. A. NASH-WILLIAMS, J. SHEEHAN) Utilitas Math., Winnipeg 1976, pp. 489–496.

[92] POLANSKY, O. E., and I. GUTMAN, On the calculation of the largest eigenvalue of molecular graph. match **5** (1979), 149–159.

[93] LE CONTE DE POLY, C., Graphes d'amitié et plans en blocs symétriques. Math. Sci. Humaines **51** (1975), 25–33, 87.

[94] PORCU, L., Sul radoppio di un grafo. Ist. Lombardo (Rend. Sci.) A **110** (1976), 453–480. 453–480.

[95] POUZET, M., Quelques remarques sur les resultats de Tutte concernant le problème de Ulam. To appear.

[96] READ, R. C., and D. G. CORNEIL, The graph isomorphism disease. J. Graph Theory **1** (1977), No. 4, 339–363.

[97] RIGBY, M. J., R. B. MALLION and A. C. DAY, Comment on a graph-theoretical description of heteroconjugated molecules. Chem. Phys. Letters **51** (1977), No. 1, 178–182.

[98] RIGBY, M. J., R. B. MALLION and D. A. WALLER, On the Quest for an isomorphism invariant which characterises finite chemical graphs. Chem. Phys. Letters **59** (1978), No. 2, 316–320.

[99] ROLLAND, P. T., On the uniqueness of the tetrahedral association scheme. Ph. D. Thesis, City University of New York, 1976.

[100] SACHS, H., and M. STIEBITZ, Automorphism group and spectrum of a graph (abridged version of [SaSt]). Proc. Internat. Coll. on Algebraic Methods in Graph Theory, Szeged, August 25–31, 1978, pp. 657–670.

[101] SCHRIJVER, A., Association schemes and Shannon capacity: Eberlein polynomials and Erdős-Ko-Rado theorem. Proc. Internat. Coll. on Algebraic Methods in Graph Theory, Szeged, August 25–31, 1978, pp. 671–688.

[102] SCHRIJVER, A., A comparison of the Delsarte and Lovász bounds. IEEE Trans. Inform. Theory, to appear.

[103] SCHULZ, M., Automorphismen und Eigenwerte von Graphen. Diplomarbeit, Humboldt-Univ. Berlin, Sektion Mathematik, 1979. See also: Discrete Math. **31** (1980), 221–222.

[104] SCHWENK, A. J., Spectral reconstruction problems. To appear.

[105] SCHWENK, A. J., W. C. HERNDON and M. L. ELLZEY, Jr., The construction of cospectral composite graphs. Ann. New York Acad. Sci. **319** (1979), 490–496.

[106] SCHWENK, A. J., and R. WILSON, On the eigenvalues of a graph. In: Selected topics in graph theory (eds. L. BEINEKE and R. WILSON), Academic Press, New York 1979.

[107] SEIDEL, J. J., Strongly regular graphs, an introduction. In: Proc. 7th British Combinatorial Conference, Cambridge 1979, to appear.

[108] SEIDEL, J. J., and D. E. TAYLOR, Two-graphs, a second survey. Proc. Internat. Coll. on Algebraic Methods in Graph Theory, Szeged, August 25–31, 1978, to appear.

[109] SHRIKHANDE, M. S., Strongly regular graphs and group divisible designs. Pacific J. Math. **54** (1974), No. 2, 199–208.

[110] SMITH, J. H., Symmetry and multiple eigenvalues of graphs. Glas. Mat., Ser. III, **12** (1977), No. 1, 3–8.

[111] TOIDA, S., A note on Ádám's conjecture. J. Combinatorial Theory (B) **23** (1977), 239–246.

[112] TRINAJSTIĆ, N., Computing the characteristic polynomial of a conjugated system using the Sachs theorem. Croat. Chem. Acta **49** (4) (1977), 593–633 .

[113] TRINAJSTIĆ, N., Hückel theory and topology. In: Semiempirical Methods of Electronic Structure Calculations — Part A: Techniques, Modern Theoretical Chemistry, Vol. 7, Plenum Press, New York 1977 (ed. G. J. SEGAL), pp. 1–27.

[114] WILSON, R. J., Graph theory and chemistry. In: Combinatorics (Colloquia Mathematica Societatis János Bolyai, **18**), Edited by A. HAJNAL and V. Sós. North-Holland Publishing Co., Amsterdam–Oxford–New York 1978, Vol. II, pp. 1147–1164.

[115] YAP, H. P., The characteristic polynomial of the adjacency matrix of a multi-digraph. Nanta Math. **8** (1975), No. 1, 41–46.

[116] ДЯДЮША, Г. Г., и А. Д. КАЧКОВСКИЙ, Применение графов к теории основности концевых групп полимесиновых красителей. Теорет. эксп. хим. **15** (1979), 152–161.

[117] ГИНЗБУРГ, Б. Д., О числе внутренней устойчивости графов. Сообщения АН СССР **85** (1977), 289–292.

[118] СУББОТИН, В. Ф., и Р. Б. СТЕКОЛЬЩИК, Спектр преобразования Кокстера и регулярность представлений графов. Тр. Мат. фак. Воронеж. ун-та 1975, вып. **16**, 62–65.

[119] SUBBOTIN, V. F., und R. B. STEKOL'ŠČIK, Jordanform, Coxetertransformation und Anwendungen auf Darstellungen endlicher Graphen (Russian). Funkcional'. Analiz Priloženija **12** (1978), 84–85 (according to „Zentralblatt").

Comment

Recent *expository papers* are [18], [19], [39], [61], [67], [106], [107], [108], [114].

Cospectral graphs (*digraphs*) appear in [25], [40], [58], [69], [70], [75], [104], [105].

The graph isomorphism problem in connection with eigenvalues is considered in [6], [7], [74], [96], [98].

The graph reconstruction problem is discussed from the view of spectral graph theory in [12], [18], [42], [94], [104].

The matching polynomial, a graph polynomial related to the characteristic polynomial, is investigated in [34], [38], [54], [68].

L. LOVÁSZ [84] developed an eigenvalue method for determining the so-called *Shannon capacity* of a graph and, in particular, proved that the capacity of C_5 is equal to $\sqrt{5}$. He solved in that way an old problem posed by C. E. SHANNON (see, for example, [Ber 1]). The paper [84] caused some further discussion: [62], [63], [101], [102].

All *graphs with the least eigenvalue greater than* -2 are determined in [29]. A characterization of *generalized line graphs* by forbidden subgraphs is given in [20], [21].

Graph equations and eigenvalues are related one to another in [18], [22], [23], [24].

Eigenvalues and groups are treated in [2], [3], [26], [80], [88], [100], [103], [110].

The *Jordan normal form* of the adjacency matrix of a digraph is investigated in [69], [70].

Applications of graph spectra to Physics and Chemistry, other than those described in Chapter 8, can be found in [30], [31], [36], [86].

Numerical tables are contained in [88].

The chemist's attention is particularly directed towards the following two books on Hückel theory, which just have appeared:

COULSON, C. A., B. O'LEARY and R. B. MALLION, Hückel Theory for Organic Chemists. Academic Press, London–New York–San Francisco 1978.

YATES, K., Hückel Molecular Orbital Theory. Academic Press, London–New York–San Francisco 1979.

Index of Symbols

Index of Names

Subject Index

Bold-face page numbers indicate definitions. For some general concepts, marked by an asterisk (*), only the place of their definition is given.